STANDARD LEVEL

Biology
for the IB Diploma Programme

Alan Damon, Randy McGonegal, William Ward

Published by Pearson Education Limited, 80 Strand, London, WC2R 0RL.

www.pearson.com/international-schools

Text © Pearson Education Limited 2023
Development edited by Judith Head and Jess White
Copy edited by Eva Fairnell
Proofread by Eva Fairnell
Indexed by Georgie Bowden
Designed by Pearson Education Limited
Typeset by Tech-Set ltd
Picture research by Integra
Original illustrations © Pearson Education Limited 2023
Cover design © Pearson Education Limited 2023

The rights of Alan Damon, Randy McGonegal, and William Ward to be identified as the authors of this work has been asserted by them in accordance with the Copyright, Designs and Patents Act 1988.

First published 2023

25 24 23
10 9 8 7 6 5 4 3 2 1

British Library Cataloguing in Publication Data
A catalogue record for this book is available from the British Library

ISBN 978 1 29242 773 7

Acknowledgements

The authors and publisher would like to thank the following individuals and organisations for permission to reproduce photographs, illustrations, and text:

Text extracts relating to the IB syllabus and assessment have been reproduced from IBO documents. Our thanks go to the International Baccalaureate for permission to reproduce its copyright.

 The "In cooperation with IB" logo signifies the content in this textbook has been reviewed by the IB to ensure it fully aligns with current IB curriculum and offers high-quality guidance and support for IB teaching and learning.

KEY (t – top, c – center, b – bottom, l – left, r – right)

Images:

123RF: Alila 176, Grigory_bruev 222, Arthit Buarapa 228c, Sebastian Kaulitzki 286, James Cumming 353br, Achim Prill 466, Anna Grigorjeva 542b, Zhijie Zhuang 567br, Oleg Beloborodov 585, Trish233 616, Veronique Martel 617, sdigitall 618; **Alamy Stock Photo:** Mint Images Limited 56, Encyclopaedia Britannica/Universal Images Group North America LLC 59, Wayne Lynch/All Canada Photos 78, Encyclopaedia Britannica/Universal Images Group North America LLC 85, GERAULT Gregory/Hemis.fr 98b, Matthew Taylor 138, Tetiana Zhabska 160l, Blickwinkel/M. Woike 199, Leo Francini 202, Nick Harrison 308, Encyclopaedia Britannica/Universal Images Group North America LLC 317, Arindam Ghosh 321, Tom Uhlman 339bl, Peter Groenendijk/Robertharding 362, Photo Researchers/Science History Images 414b, Heather Angel/Natural Visions 452, Pattarawit Chompipat/agefotostock 467, Ester van Dam 512, Steven May 529r, Pictorial Press Ltd 534, Rosanne Tackaberry 535r, Michael Lidski 536, David Tipling Photo Library 552, Enigm 571; **Alan Damon:** Alan Damon 160r, 234, 256, 393, 510; **Getty Images:** David_Slater/iStock 272, Vchal/iStock 386, Micro_photo/Istock 445c, Ryan McVay/Photodisc 567tr, Alain Denantes/Gamma-Rapho 567cl, Dorling Kindersley 576; **Google:** Screenshot from "The Enteric Nervous System", Gastroenterology 101 304; **International Union for Conservation of Nature and Natural Resources:** International Union for Conservation of Nature and Natural Resources 102, Indrayana/International Union for Conservation of Nature and Natural Resources 102; **National Human Genome Research Institute:** POINT MUTATION, Chris Gunter, Ph.D. Associate Investigator, Social and Behavioral Research Branch 109; **National Institutes of Health:** National Institutes of Health 605; **Kunstformen der Natur:** Natur,Kunstformen der, 'plate 32: Rotatoria' 572br; **Pearson Education Ltd:** Oxford Designers & Illustrators Ltd/Pearson Education Ltd 354, 418t, 428, Trevor Clifford/Pearson Education Ltd 8; **Science Photo Library:** Kateryna Kon 2, 73, 411, Ence Mate/Nature Picture Library 6, Markus Varesvuo/Nature Picture Library 11, Ron Miller 28, Steve Gschmeissner 34, 44l, 274t, 278t, 309, 445t, Niaid/CDC 37, A.B. Dowsett 39, 49l, CNRI 44b, 65, 311t,Dr Karl Lounatmaa 45l, Biology Media 45c, Dr Jeremy Burgess 46, 62, Don W. Fawcett 47r, J.C. Revy, ism 47b, 238, Biophoto Associates 71, ism 185, Sheila Terry 58, 471, Roma Uchytel 63, Javier Trueba/MSF 68, 219r, John Birdsall Social Issues Photo Library 70, Turtle Rock Scientific/Science Source 71, Nicolle Rager Fuller/National Science Foundation 84t, Mark Garlick 98t, animate4.com 112t, Dr Tim Evans 118, Alice J. Belling 122, Gary Carlson 127, Klaus Guldbrandsen 154c, Jannicke Wiik-Nielsen 173, DR David Furness, Keele University 191, David Aubrey 213, Mauricio Anton 219l, D. Roberts 221, Clay Coleman 225t, Professors P. Motta & T. Naguro 251, John Durham 258, Trevor

Clifford Photography 264, Omikron 293, Eric Grave 294, Springer Medizin 302, Eye Of Science 311b, 410t, Alfred Pasieka 314, 401, Bob Gibbons 349, Adrian Thomas 351t, Noaa 365, Ted Kinsman 371, David Parker 392b, Equinox Graphics 414t, Patrick Landmann 415, Science Photo Library 416, 584, Art for Science 429, Vivid Biology 454, 511, Dr Yorgos Nikas 455, Dr. Keith Wheeler 458, Carolina Biological Supply Company 463, Doug Allan 475, Don Fawcett 498, Art Wolfe 514, Paul Zahl 519, British Antarctic Survey 539b, Simon Fraser 543, Katey Walter Anthony/University Of Alaska Fairbanks/Nasa 544t, Science Photo Library 548, European Space Agency/Atg Media Lab 549t, Nasa/Gsgc 549b, Adam Jones 579; **Science Courseware:** Copyright © 2019, 2022, Robert A. Desharnais, Ph.D. Virtual Courseware Project, ScienceCourseware.org 603; **Shutterstock:** Corona Borealis Studio 15, Leonid Andronov 22, 110, Art of Science 23, 51t, 3d_man 25, Sirikwan Dokuta 38t, BioMedical 38b, Designua 49r, 278b, 347t, 347c, 502, Schira 51b, Jose Luis Calvo 52, 165b, 292, Yaruna 112b, Sansanorth 113, 121t, BigBearCamera 121b, Biology Education 126, Zcxes 128l, Magnetix 128c, Ali DM 133, 316, Vecton 142, VectorMine 146, Stanislaw Mikulski 153c, Chakapong 153b, Soleil Nordic 154t, W.Y. Sunshine 164, Rattiya Thongdumhyu 165t, 215, T.Thinnapat 4, Aldona Griskeviciene 7, 345b, JohnKwan 9, Polarman 12, Marina Veder 61, Chris Howey 81, Photographer DmitryStrong 83, Panaiotidi 87, Sergey Uryadnikov 89cl, Bardocz Peter 89b, Andy333 89cr, Henner Damke 93, jo Crebbin 95, Zern Liew 96, Rich Carey 99, 355, Yusnizam Yusof 100, Barbara Tripp 104, Yandong Yang 106, Christian Colista 107, Design_Cells 170, Sergio Gutierrez Getino 172, Art of Kosi 174, Pissamai Boonkane 180, Songkram Chotik-anuchit 183, Blamb 187t, SciePro 187b, Ilusmedical 188, Michael Hare 189, Willem van de Kerkhof 198, Chad Zuber 200, Cathy Keifer 206l, leungchopan 206r, Lhbllc 208t, Vinnie Lauritsen 208b, Ticiana Giehl 209, Klaus Ulrich Mueller 210, Zagorulko Inka 214, Lebendkulturen.de 216, Jolanda Aalbers 217, Achiichiii 220, Wiangya 224, Dirk Ercken 225b, Dr Ajay Kumar Singh 226, Steve Estvanik 227, Torsten ursche 228l, Systemoff 274b, Scubaluna 282, GraphicsRF.com 291, Terpsychore 297, TimeLineArtist 303, J. Marini 312, Magic mine 313, C.Lotongkum 323, Kateryna Kon 324, 82841 332, Sergey 402, 334, Samib123 335, Raquel Pedrosa 337t, Matt Tilghman 337b, Moolkum 339c, Roman Khomlyak 339br, Yumeee 343, David Osborn 344, TalyaPhoto 345t, Worachat Tokaew 346, Nwdph 348, Wolfen 351c, Critterbiz 353cr, Bildagentur Zoonar GmbH 356, Vadim Sadovski 360, Paul Aniszewski 361, Volkov Alexey 364, Solarseven 372, Kara Grubis 376, Richsouthwales 377, M. Patthawee 392t, Haroon Abbasi 404, Ellepigrafica 410b, Peter Hermes Furian 417, 572tl, Alila Medical Media 418b, 461, Vita khorzhevska 419, Elenabsl 420, Dimarion 424, 432r, OSweetNature 426, Fancy Tapis 427b, 437, 921248 432l, Yakobchuk Viacheslav 436, Steve Allen 450, Scisetti Alfio 483, Michaeljung 490, Olivier Le Moal 497, Fabiano Vieira 500, Sebastian Kaulitzki 501, Iv-olga 503t, Steven McDowell 503b, ParabolStudio 508, Francesco de marco 516t, Zhaoyan 516b, Juefraphoto 517, Szefei 518, Marktucan 523, Holly Kuchera 528t, Darkydoors 528b, Alexandru V 529b, Avigator Fortuner 531, Manishankar Patra 532, Olha1981 535c, Jeanie333 538, Ziablik 539t, Tony Skerl 542t, Condor 36 544b, Paralaxis 545t, Maksimilian 545b Anticiclo 546, Gentoo Multimedia Limited 547t, Outdoorsman 547b, Evgenii mitroshin 550, Peter Leahy 553l, 592411 553r, Callums Trees 554, ArtMediaFactory 555, Robin Nieuwenkamp 556, Ververidis Vasilis 563, Stockagogo, Craig Barhorst 564, Becris 565, IMG Stock Studio 567tl, Stillfx567tc, Matthew Jacques 567cc, Asharkyu 567cr, Ground Picture 567bl, Rawpixel.com 567bc, Julie Phipps 569, Goodluz 570, MJConline 574, MikhailSh 578, Chess Ocampo 581, Wichudapa 599, NatalieIme 600, Deyan Georgiev 602, Ket Sang 608, Christian Musat 613.

Text:

American Association for the Advancement of Science (AAAS): Changes in Melatonin Secretion during Day and Night Time (IMAGE), © THE KOREA ADVANCED INSTITUTE OF SCIENCE AND TECHNOLOGY (KAIST) 298; **American Heart Association, Inc:** Marma, A. K., & Lloyd-Jones, D. M. (2009). Systematic examination of the updated Framingham heart study general cardiovascular risk profile. Circulation, 120(5), 384-390 188; **Centers for Disease Control and Prevention:** Developed by the National Center for Health Statistics in collaboration with the National Center for Chronic Disease Prevention and Health Promotion (2000) 492; **Council on Foreign Relations:** Diana Roy, Deforestation of Brazil's Amazon Has Reached a Record High. What's Being Done?, August 24, 2022 524; **Frontiers Media S.A** © Frontiers Media S.A 327; **George E. P. Box:** George E. P. Box (innovator in statistical analysis) Box and Draper, 1987. Copyright © 1987 by John Wiley & Sons, Inc. 575; **Intergovernmental Science-Policy Platform on Biodiversity and Ecosystem Services (IPBES):** SUMMARY FOR POLICYMAKERS OF THE IPBES GLOBAL ASSESSMENT REPORT ON BIODIVERSITY AND ECOSYSTEM SERVICES Copyright © 2019, Intergovernmental Science-Policy Platform on Biodiversity and Ecosystem Services (IPBES) ISBN No: 978-3-947851-13-3 101; **J. Robert Oppenheimer:** Robert Oppenheimer Barnett 1949 572; **John N. Bahcall:** John Bahcall (commenting on the Hubble space telescope's capabilities) 585; **John Wiley and Sons:** Boisvert-Marsh, L., Périé, C., & de Blois, S. (2014). Shifting with climate? Evidence for recent changes in tree species distribution at high latitudes. Ecosphere, 5(7), 1-33. 551t, 551b; **Karl Popper:** Popper, K. R. (1992). The logic of scientific discovery. United Kingdom: Routledge. 579, Karl Popper Popper 1992b 583; **Louis Pasteur:** Louis Pasteur 573; **Marie Curie:** Marie Curie 584; **Mr G's Environmental Systems:** Why are Biomes where they are?, Copyright © 2007-2022 Mr G's Environmental Systems 205; **National Academy of Science:** Freeman, B. G., & Class Freeman, A. M. (2014). Rapid upslope shifts in New Guinean birds illustrate strong distributional responses of tropical montane species to global warming. Proceedings of the National Academy of Sciences, 111(12), 4490-4494. 552; **National Library of Medicine:** Rico, A., Brody, D., Coronado, F., Rondy, M., Fiebig, L., Carcelen, A., … Dahl, B. A. (2016). Epidemiology of Epidemic Ebola Virus Disease in Conakry and Surrounding Prefectures, Guinea, 2014–2015. Emerging Infectious Diseases, 22(2), 178–183. 331; **National Oceanic and Atmospheric Administration (NOAA):** Trends in Atmospheric Carbon Dioxide, Monthly Average Mauna Loa CO2, National Oceanic and Atmospheric Administration (NOAA) 379t, 379b; **Niels Bohr:** Niels Bohr (a Danish physicist who helped us understand how atoms work) Ellis 1970 580; **Our World In Data:** Two centuries of rapid global population growth will come to an end, © Our World In Data 103; **Oxford University Press:** Phenotypic Plasticity: Functional and Conceptual Approaches. (2004). United Kingdom: Oxford University Press. 478, John A. Endler (1980). Natural Selection on Color Patterns in Poecilia reticulata. Evolution, 34(1), 76–91. doi:10.2307/2408316 520; **PMEL Carbon Group:** Hawaii Carbon Dioxide Time-Series; © PMEL Carbon Group 204; **The University of Waikato Te Whare Wānanga o Waikato:** How the distribution of a species is affected by limiting factors; © The University of Waikato Te Whare Wānanga o Waikato 201; **Thomas Henry Huxley:** Thomas Henry Huxley 583, 585; **Uang Tzu Taois:** Chuang Tzu Taoist text (written more than 2,000 years ago) 576; **World Health Organisation:** Situation by WHO Region, WHO Coronavirus (COVID-19) Dashboard, © WHO 327.

Contents

Syllabus roadmap   xi

Author's introduction to the third edition   xii

Theme A

A Unity and diversity – Molecules   2

A1.1 Water   3
   A1.1.1 – The medium of life   3
   A1.1.2 – The structure and polarity of water molecules   5
   A1.1.3 – Cohesion of water molecules   6
   A1.1.4 – Adhesion between water and other polar substances   7
   A1.1.5 – The solvent properties of water   8
   A1.1.6 – The physical properties of water   10

A1.2 Nucleic acids   14
   A1.2.1 – DNA is the universal genetic material   14
   A1.2.2 – The structure of nucleotides   15
   A1.2.3 – Sugar to phosphate "backbone" of DNA and RNA   16
   A1.2.4 – Nitrogenous bases within nucleic acids   17
   A1.2.5 – The structure of RNA   18
   A1.2.6 – The structure of DNA   19
   A1.2.7 – Distinguishing between DNA and RNA   20
   A1.2.8 – The importance of complementary base pairing   22
   A1.2.9 – Storage of genetic information   23
   A1.2.10 – Genetic uniqueness   24

A Unity and diversity – Cells   28

A2.2 Cell structure   29
   A2.2.1 – Cells and the functions of life   29
   A2.2.2 – Cells and the microscope   30
   A2.2.3 – Advanced microscopy   33
   A2.2.4 – Structures common to all cells   35
   A2.2.5 – The prokaryote cell   36
   A2.2.6 – The eukaryote cell   40
   A2.2.7 – Unicellular organisms   48
   A2.2.8 – Different types of eukaryotic cells   49
   A2.2.9 – Atypical eukaryotes   50
   A2.2.10 and A2.2.11 – Electron micrograph skills 52

A Unity and diversity – Organisms   56

A3.1 Diversity of organisms   57
   A3.1.1 – Variation between organisms   57
   A3.1.2 – Species as groups of organisms   59
   A3.1.3 – The binomial naming system   59
   A3.1.4 – Biological species   61
   A3.1.5 – Distinguishing between populations and species   62
   A3.1.6 – Diversity in chromosome numbers   64
   A3.1.7 – Karyotypes   65
   A3.1.8 – Unity and diversity of genomes   68
   A3.1.9 – Eukaryote genomes   70
   A3.1.10 – Genome sizes   73
   A3.1.11 – Whole genome sequencing   74

A2.1 is not included as it is for HL students only.

A Unity and diversity – Ecosystems   78

A4.1 Evolution and speciation   79
A4.1.1 – Evolution   79
A4.1.2 – Biochemical evidence for evolution   81
A4.1.3 – Selective breeding   83
A4.1.4 – Homologous and analogous structures   84
A4.1.5 – Convergent evolution   86
A4.1.6 – Speciation   87
A4.1.7 – Reproductive isolation and differential selection   88

A4.2 Conservation of biodiversity   93
A4.2.1 – Biodiversity exists in many forms   93
A4.2.2 – Has biodiversity changed over time?   95
A4.2.3 – Human activities and the rate of species extinction   97
A4.2.4 – Human activities and ecosystem loss   99
A4.2.5 – A biodiversity crisis   100
A4.2.6 – Causes of the biodiversity crisis   102
A4.2.7 – Conservation of biodiversity   104
A4.2.8 – The EDGE of Existence programme   107

Theme B

B Form and function – Molecules   110

B1.1 Carbohydrates and lipids   111
B1.1.1 – The variety of compounds containing carbon   111
B1.1.2 and B1.1.3 – Condensation and hydrolysis   115
B1.1.4 – Monosaccharides   118
B1.1.5 – Polysaccharides and energy storage   119
B1.1.6 – Cellulose as a structural polysaccharide   121
B1.1.7 – Conjugated carbon molecules   123
B1.1.8 – Lipid solubility   124
B1.1.9 – Triglycerides and phospholipids   124
B1.1.10 – Properties of fatty acids   126
B1.1.11 – Adipose tissue   127
B1.1.12 – Phospholipid bilayers   127
B1.1.13 – Steroid hormones   128

B1.2 Proteins   131
B1.2.1 – The common structure of amino acids   131
B1.2.2 – Condensation reactions bond amino acids together   132
B1.2.3 – Essential amino acids   133
B1.2.4 – The vast variety of polypeptides   133
B1.2.5 – The effect of pH and temperature   134

B Form and function – Cells   138

B2.1 Membranes and membrane transport   139
B2.1.1 and B2.1.2 – Membrane structure   139
B2.1.3 – Diffusion across cellular membranes   141
B2.1.4 – Membrane proteins   142
B2.1.5 and B2.1.6 – Membrane transport   143
B2.1.7 – Active transport and pump proteins   146
B2.1.8 – Membrane permeability   146
B2.1.9 – Glycoproteins and glycolipids   147
B2.1.10 – The fluid mosaic model   148

B2.2 Organelles and compartmentalization 152
    B2.2.1 – Cell compartmentalization 152
    B2.2.2 – The nucleus and cytoplasm 155
    B2.2.3 – Compartmentalization of the cytoplasm 156

B2.3 Cell specialization 159
    B2.3.1 – Cell reproduction and organism development 159
    B2.3.2 – Stem cells 160
    B2.3.3 – Stem cell niches 161
    B2.3.4 – Types of stem cell 162
    B2.3.5 – Cell size and specialization 163
    B2.3.6 – Constraints on cell size 166

B Form and function – Organisms 170

B3.1 Gas exchange 171
    B3.1.1 – The exchange of gases between organisms and their environment 171
    B3.1.2 – Gas exchange surfaces 172
    B3.1.3 – Concentration gradients at exchange surfaces in animals 172
    B3.1.4 – Gas exchange in mammalian lungs 174
    B3.1.5 – Lung ventilation 175
    B3.1.6 – Lung volume 177
    B3.1.7 – Gas exchange in leaves 178
    B3.1.8 – Leaf tissue distribution 179
    B3.1.9 – Transpiration 179
    B3.1.10 – Stomata 180

B3.2 Transport 183
    B3.2.1 – Capillaries and chemical exchange 183
    B3.2.2 – Arteries and veins 184
    B3.2.3 – Adaptations of arteries 185
    B3.2.4 – Measuring the pulse rate 186
    B3.2.5 – Adaptations of veins 186
    B3.2.6 – Occlusion of coronary arteries 187
    B3.2.7 – Water transport from roots to leaves 189
    B3.2.8 – Adaptations of xylem vessels 190
    B3.2.9 – Tissues in a dicotyledonous stem 191
    B3.2.10 – Tissues in a dicotyledonous root 192

B Form and function – Ecosystems 196

B4.1 Adaptation to environment 197
    B4.1.1 – What is a habitat? 197
    B4.1.2 – Adaptation to the abiotic environment 198
    B4.1.3 – Abiotic variables 200
    B4.1.4 – Limiting factors 201
    B4.1.5 – Coral reef formation 203
    B4.1.6 – Terrestrial biomes 204
    B4.1.7 – Biomes, ecosystems and communities 205
    B4.1.8 – Hot deserts and tropical rainforests 207

B4.2 Ecological niches 212
    B4.2.1 – Species and ecosystems 212
    B4.2.2 – Obligate anaerobes, facultative anaerobes and obligate aerobes 213
    B4.2.3 – Photosynthesis 215
    B4.2.4 – Holozoic nutrition 216
    B4.2.5 – Mixotrophic nutrition 216
    B4.2.6 – Saprotrophic nutrition 217
    B4.2.7 – Diversity of nutrition in archaea 218

B4.2.8 – The relationship between dentition and diet 218
B4.2.9 – Adaptations of herbivores and plants 222
B4.2.10 – Adaptations of predators and prey 223
B4.2.11 – Harvesting light 227
B4.2.12 – Ecological niches 228
B4.2.13 – Competitive exclusion 229

Theme C

C Interaction and interdependence – Molecules 234

C1.1 Enzymes and metabolism **235**
C1.1.1 – Enzymes as catalysts 235
C1.1.2 – Metabolism 236
C1.1.3 – Anabolism and catabolism 236
C1.1.4 – Globular proteins and active sites 238
C1.1.5 and C1.1.10 – Enzyme activation 239
C1.1.6 – Molecular motion 241
C1.1.7 – Mechanism of enzyme action 242
C1.1.8 – Factors affecting enzyme-catalysed reactions 242
C1.1.9 – Measuring enzyme-catalysed reactions 245

C1.2 Cell respiration **248**
C1.2.1 – ATP structure and function 248
C1.2.2 – Life processes within cells require ATP 249
C1.2.3 – ATP and ADP 253
C1.2.4 and C1.2.5 – ATP and cell respiration 256
C1.2.6 – The rate of cell respiration 253

C1.3 Photosynthesis **256**
C1.3.1 – Light energy and life processes 256
C1.3.2 and C1.3.3 – The equation for photosynthesis 257
C1.3.4 – Photosynthetic pigments and light absorption 258
C1.3.5 and C1.3.6 – Absorption and action spectra 261
C1.3.7 – Measuring the rate of photosynthesis 263
C1.3.8 – Carbon dioxide levels and future rates of photosynthesis 267

C Interaction and interdependence – Cells 272

C2.1 is not included as it is for HL students only.

C2.2 Neural signalling **273**
C2.2.1 – The role of neurons 273
C2.2.2 and C2.2.3 – Generation and transmission of an impulse along a neuron 275
C2.2.4 – The speed of nerve impulses 277
C2.2.5 and C2.2.6 – Synapses, neurotransmitters, and their actions 280
C2.2.7 – Acetylcholine and the generation of a postsynaptic potential 281

C Interaction and interdependence – Organisms 286

C3.1 Integration of body systems **287**
C3.1.1 – Coordinating systems 287
C3.1.2 – Hierarchy of body subsystems 288
C3.1.3 – Integration of organs in animals 289
C3.1.4 – The brain and information processing 290
C3.1.5 – The spinal cord and unconscious processes 292
C3.1.6 – Sensory neurons and conveying information 292
C3.1.7 – Motor neurons and muscle stimulation 293
C3.1.8 – Nerve fibres 294
C3.1.9 – Pain reflex arcs 295

C3.1.10 – The cerebellum and skeletal muscle coordination 296
C3.1.11 – Melatonin secretion and sleep patterns 297
C3.1.12 – Epinephrine and vigorous activity 298
C3.1.13 – The hypothalamus, pituitary gland and endocrine system 299
C3.1.14 – Feedback control of heart rate 300
C3.1.15 – Feedback control of ventilation rate 301
C3.1.16 – Control of peristalsis in the alimentary canal 303

C3.2 Defence against disease 307
C3.2.1 – Infectious diseases are caused by pathogens 307
C3.2.2 – Skin and mucous membranes as the first line of defence 309
C3.2.3 – Blood clotting minimizes blood loss and infection 310
C3.2.4 – A two-layered immune system: innate and adaptive 311
C3.2.5 – The role of phagocytes 312
C3.2.6 – The role of lymphocytes 313
C3.2.7 – Antigens trigger antibody production 314
C3.2.8 – The role of helper T-lymphocytes 316
C3.2.9 – Activation of a B-lymphocyte results in cloning 317
C3.2.10 – The role of memory cells 318
C3.2.11 – HIV transmission 319
C3.2.12 – The result of HIV infection 319
C3.2.13 – Antibiotics against bacterial infections 320
C3.2.14 – Pathogenic resistance to antibiotics 321
C3.2.15 – Zoonotic diseases 322
C3.2.16 – Vaccines and immunity 324
C3.2.17 – The role of herd immunity 325
C3.2.18 – Evaluating COVID-19 data 325

C Interaction and interdependence – Ecosystems 332

C4.1 Populations and communities 333
C4.1.1 – Populations 333
C4.1.2 – Estimating population size 334
C4.1.3 – Sampling sessile organisms 335
C4.1.4 – Sampling motile organisms 336
C4.1.5 – Carrying capacity 337
C4.1.6 – Negative feedback 338
C4.1.7 – Population growth 339
C4.1.8 – Modelling population growth 342
C4.1.9 – Communities 342
C4.1.10 – Intraspecific relationships 344
C4.1.11 – Interspecific relationships 344
C4.1.12 – Mutualism 345
C4.1.13 – Endemic and invasive species 348
C4.1.14 – Interspecific competition 349
C4.1.15 – The chi-squared test 351
C4.1.16 – Predator–prey relationships 353
C4.1.17 – Control of populations 354
C4.1.18 – Allelopathy and antibiotic secretion 356

C4.2 Transfers of energy and matter 359
C4.2.1 – Ecosystems are open systems 359
C4.2.2 – Sunlight sustains most ecosystems 360
C4.2.3 – The flow of energy 362
C4.2.4 – Food chains and food webs 362
C4.2.5 – Decomposers 364
C4.2.6 – Autotrophs 364
C4.2.7 – Energy sources 365

C4.2.8 – Heterotrophs	366
C4.2.9 – The release of energy by cell respiration	367
C4.2.10 – Trophic levels	367
C4.2.11 – Energy pyramids	368
C4.2.12 – Energy loss between trophic levels	370
C4.2.13 – Heat loss from cell respiration	371
C4.2.14 – The number of trophic levels	372
C4.2.15 – Primary production	373
C4.2.16 – Secondary production	374
C4.2.17 – The carbon cycle	374
C4.2.18 – Carbon sinks and sources	376
C4.2.19 – The release of carbon dioxide during combustion	376
C4.2.20 – The Keeling Curve	378
C4.2.21 – The dependence on atmospheric oxygen and carbon dioxide	380
C4.2.22 – The recycling of chemical elements	381

Theme D

D Continuity and change – Molecules	**386**
D1.1 DNA replication	**387**
D1.1.1 – The role of DNA replication	387
D1.1.2 and D1.1.3 – Semi-conservative replication and complementary base pairing	388
D1.1.4 – Amplifying and separating DNA	390
D1.1.5 – Applications of amplifying and separating DNA	392
D1.2 Protein synthesis	**395**
D1.2.1 – The synthesis of RNA	395
D1.2.2 – Hydrogen bonding and complementary base pairing	396
D1.2.3 – DNA templates	397
D1.2.4 – The expression of genes	397
D1.2.5 – The synthesis of polypeptides	398
D1.2.6 – RNA and ribosomes	398
D1.2.7 – RNA complementary base pairing	400
D1.2.8 – The genetic code	401
D1.2.9 – mRNA codons	402
D1.2.10 – Producing a polypeptide chain	403
D1.2.11 – Changing the protein structure	404
D1.3 Mutation and gene editing	**406**
D1.3.1 – Gene mutations	406
D1.3.2 – Base substitutions	409
D1.3.3 – Insertions and deletions	411
D1.3.4 – Mutagens and replication errors	413
D1.3.5 – Location of mutations	416
D1.3.6 – Mutations in germ cells and somatic cells	417
D1.3.7 – Genetic variation	419
D Continuity and change – Cells	**424**
D2.1 Cell and nuclear division	**425**
D2.1.1 – Generating new cells	425
D2.1.2 – Cytokinesis	426
D2.1.3 – Cytoplasm division	427
D2.1.4 – Nuclear division	428
D2.1.5 – DNA replication	428

D2.2 is not included as it is for HL students only.

D2.1.6 – DNA condensation and chromosome movement   429
D2.1.7 – Mitosis   430
D2.1.8 – Identifying the phases of mitosis   432
D2.1.9 – Meiosis   433
D2.1.10 – Non-disjunction   435
D2.1.11 – Genetic diversity   437

D2.3 Water potential   **440**
D2.3.1 – Water as a solvent   440
D2.3.2 – Water movement in relation to solute concentration   442
D2.3.3 and D2.3.4 – Hypotonic and hypertonic solutions and osmosis   443
D2.3.5 – Water movement without cell walls   444
D2.3.6 – Water movement with cell walls   446
D2.3.7 – Isotonic solutions   447

D Continuity and change – Organisms   450

D3.1 Reproduction   **451**
D3.1.1 – Sexual and asexual reproduction   452
D3.1.2 – The role of meiosis and gametes   453
D3.1.3 – Male and female gametes   455
D3.1.4 – Male and female reproductive systems   456
D3.1.5 – Hormonal control of the menstrual cycle   457
D3.1.6 – The process of fertilization   460
D3.1.7 – In vitro fertilization   462
D3.1.8 – Sexual reproduction in plants   462
D3.1.9 – Insect pollination   463
D3.1.10 – Cross-pollination in plants   464
D3.1.11 – Self-incompatibility mechanisms   465
D3.1.12 – The role of seeds   465

D3.2 Inheritance   **470**
D3.2.1 – Haploid gametes and diploid zygotes   470
D3.2.2 – Genetic crosses in flowering plants   471
D3.2.3 – Combinations of alleles   474
D3.2.4 – Phenotype   476
D3.2.5 – Dominant and recessive alleles   476
D3.2.6 – Phenotypic plasticity   477
D3.2.7 – Recessive genetic conditions   478
D3.2.8 – Single-nucleotide polymorphisms and multiple alleles   480
D3.2.9 – ABO blood groups   481
D3.2.10 – Intermediate and dual phenotypes   483
D3.2.11 – Sex determination   484
D3.2.12 – Haemophilia   485
D3.2.13 – Pedigree charts   486
D3.2.14 – Continuous variation   489
D3.2.15 – Box-and-whisker plots   491

D3.3 Homeostasis   **496**
D3.3.1 – Maintaining the body's internal environment   496
D3.3.2 – Negative feedback mechanisms   497
D3.3.3 – The role of hormones   498
D3.3.4 – Type 1 and type 2 diabetes   499
D3.3.5 and D3.3.6 – Body temperature control   500

D Continuity and change – Ecosystems 508

D4.1 Natural selection 509
    D4.1.1 – Evolutionary change 509
    D4.1.2 – Sources of variation 511
    D4.1.3 – Overproduction and competition 513
    D4.1.4 – Selection pressure 513
    D4.1.5 – Intraspecific competition 515
    D4.1.6 – Heritable traits 516
    D4.1.7 – Sexual selection 517
    D4.1.8 – Modelling selection pressures 518

D4.2 Stability and change 522
    D4.2.1 – Sustainability of natural ecosystems 522
    D4.2.2 – Requirements for sustainability 523
    D4.2.3 – Tipping points 524
    D4.2.4 – Mesocosms 525
    D4.2.5 – Keystone species 527
    D4.2.6 – Sustainable harvesting of natural resources 528
    D4.2.7 – Sustainability of agriculture 530
    D4.2.8 – Eutrophication 532
    D4.2.9 – Biomagnification 532
    D4.2.10 – Microplastic and macroplastic pollution 534
    D4.2.11 – Rewilding 535

D4.3 Climate change 538
    D4.3.1 – Human activity and climate change 538
    D4.3.2 – Global warming 541
    D4.3.3 – Tipping points 545
    D4.3.4 – Polar habitat change 547
    D4.3.5 – Ocean current change 548
    D4.3.6 – Range shifts 550
    D4.3.7 – Ecosystem collapse 553
    D4.3.8 – Carbon sequestration 554

Theory of Knowledge in biology 563

Internal Assessment 586

Skills in the study of biology 598

Extended Essay 621

Index 628

Syllabus roadmap

The aim of the syllabus is to integrate concepts, topic content and the nature of Science through inquiry. Students and teachers are encouraged to personalize their approach to the syllabus to best fit their interests.

Theme	Level of organization			
	1. Molecules	2. Cells	3. Organisms	4. Ecosystems
A **Unity and diversity**	Common ancestry has given living organisms many shared features while evolution has resulted in the rich biodiversity of life on Earth.			
	A1.1 Water **A1.2** Nucleic acids	**A2.2** Cell structure	**A3.1** Diversity of organisms	**A4.1** Evolution and speciation **A4.2** Conservation of biodiversity
B **Form and function**	Adaptations are forms that correspond to function. These adaptations persist from generation to generation because they increase the chances of survival.			
	B1.1 Carbohydrates and lipids **B1.2** Proteins	**B2.1** Membranes and membrane transport **B2.2** Organelles and compartmentalization **B2.3** Cell specialization	**B3.1** Gas exchange **B3.2** Transport	**B4.1** Adaptation to environment **B4.2** Ecological niches
C **Interaction and interdependence**	Systems are based on interactions, interdependence and integration of components. Systems result in emergence of new properties at each level of biological organization.			
	C1.1 Enzymes and metabolism **C1.2** Cell respiration **C1.3** Photosynthesis	**C2.2** Neural signalling	**C3.1** Integration of body systems **C3.2** Defence against disease	**C4.1** Populations and communities **C4.2** Transfers of energy and matter
D **Continuity and chance**	Living things have mechanisms for maintaining equilibrium and for bringing about transformation. Environmental change is a driver of evolution by natural selection.			
	D1.1 DNA replication **D1.2** Protein synthesis **D1.3** Mutations and gene editing	**D2.1** Cell and nuclear division **D2.3** Water potential	**D3.1** Reproduction **D3.2** Inheritance **D3.3** Homeostasis	**D4.1** Natural selection **D4.2** Stability and change **D4.3** Climate change

Authors' introduction to the third edition

Welcome to your study of IB Diploma Programme (DP) biology. This is the third edition of Pearson's highly successful Standard Level (SL) biology book, first published in 2007. It has been rewritten to match the specifications of the new IB biology curriculum for first assessments in 2025 and provides comprehensive coverage of the course. It is our intention as authors of this textbook to open a door to biological knowledge that will provide a pathway towards an ever-present curiosity of life, the factors that affect it today, and the factors that may affect it in the future.

While there is much new and updated material in this textbook, we have kept and refined the features that made the previous editions so successful and effective. We hope our knowledge and enthusiasm for biology as well as our understanding of the IB biology requirements will be passed onto you.

Content

This book covers the content that is set out in the IB DP biology subject guide for first assessments in 2025. It utilizes the overarching theme of Nature of Science (NOS) to provide the means for you to accomplish the following aims:

1. to develop conceptual understanding that allows connections to be made between different areas of the subject, and to other DP science subjects
2. to acquire and apply a body of knowledge, methods, tools and techniques that characterize science
3. to develop the ability to analyse, evaluate and synthesize scientific information and claims
4. to develop the ability to approach unfamiliar situations with creativity and resilience
5. to design and model solutions to local and global problems in a scientific context
6. to develop an appreciation of the possibilities and limitations of science
7. to develop technology skills in a scientific context
8. to develop the ability to communicate and collaborate effectively
9. to develop awareness of the ethical, environmental, economic, cultural and social impact of science.

Chapters are presented in the same sequence as provided in the subject guide. There are four main themes:

A. Unity and diversity
B. Form and function
C. Interaction and interdependence
D. Continuity and change

Each theme is then discussed at four different levels of organization. They are:

1. Molecules
2. Cells
3. Organisms
4. Ecosystems

Each topic begins with an introductory image and caption supplying a brief entry point into its content. Guiding Questions are then presented for further clarification of chapter content.

Guiding Questions

What plausible hypothesis could account for the origin of life?

What intermediate stages could there have been between non-living matter and the first living cells?

The text covers the course content with all scientific terms explained. We have been careful to apply the same terminology you will see in IB assessments.

Linking Questions that relate topics to one another can be found in each chapter. When encountered, Linking Questions should be considered in order to understand how other concepts from within the course relate to those currently being discussed. When used effectively, Linking Questions can provide an excellent tool for revision.

For what reasons is heredity an essential feature of living things?

Each chapter concludes with Guiding Questions revisited and a summary of the chapter. The summary presents key points from the chapter you should be especially aware of.

Guiding Question revisited

How can viruses exist with so few genes?

Nature of Science

Throughout the course you are encouraged to think about the nature of scientific knowledge and the scientific process as it applies to biology. Examples are given of the evolution of biological theories as new information is gained, the use of models to conceptualize our understandings, and the ways in which experimental work is enhanced by modern technologies. Ethical considerations, environmental impacts, the importance of objectivity, and the responsibilities regarding scientists' code of conduct are also considered here. The emphasis is on appreciating the broader conceptual themes in context. We recommend that you familiarize yourself with these examples to enrich your understanding of biology.

Throughout the book you will find NOS themes and questions emerging across different topics. We hope they help you to develop your own skills in scientific literacy.

Nature of Science

Science has progressed and continues to progress with the development of new study techniques. Not only has the microscope increased our knowledge of the cell, but ultracentrifuges and fractionation of cells have also greatly enhanced our understanding of the cell and its organelles.

Key to feature boxes

A popular feature of our past editions is maintained in this book, that is the different coloured boxes interspersed throughout each chapter. These boxes can be used to enhance your learning.

 Global context

The impact of the study of biology is global, and includes environmental, political and socio-economic considerations. Examples of these are given to help you see the importance of biology in an international context. These examples also illustrate some of the innovative and cutting-edge aspects of research in biology.

 Thanks to modern communication technologies, it is possible for scientists working all over the world to collaborate and contribute to a scientific endeavour such as sequencing the genome of plants that help feed the world. Rice is one example: biologists from 10 countries contributed to sequencing the first rice genome.

 Skills in the study of biology

Surface area-to-volume ratio. Full details on how to carry out this activity with a worksheet are available in the eBook.

These boxes indicate links to the skills section of the course, including ideas for laboratory work and experiments that will support your learning and help you prepare for the Internal Assessment. These link to further resources in the eBook (look out for the grey icon).

 Theory of Knowledge

When you study the action of sarcomeres, how much is your knowledge limited by two-dimensional models, such as Figure 3?

These questions, which are mostly from the Theory of Knowledge (TOK) guide, stimulate thought and consideration of knowledge issues as they arise in context. The questions are open-ended and will help trigger critical thinking and discussion.

 Key fact

The sequence of nitrogenous bases in DNA, later transcribed into RNA, forms the basis of the genetic code.

Key facts are drawn out of the main text and highlighted in bold. These boxes will help you to identify the core learning points within each section. They also act as a quick summary for review.

 Hint for success

You are not required to know all the names of the intermediate molecules of the respiration process. However, you must understand the steps and the overall products.

These boxes give hints on how to approach questions, and suggest approaches that examiners like to see. They also identify common pitfalls in understanding, and omissions made in answering questions.

Challenge yourself

These boxes contain probing questions that encourage you to think about the topic in more depth, and may take you beyond the syllabus content. They are designed to be challenging and to make you think.

Challenge yourself

1. Using Figure 8, showing the DNA profiles from six suspects, can you identify which one matches the DNA profile of the blood stain found at the crime scene?

 Interesting fact

These give background information that will add to your wider knowledge of the topic and make links with other topics and subjects. Aspects such as historic notes on the life of scientists and origins of names are included here.

Where does the term gene knockout come from? In contact sports such as boxing, a knockout marks the end of the combat, because the boxer who has been knocked out is no longer able to stand and fight. A gene that has been knocked out will no longer be able to make the protein that produced the original effect or trait

Questions

There are three types of question in this book.

1. Worked examples with solutions

Worked examples appear at intervals in the text and are used to illustrate the concepts covered. They are followed by the solution, which shows the thinking and the steps used in solving the problem.

Worked example

The length of an image you are looking at is 50 mm. If the actual length of the subject of the image is 5 μm, what is the magnification of the image?

Solution

Magnification = 50 mm/5 μm = 50,000 μm/5 μm = 10,000×

Or

Magnification = 50 mm/5 μm = 50×10^{-3} m/1×10^{-6} m = 10,000×

2. Exercises

These questions are found at the end of each chapter. They allow you to apply your knowledge to test your understanding of what you have just been reading. The answers to these are accessed via icons on the first page of each chapter in the eBook. Exercise answers can also be found at the back of the eBook.

Exercises

Q1. Explain why the obligate parasitism shown by viruses may have been a major factor in convergent evolution within the group.

3. Practice questions

These questions are found at the end of each group of chapters displaying a common theme and level of organization. The significance of these questions is that they are IB exam-style questions. The mark schemes used by examiners when marking these questions are accessed via icons in the eBook next to the questions. These questions and mark schemes are essential in providing insight into the depth of comprehension necessary to achieve success in an IB exam.

A2 Practice questions

1. **(a)** An organelle is a discrete structure within a cell with a specific function. In the table below, identify the missing organelles and outline the missing functions.

Name of organelle	Structure of organelle	Function of organelle
Nucleus	Region of the cell containing chromosomes, surrounded by a double membrane, in which there are pores.	Storage and protection of chromosomes.
Ribosome	Small spherical structures, consisting of two subunits.	
	Spherical organelles, surrounded by a single membrane and containing hydrolytic enzymes.	Digestion of structures that are not needed within cells.
	Organelles surrounded by two membranes, the inner of which is folded inwards.	

(4)

(b) The table above shows some of the organelles found in a particular cell. Discuss what type of cell this could be.

(2)

(Total 6 marks)

eBook

In your eBook you will find more information on the Skills section of the course, including detailed suggestions for laboratory work, and the answers to the exercises and practice questions found in the text. You will also find links to videos and command term worksheets in the Resources tab of your eBook account. In addition, there are auto-marked quizzes in the Exercises tab of your eBook account (see screenshot below).

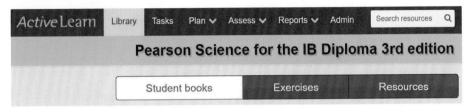

We truly hope that this book and the accompanying online resources help you enjoy the fascinating subject of IB biology. We wish you success in your studies.

Alan Damon, Randy McGonegal and William Ward

THEME **A** **Unity and diversity**
1 Molecules

This is DNA, one of the molecules classified as a nucleic acid and a molecule that is integral to life on Earth. The molecules that are important to life are diverse and complex. Yet their basic structures are largely consistent from species to species. This allows us to study the fundamental structures and functions of these molecules and apply that knowledge to all living organisms. In this chapter, you will first study the solvent of all biochemically important molecules, water. Later, you will consider the structure of nucleic acids.

A1.1 Water

Guiding Questions

What physical and chemical properties of water make it essential for life?

What are the challenges and opportunities of water as a habitat?

What makes water essential for living organisms? What physical and chemical properties does water have that provide essential benefits to aquatic, marine and terrestrial organisms? What opportunities and challenges does water pose for life? These are not questions designed to be answered in one or more short statements. They are questions that deserve to be explored. A portion of this chapter will attempt to begin that exploration.

Life first evolved in water and all living things are still dependent on this amazing molecule. Fortunately, we live on a planet where water exists in all three states: there is abundant liquid water, water vapour and ice. Water, as a polar molecule, is an excellent solvent for the vast majority of elements and compounds necessary for life. Water molecules are found inside and outside cells, and chemical communication in and out of cells must occur in a water environment.

Water has both advantages and disadvantages for the aquatic and marine organisms that use it as a habitat. Advantages include the fact that water provides buoyancy and stable thermal properties for these organisms. Disadvantages include its relatively high viscosity compared to air. This means that many organisms living in water have adapted their body shape and propulsion mechanisms in order to move easily through an aquatic environment.

A1.1.1 – The medium of life

A1.1.1 – Water as the medium for life

Students should appreciate that the first cells originated in water and that water remains the medium in which most processes of life occur.

Life on Earth has never been possible without water. Imagine a primordial planet slowly cooling from its original molten mass. That primitive Earth would not have had any water because of the extremely high temperatures at its centre *and* on its surface.

The surface of the Earth may have looked like this early in its history, with magma giving off tremendous heat at the surface.

Approximately 70% of our planet's surface is covered by water. The deepest parts of the Pacific Ocean are deeper than the height of the highest land peaks.

Every solution where water is the solvent is called an aqueous solution. Thus, cytoplasm, rivers, blood and oceans are all aqueous solutions.

Challenger Deep (the lowest known portion of the Mariana Trench) is 10,984 m below the surface of the Pacific Ocean. Mount Everest (the tallest known land mass) is 8,848 m above sea level. The difference between those points is over 19 km or 12 miles.

The origin and evolution of the first cells could not begin until temperatures cooled enough for water to form and, later, for the water cycle to begin. We take for granted the changes that water makes as it goes between its solid, liquid and gaseous phases. Earth's varied temperatures allow these changes. That was not the case in our planet's early history.

It is thought that the first cells formed and slowly evolved in the oceans. Cells require a complex series of biochemical reactions. This means a **solvent** is needed for reactions to occur. Ocean water provided the source for that solvent. The first cells evolved a membrane to separate the water in the cytoplasm from the "ocean water".

When most people think of water, their first thoughts are about the water they drink and bathe or swim in. But water is more widespread than that. Below are a few examples of where the importance of water as a solvent is vital to living organisms.

Water is the solvent that:

- makes up the fluid (cytoplasm) in all cells where all cellular reactions occur
- makes up the fluid inside all organelles in cells
- is found between cells of multicellular organisms (intercellular or tissue fluid)
- permits transport of substances into and out of cells
- is essential to blood and many other body fluids in humans and other organisms
- provides the medium in which all organisms in oceans, lakes and rivers live.

Nature of Science

Measurements in science often change over time. If you research the world's deepest and tallest points you may find slightly different numbers (meters below and above sea level). There are various possible reasons including: how recently the data point was taken; what method was used to obtain the data; whether or not the data change over time due to natural causes. Can you think of other reasons for the data to vary?

A1.1.2 – The structure and polarity of water molecules

A1.1.2 – Hydrogen bonds as a consequence of the polar covalent bonds within water molecules

Students should understand that polarity of covalent bonding within water molecules is due to unequal sharing of electrons and that hydrogen bonding due to this polarity occurs between water molecules.

Students should be able to represent two or more water molecules and hydrogen bonds between them with the notation shown below to indicate polarity.

To understand the properties of water and its importance to living organisms, it is necessary to understand the molecular structure of water molecules.

◀ **A1.1 Figure 1** This image shows the covalent bonds in a water molecule. Each of two hydrogen atoms is bonded at an angle to a single oxygen atom. Remember that each of the two covalent bonds is a pair of shared electrons.

The covalent bonds between the oxygen atom and the two hydrogen atoms of a water molecule are categorized as **polar covalent bonds**.

You may remember from fundamental chemistry that covalent bonds form when two atoms share electrons. Electrons are negatively charged and the nucleus of an atom is positively charged (because of the protons). So, any equally shared electrons create a **non-polar covalent bond**. This is because neither of the atoms has a higher density of electrons than the other. Good examples of non-polar covalent bonds include the covalent bond between two carbons and the covalent bond between two hydrogens.

Polar covalent bonding results from an unequal sharing of electrons. In water, the single oxygen atom is bonded to two different hydrogen atoms. Each oxygen–hydrogen bond is a polar covalent bond. This results in a slight negative charge at the oxygen end of the molecule and a slight positive charge at the end with the two hydrogens.

Because of the open triangular shape of a water molecule, the two "ends" of each molecule have opposite charges. The oxygen side is slightly negative and the hydrogen side is slightly positive. This is why water is a polar molecule: it has different charges at each end. Because of this, water molecules interact with each other and other molecules in very interesting ways. Many of these interactions are explained by the usually short-lived (ephemeral) attractions between either two water molecules or between a water molecule and another type of charged atom (or ion). These ephemeral attractions are called **hydrogen bonds** and will be explained further in the following sections.

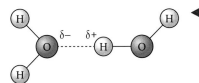

◀ **A1.1 Figure 2** Two water molecules showing a single hydrogen bond between them. The bonding force of each hydrogen bond (indicated by the dotted line) is weak. In liquid water, the bond is ephemeral because the water molecules continue to move around.

You may be used to seeing the Greek symbol Δ called delta. Δ is the capital letter symbol and δ is the corresponding small case letter symbol for delta.

The electrons being shared to create the covalent bonds within a water molecule are not being shared equally between the two atoms. In Figure 1, you see the symbols δ+ and δ- (delta positive and delta negative). These symbols represent areas of low or high electron density in the sharing of electrons to create a covalent bond. Each hydrogen atom is assigned a δ+ because that is an area of lesser electron density (thus a small positive charge due to the single proton of the hydrogen atom). The oxygen atom is assigned a δ- charge due to its high electron density.

Practise sketching from memory a diagram similar to the one shown in Figure 2. Include the hydrogen bond and the delta symbols and charges as shown. Practise adding a third and fourth water molecule with the same symbolism and orientation.

A1.1.3 – Cohesion of water molecules

A1.1.3 – Cohesion of water molecules due to hydrogen bonding and consequences for organisms

Include transport of water under tension in xylem and the use of water surfaces as habitats due to the effect known as surface tension.

Water molecules are highly cohesive. **Cohesion** occurs when *molecules of the same type are attracted to each other*. As you have seen, water molecules have a slightly positive end and a slightly negative end. Whenever two water molecules are near each other, the positive end of one attracts the negative end of another – this is hydrogen bonding. When water cools below its freezing point, the molecular motion of the water molecules slows to the point where the hydrogen bonds become locked into place and an ice crystal forms. Liquid water has molecules with a faster molecular motion, and the water molecules are able to influence each other, but not to the point where molecules stop their motion. This influence is highly important and leads to many of the physical and chemical properties of water. The ephemeral hydrogen bonding between liquid water molecules explains a variety of events, including the following.

You can float a paperclip on water because of the surface tension of the water. Make sure you maximize the surface area of the paperclip on the water if you try this.

- Why water has a **surface tension**. Surface tension is due to the fact that the layer of water molecules at the surface of a body of water does not have molecules of water above it. Because of this, the water molecules show a relatively strong cohesive force to the molecules immediately around and below them (no molecules are pulling upwards). This surface tension must be broken in order for an object to move through the surface from above. It is surface tension that causes you pain when you do a "belly flop" into a body of water. It is also surface tension that creates a habitat for some animals such as water striders and basilisk lizards.

A green basilisk (*Basiliscus plumifrons*) (found in Central America) running across the surface of water. Aided by its webbed feet to increase the surface area in contact with the water, the lizard must keep running in order to not break through the surface tension.

- How water is able to move as a "water column" in the vascular tissues of plants. The majority of water moving upwards in a plant moves within small tubes called **xylem**. Think of xylem as being similar to numerous tiny straws. When water evaporates from a leaf (in a process called transpiration) the water that evaporates in order to exit the leaf has cohesion to the water in a xylem tube that adjoins the exit point. The evaporation with corresponding cohesion creates a low pressure in this area called **tension**. This tension pulls on the other water molecules in the xylem tube so they all move upwards towards the leaf. The molecules are all cohesive to each other and all move up collectively. This evaporation occurs in small, controlled openings called stomata, which are usually found on the underside of leaves. The water that transpires from the leaf is replaced in the xylem in the root system of the plant.

Think of a xylem tube and the upwards movement of water as being similar to what happens when you use a straw in a drink. The suction you provide creates tension (low-pressure area at the top of the straw) and the fluid is moved upwards along the straw. The bottom of the straw in your drink is similar to the bottom of the xylem tubes found in the root system of a plant.

An example of the importance of cohesion. At the top, water is evaporating from a stoma (singular of stomata). Stomata are very small openings that can be opened or closed and are found primarily on the under surface of leaves. The evaporation of water from open stomata is called transpiration. The water is provided to the leaf by many xylem tubes. The transpiration of water creates tension (a low-pressure area in the leaf and xylem tube) and the polarity of water molecules pulls the entire water column to move towards the low-pressure area. The xylem tube within the leaf is continuous with the xylem in the stem and root. The water moving upwards is replaced by ground water moving into the root system.

A1.1.4 – Adhesion between water and other polar substances

A1.1.4 – Adhesion of water to materials that are polar or charged and impacts for organisms

Include capillary action in soil and in plant cell walls.

Water molecules are certainly not the only molecules in nature that exhibit polarity. An attraction between two *unlike* molecules due to hydrogen bonding is called **adhesion**. When water molecules are attracted to cellulose molecules by hydrogen bonding, the attraction is an example of adhesion because the hydrogen bonding is between two different kinds of molecule. Where is this important in nature?

Cohesion and adhesion are both a result of the polarity of water molecules. Cohesion is an attraction between two water molecules and adhesion is an attraction between a water molecule and another polar molecule that is not water.

- Water within the xylem. Cohesion and adhesion are both at work in this example. When the column of water is "pulled up", cohesion moves each molecule up; when the column is not being "pulled up", adhesion keeps the entire column from dropping down within the tube. The same phenomenon occurs when water is placed in a capillary tube – you can think of the xylem tissue in plants as being biological capillary tubes.

A capillary tube is a glass tube (similar to a straw) that has a very narrow inside opening. In this photo, a capillary tube has been inserted into a vessel filled with water with a red dye. The liquid will spontaneously climb upwards into the capillary tube due to adhesion and remain in a fixed position within the tube. The adhesion is the attraction between the inside surface of the glass tube and water molecules.

How do the various intermolecular forces of attraction affect biological systems?

- Capillary action in soil. Even soil that appears to be dry contains water in microscopic channels. These small channels act in a similar way to capillary tubes. Water molecules adhere to the polar molecules making up the soil and other water molecules are then sometimes moved by cohesion. The small root hairs of plants intrude into the water-filled spaces and water is taken into the root.

A1.1.5 – The solvent properties of water

A1.1.5 – Solvent properties of water linked to its role as a medium for metabolism and for transport in plants and animals

Emphasize that a wide variety of hydrophilic molecules dissolve in water and that most enzymes catalyse reactions in aqueous solution. Students should also understand that the functions of some molecules in cells depend on them being hydrophobic and insoluble.

As you have seen, water is a polar molecule and thus a polar solvent. In nature, water is almost always found as a solvent carrying one or more of a wide variety of other substances as solutes. Any solution that has water as the solvent is called an **aqueous solution**. Any substance that dissolves readily in water is described as **hydrophilic** (water loving) and any substance that does not dissolve easily is called **hydrophobic** (water fearing).

Hydrophilic molecules

The cytoplasm of a cell is a good example of an aqueous solution and contains a wide variety of water-soluble substances. These hydrophilic solutes include (among others) glucose, ions, amino acids and proteins. Some of the dissolved proteins in cells are the biological catalysts called enzymes. Reactions within the cytoplasm depend on enzymes to proceed at a rate necessary for life and at a temperature tolerated by that type of cell.

Water is an excellent medium for transporting dissolved substances. The water contained in xylem vessels of plants is not pure water. It is an aqueous solution that transports inorganic ions such as sodium, potassium and calcium. These and many other essential substances are hydrophilic; they dissolve easily in water and are transported upwards from the root system to the leaves.

The blood of many animals, including humans, is also an aqueous solution. The red and white blood cells are suspended in plasma. Plasma is an aqueous solution of an incredible array of molecules. Anyone looking at the results of a typical medical blood test can see the variety of solutes in this solution.

> The biochemistry of a cell occurs in its cytoplasm and also within membrane-bound organelles such as the nucleus and mitochondria. The fluids of these cellular environments use water as a solvent because most biochemically active molecules are polar and dissolve easily in an aqueous solvent.

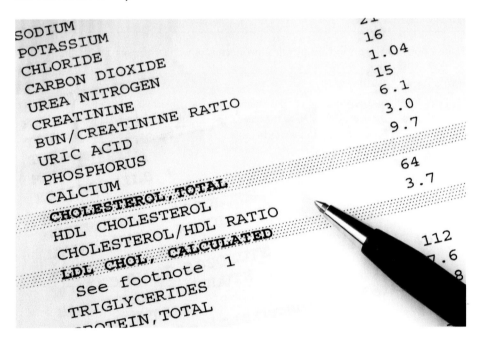

A small section of the results of a human blood test showing some of the dissolved substances in the aqueous portion of blood called plasma.

Hydrophobic molecules

Some non-polar (insoluble) molecules found in nature are important to living organisms. Here are some examples.

- Steroid hormones, such as oestradiol and testosterone, are able to pass directly through the plasma membrane and nuclear membrane of a cell. Steroid hormones can do this because they are hydrophobic and are able to pass directly through the hydrophobic layers of cell membranes.
- Many proteins have some sections that are hydrophilic and other sections that are hydrophobic. Membrane-bound proteins may use one or more hydrophobic areas to embed into the hydrophobic layers of a membrane while their hydrophilic section(s) extends into either the intercellular fluid or cytoplasm. This enables the protein to stay attached to the membrane but still interact with soluble substances in the surrounding cell fluids.

- The epidermal cells of leaves are capable of secreting a wax that is used to coat the leaves and is called the **cuticle**. This wax cuticle is hydrophobic and acts as a barrier to water entering and especially exiting the leaf by evaporation. Without this cuticle, leaves would quickly dehydrate because their function requires a thin, broad surface area exposed to the Sun.

A1.1.6 – The physical properties of water

A1.1.6 – Physical properties of water and the consequences for animals in aquatic habitats
Include buoyancy, viscosity, thermal conductivity and specific heat. Contrast the physical properties of water with those of air and illustrate the consequences using examples of animals that live in water and in air or on land, such as the black-throated loon (*Gavia arctica*) and the ringed seal (*Pusa hispida*).
Note: When students are referring to an organism in an examination, either the common name or the scientific name is acceptable.

Table 1 outlines the important physical properties of water compared with air.

Property	**Water**	**Air**
Buoyancy or buoyant force (an upwards force exerted on an object placed in the medium – either water or air)	Buoyant force equals the weight of the water displaced by the object. The buoyant force is upwards because there is more pressure from below (in the water) than above (in the air).	An object placed in air has an almost insignificant buoyant force. This force is equal to the weight of the air displaced by the object.
Viscosity	Water's resistance to an object moving through it.	Air's resistance to an object moving through it. Since air is far less dense than water, air's viscosity is far less.
Thermal conductivity	The ability of a substance to transfer heat. Water has a high thermal conductivity.	The thermal conductivity of air is very low compared to water.
Specific heat capacity	In simplest terms, water can absorb or give off a great deal of heat without changing temperature very much. Think of a body of water on a very cold night: even though the air may be very cold, a nearby body of water is relatively stable in temperature.	Air's ability to absorb or give off heat without changing temperature is very low compared to that of water. The temperature of the air changes easily and rapidly due to weather events.

A1.1 Table 1 Physical properties of water

The physical properties of water have important consequences for animals that live in aquatic habitats, such as the black-throated loon (*Gavia arctica*) and the ringed seal (*Pusa hispida*).

The black-throated loon is a beautiful bird that lives primarily in very cold regions of the Northern Hemisphere. As with most aquatic birds, the loon transfers regularly between land (for nesting), water (for feeding) and air (for flying). Even though this bird is capable of diving for food, it spends much of its time in water on the surface relying on the buoyant force of the water to float. The bird requires energy to overcome the viscosity of water to move across the water surface and even more when it dives for fish and other food sources below the surface. Webbed feet and efficient, streamlined body shape aid the loon in this movement. When the bird is in water, the high thermal conductivity of the water would cause the loon to lose more body heat than when it is in the air. Like many waterbirds, loons use an adaptation to prevent this. They have an oil gland near their tail and they use their beaks to rub this oil over their feathers to make them waterproof. When the air is very cold (below 1°C) the surrounding water is likely to be warmer than the air because the high specific heat of water allows its temperature to remain relatively stable in comparison to air.

Black-throated loon
(*Gavia arctica*)

You are not required to memorize the scientific names (genus and species) of example organisms.

The ringed seal is another animal that is common in cold environments of the Northern Hemisphere. This small seal is buoyant, although not as buoyant as a loon – less of its body is above the surface of the water when resting. It is buoyant enough to keep its snout above water easily and thus has an easily available supply of air. Seals spend a great deal of time swimming in and under the water to catch food (fish and invertebrates) and occasionally to escape a predator such as an orca. Their streamlined shape and paddle-like feet are great assets in overcoming the viscosity of water. But water has high thermal conductivity compared to air, so ringed seals need to minimize body heat loss. They do this by having a thick blubber under their skin. The blubber is insulation and reduces heat loss from the seals' internal organs. Like the black-throated loon, ringed seals are protected from very low air temperatures by the relatively high temperature of arctic water (compared to arctic air) which is due to the high specific heat of water.

Melting sea ice due to global warming is threatening many species, including seals, because their habitats are fundamentally changing in a very short period of time. No one country by itself can solve the problem of global warming.

What biological processes only happen at or near surfaces?

Ringed seal (*Pusa hispida*) ▶

Guiding Question revisited

What physical and chemical properties of water make it essential for life?

In this chapter we have described how and why water has:

- polar covalent bonds due to an unequal sharing of electrons between oxygen and hydrogen
- cohesive forces attracting one molecule of water to another
- adhesive forces attracting molecules of water to other types of polar molecules
- excellent solvent properties for other polar molecules (solutes)
- properties making water the "solvent of life" as exhibited by cytoplasm, intercellular fluids, blood and many other solutions that are vital to living organisms.

Guiding Question revisited

What are the challenges and opportunities of water as a habitat?

In this chapter we have investigated:

- physical and chemical properties of water that provide both opportunities and challenges for living organisms
  - buoyancy – important to all aquatic and semi-aquatic organisms to keep them at or near the water surface
  - viscosity – the body shape and propulsion mechanisms of animals have become adapted to overcome this resistance that water has for objects moving through it
  - thermal conductivity – organisms living in cold-water environments must have either a physiology adapted for that water temperature or a means of insulation from the cold because water readily conducts heat away from an organism's body
  - specific heat – water in oceans, lakes and rivers has a very high specific heat that protects many aquatic organisms from much colder surrounding air temperatures.

Exercises

Q1. Describe how a polar covalent bond differs from a non-polar covalent bond.

Q2. Describe the pathway and the forces involved in getting water from the soil surrounding a large tree to a leaf in one of the uppermost branches of that tree (hint: start with the leaf).

Q3. State:

    (a) an example of a molecule that is soluble in the cytoplasm of a cell

    (b) the function of that same molecule.

Q4. State:

    (a) an example of a molecule that is insoluble in the cytoplasm of a cell

    (b) the function of that same molecule.

Q5. Describe two adaptations that the black-throated loon (*Gavia arctica*) has evolved for overcoming the viscosity of water.

A1.2 Nucleic acids

The organisms alive on Earth today have a long history and a very long family tree. Living things do not just appear, rather they are descended from previous generations. This is based on genetics. The information that is being passed from one generation to the next is in the form of DNA. Humans have 46 DNA molecules in each cell in the form of chromosomes. Written in the genetic code of DNA is information that makes a blue whale what it is and makes you what you are.

Along the length of DNA molecules there are chemical messages that code for specific proteins. Most of these protein messages are common to a species, but a few are individual to one single individual of that species. Thus, each living organism is unique. Preceding every cell division, the DNA replicates in an amazingly accurate series of steps that produces two DNA molecules where there was once one. Life has continued in this way for millions of years.

This chapter will introduce you to DNA and other molecules termed nucleic acids. Nucleic acids include DNA and three types of RNA that are all involved in the synthesis of proteins in cells.

A1.2.1 – DNA is the universal genetic material

A1.2.1 – DNA as the genetic material of all living organisms
Some viruses use RNA as their genetic material but viruses are not considered to be living.

Deoxyribonucleic acid (DNA) is the molecule that provides the long-term stored genetic information for all organisms on Earth. When mutations occur that influence evolution, they happen within DNA and are passed on to the next generation. The fact that DNA is universal to all living organisms is evidence of our common ancestry, even back to when the most complex life forms were single cells living in the oceans.

In addition to sugars and phosphate groups acting as a structural framework, DNA has within it four **nitrogenous bases**: adenine, thymine, cytosine and guanine, which are found along the length of the very long molecule. These four bases can be combined in a tremendous variety of orders and lengths. The sequences of nitrogenous bases are the genetic messages or **genes**. The messages are codes for **amino acids**. Amino acids are the "building blocks" of proteins, and a cell's identity and function is determined by the proteins it is able to synthesize. Every cell in a multicellular organism has the same DNA, but each different type of cell only uses the genetic information that is appropriate for that cell.

Viruses are not living organisms. Some viruses contain RNA as their genetic information and some contain DNA. No matter which nucleic acid acts as the genetic code for viral proteins, viruses are not considered to be alive because they cannot survive without a cell of a living organism, and they have no internal biochemistry when they exist as a separate particle. Only when they infect a cell will their nucleic acid (RNA or DNA) become active and use the internal biological components of the cell for their own uses. A virus has absolutely no other function other than to reproduce itself: viruses exist to reproduce. Sometimes that reproduction damages cells to the point of causing great harm to the host organism.

A1.2.2 – The structure of nucleotides

A1.2.2 – Components of a nucleotide

In diagrams of nucleotides use circles, pentagons and rectangles to represent relative positions of phosphates, pentose sugars and bases.

Both DNA and RNA are **polymers** of **nucleotides**. This means that both DNA and RNA have repeating units called nucleotides within the much larger molecule. So, in order to understand the structure of these two molecules important to life, we must first start with the structure of the nucleotides. Individual nucleotides consist of three major parts: one phosphate group, one five-carbon monosaccharide (also called a pentose sugar) and a nitrogenous base. Covalent bonds occur at specific locations in order to produce a functional unit.

It is important to note that in Figure 1 a circle is used to represent a phosphate, a pentagon is used to represent a pentose sugar, and a rectangle is used to represent a nitrogenous base.

▲
A1.2 Figure 1 Two representations of a single nucleotide are shown in the diagram. The upper drawing shows more detail, although not every atom and bond are shown of the pentose sugar and only a bonding location is shown for a nitrogenous base. The lower drawing shows the level of detail the IB requires you to draw from memory.

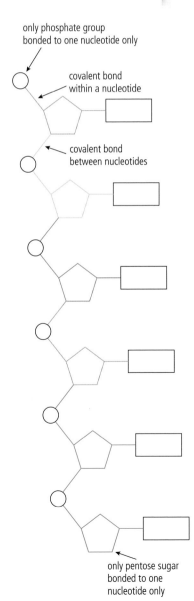

A1.2 Figure 2 Some nucleic acids are formed from a single chain of nucleotides.

A1.2.3 – Sugar to phosphate "backbone" of DNA and RNA

A1.2.3 – Sugar–phosphate bonding and the sugar–phosphate "backbone" of DNA and RNA

Sugar–phosphate bonding makes a continuous chain of covalently bonded atoms in each strand of DNA or RNA nucleotides, which forms a strong "backbone" in the molecule.

Nucleotides in both DNA and RNA bond together to produce long chains or polymers. In order to form a chain of nucleotides, the pentose sugar of one nucleotide is covalently bonded to the phosphate group of the next nucleotide. This means that there will always be one phosphate group with only one bond to a sugar at one end of the nucleic acid polymer, and a pentose sugar with only one bond to a single phosphate at the other end.

Challenge yourself

Examine Figure 1 on the previous page. Notice that the carbons of the pentose sugar are numbered. Now look at Figure 2, showing six nucleotides bonded together as a single-stranded polymer. Answer the following.

1. Within the polymer of six nucleotides, which *sugar* carbons are bonded to phosphate groups? (Do not consider the first nucleotide.)

2. Within a *single* nucleotide, what number carbon is always attached to the phosphate group?

3. Which carbon number is always attached to the nitrogenous base?

Nucleotides bond to one another to form a chain or polymer as a result of **condensation reactions** forming covalent bonds between the sugar of one nucleotide and the phosphate group of the next nucleotide. The fact that covalent bonds hold the chain together is important as covalent bonds are relatively strong (require a great deal of energy to break) and thus a nucleic acid polymer made of nucleotides is quite stable.

A1.2.4 – Nitrogenous bases within nucleic acids

A1.2.4 – Bases in each nucleic acid that form the basis of a code
Students should know the names of the nitrogenous bases.

In total, there are five possible **nitrogenous bases** in RNA and DNA. Four are found within RNA, and four are found in DNA. Only one of the bases differs in the two types of polymers, as shown in Table 1.

RNA nitrogenous bases	DNA nitrogenous bases
Adenine (A)	Adenine (A)
Uracil (U)	Thymine (T)
Cytosine (C)	Cytosine (C)
Guanine (G)	Guanine (G)

A1.2 Table 1 The five nitrogenous bases found in RNA and DNA

It may look like some of the nucleotides found in RNA and DNA are identical, for example because they both contain the base adenine. However, they are not identical because all the nucleotides found in RNA contain ribose as their pentose sugar, and all the nucleotides in DNA contain deoxyribose. In addition, the base uracil only occurs in RNA, not DNA, and the base thymine only occurs in DNA, not RNA. Thus, there are eight different nucleotides in total. When drawing nucleotides, it is common practice to put the capitalized first letter of the base inside the rectangle, as used by the IB.

Make sure you know the names of the five nitrogenous bases found in RNA and DNA, and do not just rely on the abbreviated form of a capital letter.

The sequence of nitrogenous bases in DNA, later transcribed into RNA, forms the basis of the genetic code.

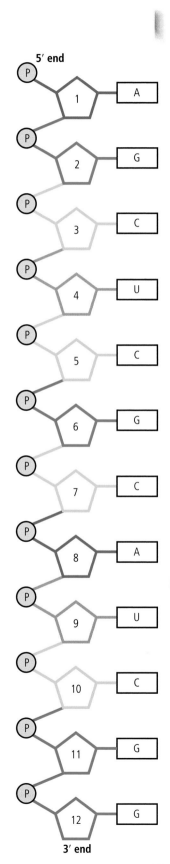

5' end

3' end

Challenge yourself

4. Use the geometric symbols required by the IB (see Figure 1) to represent all the possible separate nucleotides of DNA. Once you have sketched the four for DNA, do the same for RNA. To remind yourself of the fundamental pentose sugar difference between RNA and DNA nucleotides, you might want to put the letter "R", for ribose, inside the pentose shape of all RNA nucleotides. Then put "DR", for deoxyribose, inside all of the four DNA nucleotides. Make sure you end up with eight different nucleotides in total, one containing uracil and one containing thymine.

A1.2.5 – The structure of RNA

A1.2.5 – RNA as a polymer formed by condensation of nucleotide monomers

Students should be able to draw and recognize diagrams of the structure of single nucleotides and RNA polymers.

RNA is formed when nucleotides become bonded together in very specific sequences. The nucleotides are joined together by a **condensation reaction** between the pentose sugar of one nucleotide and the phosphate group of the next nucleotide. This reaction releases a water molecule (which is why this is called a "condensation" reaction). If an RNA molecule contains 322 nucleotides, 321 molecules of water would have been produced during its **synthesis**, as it would have required 321 condensation reactions to form.

Challenge yourself

5. How many water molecules would have been produced when the condensation reactions occurred that produced the 12 nucleotide RNA sequence shown in Figure 3?

 Even though the RNA depiction in Figure 3 has only 12 nucleotides shown, the actual RNA may have as many as a few thousand nucleotides.

◀ **A1.2 Figure 3** Twelve nucleotides bonded to form a very small section of a strand of RNA. The molecule is recognized readily as RNA because of the presence of uracil and because it is a single strand. Each adjoining nucleotide has been drawn in a different colour to emphasize the nucleotide structures. Notice that the chain has an alternating pentose–phosphate backbone, with the nitrogenous bases extending outwards from the backbone.

A1.2.6 – The structure of DNA

A1.2.6 – DNA as a double helix made of two antiparallel strands of nucleotides with two strands linked by hydrogen bonding between complementary base pairs

In diagrams of DNA structure, students should draw the two strands antiparallel, but are not required to draw the helical shape. Students should show adenine (A) paired with thymine (T), and guanine (G) paired with cytosine (C). Students are not required to memorize the relative lengths of the purine and pyrimidine bases, or the numbers of hydrogen bonds.

RNA is composed of a single chain or strand of nucleotides, while DNA consists of two chains or strands of nucleotides connected to one another by hydrogen bonds. The strands of both DNA and RNA may involve very large numbers of nucleotides. To visualize DNA, imagine the double-stranded molecule as a ladder (see Figure 4). The two sides of the ladder are made up of the phosphate and deoxyribose sugars. The rungs of the ladder (what you step on) are made up of the nitrogenous bases. Because the ladder has two sides, there are two bases making up each rung. The two bases making up one rung are said to be complementary to each other. Notice that the base pairs are always adenine (A) bonded to thymine (T) and cytosine (C) bonded to guanine (G). There are no exceptions to this in DNA, and these base pairings are known as the **complementary base pairs**. Because the two strands are upside down in comparison to each other, but parallel, they are said to be **antiparallel** to each other.

The nitrogenous bases adenine and thymine are always paired with each other in the double-stranded DNA molecule. Likewise, cytosine and guanine are always paired. These pairings are called the complementary base pairs.

Two antiparallel strands of DNA

◄ **A1.2 Figure 4** A small section of a double-stranded DNA molecule showing hydrogen bonds (dotted lines) between complementary base pairs. This type of representation of DNA is known as a "ladder diagram" and does not attempt to show the helical shape of the molecule.

Always attempt to view DNA and RNA molecules as chains of nucleotides. Identify the first nucleotide with its own phosphate, sugar and nitrogenous base and then visually move to the next, and so on. In Figure 4 you would visually start in the upper left corner for the left strand, and you would start in the lower right corner for the right strand.

Challenge yourself

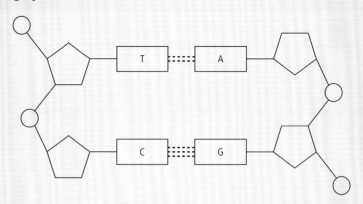

6. On your own paper and using the figure above as a guide, sketch and label the geometric shapes as shown to represent this four-nucleotide section of DNA.

7. Add four more nucleotides to each side by adding to the bottom of your sketch so that you end up with a 12-nucleotide section of antiparallel DNA. Remember to use complementary base pairs, although you can choose the base sequence.

8. Circle two *complete* nucleotides of your *added* nucleotides, one on each side, but do not circle any of the nucleotides in the corners of the figure. Check to make sure that your circles include one phosphate group, one deoxyribose sugar and one nitrogenous base, and that there are no uncircled nucleotides that are incomplete.

A1.2.7 – Distinguishing between DNA and RNA

A1.2.7 – **Differences between DNA and RNA**
Include the number of strands present, the types of nitrogenous bases and the type of pentose sugar.
Students should be able to sketch the difference between ribose and deoxyribose. Students should be familiar with examples of nucleic acids.

DNA and RNA are both linear polymers, consisting of sugars, phosphates and bases, but there are some important differences between the two molecules. Table 2 summarizes these differences.

DNA	RNA
Double-stranded molecule	Single-stranded molecule
All nucleotides contain deoxyribose sugar	All nucleotides contain ribose sugar
Thymine is one of the four nitrogenous bases	Uracil is one of the four nitrogenous bases
Shaped into a double helix	Variety of shapes depending on type of RNA
Acts as the permanent genetic code of a cell/organism	Does not contain a permanent genetic code, except in RNA viruses

A1.2 Table 2 A comparison of DNA and RNA molecules

Distinguishing between deoxyribose and ribose

Ribose has a molecular formula of $C_5H_{10}O_5$, whereas deoxyribose has a formula of $C_5H_{10}O_4$. Notice that the only difference in the molecular (chemical) formulas is that ribose has one more oxygen compared to deoxyribose. A side-by-side comparison shows where the difference occurs (see Figure 5). In organic chemistry an –OH group bonded to a carbon is called an **alcohol** or **hydroxyl group**. If you remove the oxygen from the hydroxyl group, it simply leaves a hydrogen. This may not look like much, but it is the common difference in all nucleotides of RNA versus DNA.

SKILLS

Practise sketching each of the two molecules shown in Figure 5. Learn the pattern that is common to both molecules and then modify for the single difference between deoxyribose and ribose.

deoxyribose   ribose

▲ **A1.2 Figure 5** A molecular sketch showing the deoxyribose sugar of DNA compared to the ribose sugar found in RNA molecules. Notice the difference in the lower right corners of the two molecules. Ribose has one more oxygen in its structure compared to deoxyribose.

The single "missing" oxygen in the pentose sugar of DNA leads to the name *deoxy*ribose within the full name for DNA (deoxyribose nucleic acid). The full name of RNA is ribonucleic acid.

Specific examples of nucleic acids

All living organisms use DNA as their long-term hereditary storage molecule. DNA stores genetic information as genes, but for that information to become useful to a cell there must be other nucleic acids at work. Here are four of the other nucleic acids as examples.

- **Messenger RNA (mRNA)** – This is an RNA molecule that is synthesized from an area of DNA called a **gene**. In a cell with a nucleus, the mRNA then leaves the nucleus and represents the genetic information necessary to make a protein. This is where it gets its name "messenger" RNA.

- **Transfer RNA (tRNA)** – Special genes of DNA code for tRNA molecules. When a specific protein is synthesized, specific amino acids must be added to the amino acid chain in a specific order. The function of tRNA is to transfer the correct amino acid into a growing chain of amino acids. This is the reason for its name "transfer" RNA.

- **Ribosomal RNA (rRNA)** – Again, special genes of DNA code for rRNA molecules. Along with some previously synthesized proteins, rRNA is used to create an organelle in cells called ribosomes. Cells typically have many thousands of ribosomes, and they are the cellular location where proteins are synthesized.

- **Adenosine triphosphate (ATP)** – This is a single-nucleotide nucleic acid. There are many other single-nucleotide nucleic acids in cells, but we are going to use this one as an example. ATP is used in cells as a type of chemical energy. When a muscle contracts, many ATP molecules are used as an energy source for the movement. The ultimate purpose of cellular respiration is to convert the energy contained within food molecules into the energy of ATP.

Do not concern yourself at this point with the details of these examples of nucleic acid molecules, beyond what is summarized in this section. The function of each of these molecules is explained in much greater detail in other chapters.

Challenge yourself

The figure below shows a molecular diagram of an ATP molecule. You do not need to memorize it, but based on what you have read earlier in this chapter you should be able to look at the diagram and answer the following questions.

9. Why is this molecule called a "triphosphate"?

10. Is the pentose sugar in this molecule ribose or deoxyribose?

11. The "adenosine" portion of the molecule's name comes from the nitrogenous base bonded to the pentose sugar. What is that nitrogenous base?

A1.2.8 – The importance of complementary base pairing

A1.2.8 – Role of complementary base pairing in allowing genetic information to be replicated and expressed

Students should understand that complementarity is based on hydrogen bonding.

As you recall, adenine and thymine are complementary to each other in DNA, and cytosine and guanine are complementary as well. This complementarity is based on hydrogen bonding. Adenine and thymine only form hydrogen bonds with each other; adenine does not form hydrogen bonds with any other DNA nucleotide. The same is true for cytosine and guanine.

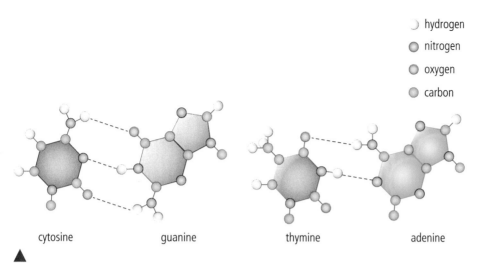

hydrogen
nitrogen
oxygen
carbon

cytosine   guanine   thymine   adenine

▲ Hydrogen bonding (shown in dotted red lines) between the complementary base pairs within DNA. It is this hydrogen bonding that holds the two antiparallel strands together and ultimately results in the double helix shape.

Complementary base pairing is important in DNA replication. Imagine that an area of DNA has been unzipped (opened up into two single strands). If free-floating individual nucleotides in solution begin pairing with the unmatched nucleotides, an exact copy of the original molecule can be made. In fact, if both sides of the original DNA are used as a template, then two molecules of DNA can be synthesized, each a duplicate of the original. In a simplified form, this is how DNA replication occurs.

A1.2.9 – Storage of genetic information

A1.2.9 – Diversity of possible DNA base sequences and the limitless capacity of DNA for storing information

Explain that diversity by any length of DNA molecule and any base sequence is possible. Emphasize the enormous capacity of DNA for storing data with great economy.

DNA stores genetic information in its sequence of nitrogenous bases. Every three bases represents a meaningful piece of information called a triplet or, more specifically, a **triplet codon**. Many triplets within DNA code for one of the 20 amino acids. There are four different DNA nucleotides that can be arranged as sequenced triplets. So, what are the odds of DNA containing any one triplet in any one gene location? Consider the odds of having G–G–G in one triplet area of DNA. If it was by random chance (although it is not) the odds would be:

$$\frac{1}{4} \times \frac{1}{4} \times \frac{1}{4} = \frac{1}{64}$$

Why? Because there is a one in four chance of the nitrogenous base being guanine, and it occurs in our example three times.

This computation also means that there are 64 combinations of nucleotides within the triplet code system. All of those 64 combinations are used in the genetic code for some purpose, most of them coding for amino acids.

Think about all the ways that the four nitrogenous bases of DNA can be grouped. If DNA was a short molecule (say around 1,000 nucleotides), the number of groupings would be large, but still not unlimited. Now consider that the length of DNA (the number of nucleotides in one strand) is only limited by the amount that will fit efficiently into a cell. The shortest DNA molecule in the human genome is about 50 million base pairs, and the longest about 260 million base pairs.

As you can see, the likelihood that two DNA molecules are identical as a result of random chance approaches zero. DNA can contain a nearly limitless amount of genetic information.

A1.2.10 – Genetic uniqueness

A1.2.10 – Conservation of the genetic code across all life forms as evidence of universal common ancestry

Students are not required to memorize any specific examples.

Imagine a section of DNA that contains the triplet code C–G–A. If that triplet code is used to synthesize a protein, the amino acid that will be produced will be alanine. If the triplet code is A–G–A, the amino acid is serine. A chart listing the triplet codes can provide this information.

It does not matter whether the organism is a species of fungus, an oak tree, or a human being. All living organisms use the same genetic code. The genetic code is therefore said to be universal.

So why are organisms different from each other? The answer to that is the DNA base sequences are different even though the code to read the sequences is the same. Your best friend, although not directly related to you, is related to you by evolution. The two of you share more than 99% of the same gene sequences. If it was 100%, you would not be the unique and different people you are.

A conserved genetic code

Why has the genetic code remained unchanged? The answer to this question lies in the process of evolution. The evolution of living organisms has been occurring for over 3.5 billion years. If you could go back in time far enough you would probably not see any organisms that you recognize today, although some of the organisms you would see will be the ancestors of today's organisms. If you were to keep moving back through time, the organisms would become even less familiar, and eventually they would be nothing more than single-celled organisms living in water.

Researchers are working on ways to store data (text files, photos, books, maps) within artificially created DNA molecules. DNA stores information using the very efficient code of four nitrogenous bases, compared to the less efficient 0 and 1 binary code used by computers.

How can polymerization result in emergent properties?

Identical twins develop when a single fertilized egg or early embryo splits into two portions. Each grows to become a separate person and shares exactly the same DNA sequences.

What makes RNA more likely to have been the first genetic material, rather than DNA?

These single-celled organisms are the ancestors of all life on Earth today. This is also postulated to be the time period in which the biochemistry of DNA and RNA evolved. All life forms from that point on used DNA to store their genetic information, and RNA to transfer that information to the order of amino acids in their proteins. Evolution changes the DNA sequences slowly, but it always has continued to use the same mechanisms of genetic coding.

Nature of Science

The theory of evolution by natural selection as proposed by Charles Darwin and independently by Alfred Wallace was based primarily on their observations of physical traits. It appeared to them that organisms developed adaptations to fit different ecological niches in the area that they lived in. In 1859, when Darwin published his famous book *On the Origin of Species*, there was absolutely no knowledge of DNA or the molecular basis of heredity and evolution. Scientific ideas that originate in one form can be corroborated by later scientific work if the ideas are sound. Today there is a mountain of evidence supporting evolutionary principles, including a vast amount of information from **molecular genetics**.

Guiding Question revisited

How does the structure of nucleic acids allow hereditary information to be stored?

In this chapter we have described how RNA and DNA are structured:

- each is composed of subunits called nucleotides
- nucleotides exist in eight types, four types in RNA and four types in DNA
- each nucleotide contains one of five possible nitrogenous base, adenine, thymine, cytosine, guanine and uracil

- in DNA, the two strands are held together by complementary base pairing between the nitrogenous bases
- the sequence of the nucleotides in sections of DNA called genes allows long-term storage of the genetic code
- RNA molecules are complementary copies of genes of DNA transcribed by using RNA nucleotides.

Guiding Question revisited

How does the structure of DNA facilitate accurate replication?

In this chapter we have described how:

- DNA exists as a double-stranded molecule
- DNA makes copies of itself
- this unwinding allows the nitrogenous bases to make new complementary pairings using the exposed nitrogenous bases as a template
- the pairings are adenine with thymine, and cytosine with guanine
- two DNA molecules are created from one during DNA replication, although neither is completely "new".

Exercises

Q1. State how many nucleotide types exist within the structures of DNA and RNA.

Q2. Suggest a reason why researchers often give DNA information:

    (a) as the sequence of nitrogenous bases without indicating the presence of the phosphate group and sugar component of each nucleotide (for example 5'ATTCCGTGTACGT3')

    (b) from one strand of DNA only.

Q3. You are visualizing a single sequence of nitrogenous bases and you see multiple uracil bases. What does that tell you about the molecule?

Q4. Which of these is not a nucleic acid?

    A DNA   **B** ATP   **C** PCR   **D** RNA

Q5. A measurement of a sample of DNA showed that 22% of the nitrogenous bases were cytosine. Calculate the expected percentage of the following nitrogenous bases:

    (a) guanine

    (b) adenine

    (c) thymine.

A1 Practice questions

1. Describe the importance of water to living organisms.

   (Total 5 marks)

2. Draw a labelled diagram showing the structure of three water molecules and how they interact.

   (Total 4 marks)

3. Draw a labelled diagram of a section of DNA showing four nucleotides.

   (Total 5 marks)

4. Distinguish between the structures of DNA and RNA.

   (Total 3 marks)

5. Where do hydrogen bonds form?

   A Between the slight negative charge of hydrogen and the slight positive charge of oxygen within a water molecule.

   B Between the slight positive charge of hydrogen and the slight negative charge of oxygen within a water molecule.

   C Between the slight positive charge of hydrogen and the slight negative charge of oxygen in different water molecules.

   D Between the slight negative charge of hydrogen and the slight positive charge of oxygen in different water molecules.

   (Total 1 mark)

THEME **A Unity and diversity**
2 Cells

◀ Early Earth provided an environment that was extremely inhospitable to life as we know it today. Yet, over long periods of time, conditions changed allowing the building blocks of life to form. Once the building blocks were in place, a slow but steady development occurred. Ultimately, the complexity of life has led to the estimated 8.7 million different species that exist today.

 A2.1 is not included as it is for HL students only.

A2.2 Cell structure

Guiding Questions

What are the features common to all cells and the features that differ?

How is microscopy used to investigate cell structure?

In the 1660s, Antonie van Leeuwenhoek became interested in the early microscopes being developed by Robert Hooke. The Dutch businessman and scientist used mostly blown-glass lenses to produce his own microscopes, which opened a completely new world to all. His powers of observation led to the first recorded descriptions of bacteria and protozoa. From van Leeuwenhoek's work the science of microbiology took form.

Countless improvements in microscopy since these simple beginnings have led to an understanding of the features common to all cells. We have also learned of the tremendous diversity that exists not only in cells but in all life.

A2.2.1 – Cells and the functions of life

A2.2.1 – Cells as the basic structural unit of all living organisms

NOS: Students should be aware that deductive reason can be used to generate predictions from theories. Based on cell theory, a newly discovered organism can be predicted to consist of one or more cells.

Whether organisms are extremely small or extremely large, understanding their smallest functional units is imperative. These units are known as cells. Organisms range in size from a single cell upwards to trillions of cells. To better understand all the organisms around us we must study their cells.

Cytology is the branch of biology that studies all facets of the cell. As our understanding of the cell has increased, so has our ability to understand all forms of life and diseases that occur on planet Earth. This area of research is extremely active in laboratories all over the world.

The cell theory states:

 What are the features of a compelling theory?

- all organisms are composed of one or more cells
- cells are the smallest units of life
- all cells come from pre-existing cells.

Nature of Science

Inductive reasoning utilizes specific observations to arrive at broader generalizations. Deductive reasoning works in the opposite direction. It allows you to make an inference using widely accepted facts or premises. Using deductive reasoning, a newly discovered organism can be predicted to carry out the functions of life and demonstrate the principles of cell theory.

A2.2.2 – Cells and the microscope

A2.2.2 – Microscopy skills

Application of skills: Students should have experience of making temporary mounts of cells and tissues, staining, measuring sizes using an eyepiece graticule, focusing with coarse and fine adjustments, calculating actual size and magnification, producing a scale bar and taking photographs.

NOS: Students should appreciate that measurement using instruments is a form of quantitative observation.

Cells are made up of many different subunits. These subunits are often of a particular size, but most are microscopically small.

Unit	Equivalent measurement
1 metre (m)	100 cm = 1,000 mm
1 centimetre (cm)	10^{-2} m (0.01 m)
1 millimetre (mm)	10^{-3} m (0.001 m)
1 micrometre (μm)	10^{-6} m (0.000001 m)
1 nanometre (nm)	10^{-9} m (0.000000001 m)

Commonly used microscope metric equivalents

Microscopes with a high **magnification** and **resolution** are needed to observe cells and especially their subunits. Magnification is the increase in an object's image size compared to its actual size. Pictures or drawings of an image from a microscope include the number of times larger than the actual object they are, for example 500× or 100,000×.

Resolution refers to the minimal distance between two points or objects at which they can still be distinguished as two. As the resolution of a microscope increases, the greater the detail that microscope will reveal. Some like to explain resolution in terms of clarity, with greater resolution providing greater clarity.

Light microscopes use light, passing through living or dead specimens, to form an image. Stains may be used to improve the visibility of structures. **Electron microscopes (EMs)** provide the greatest magnification (over 100,000×) and resolution. These use electrons passing through a specimen to form an image.

ocular lens (eyepiece)

body tube

arm

revolving nosepiece

objectives

stage

stage clip

condenser

diaphragm

coarse adjustment knob

light source

base

A light microscope (above) and an electron microscope (below).

Light microscope	Electron microscope
Inexpensive to purchase and operate	Expensive to purchase and operate
Simple and easy specimen preparation	Complex and lengthy specimen preparation
Magnifies up to 2,000×	Magnifies over 500,000×
Specimens may be living or dead	Specimens are dead, and must be fixed in a plastic material

▲

A comparison of the light microscope and the electron microscope

Most cells can be up to 100 micrometres (100 µm) in size. Organelles are up to 10 µm in size. Bacteria are between 1 and 10 µm in size. Viruses are up to 100 nanometres (nm) in size. Cell membranes are 10 nm thick, while molecules are about 1 nm in size. All these structures are three-dimensional.

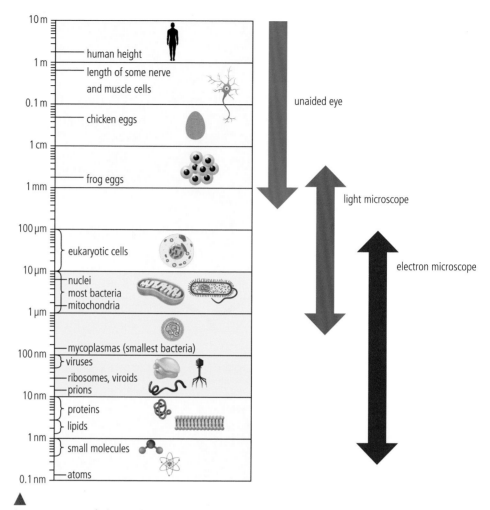

A representation of what can be used to visualize various structures important in biology.

Cells and their subunits are so small they are hard to visualize, so it is important to appreciate relative sizes. Cells are relatively large, and then in decreasing order of size are:

- organelles
- bacteria (some bacteria cells are as large as organelles)
- viruses
- membranes
- molecules.

If you want to calculate the actual size of a specimen seen with a microscope, you need to know the diameter of the microscope's **field of vision**, also called the **field of view**. The field of vision is the total area visible when looking through a microscope's **ocular** or eyepiece, and the diameter can be calculated using special **micrometers**. There are two general types of micrometers: ocular and stage. The **ocular micrometer**, also called a **graticule**, is located in the eyepiece and is engraved with equal units. It is important to note that the units on this micrometer are arbitrary. They are calibrated using a **stage micrometer**. This calibration is often done using a simple ruler or a special slide with defined units, usually millimetres. By comparing the units of the graticule to the known unit size of the stage micrometer, you can determine the size of the image being examined. The ocular micrometer has to be calibrated in this way with each objective power of the microscope. The size of the specimen can then be calculated.

A simple formula can be used to calculate the magnification being used:

$$\text{magnification} = \frac{\text{measured size of image}}{\text{actual size of specimen}}$$

Scale bars are often used with a micrograph or drawing so that the actual size can be determined. Scale bars and magnification will be addressed in more detail in a later practical activity.

Worked example

The length of an image you are looking at is 50 mm. If the actual length of the subject of the image is 5 μm, what is the magnification of the image?

Solution

Magnification = 50 mm/5 μm = 50,000 μm/5 μm = 10,000×

Or

Magnification = 50 mm/5 μm = 50×10^{-3} m/1×10^{-6} m = 10,000×

SKILLS

Use of a light microscope to investigate cells and cell structure sizes. Full details of how to carry out this activity with a worksheet are available in the eBook.

Nature of Science

Scientists need to accumulate data when conducting experiments using scientific methods. Two types of data can be collected. Qualitative data is non-numerical but descriptive. It includes attributes such as colour, presence of a structure or feature (or not) or sex. Quantitative data involves numerical values collected by a specific type of instrument. Examples of quantitative data are mass measured by a laboratory balance or length measured by a ruler. These two types of data, when collected properly, allow meaningful conclusions to be made.

A2.2.3 – Advanced microscopy

A2.2.3 – Developments in microscopy

Include the advantages of electron microscopy, freeze fracture, cryogenic electron microscopy, and the use of fluorescent stains and immunofluorescence in light microscopy.

The microscope has undergone tremendous advancement since the one used by Robert Hooke in 1665. Early microscopes were pivotal in the development of the cell theory, even though they were extremely simple by today's standards. Scientists have also perfected many new techniques in the preparation of materials for study involving the microscope. In this section we will examine some of these developments and techniques.

The top image of a leukaemia cell is from a scanning electron microscope (SEM). The bottom image is of the same cell but from a transmission electron microscope (TEM).

Both fluorescent staining and immunofluorescence have been extensively used in the study of severe acute respiratory syndrome coronavirus 2 (SARS-CoV-2) and related viruses. They have provided valuable information about the life cycle of the virus as it attacks living cells.

One significant advancement in microscopy was the development of the electron microscope (EM). The EM utilizes a beam of electrons rather than a beam of light, which the light microscope uses. Electrons have a much shorter wavelength than light. The benefits of the shorter wavelength include a 1,000 times greater resolving power than the light microscope, and the ability to magnify objects over 500,000× compared to the maximum magnification of 2,000× for a light microscope.

There are two general types of EMs – the **scanning electron microscope (SEM)** and **transmission electron microscope (TEM)**. The SEM uses a beam of electrons to scan the surface of a specimen. The TEM aims a beam of electrons through a very thin section of a specimen, allowing its inner structure to be viewed. Both SEM and TEM images provide essential information in cytology investigations.

Techniques employed when working with an EM include **freeze fracture** and **cryogenic electron microscopy**. Freeze fracture is a process of preparing a sample for observation with an EM. It involves the rapid freezing of a biological specimen followed by physically breaking the specimen apart (fracturing). This technique reveals a plane through the sample that can then be examined. Our understanding of the cell membrane has been greatly enhanced using this technique.

Cryogenic electron microscopy is a recent advancement in EM that has furthered our knowledge of structural biology. It enables an image to be formed using computer enhancement that shows the three-dimensional framework of proteins involved with the function of a cell. It utilizes low temperatures to freeze specimens in ice. Many advances in our understanding of virus composition and structure, cell membrane components and their arrangement, cellular protein synthesis, and even hereditary expression and regulation, are the result of using this technique. New applications of cryogenic electron microscopy are being developed at an amazing pace with enlightening results.

It is obvious that the EM offers tremendous advantages over the light microscope in the study of cells and their structures. However, it is important to note that EMs are expensive, require extensive training to operate, and involve non-living specimens embedded in some sort of matrix such as plastic. Often, structural features called **artifacts** are seen in the pictures produced by an EM. These artifacts do not actually exist in the cell but are produced during the preparation of the samples for an EM.

When living samples are to be studied, the light microscope must be used. Two preparation techniques developed recently for the study of cells using light microscopy involve the use of **fluorescent stains** and **immunofluorescence**. Fluorescent stains are substances or dyes that combine with specific cellular components. When these living samples are then irradiated with ultraviolet or violet-blue light, the parts that accepted the dye will fluoresce. When fluorescence occurs, assorted colours are produced, allowing more detailed visibility. Immunofluorescence also allows greater visibility of living tissue. Immunofluorescence involves antibodies that have dyes already combined with them. Specific antibodies combined with unique coloured dyes recognize and combine with target molecules. This allows the target, usually a protein, to be detected. This technique is often used to detect viral proteins that have infected cells.

Fluorescence-based methods have recently been developed to target RNA. We are now able to visualize single RNA molecules within single cells and viruses.

The light microscope has gone through many developments to improve its ability to produce images of living cells and their internal structures. One area of development has involved the part of the microscope called the **condenser**. The condenser is located between the stage and the light source. It possesses a lens that directs light rays from the light source through the specimen. From the specimen, the light rays pass through the objective lens to the ocular lens, where the image is viewed by the researcher. By changing the capabilities of the condenser, we now have some microscope types with unique and valuable features.

Type of microscope	Feature
Brightfield	Visible light is used; the specimen is viewed against a light background; it is the most common and easy to use light microscope
Darkfield	A special opaque lens is used in the condenser, that blocks direct light from entering the specimen; the specimen appears light against a dark background
Phase-contrast	A special condenser with a circular diaphragm and a modified objective lens are used to reveal detailed images of specimens without staining

Types of light microscope

Each advance of the microscope, whether light or electron, leads to a corresponding increase in our understanding of the cell's structures and functions.

A2.2.4 – Structures common to all cells

A2.2.4 – Structures common to cells in all living organisms
Typical cells have DNA as genetic material and a cytoplasm composed mainly of water, which is enclosed by a plasma membrane composed of lipids. Students should understand the reasons for these structures.

As all organisms are composed of one or more cells and demonstrate common functions, all cells possess certain common structures. These include:

- DNA, as their genetic material
- a cytoplasm, composed of mainly water
- a plasma membrane, composed of lipids that surrounds the cytoplasm.

cytoplasm

DNA

plasma membrane

All cells possess three common structures: DNA, cytoplasm and a plasma membrane. Cells usually demonstrate greater complexity than this, with many more structures. However, greater complexity is not required for a cell to carry out the functions of life.

For new cells to be formed from pre-existing cells, there must be a means to store and transfer information. DNA fulfils this role because of its ability to form large molecules from small building blocks called **nucleotides**. Four different nucleotides make up DNA. It is the specific sequence of these unique nucleotides, and their ability to combine to form huge chains, that results in the production of the exact proteins essential for passing on distinctive characteristics from cell to cell and even from organism to organism. DNA also controls the production of enzymes within an organism, which serve a controlling role in chemical reactions.

The cytoplasm is found within the boundary of a cell. This region of a cell consists of a matrix composed mainly of water called **cytosol.** Cytosol contains all the ingredients necessary for a cell to conduct its day-to-day activities. These ingredients include many different carbon compounds, as well as **ions,** which are atoms with a charge, and other inorganic compounds. The cytoplasm of a cell is the location where most chemical reactions take place.

The plasma membrane encloses the cell and protects its contents from the surrounding environment. Its major component is two layers of lipids combined as a **bilayer.** Proteins and the element phosphorus are also associated with this bilayer. The membrane controls interactions between a cell's contents and the exterior. Materials needed by the cell are transported into the cell through the membrane, while waste material is transported out of the cell. Membrane proteins provide identity properties to the cell, which is especially important in multicellular organisms. The membrane proteins in multicellular organisms also engage in communication and transport between cells.

A matrix is an unstructured semi-fluid region within a boundary. The cytosol is a matrix with a gel-like consistency in which other cell structures may be suspended.

What explains the use of certain molecular building blocks in all living cells?

A2.2.5 – The prokaryote cell

A2.2.5 – Prokaryote cell structure

Include these cell components: cell wall, plasma membrane, cytoplasm, naked DNA in a loop and 70S ribosomes. The type of prokaryotic cell structure required is that of Gram-positive eubacteria such as *Bacillus* and *Staphylococcus*. Students should appreciate that prokaryote cell structure varies. However, students are not required to know details of the variations such as the lack of cell walls in phytoplasmas and mycoplasmas.

Two types of organism, bacteria (members of the domain Eubacteria) and archaea (members of the domain Archaea), are made up of prokaryotic cells and are called prokaryotes. Most of these organisms do not cause disease and are not pathogenic (disease-causing). They are an extremely diverse group occupying air, water and soil environments. Prokaryotes are a very successful group of organisms.

What is a prokaryotic cell?

After extensive studies of cells, it has become apparent that all cells use some common molecular mechanisms. There are huge differences between forms of life, but cells are the basic unit and different cells have many characteristics in common. Cells are often divided into groups based on major characteristics. One such division separates cells into **prokaryotic** and **eukaryotic cells**. Prokaryotic cells are much smaller and simpler than eukaryotic cells. In fact, most prokaryotic cells are less than 1 μm in diameter. As bacteria are prokaryotic cells, you can see that such cells play a large role in the world today.

Prokaryotic organisms include bacteria and archaea. Bacteria and archaea appear to have followed different branches to eukaryotes (in the domain Eukarya) in the evolution of life. Prokaryotes are mostly small and unicellular. There are thousands of distinct types differentiated by many factors, including nutritional requirements, sources of energy, chemical composition and morphology (shape).

A domain is the highest classification rank of all organisms. Three domains of life are recognized. They are the Eubacteria, the Archaea, and the Eukarya.

Features of prokaryotic cells

Study the diagram of a prokaryotic cell (Figure 1) and make sure you can identify:

- the cell wall
- the plasma membrane
- flagella
- pili
- ribosomes
- the nucleoid (a region containing free DNA).

Becoming familiar with common prefixes, suffixes and word roots will help you understand biological terms. For example, the word prokaryotic is from the Greek word "pro", which means "before", and "karyon" which means "kernel", referring to the nucleus.

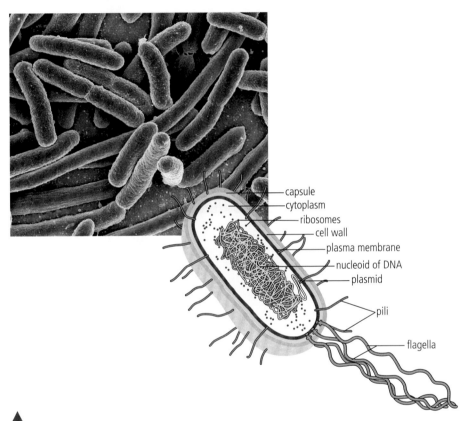

▲
A2.2 Figure 1 A false-colour scanning electron micrograph (SEM) of the bacterium *Escherichia coli*. Below it is a drawing of a prokaryotic cell.

Cell wall and plasma membrane

The prokaryotic cell wall protects and maintains the shape of the cell. It also keeps the bacterial cell from rupturing when water pressure is greater inside the cell than outside. In most prokaryotic cells this wall is composed of a carbohydrate–protein complex called **peptidoglycan**. Some bacteria have an additional layer of a type of polysaccharide outside the cell wall. This layer, called the **capsule**, makes it possible for some bacteria to adhere to structures such as teeth, skin and food.

The plasma membrane is found just inside the cell wall and is similar in composition to the membranes of eukaryotic cells. To a considerable extent, the plasma membrane controls the movement of materials into and out of the cell, and it plays a role in binary fission of the prokaryotic cell.

Antibiotics used to treat infections caused by bacteria can attack two areas of the bacterial cell. They may interfere with the proper development of the cell wall, resulting in a weakened outer protective wall. They may also act on ribosomes, to prevent the synthesis of the cell's required proteins. These same antibiotics do not act on eukaryote cell walls or ribosomes, so they can be used to successfully treat bacterially caused infections without harming the cells of the affected eukaryotic organism.

SKILLS

Follow the Gram-staining procedure accessed from this page of your eBook.

One major way to classify bacteria is by their ability to retain a dye called crystal violet. Bacteria that are "Gram-positive" have cell walls that, when exposed to crystal violet, take on a violet or blue appearance. "Gram-negative" bacteria do not retain this dye and do not appear violet or blue when examined with a microscope. *Bacillus* and *Staphylococcus* are examples of Gram-positive bacteria.

▲

A transmission electron micrograph (TEM) of *Bacillus subtilis* bacteria. Notice the violet-blue colour indicating that this bacterium is Gram-positive. Had this bacterium been Gram-negative, there would be a pink colour present because of the addition of Gram's safranin, as mentioned in the Gram-staining procedure.

Pili and flagella

Some bacterial cells contain hair-like growths on the outside of the cell wall. These structures are called **pili** and can be used for attachment. However, their main function is in joining bacterial cells in preparation for the transfer of DNA from one cell to another (sexual reproduction).

Some bacteria have flagella (plural) or a flagellum (singular), which are always longer than pili. Flagella allow a cell to move and are anchored to the cell wall and plasma membrane.

Cytoplasm

The cytoplasm occupies the complete interior of the cell. Using a microscope capable of high magnification, the most visible structure of the cytoplasm is the chromosome or a molecule of DNA. There are no specialized areas within the cytoplasm because internal membranes do not exist. All the cellular processes taking place within prokaryotic cells occur within the cytoplasm, without the existence of specialized compartments.

Because there are no specialized areas within prokaryotic cells, chemical reactions are not isolated from one another. This may limit the cell's development and efficiency because of possible interference between the reactions. It is interesting that without this separation of specialized areas, prokaryotes have the most diverse metabolic reactions of all organisms. When areas within a cell take on specific functions and are separated from the surrounding cytoplasm, the cell is said to show **compartmentalization**. Compartmentalization was a major development as prokaryotic cells gave rise to eukaryotic cells.

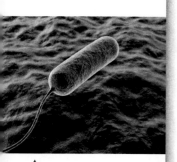

▲

A scanning electron micrograph (SEM) of a bacterial cell with a single flagellum. When flagella are present on a bacterial cell, they are usually involved in movement. Many bacteria have more than one flagellum attached.

Ribosomes

Ribosomes occur in all prokaryotic cells, and they function as sites of protein synthesis. These small structures occur in large numbers in cells that produce a substantial amount of protein, and, when numerous, they give a granular appearance to a TEM of a prokaryotic cell. Ribosomes are composed of two subunits, a protein and a type of RNA called ribosomal RNA. The structure of prokaryotic ribosomes will be explained further in the context of eukaryotic cell structures (Section A2.2.6).

The nucleoid region

The nucleoid region of a bacterial cell is non-compartmentalized and contains a single, long, continuous, circular thread of DNA, the bacterial chromosome. The nucleoid region is not surrounded by a membrane. Prokaryotic cell DNA is not associated with proteins called histones, as the DNA of eukaryotes is; hence bacterial chromosomes can be described as naked loops. This nucleoid region is involved with cell control and reproduction.

In addition to the bacterial chromosome, bacteria may also contain **plasmids**. These small, circular, DNA molecules are not connected to the main bacterial chromosome. The plasmids replicate independently of the chromosomal DNA. Plasmid DNA is not required by the cell under normal conditions, but it can help the cell adapt to unusual circumstances.

Binary fission

Prokaryotic cells divide by a very simple process called **binary fission**. During this process, the DNA is copied, resulting in two daughter chromosomes. These daughter chromosomes become attached to different regions on the plasma membrane, and the cell divides into two genetically identical daughter cells. This divisional process includes an elongation of the cell and a partitioning of the newly produced DNA by specialized fibres.

◀ A false-colour transmission electron micrograph (TEM) showing *Escherichia coli* dividing by binary fission.

Plasmids have especially important roles to play in some techniques involving genetic engineering/ modification. Current research into genetic modification is progressing rapidly with the use of a recently discovered biological scalpel called CRISPR. It is hoped that CRISPR will provide a future cure for some genetic diseases.

Some types of bacteria go through binary fission every 20 minutes when conditions are ideal. This results in huge populations and greater potential for infections. Refrigeration of foods is often used to reduce ideal conditions for bacteria, resulting in lower bacteria counts in our food and less chance of infection/ food poisoning.

Challenge yourself

1. Prepare a drawing of the ultrastructure of a prokaryotic cell based on electron micrographs. Remember to use a sharp pencil; use simple, narrow lines, and do not use shading. Label each of the structures, including their function.

A2.2.6 – The eukaryote cell

A2.2.6 – Eukaryote cell structure

Students should be familiar with features common to eukaryote cells: a plasma membrane enclosing a compartmentalized cytoplasm with 80S ribosomes; a nucleus with chromosomes made of DNA bound to histones, contained in a double membrane with pores; membrane-bound cytoplasmic organelles including mitochondria, endoplasmic reticulum, Golgi apparatus and a variety of vesicles or vacuoles including lysosomes; and a cytoskeleton of microtubules and microfilaments.

What is a eukaryotic cell?

The term "eukaryote" comes from the Greek words meaning "true kernel" or nucleus.

Whereas prokaryotic cells occur in bacteria and archaea, eukaryotic cells occur in organisms such as algae, protozoa, fungi, plants and animals. Eukaryotic cells range in diameter from 5 to 100 μm. A "kernel" or nucleus is usually noticeable in the cytoplasm. Other **organelles** may be visible within the cell if you have a microscope with a high enough magnification and resolution. Organelles are non-cellular structures that carry out specific functions (a bit like organs in multicellular organisms); different types of cells may have different organelles. These structures enable compartmentalization in eukaryotic cells, which is not a characteristic of prokaryotic cells. Compartmentalization enables different chemical reactions to be separated, which is especially important when adjacent chemical reactions are incompatible. Compartmentalization also allows chemicals for specific reactions to be isolated; this isolation results in increased efficiency.

Examine Figures 2 and 3, illustrating typical animal and plant eukaryotic cells.

Plasma membrane

Cytoplasm

Mitochondria are organelles that carry out respiration.

Golgi apparatus stores, modifies and packages proteins.

Nuclear pore allows communication between the nucleus and the rest of the cell.

Nucleolus is a dense, solid structure involved in ribosome synthesis.

Ribosomes are small structures, free in the cytoplasm or associated with the endoplasmic reticulum (ER).

Rough endoplasmic reticulum is a site of protein synthesis.

Smooth endoplasmic reticulum is ER without ribosomes.

Nucleus

Centrioles are associated with nuclear division. They are composed of microtubules. The area in which centrioles are found is called the centrosome. It is present in all eukaryotic cells, but centrioles are absent from higher plant cells.

Nuclear membrane

Lysosomes are sacs bounded by a single membrane. They contain and transport enzymes. Lysosomes show little internal structure. Lysosomes are usually absent from plant cells.

Vacuoles are smaller than those found in plant cells.

▲ **A2.2 Figure 2** Look at this drawing of a typical animal cell and compare it with Figure 3.

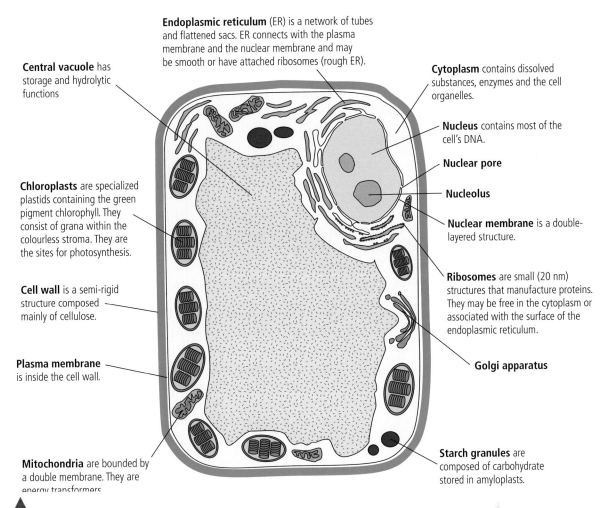

Endoplasmic reticulum (ER) is a network of tubes and flattened sacs. ER connects with the plasma membrane and the nuclear membrane and may be smooth or have attached ribosomes (rough ER).

Central vacuole has storage and hydrolytic functions

Cytoplasm contains dissolved substances, enzymes and the cell organelles.

Nucleus contains most of the cell's DNA.

Nuclear pore

Chloroplasts are specialized plastids containing the green pigment chlorophyll. They consist of grana within the colourless stroma. They are the sites for photosynthesis.

Nucleolus

Nuclear membrane is a double-layered structure.

Cell wall is a semi-rigid structure composed mainly of cellulose.

Ribosomes are small (20 nm) structures that manufacture proteins. They may be free in the cytoplasm or associated with the surface of the endoplasmic reticulum.

Plasma membrane is inside the cell wall.

Golgi apparatus

Mitochondria are bounded by a double membrane. They are energy transformers.

Starch granules are composed of carbohydrate stored in amyloplasts.

A2.2 Figure 3 What is different and what is similar between this typical plant cell and the animal cell in Figure 2?

As you read about the organelles of eukaryotic cells, refer to Figures 2 and 3.

Organelles of eukaryotic cells

Common organelles include the following (see Figures 2 and 3):

- endoplasmic reticulum
- ribosomes
- lysosomes (not usually found in plant cells)
- Golgi apparatus
- mitochondria
- nucleus
- chloroplasts (only in plant and algal cells)
- centrosomes (present in all eukaryotic cells, but centrioles are not found in most plant and fungal cells)
- vacuoles.

The microscope has given us an insight into the structure and function of eukaryotic cell organelles and characteristics.

Cytoplasm

All eukaryotic cells have a region called the **cytoplasm** that occurs inside the plasma membrane and outside the nucleus. It is in this region that the organelles are found. The fluid portion of the cytoplasm around the organelles is called the **cytosol**. Eukaryotic cytoplasm includes small fibres and rods called a cytoskeleton, which creates a complex internal structure. Prokaryotic cytoplasm lacks a cytoskeleton.

Cytoskeleton

The eukaryotic cell cytoplasm contains a network of fibres collectively called the **cytoskeleton**. These fibres are composed of protein and provide the following functions within the cell:

- maintaining cell shape
- anchoring some organelles
- aiding cellular movements
- providing a means for some organelles to move within the cell.

The cytoskeleton contains actin filaments, intermediate filaments and microtubules. These fibres can rearrange their protein components so that the cell can respond to changes in both internal and external environments. Actin filaments are also called microfilaments, and function in cell division and cell movement, especially involving contractions, as in muscle cells. Intermediate filaments are found in most animal cells and reinforce cell shape as well as anchoring some organelles. Microtubules shape and support the cell. They also function as movement paths or tracks through the cell for some organelles.

Endoplasmic reticulum

The **endoplasmic reticulum (ER)** is an extensive network of tubules or channels that extends most everywhere in the cell, from the nucleus to the plasma membrane. Its structure enables its function, which is the transportation of materials throughout the internal region of the cell. There are two general types of ER: **smooth ER** and **rough ER**. Smooth ER does not have any of the organelles called ribosomes on its exterior surface. Rough ER has ribosomes on its exterior.

Smooth endoplasmic reticulum (ER) and rough endoplasmic reticulum (ER).

Smooth ER has many unique enzymes embedded on its surface. Its functions include:

- the production of membrane phospholipids and cellular lipids
- the production of sex hormones such as testosterone and oestradiol
- detoxification of drugs in the liver
- the storage of calcium ions in muscle cells, needed for contraction
- transportation of lipid-based compounds
- helping the liver to release glucose into the bloodstream when needed.

Rough ER has ribosomes on the exterior of its channels. The ribosomes participate in protein synthesis, so this type of ER engages in protein development and transport. These proteins may become parts of membranes, enzymes or even messengers between cells. Most cells contain both types of ER, with the rough ER being closer to the nuclear membrane.

Ribosomes

Ribosomes are unique structures that do not have an exterior membrane. They conduct protein synthesis within the cell. These structures may be found free in the cytoplasm, or they may be attached to the surface of ER. They are always composed of a type of RNA and protein. You will recall that prokaryotic cells also contain ribosomes. However, the ribosomes of eukaryotic cells are larger and denser than those of prokaryotic cells. Ribosomes are composed of two subunits. In eukaryotic cells these subunits together equal 80S. The ribosomes in prokaryotic cells are also of two subunits, but they only equal 70S.

Lysosomes

Lysosomes are intracellular digestive centres that arise from the Golgi apparatus. A lysosome does not have any internal structures. Lysosomes are **vesicles** (sacs) bounded by a single membrane that contains as many as 40 different enzymes. The enzymes are all **hydrolytic** and catalyse the breakdown of proteins, nucleic acids, lipids and carbohydrates. Lysosomes fuse with old or damaged organelles within the cell to break them down, so that recycling of the components can occur. Lysosomes are also involved in the breakdown of material that is brought into the cell by **phagocytosis**. Phagocytosis is a type of **endocytosis** and is a means by which materials can enter a cell.

The interior environment of a functioning lysosome is acidic; acidic conditions are necessary for the enzymes to hydrolyse large molecules. When **hydrolysis** occurs, large molecules are broken down with the addition of water.

Golgi apparatus

The **Golgi apparatus** consists of flattened sacs called **cisternae**, which are stacked one on top of another. This organelle functions in the collection, packaging, modification and distribution of materials synthesized in the cell. One side of the apparatus is near the rough ER, called the *cis* side. It receives products from the ER. These products then move into the cisternae of the Golgi apparatus. They continue to move to the discharging or opposite side, the *trans* side. Small sacs called **vesicles** can then be seen coming off the trans side. Lysosomes are an important example of vesicles produced by the Golgi apparatus. The vesicles carry modified materials to wherever they are needed inside or outside the cell. The Golgi apparatus is especially prevalent in glandular cells, such as those in the pancreas, which manufacture and secrete substances.

The letter S used in the measurement of ribosomes refers to Svedberg units, which indicate the relative rate of sedimentation during high-speed centrifugation. The higher the S value, the quicker the structure will become part of the sediment and the more mass it will have.

Endocytosis is the uptake of new materials into the cell by invagination of the plasma membrane. If the material entering the cells is solid, the process is known as phagocytosis. When liquid containing dissolved materials enters the cell, it is known as pinocytosis.

A transmission electron micrograph (TEM) of a pancreatic exocrine cell. Can you tell that this is an animal cell? Locate as many of the structures of an animal cell as you can. How do the structures of this cell reflect the overall functions of the pancreas?

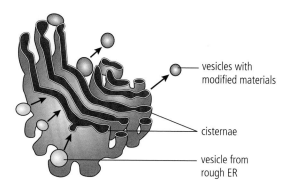

vesicles with modified materials

cisternae

vesicle from rough ER

In this drawing of the Golgi apparatus, the movement of the vesicles is shown by arrows. Can you identify which side is the *cis* side and which is the *trans* side?

Mitochondria

Mitochondria (singular mitochondrion) are rod-shaped organelles that appear throughout the cytoplasm. They are close in size to a bacterial cell. Mitochondria have their own DNA, a circular chromosome like that in bacterial cells, allowing them some independence within the cell. They have a double membrane: the outer membrane is smooth, but the inner membrane is folded into **cristae** (singular crista). Inside the inner membrane is a semi-fluid substance called the **matrix.** An area called the **inner membrane space** lies between the two membranes.

The cristae provide a huge surface area within which the chemical reactions characteristic of mitochondria occur. Most mitochondrial reactions involve the production of usable cellular energy called **adenosine triphosphate (ATP)**. Because of this, the mitochondria are often called the powerhouse of a cell. This organelle also produces and contains its own ribosomes. These ribosomes are of the 70S type. Cells that have high energy requirements, such as muscle cells, have large numbers of mitochondria.

Compare this drawing of a mitochondrion with the corresponding false-colour transmission electron micrograph (TEM) below it.

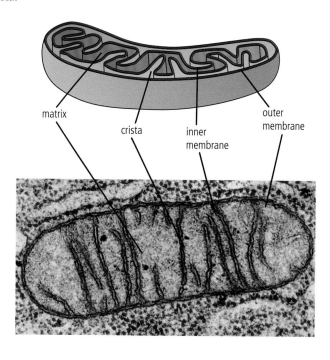

matrix

crista

inner membrane

outer membrane

Nucleus

The **nucleus** in eukaryotic cells is an isolated region where DNA resides. It is bordered by a double membrane referred to as the **nuclear envelope**. This membrane allows compartmentalization of the eukaryotic DNA, thus providing an area where DNA can conduct its functions without being affected by processes occurring in other parts of the cell. The nuclear membrane does not result in complete isolation, because it has numerous pores that allow communication with the cell's cytoplasm.

▲ False-colour TEM showing nucleus and nucleolus.

◀ False-colour TEM showing pores in the nuclear membrane.

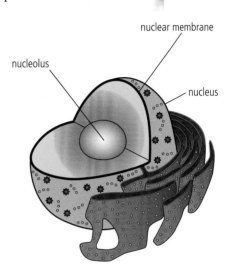

▲ The nucleus has a double membrane with pores and contains a nucleolus.

The DNA of a eukaryotic cell often occurs in the form of chromosomes; chromosomes vary in number depending on the species. Chromosomes carry all the information that is necessary for the cell to exist, thus allowing an organism to survive, whether it is unicellular or multicellular. The DNA is the genetic material of the cell. It enables certain traits to be passed on to the next generation. When the cell is not in the process of dividing, the chromosomes are not present as visible structures. During this phase, the cell's DNA is in the form of **chromatin.** Chromatin is formed of strands of DNA and proteins called **histones**. The DNA and histone combination often results in structures called a **nucleosome.** A nucleosome consists of eight spherical histones with a strand of DNA wrapped around them and secured with a ninth histone. This produces a structure that resembles a string of beads. A chromosome is a highly coiled structure of many nucleosomes.

◀ This drawing shows how DNA is packaged into chromosomes.

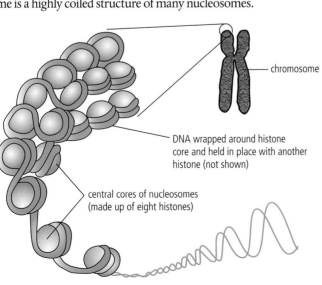

chromosome

DNA wrapped around histone core and held in place with another histone (not shown)

central cores of nucleosomes (made up of eight histones)

The nucleus is often located centrally within the cell's cytoplasm, although in some cell types it is pushed to one side or the other. The side position is characteristic of plant cells, because these cells often have a large central vacuole. Most eukaryotic cells possess a single nucleus, but some do not have a nucleus at all, and others have multiple nuclei. Without a nucleus, cells cannot reproduce. The loss of reproductive ability is often paired with increased specialization to carry out certain functions. For example, human red blood cells do not have nuclei: they are specialized to transport respiratory gases. Most nuclei also include one or more dark areas called **nucleoli** (singular nucleolus). Ribosome molecules are manufactured in nucleoli. The molecules pass through the nuclear envelope before assembling as ribosomes.

Chloroplasts

Chloroplasts occur only in algae and plant cells. The chloroplast contains a double membrane and is about the same size as a bacterial cell. Like the mitochondrion, a chloroplast contains its own DNA and 70S ribosomes. The DNA of the chloroplast takes the form of a ring.

You should note all the characteristics that chloroplasts and mitochondria have in common with prokaryotic cells.

As well as DNA and ribosomes, the interior of the chloroplast includes **grana** (singular granum), **thylakoids** and the **stroma**, which are labelled in Figure 4. A granum is made up of numerous thylakoids stacked like a pile of coins. The thylakoids are flattened membrane sacs with components necessary for the absorption of light. Absorption of light is the first step in **photosynthesis**. Photosynthesis is a process that converts light energy into chemical energy. The chemical energy is then stored in sugars made from carbon dioxide. The fluid stroma is like the cytoplasm of the cell. It occurs outside the grana but within the double membrane. The stroma contains many enzymes and chemicals necessary to complete the process of photosynthesis. Like mitochondria, chloroplasts can reproduce independently of the cell.

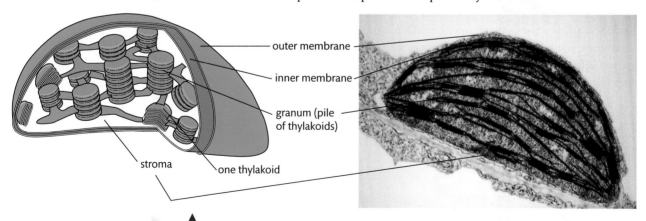

outer membrane

inner membrane

granum (pile of thylakoids)

stroma

one thylakoid

▲
A2.2 Figure 4 Compare the drawing of a chloroplast with the corresponding transmission electron micrograph (TEM) of a chloroplast.

Centrosome

The **centrosome** occurs in all eukaryotic cells. In animal cells it consists of a pair of **centrioles** that are often at right angles to one another. The centrioles are involved with the assembly of **microtubules**, which are important to a cell because they provide structure and allow movement. Microtubules are also important for cell division. Plant and fungal cells do not have centrioles. However, they are able to produce microtubules from their centrosome-like regions, which suggests that centrioles are not necessary for producing microtubules.

The centrosome is located at one end of the cell, close to the nucleus. **Basal bodies** are structures related to the centrosome of eukaryotic cells and are located at the base of cilia and flagella. Not all eukaryotic cells have cilia or flagella, therefore not all eukaryotic cells have basal bodies. The basal bodies are thought to direct the assembly of microtubules within the associated cilia or flagella. When present, centrioles appear to produce basal bodies.

Vacuoles

Vacuoles are storage organelles that are usually formed from the Golgi apparatus. They are membrane-bound and have many possible functions. They occupy a very large space inside the cells of most plants. In animal cells, vacuoles are small and may be numerous. Vacuoles may store several different substances, including potential food (to provide nutrition, as in plant cells), metabolic waste and toxins (to be expelled from the cell) and water. Vacuoles enable cells to have higher surface area-to-volume ratios even at larger sizes. In plants, they allow an uptake of water, which provides rigidity to the organism. When a large vacuole occurs in the central area of a plant cell, it is called a **central vacuole**. Vacuoles are like vesicles except that they are larger.

▲ A transmission electron micrograph (TEM) showing the two centrioles of a centrosome. The presence of centrioles indicates that the micrograph is of a eukaryotic cell, but not a plant or fungal cell.

◄ A coloured transmission electron micrograph (TEM) of a plant cell that has a central vacuole filled with water. Note the central location of the vacuole, with the cytoplasm and all the other organelles pushed to the cell margins.

When comparing
items, be certain to
state the characteristic
for each type of item,
as shown in Table 1
for prokaryotic and
eukaryotic cells.

A comparison of prokaryotic and eukaryotic cells

A table is an effective way of summarizing the differences between prokaryotic and eukaryotic cells.

Prokaryotic cells	Eukaryotic cells
DNA in a ring form without protein	DNA with proteins as chromosomes/chromatin
DNA free in the cytoplasm (nucleoid region)	DNA enclosed within a nuclear envelope (nucleus)
No mitochondria	Mitochondria present
70S ribosomes	80S ribosomes
No internal compartmentalization to form organelles	Internal compartmentalization present, forming many types of organelles
Size less than 10 μm	Size more than 10 μm

▲ **A2.2 Table 1** Comparing prokaryotic and eukaryotic cells

If asked to state the similarities between the two types of cells, make sure you include the following:

- both types of cells have some outside boundary that always involves a plasma membrane
- both types of cells conduct all the functions of life
- DNA is present in both cell types.

A2.2.7 – Unicellular organisms

A2.2.7 – **Processes of life in unicellular organisms**
Include these functions: homeostasis, metabolism, nutrition, movement, excretion, growth, response to stimuli and reproduction.

All organisms, whether unicellular or multicellular, carry out all the functions of life. The functions of life are summarized in Table 2.

Metabolism	The sum of all the chemical reactions that occur within an organism
Growth	The development of an organism
Reproduction	The ability to produce offspring
Response to stimuli	As the environment changes, the organism adapts
Homeostasis	Maintenance of a constant internal environment
Nutrition	The ability to acquire the energy and materials needed to maintain life
Excretion	The ability to release materials not needed or harmful into the surrounding environment
Movement	The ability to move or change position

▲ **A2.2 Table 2** The functions of life

It is important to note that if the functions of life are evident, then life is said to be present.

Unicellular organisms have unique ways of carrying out the life functions compared to **multicellular** organisms.

- The cell membrane controls the movement of materials in and out of the cell, to help maintain homeostasis.
- Vacuoles isolate and store waste so that it does not harm the organism.
- Cells often possess cilia or flagella that allow movement in response to changes in the environment.
- Vacuoles carry out digestion, to provide nutrition for the organism.
- Mitochondria or areas of enzymes allow energy production to continue for all the functions of life.
- Ribosomes provide the building blocks for growth and repair.

Multicellular organisms often have whole groups of cells called **organs** carrying out these functions.

A2.2.8 – Different types of eukaryotic cells

A2.2.8 – Differences in eukaryotic cell structure between animals, fungi and plants

Include presence and composition of cell walls, differences in size and function of vacuoles, presence of chloroplasts and other plastids, and presence of centrioles, cilia and flagella.

The eukaryotic cells of different types of organisms can vary. Three types of organisms with eukaryotic cells are plant cells, animal cells and fungal cells. There are over 14,000 species of fungi, and it is believed that they were the first eukaryotes to live on land.

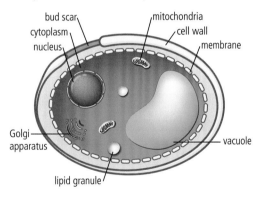

This drawing of a yeast cell illustrates some of the major cell organelles common to fungi.

This transmission electron micrograph (TEM) of a yeast cell represents one of the many species of fungi. From our previous work with organelles, identify as many as possible.

Fungi can be unicellular or multicellular. They include yeasts, mushrooms, truffles and bread moulds, plus many more. No fungus can produce its food. Fungi secrete (release into the surrounding environment) digestive enzymes and then absorb the externally digested nutrients as their source of energy. They have major roles in our planet, including decomposing organic debris to enable the recycling of nutrients, being a source of food, being used in medicines, and even controlling many harmful insects.

Most believe fungi are more closely related to animals than to plants. Table 3 summarizes the differences between plant, animal and fungal eukaryotic cells. However, do not forget the similarities between these three cell types as well.

Plant cells	Animal cells	Fungal cells
Exterior of cell includes an outer cell wall composed of cellulose, with a plasma membrane just inside	Exterior of cell includes a plasma membrane. There is no cell wall	Exterior of cell includes an outer cell wall composed of chitin, with a plasma membrane just inside
Chloroplasts are present in the cytoplasm area, enabling the production of carbohydrates	There are no chloroplasts for carbohydrate production	There are no chloroplasts for carbohydrate production
Possess large centrally located vacuoles for the storage of carbohydrates	Vacuoles are generally small and numerous, when present, with many unique functions	Vacuoles are generally small and numerous, with many unique functions
Store carbohydrates as starch	Store carbohydrates as glycogen	Store carbohydrates as glycogen
Usually do not contain cilia, flagella or basal bodies	May have cilia or flagella, with associated basal bodies	May have cilia or flagella, but do not have associated basal bodies
Because a rigid cell wall is present, this cell type has a fixed, often angular, shape	Without a cell wall, this cell is flexible and more likely to be a rounded shape	The cell wall allows a degree of flexibility, along with support for the cell; the cell shape may vary
Possess centrosomes but no centrioles	Possess both centrosomes and centrioles	Possess centrosomes but no centrioles

A2.2 Table 3 Differences between plant, animal and fungal cells

Most of the organelles discussed are present in all eukaryotic cells. When an organelle is present in each of the eukaryotic cell types, it usually has the same structure and function. For example, all three cell types contain mitochondria that possess cristae, a matrix and a double membrane. Also, in all three cell types, the mitochondria function in the production of ATP for use by the cell.

A2.2.9 – Atypical eukaryotes

A2.2.9 – Atypical cell structure in eukaryotes

Use numbers of nuclei to illustrate one type of atypical cell structure in aseptate fungal hyphae, skeletal muscle, red blood cells and phloem sieve tube elements.

The structure of some eukaryotic cells is unique or atypical, which allows them to carry out specialized functions. One example of this atypical structure involves cell nuclei.

Some multicellular fungi produce filaments called **hyphae**. Most of these hyphae consist of chains separated by cross-walls that have pores to allow various organelles and cytoplasm to flow from cell to cell. However, some fungi produce hyphae that lack cross-walls. The result of this is a single mass of cytoplasm (one cell) with more than one nucleus.

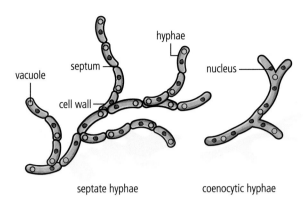

Notice the two types of hyphae shown in this image. The hyphae on the right do not contain cross-walls, while cross-walls are present in the hyphae on the left.

Phloem sieve tube elements, shown in Figure 5, have a specialized function allowing transportation within a multicellular plant. These unique elements/cells have end walls with pores and minimal cellular components such as nuclei, ribosomes, cytoskeleton and cytoplasm. They are connected end to end, forming tube structures. These cells can only remain alive with the help of **companion cells**, which maintain a close connection with the sieve tube elements.

Figure 6 shows a micrograph of human red blood cells. They have the specialized function of carrying oxygen throughout the body. They contain substantial amounts of a molecule called haemoglobin, which easily combines with oxygen. They are shaped to allow a large surface area for the absorption and release of oxygen. They do not have a nucleus, which allows them to carry even more oxygen.

▲
A2.2 Figure 5 A transmission electron micrograph (TEM) of a plant's sieve tube elements and associated companion cells. Notice the lack of substance in the sieve tube elements and the pores in the end wall.

▲
A2.2 Figure 6 A micrograph of a human blood smear.

Figure 7 shows an electron micrograph of human skeletal muscle. This muscle type specializes in allowing body movement. It can carry out this function because of the presence of specialized proteins arranged in bands that contract and relax. The presence of cell membranes is limited, resulting in large, tubular cells with multiple nuclei, allowing more coordinating protein molecules.

A2.2 Figure 7 An electron micrograph (EM) of human skeletal muscle. Note the large, continuous cells.

Other cells with specialized structures to enable unique functions include:

- nerve cells, which are long and thin with branched connections at each end to transmit electrical impulses
- sperm cells, with many mitochondria and a tail allowing movement and a head with a tip capable of producing an enzyme that facilitates penetration of an egg cell
- cells found in the tubes associated with lungs, which have many tiny hairs called cilia on their exterior that work in unison to move mucus and other particles up and out of the airways.

Draw and annotate diagrams of organelles and other cell structures shown in electron micrographs. Full details of how to carry out this activity with a worksheet are available in the eBook.

A2.2.10 and A2.2.11 – Electron micrograph skills

A2.2.10 – Cell types and cell structures viewed in light and electron micrographs

Application of skills: Students should be able to identify cells in light and electron micrographs as prokaryote, plant or animal. In electron micrographs, students should be able to identify these structures: nucleoid region, prokaryotic cell wall, nucleus, mitochondrion, chloroplast, sap vacuole, Golgi apparatus, rough and smooth endoplasmic reticulum, chromosomes, ribosomes, cell wall, plasma membrane and microvilli.

A2.2.11 – Drawing and annotation based on electron micrographs

Application of skills: Students should be able to draw and annotate diagrams of organelles (nucleus, mitochondria, chloroplasts, sap vacuole, Golgi apparatus, rough and smooth endoplasmic reticulum and chromosomes) as well as other cell structures (cell wall, plasma membrane, secretory vesicles and microvilli) shown in electron micrographs. Students are required to include the functions in their annotations.

Utilizing the text, diagrams and pictures presented in this chapter, you should be able to differentiate between prokaryotic and eukaryotic cells when presented with light or electron micrographs. You must be able to identify the following cell structures: nucleoid region, prokaryotic cell wall, nucleus, mitochondrion, chloroplast, sap vacuole, Golgi apparatus, rough and smooth endoplasmic reticulum, chromosomes, ribosomes, cell wall, plasma membrane and microvilli. The internet has many sites that show cells of various types, which you can use to develop your skills in this identification process.

 It is important that you practise the skills necessary to produce informative drawings throughout the course. Actual laboratory observation of cells using prepared slides and a microscope will help you develop your skills. Draw what you can see in the field of view, and compare your drawings, labels and explanations with those found on appropriate internet sites.

 Guiding Question revisited

What are the features common to all cells and the features that differ?

In this chapter, we have discovered the following about cells:

- whether unicellular or multicellular, all organisms are composed of cells
- features common to all cells include DNA, cytoplasm and a plasma membrane forming an exterior boundary
- prokaryotic cells display a simple composition, lacking membrane-bounded organelles in their cytoplasm
- eukaryotic cells are compartmentalized, with isolated areas carrying out specialized tasks
- the cytoplasm of eukaryotic cells has many unique organelles working together, exhibiting all the life functions of the cell/organism
- variations of the cell structure result in some unique cellular compositions, such as cells with multiple nuclei and cells with no nuclei.

 Guiding Question revisited

How is microscopy used to investigate cell structure?

In this chapter, we have discovered the following about microscopes:

- magnification and resolution are two properties of microscopes that are essential for the study of cells
- light microscopes have the advantage that living cells and tissue can be viewed
- EMs have increased the limits of magnification and resolution, allowing views of cells never thought possible even 50 years ago
- freeze fracture and fluorescent stains have furthered the study of cells via microscopy
- immunofluorescence using antibodies and specialized dyes has allowed visualization of the specific tissues viruses attack.

Exercises

Q1. Which pair of organelles is present in plant cells but not in animal cells?

   A Chloroplasts and mitochondria.

   B Centrioles and central vacuole.

   C Chloroplasts and cell wall.

   D Lysosomes and plasma membrane.

Q2. What carbon compound is most likely to be transported by rough endoplasmic reticulum?

Q3. Which of the following is not found in eukaryotic cells?

   A Microtubules.

   B Mitochondria.

   C Nucleus.

   D Chloroplasts.

Q4. Which cell type is the most likely to possess a capsule?

   A Red blood cell.

   B Prokaryotic cell.

   C Sieve tube element.

   D Eukaryotic cell.

Q5. What structure is directly related to prokaryotic cell reproduction?

   A Cilia.

   B Basal body.

   C Centriole.

   D Pili.

Q6. Which association is most accurate?

   A Red blood cell: nucleus.

   B Nucleus: mitochondrion.

   C Basal body: ribosome.

   D Golgi apparatus: vesicle.

Q7. Match the following features and organelles.

A	mitochondrion	**1**	food storage
B	cytoskeleton	**2**	cristae
C	ER	**3**	contains hydrolytic enzymes
D	lysosome	**4**	microtubules
E	vacuole	**5**	rough or smooth

A2 Practice questions

1. Describe two examples of a typical cell structure involving number of nuclei.

    (Total 4 marks)

2. List three structures common to all cells.

    (Total 3 marks)

3. Explain two advantages as well as two disadvantages concerning the use of electron microscopy.

    (Total 4 marks)

4. Compare and contrast the general features of prokaryotic and eukaryotic cells.

    (Total 4 marks)

THEME **A Unity and diversity**
3 Organisms

From single-celled organisms to coral reefs to trees, life on Earth shows a remarkable degree of variation. For centuries, physical characteristics have been used to name organisms and to put similar organisms into categories. More recently, thanks to DNA sequencing, we can use the genetic code of an organism (its genome) to help show how closely it is related to other species.

A3.1 Diversity of organisms

Guiding Questions

What is a species?

What patterns are seen in the diversity of genomes within and between species?

Although there are at least two dozen definitions for the concept of species in biology, we will examine two: the morphological definition that has been used for hundreds of years, and the biological species concept definition, which has only existed in the past few decades. The first looks at what physical features organisms have, while the second considers whether or not individuals can breed to produce fertile offspring. Each definition has its strengths and weaknesses. No single definition can encompass all living organisms as well as extinct species, because such an astoundingly large diversity exists among the various forms of life on Earth.

When DNA sequences of organisms are compared, it is possible to see that, between individuals of the same species, there are remarkably few differences compared to the differences between individuals belonging to two different species. A single-celled organism with no specialized tissue is likely to have a much smaller quantity of DNA than a multicellular organism with hundreds of different specialized tissues.

A3.1.1 – Variation between organisms

A3.1.1 – Variation between organisms as a defining feature of life

Students should understand that no two individuals are identical in all their traits. The patterns of variation are complex and are the basis for naming and classifying organisms.

If you have pigeons where you live, you might think that they all look the same. But ask pigeon experts and they will tell you that the level of diversity and variation among pigeons is equivalent to the level of diversity and variation in humans. Animal breeders such as pigeon fanciers recognize each individual in the population they are raising, just as you would recognize your dog in a group of similar dogs. No two individuals in a population share all the same traits. Even identical twins have slight differences.

Observing the differences between individuals within one species and observing the differences between one species and another is a daunting task, especially when we consider that there are millions of species on Earth to observe, from invisible microbes to mighty redwood trees over 100 m tall.

How can we classify organisms? There are countless possible ways; a few examples are listed below.

- By feeding habits: it makes its own food/it is a carnivore or herbivore.
- By habitat: land-dwelling/aquatic.
- By movement: sessile (stuck in one place)/free moving.
- By daily activity: nocturnal/diurnal.
- By risk: harmless/venomous.
- By anatomy: plant/animal/vertebrate/invertebrate.

We generally start by categorizing organisms based on **morphology** (the physical appearance of an organism). Is the organism made of a single cell without a nucleus, or does it have a nucleus? If it has a nucleus, is it single-celled or multicellular? Think of these categories as boxes into which the organisms are placed. Each category is called a **taxon** (plural taxa). The biggest taxa are very broad and encompass many species, but as the defining features used become more and more detailed and specific, smaller and smaller boxes are used, containing fewer and fewer species per taxa, until we arrive at a single species. The largest taxon is a "domain" and it contains all the more specific taxa, from "kingdom" down to "species".

Table 1 illustrates the identification of two species from very different kingdoms: one species is an animal, humans, and the other is a plant, garden peas. The science and skill of categorizing life is called **taxonomy** and specialists who do it are called **taxonomists**.

The garden pea (*Pisum sativum*) is the plant Gregor Mendel studied.

How do species exemplify both continuous and discontinuous patterns of variation?

To help remember the order of the taxa, a mnemonic (memory trick) is helpful. Make a sentence using the first letters of each level, such as "King Philip Came Over For Good Soup". The human brain is very poorly adapted for remembering lists of words but very highly adapted for remembering stories. Transforming lists into stories is a good example of a mnemonic.

Taxa	Human	Garden pea
Kingdom	Animalia	Plantae
Phylum	Chordata	Angiospermophyta
Class	Mammalia	Dicotyledoneae
Order	Primate	Rosales
Family	Hominidae	Papilionaceae
Genus	*Homo*	*Pisum*
Species	*sapiens*	*sativum*

▲
A3.1 Table 1 The classification of two species

The variations in characteristics for sorting species into their designated taxon might be obvious (plants have leaves and roots, whereas humans have limbs and a head), but can sometimes be very subtle. Two species of frog might look identical on the outside but can be distinguished by different mating calls. In such a case, the patterns of variation in morphology are not sufficient for classification.

When variation can be placed into distinct categories (type A blood versus type B, for example), we say it is **discontinuous**. When variation has a wide range of possibilities (how tall a tree can grow, for example), we say it is **continuous**. Sometimes we impose categories such as eye colour as if it is an example of discontinuous variation when, in fact, a hundred people who have blue eyes will show a certain amount of continuous variation, from deep blue to very light blue.

A3.1.2 – Species as groups of organisms

A3.1.2 – Species as groups of organisms with shared traits
This is the original morphological concept of the species as used by Linnaeus.

Carolus Linnaeus, an 18th century professor of medicine and botany in Sweden, had difficulty identifying the plants he found on his travels because different botanists used different systems for naming them. This made it difficult to categorize the organisms. Linnaeus then had a remarkable idea: what if we take all the known living organisms, put them into categories, and give them a name using a uniform system? Not just plants, but animals, too. By creating the names using Latin or Greek, no matter what anyone calls the organism in their native language (such as Swedish), it will always have a universally known name.

Linnaeus based the classification system, as well as the names, on the physical features of the organisms. This **morphological classification**, first published in his book *Systema Naturæ* in 1735, was used by generations of botanists and zoologists, and the naming system he created is still used today. Thousands of organisms still carry the scientific name that Linnaeus gave them over two-and-a-half centuries ago, such as the Asian elephant, which he named *Elephas maximus* in 1758.

A3.1.3 – The binomial naming system

A3.1.3 – Binomial system for naming organisms
Students should know that the first part of the name is the genus, the second part of the name is the species. Species in the same genus have similar traits. The genus name is given an initial capital letter but the species name is lowercase.

You have a scientific name based on your species: *Homo sapiens*. This system of naming organisms using two names is called **binomial nomenclature**. "Bi" means two, "nomial" means name and "nomenclature" refers to a system used to name things.

Myrmecophaga tridactyla is a name that literally means "eater of ants" plus "with three fingers". This name refers to the giant anteater of Central and South America. In fact, the animal really has five fingers, but they are hard to see because the animal walks on its front knuckles.

▲
The giant anteater (*Myrmecophaga tridactyla*)

30 cm

In the early days of classification, all known organisms were classified into only two kingdoms: plants and animals. With the invention of the microscope in the mid-1600s, many new creatures were discovered that were nothing like plants or animals. In effect, the microscope revealed that there is an entire world of invisible organisms living throughout the world's ecosystems.

The first name in the binomial nomenclature system is always capitalized and it refers to the **genus**; the second name always begins with a small letter and refers to the **species**. Both are always written in italics when typed, or underlined when written by hand. Organisms in the same genus will have a higher number of similar characteristics compared to organisms in a different genus.

There are three main objectives and associated rules to using binomial nomenclature:

1. each organism has a unique name that cannot be confused with another organism
2. the names can be universally understood, no matter what nationality or culture is using the name
3. there is some stability in the system, so that people cannot change the names of organisms without valid reasons.

Examples of binomial nomenclature

Sometimes scientific names for organisms are relatively easy to decipher because they contain their common names:

- *Amoeba amazonas*
- *Equus zebra*
- *Gekko gecko* (this lizard gets its name from the sounds it makes)
- *Gorilla gorilla*
- *Paramecium caudatum* (caudate means having a tail).

Sometimes, it is more difficult to guess their common name:

- *Apis mellifera* (honeybee, although you might have guessed this if you know that beekeeping is also called apiculture)
- *Aptenodytes patagonicus* (king penguin, although you can probably guess where it lives from its species name)
- *Loxodonta cyclotis* (African forest elephant)
- *Malus domestica* (apple tree).

Homo sapiens

Genus   species

The rules about writing binomial nomenclature names are that:
- the genus name is capitalized but the species name is not
- both are written in italics when typed, or underlined when handwritten.

SKILLS

In taxonomy, there are two opposing philosophies concerning what to do when an organism does not fit easily into existing categories: (1) broaden the definition of an existing category to include the new organism; or (2) invent a new category or subcategory. Specialists who take the first approach are referred to as **lumpers**, while those who take the second approach are referred to as **splitters**.

Scientists naming organisms sometimes have a sense of humour. Here are some examples.
- *Agra schwarzeneggeri* Erwin, 2002. This Costa Rican ground beetle was named after Arnold Schwarzenegger because of the insect's large biceps.
- *Dracula vampira* Luer, 1978. This orchid in Ecuador got its name from the fact that the petals on the flower look like a bat's wings.

Challenge yourself

1. Look up the following to find out what their scientific names are:
   - your favourite animal
   - your favourite fruit or vegetable
   - your favourite flower, tree or house plant.

A3.1.4 – Biological species

A3.1.4 – Biological species concept
According to the biological species concept, a species is a group of organisms that can breed and produce fertile offspring. Include possible challenges associated with this definition of a species and that competing species definitions exist.

Another definition of a species that is now often preferred over Linnaeus' morphological definition is the **biological species concept**. This was proposed by Ernst Mayr in 1942. Using this definition, in order to be classified as the same species, individuals must be able to breed together and produce fertile offspring. All modern dogs, *Canis familiaris*, can interbreed to produce fertile offspring, so they are considered to be one species.

Not every biologist is happy with this definition, however. How can this definition apply to organisms that reproduce asexually and therefore do not breed? Hybrids produced from parents of closely related but separate species are usually infertile, but not always. Some species are made up of a mosaic of DNA from multiple species. How should they be classified? Should they receive multiple species names if they are composed of more than one? How can we apply the concept to extinct species such as velociraptors when we cannot know from skeletons whether members of a population could interbreed?

▲
All domestic dogs are of the same species.

Depending on which expert you ask, there are dozens of definitions of the word "species". We have discussed two so far: the morphological definition used in the 18th century, and a more recent definition, the biological species concept, involving the ability to breed and produce fertile offspring. But other characteristics can also be taken into account when deciding on what counts as a species, such as the following.

- The ecological niche of an organism. Because microbes are single-celled, it is challenging to use just morphology to determine what species they belong to. Where they live and what they eat can help classify microbes into different species.
- Genetics. When a sequence of DNA found in a sample of soil from a forest does not match any known sample, it suggests that it is from a species that has not been catalogued yet.
- The types of molecules an organism can produce. This is also useful when classifying microscopic organisms that do not have easily observable features, unlike birds and primates, for example. It is common to find microbes that produce carbon dioxide, but some can make methane or hydrogen gas.
- For extinct species, their lineage. If we find a fossil of an extinct snail that has a shell similar to a modern species, we can use the similarities to assign it a species name based on its position on the same part of the evolutionary tree as the existing species.

Microscopic soil organisms can be challenging to identify because morphology is insufficient as a criterion to differentiate species.

Nature of Science

To some extent, the debate about what a species really is becomes just as philosophical as biological. "Is all we are doing simply naming things?" "Do the categories we use actually exist in reality or just in our minds?" "Is the difficulty of agreeing on a definition a fault of the limitations of language?" "Is it possible to use the same term (species) for organisms that exist today and to express how their populations evolved over time?" These questions are currently being debated by biologists and, because the variety of life is so diverse, it is difficult to find a consensus.

A3.1.5 – Distinguishing between populations and species

A3.1.5 – Difficulties distinguishing between populations and species due to divergence of non-interbreeding populations during speciation

Students should understand that speciation is the splitting of one species into two or more. It usually happens gradually rather than by a single act, with populations becoming more and more different in their traits. It can therefore be an arbitrary decision whether two populations are regarded as the same or different species.

Speciation, as explored in more detail in Chapter A4.1, is the process by which a population is separated into two groups that can no longer reproduce together. One part of the population evolves one way and the other, living with different selection pressures and producing different sets of mutations, evolves in a different way. The two populations become different enough over time that they can no longer interbreed to produce fertile offspring. As a result, a new species has branched off from the previous one, resulting in two species that have a common ancestor.

Lake Victoria in East Africa is, geologically speaking, a young lake, being only about 400,000 years old. Any fish species that live there have arrived since then. African cichlid fishes, of which there are over 200 species in the lake, all appear to have evolved from a single species introduced about 200,000 years ago. Each one has evolved in its own niche and as a result split off from the others. Some specialize in eating algae, some eat plankton and others eat snails. But each split would have taken many generations and, during those generations, the population that started to explore the new source of food would have continued to interbreed with some success with the original population. As the two populations became more different from each other, the success rates of interbreeding would have diminished until it was no longer possible. It is difficult for specialists to decide when the speciation occurred. When a cut-off point is chosen, it has an arbitrary and subjective aspect to it.

The last woolly mammoth became extinct thousands of years ago. It appeared to share many similar characteristics with today's Asian elephants (*Elephas maximus*), which is why it was originally classified in 1799 in the same genus, as *Elephas primigenius*. Because of the gap in time, it is difficult to apply the biological species concept to decide whether or not the two populations are one and the same species, because there are no living mammoths to test the hypothesis by breeding them with elephants. The mammoth's scientific name has since been changed to *Mammuthus primigenius*, without knowing for sure whether they could breed together or not, so it is a relatively arbitrary decision from the point of view of the biological species concept.

What might cause a species to persist or go extinct?

▲ The woolly mammoth went extinct thousands of years ago. We cannot test whether it was able to breed with modern elephants or not.

Figure 1 shows a common ancestor giving rise to four species. The first speciation event shown happened earlier in time, then the split that generated species B occurred, and, finally, D split from C. Although this type of diagram helps illustrate the sequence of events, it gives the impression that the splits occurred suddenly, which is not always the case.

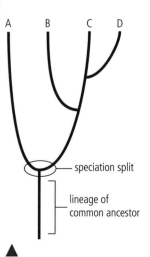

▲ **A3.1 Figure 1** Species A, B, C and D evolved from a common ancestor. Three speciation splits led to the generation of these species, the first of which is circled.

63

A3.1.6 – Diversity in chromosome numbers

A3.1.6 – Diversity in chromosome numbers of plant and animal species
Students should know in general that diversity exists. As an example, students should know that humans have 46 chromosomes and chimpanzees have 48. Students are not required to know other specific chromosome numbers but should appreciate that diploid cells have an even number of chromosomes.

Diploid and haploid cells

The term **diploid** is used to describe a nucleus that has chromosomes organized into homologous pairs. Most cells in the human body are diploid cells, and in such cells the nucleus contains a set of 23 chromosomes from the mother and 23 from the father. There is a category of cells that only contain 23 chromosomes in total: the sex cells, also called **gametes**. Because the chromosomes in sperm and egg cells do not come in pairs, but rather only have a single chromosome from each pair, they are said to be **haploid**. The adult form of animal cells is rarely haploid, but there are exceptions, for example adult male bee, wasp and ant cells are haploid. Generally speaking, the vast majority of cells in sexually reproducing organisms are diploid, and only the gametes are haploid.

The variable n represents the **haploid number**, and it refers to the number of sets of chromosomes that a nucleus can have. For a human egg cell, $n = 23$. When an egg cell is fertilized by a sperm cell (a sperm is also haploid and therefore contains 23 chromosomes), a **zygote** is formed and the two haploid nuclei fuse together, matching up their chromosomes into pairs. Hence humans generally have a total of $23 + 23 = 46$ chromosomes. This means that in humans, $2n = 46$, so diploid cells in humans have 23 pairs of chromosomes making a total of 46 chromosomes. Compare this number with some of the other species in Table 2.

Note in Table 2 that diploid cells always have an even number of chromosomes. This is logical because one chromosome in each pair comes from one parent and the other from the other parent.

A3.1 Table 2 A comparison of types of cells and chromosome numbers ▶

	Types of cells and chromosome numbers	
Species	**Haploid = n**	**Diploid = $2n$**
Human (*Homo sapiens*)	23	46
Chimpanzee (*Pan troglodytes*)	24	48
Domestic dog (*Canis familiaris*)	39	78
Rice (*Oryza sativa*)	12	24
Roundworm (*Parascaris aquonum*)	1	2

The number of chromosomes is a characteristic of a species

As you can see from Table 2, the number of chromosomes for humans (46) is very different to the number of chromosomes for the roundworm. One of the best-studied worms in genetics laboratories is *Caenorhabditis elegans*, whose genome was first sequenced in 1998. It has six chromosomes, meaning its diploid number, $2n$, is 6, and therefore its haploid number, n, is 3. It would be expected that all the cells in *C. elegans* would have six chromosomes, and, likewise, that all cells in humans would have 46. Although this is true for most cells, we have already seen the exception of haploid cells (n). Note as well that some cells do not contain a nucleus and have no chromosomes, such as red blood cells.

A3.1.7 – Karyotypes

A3.1.7 – Karyotyping and karyograms

Application of skills: Students should be able to classify chromosomes by banding patterns, length and centromere position. Students should evaluate the evidence for the hypothesis that chromosome 2 in humans arose from the fusion of chromosomes 12 and 13 with a shared primate ancestor.

NOS: Students should be able to distinguish between testable hypotheses such as the origin of chromosome 2 and non-testable statements.

A **karyogram** is a representation of the chromosomes found in a cell arranged according to a standard format, as in the example in Figure 2. The chromosomes are placed in order according to their size and shape. The shape depends mainly on the position of the **centromere**. A karyogram is used to show a person's **karyotype**, which is the specific number and appearance of the chromosomes in their cells.

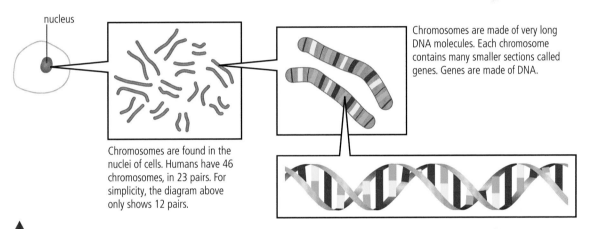

nucleus

Chromosomes are found in the nuclei of cells. Humans have 46 chromosomes, in 23 pairs. For simplicity, the diagram above only shows 12 pairs.

Chromosomes are made of very long DNA molecules. Each chromosome contains many smaller sections called genes. Genes are made of DNA.

▲
Zooming into a cell reveals where DNA is found.

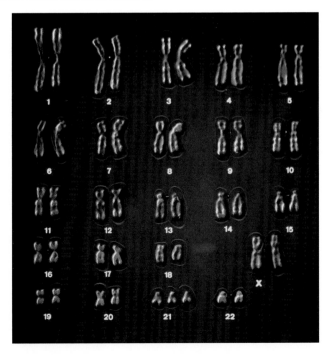

◄ **A3.1 Figure 2** This is a karyogram showing all 23 pairs of chromosomes. What can we learn about the individual's karyotype from this figure? This karyogram was prepared using false-colour imagery.

SKILLS

How is a karyogram image obtained? Once the cells of an organism have been collected and grown in culture, a karyogram is made following the steps below.

1. The cells are stained and prepared on a glass slide, to see their chromosomes under a light microscope.

2. Photomicrograph images are obtained of the chromosomes during a specific phase of cell division called the mitotic metaphase (see Chapter D2.1).

3. The images are cut out and separated, a process that can be done using a print out and scissors or on a computer.

4. The images of each pair of chromosomes are placed in order by size and the position of their centromeres. Generally speaking, the chromosomes are arranged in order by decreasing length. The exception is in the 23rd pair of chromosomes, which can contain one or two X chromosomes, which are considerably larger than the chromosomes in the 22nd pair (see the chromosome pair marked X in Figure 2). In addition, the coloured bands that show up in the image can be used to identify which chromosome it is. For example, chromosomes 3 and 4 in the image show very different banding patterns.

The evolution of human chromosome 2

Modern humans have 46 chromosomes. Other human species that no longer exist but whose preserved fossil DNA we can study, such as Neanderthals and Denisovans, also had only 46 chromosomes. Gorillas and chimpanzees are the species most closely related to humans. Our last common ancestor with gorillas existed about 9 million years ago and the speciation split with chimpanzees occurred about 6 million years ago. However, when we prepare a karyogram of the contents of their nuclei, both gorillas and chimpanzees have 48 chromosomes instead of 46. If we shared a common ancestor with them, what happened to our chromosome number?

Two possible hypotheses can be formulated:

1. a complete chromosome disappeared

2. two chromosomes from an earlier common ancestor fused to become a single chromosome.

It is unlikely that an entire chromosome was deleted and disappeared, because removing hundreds of genes in that way would cause a major threat to the viability of the species. To test the second hypothesis, we can look for evidence, and can start by examining the two characteristics that help identify a chromosome: its shape (position of the centromere) and its banding patterns. One shape a chromosome can have is the "X" shape, with the centromere close to the centre. This is called a **metacentric** chromosome. Chromosomes can also have an **acrocentric** shape, meaning the centromere is at one end, making one arm of the chromosome much shorter and the other much longer. All primates have both types.

One hypothesis is that chromosome 2 in humans arose from the fusion of chromosomes 12 and 13 in a shared ancestor. In an article from *Molecular Cytogenetics* by Paweł Stankiewicz in 2016, human chromosome 2 was compared to chimpanzee chromosomes 12 and 13. In terms of shape, these two acrocentric non-human chromosomes, when placed end to end, have a similar length to the human chromosome, although some parts overlap. The position of the centromere in human chromosome 2 lines up with the chimpanzee chromosome 12 but not with chromosome 13. This latter piece of evidence refutes the hypothesis. However, in the zone marked B on the human chromosome in Figure 3, we find the type of DNA we usually encounter in the centromere, known as **satellite DNA**, which consists of short repeating sequences of DNA. This zone corresponds to the position of the centromere in the non-human chromosome 13, giving credibility to the hypothesis. In terms of banding patterns, the long arm of chimpanzee chromosome 12 matches that of the short arm of human chromosome 2, and the long arm of chimpanzee chromosome 13 matches the banding patterns of the long arm of human chromosome 2.

chimp
12

A —

B —

human
2

chimp
13

A3.1 Figure 3 A comparison of human chromosome 2 with chimpanzee chromosomes 12 and 13.

Besides shape and banding patterns, other evidence to support the idea of fusion is the presence of telomeric DNA in the centre of human chromosome 2. The **telomeres** are caps at the tips of chromosomes that contain repeating sequences of DNA and provide protection, the same way that bumpers protect cars and aglets protect the ends of shoelaces. Such repeating telomeric DNA is not supposed to be in the centre of chromosomes, only at the tips. And yet, at position A in the human chromosome 2 shown in Figure 3, telomeric DNA is present at the position where the two chromosomes would have fused.

It is very important to understand that this evidence does not say we descended from chimpanzees. The fusion of the chromosomes would have happened after the speciation split of a common ancestor that led to the evolution of chimpanzees on one branch of the tree of life and the evolution of humans on another branch.

When asked to evaluate evidence for a claim, scientists and students need to express their opinion of whether or not the evidence is sufficient to convincingly confirm the claim. Some questions to consider asking are:
• Is the quantity of evidence sufficient to accept the claim?
• Has the method for collecting evidence been repeated and tested by other scientists, and have they found similar evidence?
• Is the method being used a reliable method?
• Are any counterclaims or refuting evidence enough to doubt the claim?
• Is there a mechanism to explain the cause, or is what we are seeing just a coincidence?

Nature of Science

Some claims are testable and others are not. The hominid fossil nicknamed Lucy, discovered in Ethiopia in 1974, is complete enough to test and confirm claims such as (1) she was a female, (2) she was not a modern human but rather an australopithecine, (3) she could walk on two legs and (4) she lived about 3.2 million years ago. There might be some debate about the details, but the challenges can also be tested. Can you think of any claims about her that would not be testable? For example: "Lucy had a great sense of humour." "Lucy had a recurring dream where she encountered a wildcat." "Lucy spoke three languages." Current tools in science have no way of testing these claims. Statements like these are speculation. What about these: "Lucy had very little meat in her diet." "Australopithecines such as Lucy had strong spiritual beliefs." Are they testable claims?

Some claims about the fossil called Lucy are testable and others are not. ▶

A3.1.8 – Unity and diversity of genomes

A3.1.8 – Unity and diversity of genomes within species

Students should understand that the genome is all the genetic information of an organism. Organisms in the same species share most of their genome but variations such as single-nucleotide polymorphisms give some diversity.

It seems counterintuitive, but it is possible to find lots of evidence to support the claim "we are all the same", and it is also possible to find lots of evidence to support the claim "we are all different". From a genetics point of view, humans share many more similarities than differences with each other, especially compared to another species.

If a chimpanzee was walking down your street, you would recognize right away that it was a non-human primate. And yet, the genetic difference between us and chimpanzees is only about 4%. That is a much bigger difference, however, than between you and other humans, which is estimated to be 0.1% to 0.6%. Why does *Homo sapiens* display so many similarities within its global population? Our unity arises largely from the fact that all humans share the same genes. We do not all have the same versions of each of the genes (called **alleles**, see Chapter D3.2); some of us have type B blood and some have type O, for example. But we all possess the genes that determine the ABO blood type.

Where do we find these small but crucial differences between humans? The estimated 3 million to 20 million **base pairs** (e.g. A–T or G–C) of our DNA sequence that can reveal the differences are found scattered all over our chromosomes. Where most people have a T (thymine) nucleotide, for example, a small portion of humans might have a G (guanine) instead at that position. Such variations can start out as mutations (see Chapter D1.3) but are then passed down from generation to generation. Such a variation involving only one base is called a **single nucleotide polymorphism** or SNP (see Figure 4). It is estimated that about every 100 to 300 bases in a human's genetic code contains an SNP. Geneticists interested in the human genome have identified millions of SNPs, and they can be used to help determine ancestry or risk of genetic diseases.

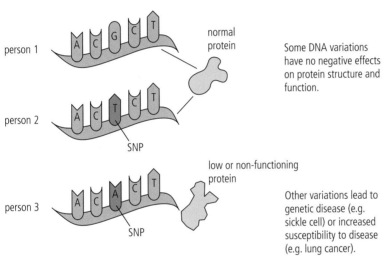

▲ **A3.1 Figure 4** Person 1 has a gene that expresses a normal protein. Person 2 has a T (thymine) nucleotide instead of a G (guanine) in the SNP, but also expresses a normal protein. Person 3, however, has an SNP that causes the protein to not form correctly.

Only about 5% of SNPs are functional, meaning they actually produce a difference in a person's body. Most are neutral, meaning that they will not affect a person's **phenotype** (the physical expression of a gene, such as blood type or colour vision, see Chapter D3.2).

The Human Genome Project

In 1990, an international cooperative venture called the Human Genome Project set out to sequence the complete human **genome**. Because the genome of an organism is a catalogue of all the bases it possesses, the Human Genome Project hoped to determine the order of all the bases A, T, C and G in human DNA. As there were approximately 3,200,000,000 to find, it took over a decade. In 2003, the Project announced that it had succeeded in achieving its goal. Now, scientists are working on deciphering which sequences represent genes and which genes do what. The human genome can be thought of as a map that can be used to show the position of any gene on any one of the 23 pairs of chromosomes.

In the 1997 science fiction film *GATTACA*, one of the main characters brings a sample of cells to a walk-up window at an establishment that provides anonymous genome services. Within seconds, she gets a full printout and analysis of the genome she is interested in. How far are we from being able to do this today? What ethical implications are there to such a service? Are there laws protecting your genome?

Thanks to modern
communication
technologies, it is
possible for scientists
working all over the
world to collaborate
and contribute to a
scientific endeavour
such as sequencing the
genome of plants that
help feed the world.
Rice is one example:
biologists from 10
countries contributed
to sequencing the first
rice genome.

The current estimate is that humans have approximately 22,000 genes, and, thanks to advances in technology, the sequencing of a person's genome can be done in hours instead of years.

Many companies offer genome sequencing for private citizens willing to pay the price. Some of the products reveal ancient family origins and risk factors for some health problems, such as the chances of developing certain types of cancer or heart disease. Would you want to know if there is a chance that your life could be suddenly shortened by the presence or absence of a certain gene? Would you tell your family and friends? Would you want your parents to have such a test? Should people tell their employer or each other about any health-related issues revealed by a genomic analysis? Or, in contrast, is this a private, personal thing that no one else needs to know about? How accurate and reliable are these analyses? Should we believe everything they say? Does all knowledge impose ethical obligations on those who know it?

A3.1.9 – Eukaryote genomes

A3.1.9 – Diversity of eukaryote genomes

Genomes vary in overall size, which is determined by the total amount of DNA. Genomes also vary in base sequence. Variation between species is much larger than variation within a species.

No humans have genes for characteristics such as bioluminescence (glowing in the dark), which many deep-sea organisms do. Although we see some diversity among humans, we do not see such huge ranges of diversity in the human population as wings for flight, gills to breathe underwater, echolocation organs for seeing without light, chloroplasts for photosynthesis, and so on. There is more unity within the human species (comparing any two humans) than diversity compared to other species (comparing humans to non-humans).

Humans are a diverse global population but there are remarkably few differences between any two humans compared to differences between humans and other species.

One major difference between genomes is their size: the quantity of DNA they have in their nuclei. As we will see in Section A3.1.10, some eukaryotic genomes only have a few thousand genes while others can have tens of thousands of genes. This means that one eukaryote will possess genes that another will not have at all. A fish does not need to have genes to produce pollen, and a rose bush does not need genes for making fins to swim. Even with closely related species that have undergone a relatively recent speciation split, they have been evolving separately to the point where the genes are now different enough that they cannot interbreed anymore.

Such differences can be seen in the sequences of base pairs in each genome. Sequencing technology along with databases and computer programs for searching and comparing large data sets have allowed biologists to compare the genomes of organisms from all over the world.

Bioinformatics is a research field that uses both computer science and information technology to help us understand biological processes. Bioinformatics has grown exponentially in recent years. The most data-rich area of bioinformatics is genomics. Genome data is now available in public databases such as The National Center for Biotechnology Information (NCBI). Genetic information can also be explored using the following databases:

- Swiss-Prot, a database of protein sequences
- Ensembl, a database and browser of genomic information about humans and other vertebrates
- GenBank, a National Institutes of Health genetic sequence database that is an annotated collection of all publicly available DNA sequences.

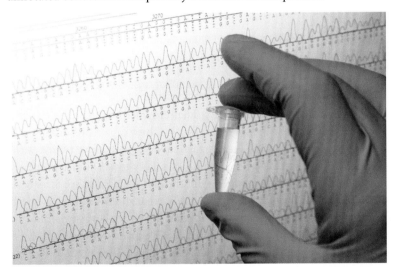

A micropipette containing a DNA sample can be sequenced and added to a database and shared worldwide thanks to web-based information technology.

Instead of sifting through the entire genome of an organism, one way to compare genetic diversity in eukaryotes is to focus on their **mitochondrial DNA**. All eukaryotes have mitochondria, and the way mitochondrial DNA, present only in the egg, not in the sperm cell, is passed down from mother to offspring, means there is not the shuffling and mixing that we see in chromosomal DNA. It is estimated that, within a species, roughly 1 in 1,000 of the genetic code letters is different between individuals' mitochondrial DNA. These genetic differences are expressed in the amino acid sequence that is coded for by the organism's DNA sequence. To see differences between individuals within a species, or to see differences between species, it is possible to look up the amino acid sequences for a particular gene in a database and match them to see if there are amino acids missing, added or modified. Instead of the DNA bases A, T, C and G being displayed, the letters in the databases correspond to the 20 possible amino acids, such as S for serine, G for glycine, A for alanine and V for valine. Some amino acids have a letter that is different from their first letter, such as E for glutamic acid, F for phenylalanine and K for lysine. You will not be asked to memorize the 20 amino acid names and their letters, but you do need to understand that, when comparing genetic differences, it is possible to either use the DNA code or the amino acid sequences.

Table 3 shows part of the sequence for a single gene selected from the online UniProt protein database. The chosen gene is one that all eukaryotes have in their DNA: *cyc1*, the gene for cytochrome c, which is a protein needed by mitochondria to perform their essential task of cellular respiration, to convert sugar into usable energy. Of the hundreds of species available in the database, four species of animal were selected and, rather than looking at all the amino acids that the gene codes for, a short sequence of 60 amino acids was selected for comparison. The differences between the first species and the three other species are highlighted in yellow.

A3.1 Table 3 Comparing a short sequence of 60 amino acids from the mitochondrial gene, *cyc1*, for cytochrome c, in four species

▼

Database codes for specific species	Fragment of the sequence of amino acids coded for in the *cyc1* gene
golden-crowned babbler: TR|A0A7K9SBC6|A0A7K9SBC6_9PASS	SL--ALALSLGGGPLSAGELELHPPNFPWSHGGPLSALDHASVRRGFQVYRQVCSACHSM
brown-headed cowbird: TR|A0A7L3VSC4|A0A7L3VSC4_MOLAT	SLAVALSLSLGGGPVSAGELELHPPGLPWSHGGFLSALDHASVRRGFQVYRQVCSACHSM
green anole: TR|H9GCG1|H9GCG1_ANOCA	GLAVALH-----SAVSAGELELHPPSFPWSHSGPLSSLDHSSVRRGYQVYKQVCSACHSM
big-headed turtle: TR|A0A4D9DRJ9|A0A4D9DRJ9_9SAUR	GLALALH-----TAVSASDLELHPPSYAWSHNGLLASLDHSSIRRGYQVYKQVCAACHSM

The first organism in Table 3 is a bird, the golden-crowned babbler (*Sterrhoptilus dennistouni*), which lives in the Philippines. The next three organisms in Table 3 are a brown-headed cowbird (*Molothrus ater*), a lizard called a green anole (*Anolis carolinensis*), and a big-headed turtle that lives in Southeast Asia (*Platysternon megacephalum*). If we look at the first amino acid in the sequence for the first species, we see S, for serine. Moving down the second column in Table 3, we see that species 2 also has an S but species 3 and 4 have G for glycine instead. Species 1 does not have any amino acids at positions three and four, while the other three do. Of those three, they all have A for alanine in the third position but not all have V for valine in the fourth.

Not surprisingly, compared to the first bird's sequence, there are more differences in the lizard and in the turtle than there are in the other bird species, because the two bird species are more closely related to each other than they are to lizards and turtles. If we looked at the whole amino acid sequence and not just the fragment of 60 amino acids used for Table 3, we would see that the three species in Table 3 have the following percentage of matches with the golden-crowned babbler: 92.9%, 84% and 76.8%, respectively.

Between any two golden-crowned babblers, we would expect more than 99% of the amino acid sequence to be identical, with only one difference every few hundred amino acids. This illustrates that there is much more diversity between organisms in different species compared to organisms within the same species.

The Human Genome Project has shown that there are only a very small number of DNA bases that make one person different from any other person in the world. This creates a feeling of unity. All humans carry inside them a common genetic heritage.

On the other hand, the Human Genome Project has shown that the small differences that do exist make each person unique in terms of skin colour, facial features and resistance to disease, for example. These differences should be appreciated and celebrated as strengths. Unfortunately, they are often the basis of discrimination and misunderstanding.

Can one group of people be considered genetically superior to another? History has shown that many people think so, yet genetics shows that this is not the case. All human populations, whatever slight differences their genomes may have, deserve equal esteem as human beings.

A3.1.10 – Genome sizes

A3.1.10 – Comparison of genome sizes

Application of skills: Students should extract information about genome size for different taxonomic groups from a database to compare genome size to organism complexity.

Using online tools, it is possible to compare the genome of an organism, such as a fruit fly, with other eukaryotes. Table 4 shows data extracted from the NCBI database at the time of writing; because the database is being continually updated, the numbers you find might be different.

Species	Genome size in millions of base pairs, Mb
Saccharomyces cerevisiae, baker's yeast	12.1
Drosophila melanogaster, fruit fly	143.7
Mus musculus, house mouse	2,500
Escherichia coli, bacterium	5.12
Homo sapiens, modern human	3,200
Neoceratodus forsteri, Australian lungfish	34,557.6
Plasmodium falciparum, a parasite that causes malaria	22.9
Oryza sativa, rice	420
Caenorhabditis elegans, a nematode worm	100

A3.1 Table 4 A comparison of genome sizes of various organisms

Escherichia coli, a bacterium that lives in your large intestine, has about 5 million letters (base pairs) in its DNA code.

73

Do you get the impression that the more complex an organism is, the bigger its genome is? For example, we think of humans as being extremely complex and advanced, so when we compare ourselves to the fungus in Table 4, the baker's yeast, we see that our genome size is hundreds of times bigger. But rice has only three times more DNA than the fruit fly. And when we compare our human genome size to the Australian lungfish, it is ten times smaller. Does that mean lungfish are more complex than we are or that we are more complex than yeast? It depends on our definition of complex. Although they may not be capable of doing creative and complex tasks such as sending a spaceship to Mars, both lungfish and yeast can survive in conditions in which humans would die. The examples given and the ones you can find on your own will often give the impression that genome size can indicate complexity, but there are enough exceptions to conclude that it is not a reliable indicator.

A3.1.11 – Whole genome sequencing

A3.1.11 – Current and potential future uses of whole genome sequencing

Include the increasing speed and decreasing costs. For current uses, include research into evolutionary relationships and for potential future uses, include personalized medicine.

Researchers are very excited about genome sequencing because it allows them to identify species and compare them to see evolutionary relationships. They can compare whole genome sequences to see how organisms are related to each other. Such a technique is known as **phylogenetics**. In general, organisms that share similar genomes tend to be more closely related than those that do not.

In Figure 5, the mouse is shown to be much more closely related to the chimpanzee than to the salamander. The DNA sequences (or corresponding amino acid sequences) of the mouse and the chimpanzee would show fewer differences between each other than if one of their DNA sequences was compared to the salamander's genome. In humans, it can tell us about our ancestry, and about possible health risks related to the genes we have inherited.

A3.1 Figure 5 A phylogenic tree of vertebrate chordates

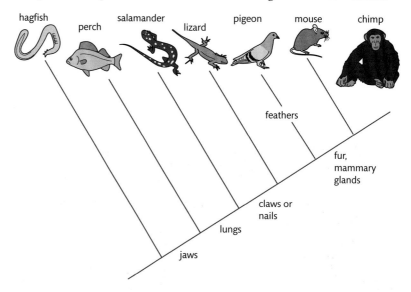

Thanks to **next-generation sequencing techniques**, which use a mix of laboratory hardware, chemical markers and powerful software to increase the speed and decrease the cost of sequencing people's genomes, it is possible for private citizens in some countries to get their genomes sequenced. Other countries have made it illegal to request genome sequencing: laws have been put in place to protect people's privacy. A parent who has put up a child for adoption and does not wish to be identified, for example, might have their identity revealed by this technology even if they do not have their own genome scanned, because a close relative's genome might be sufficient to find the match. In other countries, such services are fully legal and gaining popularity. Several companies in the United States offer genomic testing and provide detailed reports about ancestry and possible health issues related to DNA.

One potential such sequencing holds is the concept of **personalized medicine**, sometimes called precision medicine: information about a person's genetic makeup can be applied to an individual when prescribing treatments. The premise is that, if doctors know a patient's DNA profile, the best adapted treatment can be prescribed. When a doctor prescribes a drug today, the choice of molecule and the dose is based on studies involving people who might not be representative of everyone's genetic makeup. By sequencing the genomes of the participants in drug trials, patterns can be identified that suggest one drug might work better with people who possess a particular genetic sequence, but that for others, another molecule, combination of drugs or different dose would provide better results or perhaps fewer side effects.

A3.1 Figure 6 Knowing that a particular medication produces severe side effects only in people who receive the t version of an identified gene from both parents (*tt*) would allow doctors to know that four people in this group of patients should not be prescribed that medication. All the other patients have received a T from at least one parent (they are either *TT* or *Tt*) and can benefit from the medication without severe side effects.

Personalized medicine is better adapted for diseases that are dynamic, such as cancer, type 2 diabetes or cardiovascular disease, and require different treatments at different stages of the illness. Knowing more about how a patient's genome might cause new proteins to be produced in their cells or trigger certain genes to be turned on or off could lead to breakthroughs in medical treatments. By creating databases of biomarker profiles within a population (such as **TT**, **Tt** or **tt** in the example in Figure 6), researchers of personalized medicine hope to provide better diagnoses and more effective treatments with fewer undesirable side effects.

Another advantageous use of the human genome is the production of new medications. This process involves several steps:

- find beneficial molecules that are produced naturally in healthy people
- find out which gene controls the synthesis of a desirable molecule

- copy that gene and use it to instruct synthesis of the molecule in a laboratory
- distribute the beneficial therapeutic protein as a new medical treatment.

This is not science fiction: genetic engineering firms are finding such genes regularly. One current line of research is dealing with genes that control ageing. How much money do you think people would be willing to pay for a molecule that could reverse the effects of ageing and prolong life by several decades?

Guiding Question revisited

What is a species?

In this chapter we have learned that:

- there is no single definition of the term "species" because the sheer variety of currently living species and extinct species is so enormous and complex
- using morphology works up to a point, but this methodology is poorly adapted for microbes or for species that are visually very similar
- the biological species concept works most of the time but is does not work for single-celled organisms that do not breed, or for organisms that are only found in the fossil record.

Guiding Question revisited

What patterns are seen in the diversity of genomes within and between species?

In this chapter we have discussed how:

- there is some diversity in genomes of individuals of the same species
- there is much more diversity when two different species are compared, especially if they were separated in a speciation event that occurred long ago.

Exercises

Q1. The system of giving a scientific or Latin name to organisms such as *Canis familiaris* is used worldwide. State the name of this system and identify the person who perfected and popularized it.

Q2. Distinguish between the morphological definition of species and the biological species concept.

Q3. Explain the features of chromosomes that are taken into consideration when making a karyogram.

Q4. Distinguish between haploid and diploid cells.

Q5. A karyogram can be used to determine if an unborn baby will be a girl or a boy. Explain how a karyogram is analysed to do this.

Q6. Outline the evidence for a fusion of ancestral chromosomes to become human chromosome 2.

Q7. Outline the advantages of personalized medicine using genomes.

A3 Practice questions

1. In a pollen grain of a species of flower, there are 20 chromosomes.

    Which of the following is true of the species?

    A $2n = 10$

    B $2n = 20$

    C $n = 10$

    D $n = 20$ *(Total 1 mark)*

2. What determines the genomic size of a species?

    A The total amount of DNA

    B The total number of genes

    C The total number of alleles

    D The total number of chromosomes   *(Total 1 mark)*

3. The table gives common names and binomial names for some mammals.

Common name	Binomial name
Golden bamboo lemur	*Hapalemur aureus*
Golden jackal	*Canis aureus*
Grey wolf	*Canis lupus*
Red fox	*Vulpes vulpes*

    (a) Identify the **two** species most closely related.   (1)

    (b) Identify **two** species from the list that are classified in different genera. (1)

    (Total 2 marks)

THEME

A Unity and diversity

4 Ecosystems

The hand on this marine iguana from the Galápagos Islands has five digits. It shares an ancestor with other species that have limbs with five digits. Species adapt to their environment, and when a population finds itself in a unique habitat such as the volcanic beaches of the Galápagos Islands, it can develop adaptations that might transform the genetic makeup of the population enough to make it impossible to breed with other members of the original population. When this occurs, a speciation has happened: where there was once only one species, there are now two.

This process has taken place ever since life first appeared on Earth. As a result the planet is rich in species that fill every available niche. Biodiversity is the variety of life in all its forms. However, humans impact their environment in a variety of ways and many of their actions result in a loss of biodiversity. Scientists fear that we are currently in the middle of the sixth mass extinction. Conservation programmes exist to try to halt the loss of species around the globe. For example, the Galápagos Islands are recognized as an area of particular species richness and the whole area is now a national park. National park status means that the area is carefully managed to preserve the species that live there.

A4.1 Evolution and speciation

Guiding Questions

What is the evidence for evolution?

How do analogous and homologous structures exemplify commonality and diversity?

There is abundant evidence for evolution, and we will examine three types: molecular evidence from genetic data and amino acid sequences; experimental evidence from selective breeding of animals and plants; and morphological evidence from homologous structures, which are features of organisms that reveal they come from a common ancestor. Appendages with five bony digits can be found in animals as diverse as lizards, whales and bats, illustrating the diverse ways in which a limb can be used, such as for walking, swimming and flying. But the uniformity in bone structure and positions within the limbs also reveals that all these organisms had a common ancestor. In addition to homologous structures, there are analogous structures, which evolved on different branches of the tree of life but which serve the same purpose, for example wings in birds and insects. Wings allow flight in both these groups of organisms, but they have not evolved from the same body parts.

A4.1.1 – Evolution

A4.1.1 – Evolution as change in the heritable characteristics of a population

This definition helps to distinguish Darwinian evolution from Lamarckism. Acquired changes that are not genetic in origin are not regarded as evolution.

NOS: The theory of evolution by natural selection predicts and explains a broad range of observations and is unlikely ever to be falsified. However, the nature of science makes it impossible to formally prove that it is true by correspondence. It is a pragmatic truth and is therefore referred to as a theory, despite all the supporting evidence.

Evolution is defined as the process of cumulative change in the heritable characteristics of a population.

Darwin and Wallace

At the age of 22, Charles Darwin had the opportunity to travel on board the *HMS Beagle* for a scientific exploration mission that started in 1831 and lasted for 5 years. Little did he know that it would allow him to see nature in a new way and come up with what would become one of the most important, controversial and misinterpreted ideas in biology: **the theory of evolution by natural selection**.

Darwin was not the only person to develop a theory to explain evolution. He was surprised to discover in 1858 that Alfred Russel Wallace had independently developed an almost identical theory. The two men presented their ideas jointly to the Linnaean Society in 1858.

What is evolution?

Evolution is defined as the process of cumulative change in the heritable characteristics of a population. The word heritable means that the changes must be passed on genetically from one generation to the next, which implies that evolution does not happen overnight. The word cumulative is in the definition to stress the fact that one change is not usually enough to have a major impact on a species. Finally, the word population is in the definition because the changes do not affect just one individual.

Over time, if enough changes occur in a population, a new species can arise in a speciation split (explored in Chapter A3.1). The members of the new population will be different enough from the pre-existing population that they originated from that they will no longer be able to interbreed.

Once evolution by natural selection is understood, many of the mysteries of nature are revealed. When the role of DNA in inheritance (genetics) became known, decades after Darwin's theory had been published, there was a chance that it might have contradicted evolution by natural selection; contradictions often arise with new developments in science, making us rethink and revise our theories. In fact, the opposite happened. DNA evidence provided new support for natural selection beyond anything Darwin could have dreamt of, and led to the **modern synthesis** theory, or neo-Darwinism, a combination of Darwin's ideas with the newer ideas of genetics (based on work by Gregor Mendel, also in the 19th century), which was only confirmed long after Darwin and Wallace had died. One of the fundamental insights of the modern synthesis is the concept of common ancestry (which is explored in Chapter A3.1).

Lamarckism

Darwin and Wallace's theory replaced a previous idea formulated by French naturalist Jean-Baptiste Lamarck. His theory was that organisms acquired characteristics through their lifetime and then passed them on to their offspring. For example, Lamarck explained how kangaroos developed more powerful hind limbs and tails during their lifetimes by using them a lot while letting their forelimbs atrophy through underuse. These characteristics were then passed on to their offspring. That sounds plausible, but experiments designed to illustrate the passing on of acquired traits do not produce the results Lamarck expected.

One remarkable feature of kangaroos is the large discrepancy between the size of their forelimbs and hindlimbs.

Nature of Science

The theory of evolution by natural selection predicts and explains a broad range of observations and is unlikely ever to be completely falsified. Some parts of the theory have been falsified, however, such as the pace at which natural selection can work. Darwin thought it was always slow, but we have observed it happening in just a few generations. Darwin also incorrectly predicted that the fossil record would not contribute evidence to support his theory. Scientists do not throw out an entire theory just because there is some evidence against certain aspects of it. When new evidence is presented that contradicts a theory, the theory can be updated rather than being totally invalidated. The role of a theory is to explain the mechanism of how something works in nature, and the theory of natural selection explains evolution very convincingly. No theory has been developed since that has had any success replacing it. Equally, given the nature of science, it is not possible to formally prove that the theory of evolution is true, which means that it is referred to as a theory, in spite of all the evidence supporting it.

A4.1.2 – Biochemical evidence for evolution

A4.1.2 – Evidence for evolution from base sequences in DNA or RNA and amino acid sequences in proteins

Sequence data gives powerful evidence of common ancestry.

Your DNA includes genes that go back not just to your parents, grandparents and great-grandparents, but back to when we had a common ancestor with fish (roughly 400 million years ago) and beyond. Some, but not all, of those sequences are still inside you now. This explains how, during the development of human embryos, we have, for a period of time, slits in our neck that are similar to the parts of fish embryos that develop into gills.

Using modern bioinformatic tools, we can compare nucleic acid (DNA or RNA) and protein data from many organisms, including humans, to examine their evolutionary relationships. Computer software can process millions of codes in seconds, and compile the differences and similarities to show how species are related to each other.

Access the link on this page of your eBook for instructions on how to find DNA sequences in the NCBI database.

In other words, to show which species are more closely related (e.g. chimpanzees and humans) and which are more distantly related (e.g. humans and fish).

As an example, we can use a protein present in many organisms, haemoglobin, which is the oxygen-carrying molecule found in red blood. It contains four protein chains, two alpha and two beta. Using software that is easily available online, it is possible to compare the protein chains in different organisms. The results of a DNA sequence analysis are shown in Table 1.

Species	DNA nucleotide sequence	Number of base pairs
Rat	ATAATTGGCTTTCAGGCTAAGATGATA–G————GGAAATATATATTTTGCATATAAATTT	960
Mouse	ATAATTGGCTTTTATGCCAGGGTGACAGG————GGAAGAATATATTTTACATATAAATTC	1,003
Dog	ATTGTTGGAGTCA———————————————————————TATGGATTG	971
Human	AATTTCTGGGTTAAGGCAATAGCAATATCTCTGCATATAAATATTTCTGCATATAAATTG	1,192
Chimp	AATTTCTGGGTTAAGGCAATAGCAATATTTCTGCATATAAATATTTCTGCATATAAATTG	1,312
	* * * * *** ***	

A4.1 Table 1 An excerpt showing 60 DNA nucleotides of compared sequences

Notice that in any position marked with an asterisk (*) under it, all the letters match in that column. In the other positions, there is at least one nucleotide that does not match with the others. Where there is a dash (–), a sequence does not exist there in that species. Only humans and chimpanzees have the middle four letters in this part of the sequence, for example. This implies that humans and chimpanzees are more closely related to each other than they are to the other organisms, and that they have a more recent common ancestor than the others. In fact, in this 60-letter sequence, between the human and chimpanzee there is only one nucleotide that does not match.

Although this line-by-line and base-by-base comparison is helpful, it is not easy to see which organisms are more closely related and which are more distantly related. Instead, we can use **phylogenetic trees**. Figure 1 shows an example.

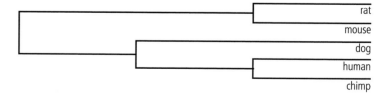

rat

mouse

dog

human

chimp

A4.1 Figure 1 An example of a phylogenetic tree.

In addition to comparing A, T, C and G in DNA sequences, it is possible to compare RNA nucleotides or amino acid sequences to reveal evolutionary changes over time.

The phylogenetic tree in Figure 1 indicates the similarities and differences between the sequences of the haemoglobin beta gene, *HBB*, in different species. It suggests that there was a speciation split between rodents and other mammals early on, then a split between dogs and primates. But we must be careful: this is only what *one* gene reveals about our past. Humans have over 20,000 genes, so just looking at evidence from one is unlikely to tell the whole story of how organisms are related.

A4.1.3 – Selective breeding

A4.1.3 – Evidence for evolution from selective breeding of domesticated animals and crop plants

Variation between different domesticated animal breeds and varieties of crop plant, and between them and the original wild species, shows how rapidly evolutionary changes can occur.

Artificial selection and evolution

The breeding of domesticated animals such as cattle, horses, dogs, sheep and pigeons, provides a good opportunity to study changes in heritable characteristics.

By controlling which males mate with which females, animal breeders can make predictions about the characteristics the offspring will have. Over the years, breeders have learned to choose the males and females with the most agriculturally desirable genetic characteristics, and breed them together. This is called **selective breeding**.

This cow has been bred to have a straight back for easier birthing and long legs for easier milking using automated mechanical pumps. She is a product of artificial selection by humans and she never existed in this form before human intervention.

After practising selective breeding for dozens and sometimes hundreds of generations, farmers and breeders realized that certain varieties of animals now had unique combinations of characteristics that did not exist before. Today, the meat or milk available to us is very different from that which was produced thousands of years ago or even only a hundred years ago. This is thanks to the accumulation of small changes in the genetic characteristics of livestock chosen by breeders.

Although selective breeding is evidence that evolution is happening as a result of an accumulation of small changes over time, the driving force is, of course, human choice. The farmers and breeders choose which animals will reproduce together and which will not. This is called **artificial selection** and it should be obvious that it is certainly not the driving force of evolution in natural ecosystems.

Plant breeding

Teosinte is a plant that you may never have heard of, but you probably consume its descendant every day. It is an ancient wild grass, from what is now Mexico, central America and the Andes region, that has small hard edible kernels. About 10,000 years ago, farmers in these regions started saving seeds from the plants that had the most desirable characteristics, and only planted those seeds the following season.

TOK Animal breeding raises ethical questions. From an animal rights activist's point of view, breeding animals involves needless suffering and cruelty, including broiler chickens that grow too quickly for their bones to support their weight, and lifelong respiratory problems in certain dog breeds. From a breeder's point of view, they are providing safe, nutritious and affordable food for billions of people, or providing adorable pets to keep us company. Whose perspective is more convincing? What counts as a good justification for a claim?

The farmers selected plants that grew successfully in varied habitats, had larger ears with more kernels on them, and ears that were better protected by the outer leaves. Over countless generations, this artificial selection led to what today we call maize or corn (*Zea mays*) one of the most successful and widely planted crops on Earth. Hundreds of millions of tons of corn are grown every year.

Thanks to Neolithic farming techniques of artificial selection, teosinte was transformed into modern corn.

Maize is an ingredient in more foods than you might think. For example, high fructose corn syrup, HFCS, is a food additive found in everything from candy to fast food, to fruit-flavoured drinks and sweet carbonated drinks, all sold worldwide. If you are eating or drinking something that has corn syrup added, you are consuming a product from *Zea mays*.

Selecting seeds with specific desirable traits generation after generation leads to small changes that accumulate over time, and results in a very different plant. The remarkable transformation from teosinte to maize is an example of evolution by artificial selection, and the changes can happen in a geologically short time. Thousands of years or even a hundred years might sound like a long time to you, but compared to the time scale of species (i.e. millions of years), these time scales are extremely short.

A4.1.4 – Homologous and analogous structures

A4.1.4 – Evidence for evolution from homologous structures
Include the example of pentadactyl limbs.

Homologous structures

Homologous structures are structures derived from the same body part of a common ancestor. One of the most striking examples of a homologous structure is the five-fingered limb found in animals as diverse as humans, whales and bats. Such limbs are called **pentadactyl limbs** because "penta" means five and "dactyl" refers to fingers. Although the shape and number of the bones may vary, the general format is the same. However, the specific functions of the limbs may be very different. Darwin explained that homologous structures were not just a coincidence but evidence that the organisms in question have a common ancestor and have therefore evolved from that common ancestor.

The front right fin of a Southern right whale (*Eubalaena australis*), showing five articulated digits.

Challenge yourself

1. **(a)** Look at the figure and complete the table.

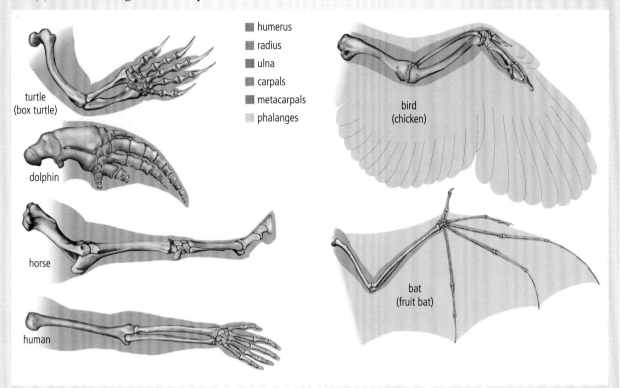

turtle
(box turtle)

dolphin

horse

human

humerus
radius
ulna
carpals
metacarpals
phalanges

bird
(chicken)

bat
(fruit bat)

Characteristic	Bat	Bird	Human	Horse	Dolphin	Turtle
Number of digits (fingers)						
Description of phalanges (finger bones) (short/long, wide/narrow)						
Type of locomotion that the limb is best adapted for						

(b) There are two species in the table that have reduced their number of digits over the course of evolution. For these two species, explain why it could have been a disadvantage to keep all 5 digits. Limit your answer to the type of locomotion.

Analogous structures

In contrast, **analogous structures** are those that may have the same function but do not necessarily come from the same body part and do not indicate a common ancestor. Wings, which have developed from different body parts in different groups of organisms, are a good example of analogous structures: eagles, mosquitoes, bats and extinct reptiles such as the pterosaurs, all use (or used) wings to fly. Although these organisms are all classified in the animal kingdom, they are certainly not placed in the same taxon simply because of their ability to fly with wings. There are many other characteristics that must be considered.

A summary of analogous versus homologous characteristics, considering form and function

	Analogous	**Homologous**
Form: from same body part	No	Yes
Function: used for the same thing	Yes	No, or at least not always
Implies recent common ancestor?	No	Yes

A4.1.5 – Convergent evolution

A4.1.5 – Convergent evolution as the origin of analogous structures

Students should understand that analogous structures have the same function but different evolutionary origins. Students should know at least one example of analogous features.

Analogous structures can also provide evidence for evolution: if a feature such as wings is seen in many different organisms, then it is clearly advantageous and could have evolved in multiple ways over time. It is possible to have two organisms with different **phylogenies** that share similar physical aspects. Phylogeny is the way a species has split from other species.

An illustration of divergent evolution (the blue lineages such as A and D, which are becoming less and less similar as time goes on) and convergent evolution (the green lineages B and C, which are becoming more and more similar over time).

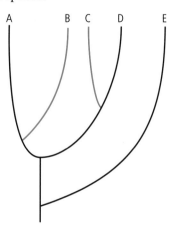

Marsupials are mammals that have a pouch instead of a placenta for nourishing their young during early development. The isolated continent of Australia is rich with examples of marsupials, which have developed in similar ways as their distant placental cousins on other continents. For example, the Tasmanian tiger (*Thylacinus*

cynocephalus), recently driven to extinction, was a marsupial that looked and behaved similarly to wolves and tigers from other continents.

A thylacine (*Thylacinus cynocephalus*), also known as the Tasmanian tiger, is classified as a marsupial. The species is believed to have been driven to extinction in the first half of the 20th century.

Convergent and divergent evolution can refer not only to entire organisms but also to physical features (such as horns, eyes or wings) and even refer to how organisms use certain molecules. The use of bioluminescent (glowing) chemicals by many deep-sea marine organisms, as well as by some bacteria and fungi, is an example of the convergent evolution of a biochemical.

In all these examples, the forces of natural selection used similar pressures on distant phylogenetic lines to favour certain characteristics over others. This process is explained in more detail in Chapter D4.1.

What counts as strong evidence in biology?

Convergent evolution means that different species start to look or behave more like each other over time. Potentially this allows them to exploit similar niches. Convergent evolution results in organisms developing analogous structures. Divergent evolution results in organisms that look less similar to each other but may have homologous structures.

A4.1.6 – Speciation

A4.1.6 – Speciation by splitting of pre-existing species

Students should appreciate that this is the only way in which new species have appeared. Students should also understand that speciation increases the total number of species on Earth, and extinction decreases it. Students should also understand that gradual evolutionary change in a species is not speciation.

Speciation by divergence of isolated populations

How is a new species formed? Recall that the biological definition of a species is based on the idea that members of the same species can produce fertile offspring together. If a subgroup of a reproducing population gets separated from the main population, it might evolve differently from the main population. For example, if a few iguanas from continental Central or South America were accidentally transported to the Galápagos Islands on a tree branch that broke off in a storm and was transported out to sea, the iguanas would find themselves in a new environment. The conditions on a remote island can be very different to those on the continent, for example the types of food available may be different, or there may be no predators. Some of the iguanas on the Galápagos Islands have adapted in ways that allow them to dive for food (algae) on the ocean floor, making them the only marine iguanas in the world.

When we observe today's mainland iguana populations and island iguana populations, we can see that they have evolved differently over millions of years because they have had to adapt to different environments. Little by little, they changed over time and the differences are now large enough such that mainland iguanas can no longer breed with the island iguanas. A speciation split has occurred. Just evolving over time is not what makes a new species. There has to be a split whereby two populations are isolated and exposed to different environments that will select for some traits and against others. Speciation events like this increase the overall number of species, whereas extinctions reduce the total number of species.

Extinction

So far, we have only talked about species that have survived. The vast majority of species that have ever lived on Earth, over 99.99%, are now extinct. Extinction happens when the last individuals of a species die out. Examples include woolly mammoths, the dodo and, of course, *Tyrannosaurus rex*. On a phylogenetic tree, branches of extinct species are cut short and do not reach the extremities of branches the way existing species continue to do.

A4.1.7 – Reproductive isolation and differential selection

A4.1.7 – Roles of reproductive isolation and differential selection in speciation

Include geographical isolation as a means of achieving reproductive isolation. Use the separation of bonobos and common chimpanzees by the Congo River as a specific example of divergence due to differential selection.

Reproductive isolation

In some situations, members of the same species can be prevented from reproducing because there is an insurmountable barrier between them. Such a barrier can be geographical, temporal or behavioural. In each case the effect is the same: over time the two populations will face different selection pressures and will change in different ways. Eventually the two populations will change so much that the individuals from the two separate populations will not be able to reproduce with each other successfully to produce fertile offspring. This is called **reproductive isolation**, and at this point they will have become two separate species, as shown in the previous example with marine iguanas on the Galápagos Islands, which were geographically isolated, and therefore reproductively isolated, from their original population on the mainland.

Geographical isolation

Geographical isolation happens when physical barriers, such as land or water formations, prevent males and females from different parts of a population finding each other, thus making interbreeding impossible. For example, a river, a mountain or a clearing in a forest can separate populations. Tree snails in Hawaii exemplify this geographical isolation: one population lives on one side of a volcano and another population lives on the other side, and they never come into contact with each other.

The barrier can be produced by humans. The Great Wall of China might prevent certain organisms such as salamanders from getting to the opposite side, although it would not stop birds or seeds that are dispersed by wind or birds. Roads and dams can have a similar effect.

Bonobos and common chimpanzees

The Congo River is an example of a physical barrier that prevents the two populations of primates from interacting or interbreeding. The primates to the north and east of the river are chimpanzees (*Pan troglodytes*), and the primates south of the river are bonobos (*Pan paniscus*).

(a) The left image shows bonobos, or pygmy chimpanzees (*Pan paniscus*).
(b) The right image shows juvenile chimpanzees (*Pan troglodytes*).

Chimpanzees are found north and east of the Congo River (A), whereas bonobos are found south of the river (B). The river is a barrier preventing the two species from encountering each other, and each species has evolved separately.

The differences in habitat, availability of food and the presence of enemies such as snakes have led to differences in traits, notably behavioural traits, between the two separated populations. Chimpanzees are considered to be more aggressive and territorial, while bonobos are more peaceful and nomadic. Chimpanzee social structure is clearly male dominated, whereas bonobo social structure tends to be matriarchal, with older males playing a role in decision making for the group. When one environment favours certain traits, and another environment favours different traits, there is differential selection. Traits such as aggression in defending a territory are selected for in places where resources are scarce, whereas there is no such selective pressure when a group can move around freely to find new food sources.

Guiding Question revisited

What is the evidence for evolution?

In this chapter we have discovered that evidence for evolution is found in multiple forms, including:

- molecular evidence from genetic data and amino acid sequences
- experimental evidence from selective breeding of animals and plants
- morphological evidence from homologous structures.

Guiding Question revisited

How do analogous and homologous structures exemplify commonality and diversity?

In this chapter we have discussed how:

- the pentadactyl limb can be found in lizards, whales and bats, showing the diverse ways in which such a limb can be used, such as walking, swimming and flying
- uniformity in bone structure and position within the limbs indicates that all these organisms had a common ancestor
- analogous structures have evolved on different branches of the tree of life but serve the same purpose, examples being the wings of birds and insects
- wings allow flight in both groups of organisms, but they have not evolved from the same body parts.

Exercises

Q1. Which of the following is an example of speciation?

  A Selective breeding to produce new varieties of the wheat *Triticum aestivum* with higher crop yields.

  B Evolution of different courtship behaviours in separate populations of the cricket *Gryllus rubens*.

  C Natural selection leading to an increase in the frequency of darker individuals of *Biston betularia*.

  D Selective feeding by koalas (*Phascolarctos cinereus*) on eucalyptus species

Q2. Which evolutionary pathway is most likely to result in the evolution of analogous structures in species W and Z?

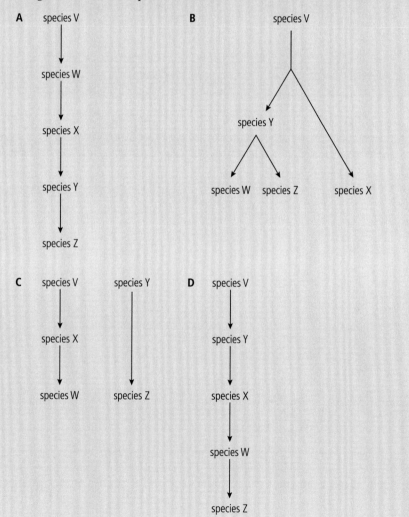

Q3. Which of the following is an example of convergent evolution?

   A   The pentadactyl limbs of bats and lizards.

   B   The opposable thumbs of humans and chimpanzees.

   C   The wings of penguins for swimming and of eagles for flying.

   D   The front fins of dolphins and sharks for swimming.

Q4. Explain how selective breeding can be a good example of evolution by selection, even though it is not natural selection.

Q5. Outline how modern maize (corn) was developed from teosinte over thousands of years of artificial selection.

A4.2 Conservation of biodiversity

Guiding Questions

What factors are causing the sixth mass extinction of species?

How can conservationists minimize the loss of biodiversity?

The amazing diversity of life on Earth allows it to withstand many environmental pressures. Living things are not designed to be unchanging. Evolution has prepared living things to be adaptable to their environment. If that was not the case, we would find almost all evolutionary genetic lineages only in the fossil record. Instead, we find living species that have changed compared to their ancestral species. Evolution allows life on Earth to adapt and respond to change. Unfortunately, a new powerful factor has been introduced in the last few centuries. That factor is the growth and influence of the human species (*Homo sapiens*).

In this chapter we will look at the influence humans have had on the diversity of life. Our population growth and negative impact on our environment have already led to the beginning of a sixth mass extinction of species, the first to be attributed to human activities. We will discuss a few examples of recent human-caused extinctions and what could be many more to come.

Fortunately, we have begun to document the loss of species richness and have initiated efforts to minimize human impacts on at least some species. These efforts are led by both government and private enterprises, whose efforts are designed to preserve as many species as possible. Humans have treated this planet as if its resources were unlimited for far too long. It is in our best interest, as well as the best interest of all species, if we reconsider the role of humans in the stewardship of our planet.

A4.2.1 – Biodiversity exists in many forms

A4.2.1 – Biodiversity as the variety of life in all its forms, levels and combinations

Include ecosystem diversity, species diversity and genetic diversity.

Biodiversity means the variety of life found in an area. A healthy coral reef has a high level of biodiversity, whereas a recently burned forest does not. Biodiversity is at its best when many types of life forms are present in reasonable numbers. This includes animals, plants, fungi and a variety of microorganisms. It is usually the interactions between these life forms that keep an ecosystem healthy and biodiversity levels high. Biodiversity can be studied at three different levels: ecosystem, species and genetics.

A young emperor angelfish (*Pomacanthus imperator*), a beautiful inhabitant of the Great Barrier Reef.

▼

Ecosystem diversity

Ecosystem diversity considers diversity from the largest overall viewpoint.

The Great Barrier Reef is made up of almost 3,000 individual reefs, along with over 1,000 islands, along the north-east coast of Australia. The entire reef system stretches for over 2,300 km, roughly north to south. This reef system is so long that the climate affecting the northern part of the reef is quite different to the climate affecting the southern part. The coral species and numerous other life forms are somewhat different from one another in the individual reefs, even though they are connected by the waters that they share.

The Great Barrier Reef is an example of one of the most ecologically diverse locations in the world. Each individual reef has its own ecosystem and a high level of biodiversity. This means that the total ecosystem biodiversity in the region is very high (i.e. there are lots of richly diverse ecosystems in one area). This diversity of ecosystems and its inhabitants provides stability in the area, and generates a great deal of species and genetic diversity.

Species diversity

Individual ecosystems have varying degrees of **species diversity**. Species diversity is sometimes known as **species richness** and is simply the number of different species in a community. **Species evenness** is a measure of the relative abundance of each of the species in a community. Some of the healthiest ecosystems have both high species richness and species evenness.

Table 1 presents the species evenness and species diversity in two hypothetical coral reef communities. Both samples were taken from the same sized area.

A4.2 Table 1 Species evenness (number of species) and diversity (types of species)

Number and type of species	
Coral reef community 1	**Coral reef community 2**
11 Hard corals	63 Hard corals
23 Fish	146 Fish
155 Sponges	64 Sponges
118 Echinoderms	21 Echinoderms
307 Total	294 Total

Look at the distribution of the species types in the two communities. Specifically take note that community 1 has a relatively high number of echinoderm and sponge species compared to hard corals and fish species. Even though community 2 has a slightly lower species richness (294 total species compared to 307 species), the species evenness evident in community 2 shows that it may be a healthier ecosystem.

Species evenness (taking into account species proportion) is often more important than species biodiversity (total number of species).

Genetic diversity

Every living organism has its own set of genes, giving that organism a unique set of characteristics within its population. All of the gene types or **alleles** found in the entire population is called the **gene pool** of that population.

Populations with greater **genetic diversity** (or a bigger gene pool) are more stable and can better withstand environmental pressures such as disease and extreme weather

Ecosystem diversity is a measure of how many types of ecosystems there are in a given location. Species diversity is how many types of species exist in single ecosystem. Genetic diversity is concerned with the diversity of the gene pool within a population.

events. That does not mean any one randomly selected individual is more likely to survive, but it does mean that at least some of the population is likely to survive.

Generally, larger populations have higher genetic diversity. One of the problems that emerges when the population of an organism falls to low levels is that the gene pool becomes very small. Any genetic diseases contained in that population are then more likely to be expressed.

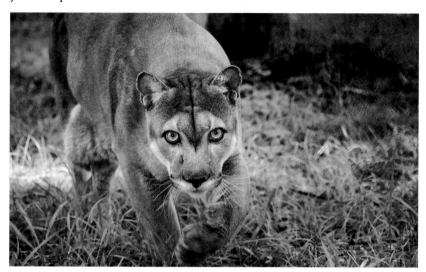

▲ **A4.2 Figure 1** In the late 20th century the Florida cougar (*Puma concolor*) population fell so low that a lack of genetic diversity became a severe problem. In the 1990s, biologists introduced eight female cougars from a population in Texas, and the genetic diversity greatly increased in the Florida population. This was a drastic step to take as biologists are reluctant to change the genetic makeup of a wild population.

A4.2.2 – Has biodiversity changed over time?

A4.2.2 – Comparisons between current number of species on Earth and past levels of biodiversity

Millions of species have been discovered, named and described but there are many more species to be discovered. Evidence from fossils suggests that there are currently more species alive on Earth today than at any time in the past.

NOS: Classification is an example of pattern recognition but the same observations can be classified in different ways. For example, "splitters" recognize more species than "lumpers" in any taxonomic group.

The rate of extinction is currently very high, and the number of species alive today is lower than it was a few hundred years ago. Most of that loss of diversity can be traced back to human activities that have resulted in extinctions.

However, the fossil record suggests that if you go back further in time there are more species alive today than at any other geological time period. It is worth noting, however, that the number of species alive today and the number of species alive in the past are both estimates. The uncertainty arises because biologists are discovering new species every day and the fossil record is incomplete.

Evolutionary theory helps explain why there are more species alive today compared to any other time period. Speciation is the formation of new species. As part of the process

The Florida cougar population in the mid-1990s suffered from a variety of genetic weaknesses, including heart failure, susceptibility to a variety of diseases, undescended testes and inability to withstand a variety of parasites.

TOK Humans have the knowledge to alter the genetic makeup of wild populations. Under what circumstances should we do so? Who should make those types of decisions?

Some of the most interesting examples of speciation occur when organisms reach one or more islands. Each island may present a different set of resources and environmental challenges. Over a long period of time the ancestral (original) organism undergoes speciation into two or more species in the different island ecosystems. This is a process called **adaptive radiation** and is a type of speciation. Charles Darwin was inspired to understand evolution better when he saw evidence of adaptive radiation while on the Galapagos Islands.

of evolution, **speciation** occurs under certain conditions. Any prolonged period of time where the rate of speciation is greater than the rate of extinction will result in a higher total number of species. Despite the last few hundred years of high extinction rates, there have been many long periods when the speciation rate was higher than the extinction rate.

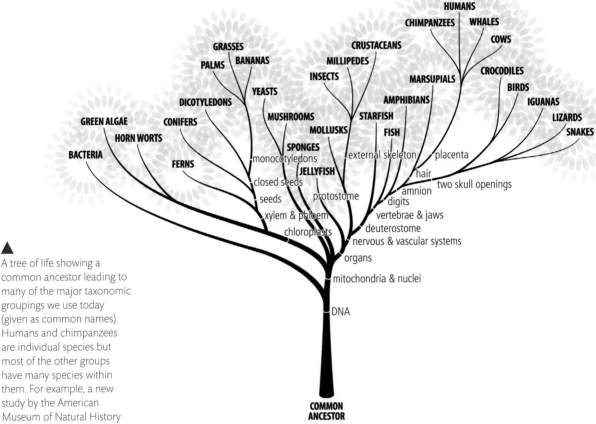

▲ A tree of life showing a common ancestor leading to many of the major taxonomic groupings we use today (given as common names). Humans and chimpanzees are individual species but most of the other groups have many species within them. For example, a new study by the American Museum of Natural History estimates that there are about 18,000 species of birds in the world.

One of the more common taxonomic grouping systems places each organism into a Kingdom, Phylum, Class, Order, Family, Genus and Species. Memorizing these taxa in sequence allows you to better understand the evolutionary relationships between two or more organisms.

SKILLS

Nature of Science

Early attempts to classify organisms were based on shared physical characteristics and there was much debate concerning the relationships between organisms. New tools for determining genetic descent are primarily based on common DNA sequences and are considered to be far more reliable than physical characteristics. However, that does not mean that there is always agreement on all taxa for every organism.

Nature of Science

Classifying organisms into categories called taxa is not always an exact science. There are two general ways of thinking about classification. One approach is represented by the "lumpers". Lumpers generally believe that similarities in organisms are more important criteria for classification than differences. Another approach is represented by the "splitters", who believe the opposite. Splitters end up with more categories of taxa than lumpers. Often a compromise is taken when deciding on taxonomic groupings.

Challenge yourself

1. Figure 1 shows a Florida cougar. Research the classification of the Florida cougar and see how it compares to chimpanzees and humans.

In what ways is diversity a property of life at all levels of biological organization?

A4.2.3 – Human activities and the rate of species extinction

A4.2.3 – Causes of anthropogenic species extinction

This should be a study of the causes of the current sixth mass extinction, rather than of non-anthropogenic causes of previous mass extinctions.

To give a range of causes, carry out three or more brief case studies of species extinction: North Island giant moas (*Dinornis novaezealandiae*) as an example of the loss of terrestrial megafauna, Caribbean monk seals (*Neomonachus tropicalis*) as an example of the loss of a marine species, and one other species that has gone extinct from an area that is familiar to students.

Note: When students are referring to an organism in an examination, either the common name or the scientific name is acceptable.

The extinction of a species caused by human activity is called **anthropogenic** species extinction. There is no doubt that most extinctions in the last few hundred years have had anthropogenic causes.

There have been five previous mass extinction events, and each occurred before humans evolved. The most recent previous extinction event was about 65 million years ago, and there is evidence that it was caused by an asteroid strike. That event killed virtually all of the dinosaurs and a great deal of other life on Earth at the time.

Many scientists propose that we are currently in the midst of a sixth extinction event and that it is the first anthropogenic event.

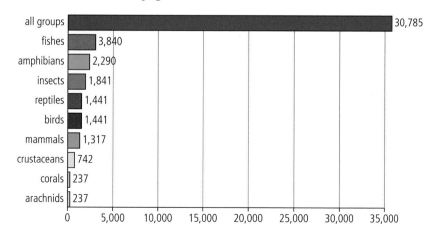

A chart based on data from the International Union for Conservation of Nature's (IUCN) Red List of threatened species, showing some of the types of organisms currently threatened by extinction. This and other data lead biologists to believe that we are now in the midst of a sixth great extinction event.

Case studies of organisms threatened by anthropogenic activities

Case study 1: the North Island giant moa

A4.2 Figure 2 An artist's interpretation of a giant moa species, which once lived in New Zealand.

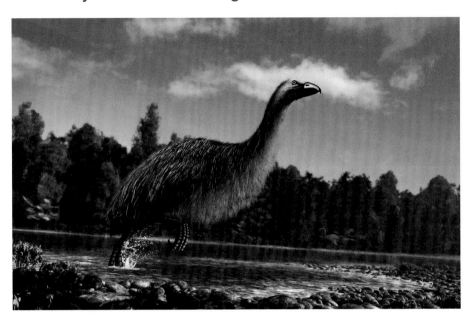

An extinct species called the North Island giant moa (*Dinornis novaezealandiae*) lived in New Zealand up until as recently as 1300 CE. Moas were extremely large herbivorous birds, which swallowed and retained stones in their gizzards in order to grind the plants in their diet to extract more nutrients. As shown in Figure 2, they had no wings. Female moas were much larger than males; skeletal remains indicate that the females reached a height of about 3 m when their necks were stretched upwards. Their bodies were covered by long feathers that were up to 18 cm in length.

New Zealand was first populated by Polynesian people around 1200–1300 CE. It is estimated that the North Island giant moa was hunted to extinction within 100 years of human arrival on the island. This shows that anthropogenic extinction has been occurring for centuries, albeit on a smaller scale than seen today.

Case study 2: the Caribbean monk seal

Caribbean monk seals (*Neomonachus tropicalis*) were declared extinct in 2008 by the US National Marine Fisheries Service, although this species may have actually gone extinct decades earlier. Caribbean monk seals were docile marine mammals living in and around the waters of the Gulf of Mexico and Caribbean islands. Prior to European colonization of the Caribbean area, this seal species was thought to have existed in at least 13 major colonies, with an overall population of approximately quarter of a million.

European colonists killed this seal for its oil, to use in lamps, and for food. The seals often hauled themselves out of water on beaches and rocks, and unfortunately showed little fear of approaching humans. Their behaviour made them easy targets for humans with guns and clubs. Some of the last Caribbean monk seals were killed to provide scientific specimens.

A photo reconstruction of a Caribbean monk seal.

Case study 3: your choice of extinct species

The IB requires you to choose and research a brief case study of a species that has become extinct as a result of anthropogenic factors. You should choose an extinct species from your area of the world, or at least an area that is familiar to you. You can refer to this species by its common or scientific name. Use the two previous case studies as a guide to decide the level of detail required.

A4.2.4 – Human activities and ecosystem loss

A4.2.4 – Causes of ecosystem loss
Students should study only causes that are directly or indirectly anthropogenic. Include two case studies of ecosystem loss. One should be the loss of mixed dipterocarp forest in Southeast Asia, and the other should, if possible, be of a lost ecosystem from an area that is familiar to students.

In recent years, human activities have not only resulted in species extinctions, but also the destruction of entire ecosystems. More often than not, ecosystem loss is caused by massive habitat destruction such as deforestation.

Case studies of ecosystem loss by anthropogenic activities

Case study 1: mixed dipterocarp forests in Southeast Asia

The dipterocarps are a family (Dipterocarpaceae) of hardwood, tropical trees comprising about 500 species. Dipterocarp forests once dominated the island nations of Southeast Asia. The ecosystems provided by dipterocarp tree species were rich and varied.

Deforestation of dipterocarp forest in Southeast Asia. The wood is sold to worldwide markets.

Southeast Asia has long been known for its incredible rainforests. Collectively, however, Southeast Asia is losing about 1% of its rainforest every year, and in some individual areas the percentage is much higher. Some regions have already lost over 50% of their dipterocarp forested area. Frequently the forested land is completely stripped of its trees (a practice called clear-cutting) and there is a total loss of the local ecosystem. There are less damaging alternatives for timber removal, but clear-cutting is the least expensive option. By clear-cutting, the land is made available for agricultural use.

In many cases the land is cleared to allow the planting of palm oil trees. The fruit of these trees is used to make an oil that is used in hundreds of products throughout the world. Many of us use products containing palm oil without even knowing it. Sometimes the product ingredient list will use one of the alternative names for palm oil.

▲ A palm oil plantation in Borneo. This area was once most likely a richly biodiverse dipterocarp forest. As much as 50% of areas that were once dipterocarp forest are now being used for agriculture or urban development. Using large areas of land to grow a single crop is called **monoculture**.

Common food products that are likely to use palm oil include margarine, instant noodles, sliced bread and ice cream. Cosmetics such as lipstick, shampoo, toothpaste and moisturizers also contain palm oil, as do cleaning products such as soaps and laundry detergents.

Many of the nations of Southeast Asia have adopted a certification system for palm oil plantations. To receive certification, a plantation must observe certain practices for growing and refining palm oil. Certification requires less forest cover to be removed.

Challenge yourself

2. Suggest specific reasons why monocultures of large land areas are harmful from an ecological perspective.

Case study 2: your choice of ecosystem loss

The IB requires you to choose and research a brief case study of an ecosystem that is under extreme stress because of anthropogenic activity. You should choose an ecosystem from your area of the world or at least an area that is very familiar to you. Use the previous case study on dipterocarp forests as a guide to decide the level of detail required.

A4.2.5 – A biodiversity crisis

A4.2.5 – Evidence for a biodiversity crisis

Evidence can be drawn from Intergovernmental Science-Policy Platform on Biodiversity and Ecosystem Services reports and other sources. Results from reliable surveys of biodiversity in a wide a range of habitats around the world are required. Students should understand that surveys need to be repeated to provide evidence of change in species richness and evenness. Note that there are opportunities for contributions from both expert scientists and citizen scientists.

NOS: To be verifiable, evidence usually has to come from a published source, which has been peer-reviewed and allows methodology to be checked. Data recorded by citizens rather than scientists brings benefits but also unique methodological concerns.

As you know, anyone can create a website or post documents that seem to support a particular position. That does not make the information reliable or correct. All of us, including policymakers who make decisions about ecologically critical issues, need valid and accurate data. Two reliable sources that you and others can use are given on the following page.

IPBES

In 2019, the Intergovernmental Science-Policy Platform on Biodiversity and Ecosystem Services (IPBES, pronounced "Ip – Bes") published a comprehensive report that provides significant and reliable scientific guidance for policymakers. The information was obtained from a wide range of habitats studied by university and governmental research projects, with the input of local sources. The data is sampled regularly in order to identify trends, and findings continue to be updated through IPBES, and are made available via its website and publications. (Search IPBES for further, updated information).

 In order to understand the nature of the current biodiversity crisis, we need reliable information. Such evidence can be drawn from IPBES reports and the ICUN's Red List.

A **Current global extinction risk in different species groups**

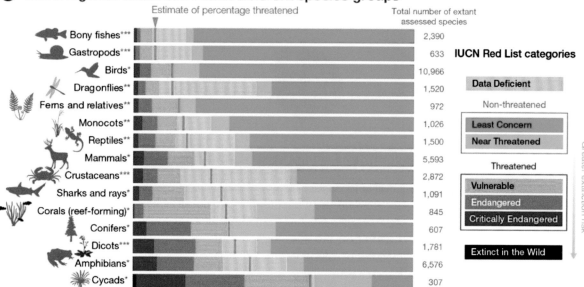

A graphic of data sampled from the 2019 IPBES report.

IUCN Red List

The International Union for Conservation of Nature's (IUCN) Red List was established in 1964. This is a continuously updated list of the world's threatened species. It has become the one of the most comprehensive and trusted sources of information on the extinction status of over 140,000 fungus, plant and animal species. Currently more than 40,000 species are listed as threatened with extinction. Each assessed species is rated on a scale (see Figure 3 on the next page) indicating its ecological health. Each entry to the list contains the details of the research papers used to make the assessment. (Search IUCN Red List for further, updated information.)

Jump to Southwest Bornean Orangutan: In detail

Southwest Bornean Orangutan

Pongo pygmaeus ssp. *wurmbii*

ABSTRACT

Southwest Bornean Orangutan *Pongo pygmaeus ssp. wurmbii* has most recently been assessed for *The IUCN Red List of Threatened Species* in 2016. *Pongo pygmaeus ssp. wurmbii* is listed as Critically Endangered under criteria A4abcd.

THE RED LIST ASSESSMENT

▶ Ancrenaz, M., Gumal, M., Marshall, A.J., Meijaard, E., Wich , S.A. & Husson, S. 2016. *Pongo pygmaeus* ssp. *wurmbii*. The ...

NOT EVALUATED	DATA DEFICIENT	LEAST CONCERN	NEAR THREATENED	VULNERABLE	ENDANGERED	CRITICALLY ENDANGERED	EXTINCT IN THE WILD	EXTINCT
NE	DD	LC	NT	VU	EN	CR	EW	EX

A4.2 Figure 3 Results from a sample search of the IUCN Red List. A rating is given specifying the status of the species. This species is currently rated as critically endangered.

Researchers who participate in creating publications such as IPBES reports and the IUCN Red List often make use of people with local knowledge concerning species in a given area. Indigenous peoples, hunters, people in fishing industries and many other local sources often have information concerning species that no one else can provide.

Nature of Science

Local individuals (citizen scientists) are an important source of information about populations, but they may not be sampling populations in a scientific manner. Data gathered from the local population must be collated by a reliable scientific organization. Such scientific organizations provide established methods of collecting data, and the results are published in peer-reviewed research papers.

A4.2.6 – Causes of the biodiversity crisis

A4.2.6 – Causes of the current biodiversity crisis

Include human population growth as an overarching cause, together with these specific causes: hunting and other forms of over-exploitation; urbanization; deforestation and clearance of land for agriculture with consequent loss of natural habitat; pollution and spread of pests, diseases and alien invasive species due to global transport.

The current estimate is that there are over 8 billion people on Earth, and our population is continuing to increase. The population of a species at any given point in time is based on the current estimated population size plus the number of births and minus the number of deaths. If the birth rate exceeds the death rate, the population is increasing. That is the situation today for the human population when the data is considered from a global perspective.

A graph showing two interrelated population factors. The shaded area represents the historic data and future predictions for the human population from 1700 through to a projected 2100. The line shows the past and predicted annual growth rate of the human population.

The annual growth rate of the human population peaked in the 1960s and has been declining since then. Based on this alone, logic would suggest that the total human population is therefore also declining. One of the primary reasons the data does not reflect this pattern is that humans are living for longer, and therefore the death rate is declining as well. This results in a pattern of continued population growth that is projected to continue until the birth rate declines and equals the death rate worldwide. Some scientific projections have this occurring around the year 2100.

The link between an increase in human population and the biodiversity crisis

Humans need resources to survive, and they produce waste and pollution. An increase in population means more resources are necessary and more pollution is produced. An increase in the human population has an impact on ecosystems because resources such as food, minerals and water must be sourced from ecosystems. Each time an ecosystem is damaged, the biodiversity within that ecosystem can be reduced. Some examples of human population growth effects on biodiversity are given below.

- Over-exploitation of resources, e.g. commercial fishing.
- Hunting, e.g. African elephant (*Loxodonta africana*) populations have decreased drastically because the animals have been hunted for their tusks, often illegally.
- Deforestation, e.g. forests have been reduced to extract minerals, hardwoods or to clear the land so that it can be used for agriculture. Often the crops planted are monocultures, which reduces the biodiversity of the area even further.
- Monoculture agriculture practices, e.g. palm oil plantations.
- Pollution, e.g. microparticles of plastics have been found in nearly every corner of the oceans.
- Increased pest species, e.g. the spruce bark beetle (*Ips typographus*) was first found in the UK in 1982. It is thought that the larvae entered the country in some untreated wood from either Europe or Asia.

The continued growth of the world's population is not evenly distributed around the globe. The continents of Africa and Asia are projected to have much greater population growth rates than other continents. Some European countries, such as Italy, already have declining populations.

- Invasive species, e.g. the Burmese python (*Python bivittatus*) was accidentally introduced into the Florida Everglades at the end of the last century. Burmese pythons have no natural predators in the Florida Everglades, and they now pose a serious risk to native wildlife, decreasing biodiversity.
- Urbanization, e.g. a growing population means that more houses and services are needed. Towns and cities are growing in size, using land that was previously unused or used for agriculture.
- Spread of disease in both humans and other organisms.

An increasing human population is resulting in increased commercial fishing, placing a strain on many populations of fish throughout the world.

A4.2.7 – Conservation of biodiversity

A4.2.7 – Need for several approaches to conservation of biodiversity

No single approach by itself is sufficient, and different species require different measures. Include in situ conservation of species in natural habitats, management of nature reserves, rewilding and reclamation of degraded ecosystems, ex situ conservation in zoos and botanic gardens and storage of germ plasm in seed or tissue banks.

As you have learned in earlier sections, human activities have resulted in biodiversity declines. There is no single solution for the loss of biodiversity. Many different approaches are required, and each biodiversity problem requires a unique approach. Fortunately, some governments and non-governmental organizations (NGOs) are making concerted and varied efforts to manage and improve biodiversity. These efforts can be classified into two categories:

- in situ conservation efforts, i.e. managing natural areas
- ex situ conservation efforts, i.e. managing one or more species outside their natural area.

In situ efforts to improve biodiversity

Establishment of national parks

A national park is an area of land established by a nation and dedicated to preserving the geology and wildlife of that area. Typically, human visitors are permitted in the park, but development and building within the park are restricted. National parks prevent ecosystems from being lost as a result of extraction of resources, urbanization or many other human activities.

A male gerenuk (*Litocranius walleri*) in Samburu National Park, Kenya.

Establishment of nature reserves

Nature reserves are smaller areas than national parks. Most attempt to provide an area where ecosystems can be protected from urbanization and uncontrolled use. Again, these steps help to maintain biodiversity.

Rewilding of areas damaged by human intervention

The aim of this private and public approach is to let nature take better care of an area than people have. Sometimes the aim is to undo previous damage by removing such things as dams and roads, and reducing active management of wildlife populations. Rewilding promotes forest and aquatic ecosystem regeneration. By leaving an area to regenerate, species that were previously lost from an area may be able to return, thereby increasing biodiversity again.

Reclamation of degraded landscapes

Human activities such as strip mining and clear-cutting of forests leave areas that have little or no possibility of natural regeneration. Reclamation projects aim to rebuild and replant as much of an ecosystem as possible.

Ex situ efforts to improve biodiversity

Breeding programmes by zoos

Some zoological parks have established **animal husbandry** facilities to promote the continuation of species that are threatened and endangered. **Artificial insemination** is a common technique used by zoos, as they typically have very small populations of captive animals. Artificial insemination facilitates the production of offspring from animals in two different zoos (possibly in different countries). This technique promotes genetic diversity within the captive population. Careful pedigrees of animals are kept in order to choose breeding pairs that will increase genetic diversity.

Botanic gardens

Botanic gardens provide a living store of plant material that helps to promote biodiversity and helps conservation efforts. Some plant species now only exist in artificial garden facilities. The plants provide a reservoir of genetic material for restoration efforts, and a source of material for scientific research of a species. Botanic gardens often exchange seeds or pollen in order to help preserve rare, threatened or endangered species.

Seed banks

How does variation contribute to the stability of ecological communities?

There are over 1,000 seed banks scattered around the globe. A seed bank is exactly what it sounds like, a place where you can safely store living seeds. The seeds can then be used to repopulate a species of plant if necessary. Seeds are ideally kept in cool, dark and dry conditions. One of the largest and most famous seed banks is located in Norway. This facility is called the Svalbard International Seed Vault (sometimes called the Doomsday Vault).

Entrance to the Svalbard International Seed Vault, Norway. Most of the facility is underground.

Animal tissue banks

The overall increase in the global population of humans has caused the current biodiversity crisis. Maintaining biodiversity is going to require a wide range of different solutions.

Two types of tissue are stored in animal tissue banks. One type is called **germplasm**, and includes sperm, eggs and embryos. The aim is to collect and store the reproductive cells of various threatened species. One of the challenges is to collect germplasm from wild populations of animals in order to have reproductive cells that can be used in captive breeding programmes. The tissue is stored cryogenically, and can be kept for a nearly indefinite period of time before use. The second type of tissue collected is called **somatic tissue**, and includes non-reproductive tissue samples. This tissue is useful for DNA research and possible cloning.

A4.2.8 – The EDGE of Existence programme

A4.2.8 – Selection of evolutionarily distinct and globally endangered species for conservation prioritization in the EDGE of Existence programme

Students should understand the rationale behind focusing conservation efforts on evolutionarily distinct and globally endangered species (EDGE).

NOS: Issues such as which species should be prioritized for conservation efforts have complex ethical, environmental, political, social, cultural and economic implications and therefore need to be debated.

In 2007 the Zoological Society of London (UK) launched a global programme with the goal of selecting evolutionarily distinct and globally endangered species. A selected species is then promoted for priority status in conservation programmes.

The selection of EDGE of Existence species is a process. First, the IUCN Red List (see Figure 3 on page 102) rating on a species is consulted. A score is then generated from this list indicating how endangered the species is. Next, the species is evaluated for its unique evolutionary history. This is done using DNA sequencing information. Those species that are the most endangered *and* the most evolutionarily distinct are given a high EDGE score, indicating that they should be prioritized.

According to the scores generated by EDGE researchers, the aardvark is the world's most evolutionary distinct mammal.

The aardvark (*Orycteropus afer*). Even though this animal is evolutionarily distinct, it is in the least concern category on the IUCN's Red List and thus does not have an overall high EDGE score.

The EDGE of Existence programme aims to inform governments, conservation organizations and local populations of the ecological peril of different species. It is hoped that this will help ensure that those that are both endangered and evolutionarily distinct will be given the highest priority when deciding which species to protect.

Nature of Science

Species selection for priority status in conservation efforts is a complex issue, with ethical, environmental, political, social, cultural and economic considerations. The selections should have input from many stakeholders.

The rapid increase in the global human population is causing the current global biodiversity crisis because humans demand increasing levels of resources. To improve biodiversity multiple approaches are needed, and organisms that are particularly at risk can be identified using the **EDGE of Existence programme.**

Guiding Question revisited

What factors are causing the sixth mass extinction of species?

This chapter has discussed how the increasing human global population is leading to:

- deforestation
- monoculture agricultural practices, such as palm oil plantations
- habitat destruction from urban development
- pollution of air, land and water
- excess commercial fishing
- unregulated hunting and poaching of wildlife.

Guiding Question revisited

How can conservationists minimize the loss of biodiversity?

This chapter has considered how the following actions can help to control or minimize the loss of biodiversity:

- establishing national parks and nature reserves
- rewilding and reclamation projects for damaged habitats
- breeding programmes in zoos
- establishing botanical gardens and seed banks
- establishing animal tissue banks
- using the research carried out by a variety of organizations to inform policymakers and the general population of biodiversity problems, including the work of the:
    - International Union for Conservation of Nature's (IUCN) Red List
    - Intergovernmental Science-Policy Platform on Biodiversity and Ecosystem Services (IPBES)
    - EDGE of Existence programme.

Exercises

Q1. Which of these events would most likely lead to adaptive radiation?

    A   Limitation of a vital resource.

    B   Immigration of an ancestral species to a diverse group of islands.

    C   A richly diverse ecosystem.

    D   Enhanced volcanic activity resulting in numerous sulfur compounds in the atmosphere.

Q2. Distinguish between genetic diversity and species diversity.

Q3. State the primary difference between the sixth (current) mass extinction and the previous five mass extinction events.

Q4. State four ways that coral reefs can be damaged (directly or indirectly) by human activity.

Q5. Briefly describe the two necessary requirements for a species to be given a high rating by the EDGE of Existence programme.

Q6. What is germplasm and why is germplasm being stored by cryogenic techniques?

Q7. Discuss why a zoo captive breeding programme might need reproductive cells extracted from a wild population of animals.

A4 Practice questions

1. Outline how reproductive isolation can occur in an animal population.

   (Total 3 marks)

2. A friend says that the environment on Earth is healthy because there are more species alive today than at any other time in history. Discuss this statement.

   (Total 4 marks)

3. Extensive areas of the rainforest in Cambodia are being cleared for large-scale rubber plantations. Distinguish between the sustainability of natural ecosystems such as rainforests and the sustainability of areas used for agriculture.

   (Total 3 marks)

4. List **two** in situ conservation efforts.

   (Total 2 marks)

5. List **two** ex situ conservation efforts.

   (Total 2 marks)

THEME

B Form and function
1 Molecules

Living things make use of molecules that are similar to each other in that they use carbon as a fundamental component. These carbon-based molecules are called organic molecules. You will study three of these organic molecule groupings in this chapter, specifically carbohydrates, lipids and proteins. Even though these molecules are varied and complex, they have patterned molecular arrangements that you will learn to recognize. At first glance, the computer-generated graphic showing the enzyme invertase (a protein) seems to be beyond comprehension. As you work your way through the study of biochemistry, the structure of this molecule and many others will become clearer to you.

B1.1 Carbohydrates and lipids

Guiding Questions

In what ways do variations in form allow diversity of function in carbohydrates and lipids?

How do carbohydrates and lipids compare as energy storage compounds?

Biochemically important organic molecules are diverse in both structure and function. Carbohydrates and lipids are two groups of vital organic molecules.

The smallest forms of carbohydrates are called monosaccharides. This group of molecules follows the formula, $C_nH_{2n}O_n$, where n equals the number of carbon atoms. Monosaccharides are the monomers of the larger carbohydrates such as disaccharides (two monosaccharides bonded together) and polysaccharides (many monosaccharides bonded together). There are numerous carbohydrates found in nature. They have a variety of different forms and they serve a multitude of functions.

Lipids are molecules that are oils at warmer temperatures and fats at cooler temperatures. Each individual lipid type has its own temperature at which that phase change occurs. The monomers of triglyceride lipids are molecules known as glycerol and fatty acids. The identity of the specific lipid is dependent on the fatty acids as they are highly variable. Lipids also have a variety of functions.

One function that carbohydrates and lipids share is to act as energy storage molecules. Per gram of substance, lipids store approximately twice the chemical energy compared to carbohydrates.

B1.1.1 – The variety of compounds containing carbon

B1.1.1 – Chemical properties of a carbon atom allowing for the formation of diverse compounds upon which life is based

Students should understand the nature of a covalent bond. Students should also understand that a carbon atom can form up to four single bonds or a combination of single and double bonds with other carbon atoms or atoms of other non-metallic elements. Include among the diversity of carbon compounds examples of molecules with branched or unbranched chains and single or multiple rings.

NOS: Students should understand that scientific conventions are based on international agreement (SI metric unit prefixes "kilo", "centi", "milli", "micro" and "nano").

The majority of molecules within all living organisms can be categorized into one of four biochemical groups: carbohydrates, lipids, proteins and nucleic acids. These four types of molecules interact with each other in a wide variety of ways in order to carry out the metabolism of each cell. All of these molecules contain carbon and this explains why life on Earth is often described as "carbon-based".

A molecular representation showing all four carbon atoms forming a covalent bond with a hydrogen atom. The resulting compound is methane (CH_4).

Carbon can share its four outer shell electrons in many diverse patterns. For example, two carbon atoms can share electrons with each other forming a carbon–carbon **covalent** bond. Carbon atoms can also share two pairs of electrons forming a double (covalent) bond. A few of the resulting patterns are shown in Figure 1.

B1.1 Figure 1 These molecules show you some of the simplest carbon compounds containing carbons and hydrogens only. You do not need to memorize these molecules. The chain of carbons can be much longer than those shown. Also notice that propene and ethene each have a double covalent bond between carbons. This reduces the number of hydrogens in the molecule. In addition, carbons can form branched formations and even ring structures.

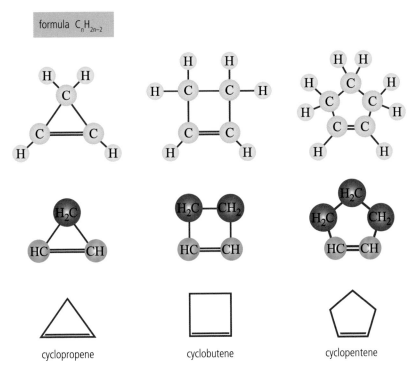

cyclopropene   cyclobutene   cyclopentene

▲ Three examples of organic molecules where the carbon atoms are formed into a ring or cyclic structure. All three happen to have a double bond between two of the carbons. There are three different ways to represent the same molecules. In the bottom row the hydrogens are omitted and are just understood to be present. Carbon atoms always have four bonds and if no other element is shown, it is assumed to be the appropriate number of hydrogens. You will see this in many representations of organic molecules. You are not required to know the names of these compounds.

The following elements are also common within the molecules of living organisms: oxygen, nitrogen and phosphorus. These elements are found in carbohydrates, proteins, lipids and nucleic acids. They often form covalent bonds with carbon, as well as forming covalent bonds with each other. Each of the elements mentioned forms a set number of covalent bonds. Memorizing the number of covalent bonds made by each of these elements will help you understand and draw the molecules that are important to living organisms. Table 1 shows the number of covalent bonds for five elements important in biochemistry.

Element	Number of covalent bonds
Hydrogen	1
Oxygen	2
Nitrogen	3
Carbon	4
Phosphorus	5

▲ **B1.1 Table 1** Each of the five most common elements in biochemically important molecules creates a unique number of covalent bonds.

In biochemistry, elements are often arranged into **functional groups**. It will help you immensely if you learn to recognize the common functional groups.

Common functional groups (highlighted in blue) shown bonded to a carbon backbone. The bonds to each of the other carbons could be all hydrogens, as shown, or there could be another functional group bonded at that location in the molecule.

Common functional groups

hydroxyl or alcohol (OH)

amino or amine (NH_2)

carboxyl (COOH)

phosphate (H_2PO_4)

As you study biochemistry, you will soon learn to recognize and categorize common biochemical molecules into appropriate categories. Table 2 shows some of the common categories of molecules.

Category	Subcategory	Example molecules
Carbohydrates	Monosaccharides	Glucose, galactose, fructose, ribose
	Disaccharides	Maltose, lactose, sucrose
	Polysaccharides	Starch, glycogen, cellulose, chitin
Proteins		Enzymes, antibodies, peptide hormones
Lipids	Triglycerides	Fat stored in adipose cells
	Phospholipids	Lipids forming a bilayer in cell membranes
	Steroids	Some hormones
Nucleic acids	Nucleotides	Deoxyribonucleic acid (DNA), ribonucleic acid (RNA), adenosine triphosphate (ATP)

B1.1 Table 2 Categories and examples of four biochemically important molecules

Nature of Science

In all scientific disciplines unit conventions are based on international agreement. This includes the use of the following metric prefixes:

kilo = 10^3 centi = 10^{-2} milli = 10^{-3} micro = 10^{-6} nano = 10^{-9}

B1.1.2 and B1.1.3 – Condensation and hydrolysis

B1.1.2 – Production of macromolecules by condensation reactions that link monomers to form a polymer

Students should be familiar with examples of polysaccharides, polypeptides and nucleic acids.

B1.1.3 – Digestion of polymers into monomers by hydrolysis reactions

Water molecules are split to provide the −H and −OH groups that are incorporated to produce monomers, hence the name of this type of reaction.

Most living organisms do not make **macromolecules** one atom at a time. Macromolecules are made up of smaller molecules called monomers. When you ingest food many of the molecules of the food are in the form of macromolecules. Digestion breaks down macromolecules as a result of chemical reactions called **hydrolysis** reactions. Hydrolysis reactions break covalent bonds between monomers.

Macromolecule	Monomer (building blocks)
Carbohydrates	Monosaccharides
Lipids	Glycerol, fatty acids, phosphate groups
Proteins (polypeptides)	Amino acids
Nucleic acids	Nucleotides

Macromolecules and their monomer subcomponents

The resulting monomers are then a suitable size to be absorbed into the bloodstream and circulated to body cells. After entering cells, often the monomers are built up into macromolecules again. This involves forming covalent bonds in reactions called **condensation** reactions.

An example of hydrolysis and condensation is as follows:

- You eat a taco containing beef, a source of protein.

- Hydrolysis reactions occur in the digestive system, resulting in amino acids.

- Amino acids are absorbed into the blood and taken to body cells.

- DNA in a body cell directs specific condensation reactions to produce a specific protein from the amino acids.

Metabolism is best thought about from a molecular perspective. Often, people think only of physiological parameters, such as heart rate and digestion, as their metabolism. But remember that metabolism is all of the reactions within all of the cells of an organism.

In a condensation reaction two products are always formed. One is a larger molecule than either of the two reactants and the other is water. In a hydrolysis reaction, water is always a reactant and two products are formed that are smaller than the initial reactant (other than water).

Condensation and hydrolysis reactions are, in many ways, the reverse of each other. In a condensation reaction, a water molecule is always formed as part of the reaction. In a hydrolysis reaction, a water molecule is split into two components and each component is added in and becomes a part of the two new (smaller) molecules. Both condensation and hydrolysis reactions require specific enzymes.

Linking monomers into polymers

Examples of condensation reactions include the following.

- Condensation reaction of two monosaccharides to form a disaccharide

  glucose + galactose → lactose + water

- Condensation of many glucose molecules to form starch (a polysaccharide)

  (many) glucose → starch + (many) water

- Condensation of amino acids to form a polypeptide

  (many) amino acids → protein + (many) water

- Condensation of nucleotide components to make a DNA or RNA nucleotide

  phosphate group + pentose sugar + nitrogenous base → nucleotide + 2 water

Notice that one water molecule is formed for every condensation reaction that occurs. We will look at the mechanism of a condensation reaction in a little more detail to see where the water molecule comes from and how the two smaller subcomponents are bonded to a larger molecule.

B1.1 Figure 2 The two amino acids shown are undergoing a condensation reaction.

It is possible to predict how many water molecules are going to be produced whenever a polypeptide is formed. You just have to know how many amino acids are being joined together. If there are 443 amino acids being joined, then 442 water molecules are formed and 442 new peptide bonds formed.

The "R" notation in Figure 2 indicates that these two amino acids could be any of the 20 different possibilities. Notice that a portion of the carboxyl group of one amino acid becomes oriented near the amine group of the other amino acid. Stress is placed on the −OH from one amino acid and the H^+ of the other. This results in the covalent bonds breaking. The released −OH (hydroxide ion) and H^+ (hydrogen ion) combine to form a water molecule. The location where the −OH and H^+ were released still contains a pair of electrons that form a new covalent bond. Whenever this occurs between two amino acids, the new covalent bond is called a **peptide bond**. The reaction is catalysed by an enzyme.

Digesting polymers into monomers

Many organisms, including all animals, rely on the foods that they eat to provide energy but also to provide monomers that can be used to make new macromolecules.

Foods are chemically digested in the alimentary canal. The digestive enzymes that accomplish this are **hydrolysing enzymes**. Each individual reaction is called a hydrolysis and requires a molecule of water as a reactant. In a **hydrolysis reaction** water is always "split" as part of the reaction. Examples of hydrolysis include the following.

- Hydrolysis of a disaccharide to two monosaccharides (see Figure 3)

    lactose + water → glucose + galactose

B1.1 Figure 3 Hydrolysis of the disaccharide lactose. Notice that water is also a reactant. The products are the two monosaccharides galactose and glucose. The shaded areas of galactose and glucose highlight the only difference between the two monosaccharides. The difference is in the orientation of the hydroxyl group; it is quite subtle, but important enough to distinguish the two monosaccharides from each other.

- Hydrolysis of a polysaccharide to many monosaccharides

    starch + (many) water → (many) glucose

- Hydrolysis of a polypeptide (protein) to amino acids

    protein + (many) water → (many) amino acids

The sum total of all the condensation reactions and all the hydrolysis reactions occurring in your body makes up a large portion of your overall metabolism.

▲ A computer graphic image of the enzyme pepsin. This enzyme is active in the acidic environment of the stomach. It is a hydrolysing enzyme that helps to digest proteins by hydrolysis reactions.

B1.1.4 – Monosaccharides

B1.1.4 – Form and function of monosaccharides

Students should be able to recognize pentoses and hexoses as monosaccharides from molecular diagrams showing them in the ring forms. Use glucose as an example of the link between the properties of a monosaccharide and how it is used, emphasizing solubility, transportability, chemical stability and the yield of energy from oxidation as properties.

You have probably already encountered several monosaccharides. Ribose and deoxyribose monosaccharides are the central components of the nucleotides of RNA and DNA, respectively (see Chapter A1.2). You have read about glucose and galactose within this chapter in reference to condensation and hydrolysis reactions. Figures 4 and 5 show the structure of ribose and glucose, respectively.

Ribose is an example of a **pentose monosaccharide**. This simply means that its carbon backbone is composed of five carbons. The chemical formula for a pentose monosaccharide is $C_5H_{10}O_5$.

▲ **B1.1 Figure 4** The structure of ribose

Glucose is an example of a **hexose monosaccharide** as its carbon backbone is composed of six carbons. Its chemical formula is $C_6H_{12}O_6$.

Notice the pattern of the chemical formula for both ribose and glucose. Both of these molecules have the general formula $C_nH_{2n}O_n$ where "n" is the number of carbon atoms within the monosaccharide. This means all you need to know is the number of carbon atoms and you can then predict the entire chemical formula. However, this formula does not apply to carbohydrates larger than monosaccharides such as disaccharides and polysaccharides, nor does it apply to modified monosaccharides such as deoxyribose.

Properties and use of glucose

Glucose is one of the most important molecules in nature. It is produced in photosynthesis and used in respiration. Glucose can also be used to make polysaccharides of various types. Some of these polysaccharides are used for structural purposes (for example cellulose) and some are used for energy storage (for example starch).

When you look at the structure of glucose (Figure 5), the alcohol (hydroxyl) functional group is found five times within the molecule. Just like in water molecules, the covalent bond between an oxygen atom and a hydrogen atom is a **polar** covalent bond (see Chapter A1.1). This leads to glucose itself being a polar molecule. Glucose molecules have the following properties.

- **Molecular stability**, because the bonds within glucose are stable covalent bonds that do not break easily.
- **High solubility in water**, because glucose is polar and dissolves readily in a polar solvent such as water.
- **Easily transportable**, because of its solubility in water, which means that glucose can easily circulate in blood and in fluids between cells.
- **Yields a great deal of chemical energy** when its covalent bonds are broken. Reactions of this type are called **oxidation reactions** and the high energy yield means that glucose is a good energy store.

B1.1 Figure 5 The structure of glucose

The monosaccharides within this section are shown as ring structures, also known as cyclic structures. Monosaccharides also have a straight-chain form that you may see in other sources. When in solution, monosaccharides switch between the ring and straight-chain form.

What are the roles of oxidation and reduction in biological systems?

B1.1.5 – Polysaccharides and energy storage

B1.1.5 – Polysaccharides as energy storage compounds

Include the compact nature of starch in plants and glycogen in animals due to coiling and branching during polymerization, the relative insolubility of these compounds due to large molecular size and the relative ease of adding or removing alpha-glucose monomers by condensation and hydrolysis to build or mobilize energy stores.

In nature, glucose in a polymer form is often used for energy storage. After glucose is synthesized by photosynthesis, a plant stores much of it as starch molecules. Starch is a polysaccharide made up of hundreds of glucose monomers. In order to make the starch as compact as possible, a plant uses two different kinds of bonds between glucose molecules. One of these bonds is called an alpha 1–4 linkage and the other is called an alpha 1–6 linkage. The numbers refer to the carbon number of the two glucose molecules that are bonded together. For example, in one type of starch called

amylose, carbon #1 is bonded to carbon #4 of the adjoining glucose. When hundreds of glucose molecules are bonded by only 1–4 linkages, the resulting molecule will be linear but in a helix shape. The 1–6 linkages are typical in another type of starch called **amylopectin** and create branches as shown in Figure 6.

(glucose − α (1-4) − glucose)

Amylose

(glucose − α (1-4) − glucose)

Branch point linkage
(glucose − α (1-6) − glucose)

Amylopectin

B1.1 Figure 6 Molecular bonding within the two carbohydrate molecules making up starch. Notice the numbering of the carbons in each glucose. These numbers are used to describe how the monomers are linked together within each polysaccharide. Amylose uses 1–4 carbon linkages and amylopectin uses 1–6 carbon linkages.

Starch is a polymer of glucose but within starch are two kinds of polysaccharides: amylose and amylopectin. A molecule of starch is a very large molecule and so is not readily soluble in water. This low solubility is important because it means that a plant can easily store the starch. Additionally, although the molecule is large it is also compact. When a plant is photosynthesizing and producing lots of glucose, it can add more glucose molecules to either amylose or amylopectin by condensation reactions. Alternatively, when a plant needs to use its reserves of glucose, hydrolysis reactions are used to break the glucose molecules away from starch.

Glycogen is a polysaccharide made of glucose monomers that are bonded in a very similar pattern as in amylopectin 1–6 linkages. The branching in glycogen is more numerous than the branching in amylopectin. Many animals, including humans, store excess glucose as glycogen. Glycogen reserves are kept within our liver and muscle tissue.

A primary advantage for organisms storing glucose as a polysaccharide is that the macromolecules are not readily soluble in cytoplasm and other fluids. This means that they do not affect the osmotic balance in living tissues, whereas individual (very soluble) glucose molecules would.

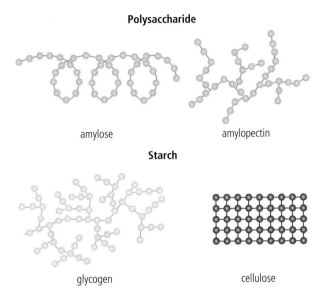

Polysaccharide

amylose

amylopectin

Starch

glycogen

cellulose

An illustration of four different polysaccharides composed of glucose monomers. Each of the smallest geometric shapes in each of the polysaccharides represents a glucose molecule. The upper two molecules, amylose and amylopectin, are both subcomponents of starch, with amylopectin typically representing a higher percentage of the overall starch molecule. Glycogen is the polysaccharide form used for energy storage in animals. Cellulose is used within the structure of the cell walls of plants and is not considered an energy storage molecule. You will read about cellulose in Section B1.1.6.

B1.1.6 – Cellulose as a structural polysaccharide

B1.1.6 – Structure of cellulose related to its function as a structural polysaccharide in plants

Include the alternating orientation of beta-glucose monomers, giving straight chains that can be grouped in bundles and cross-linked with hydrogen bonds.

alpha - glucose   **beta - glucose**

- - - reversed group - - - - -

B1.1 Figure 7 When in solution, glucose exists in two forms, alpha and beta. Both are shown here.

Notice in Figure 7 that the two forms of glucose (alpha and beta) are very similar to each other except for the reversal of the atoms shown on the right side of each molecule. This small difference is very important for the polymers that are formed from each of the two molecules. Starch and glycogen both use the alpha form of glucose, while cellulose uses the beta form. Cellulose is a primary component of the cell walls of plants. For that reason, cellulose is estimated to be the most abundant of all organic molecules on Earth.

How can compounds synthesized by living organisms accumulate and become carbon sinks?

In cellulose, the 1–4 carbon linkages are between beta-glucose molecules (whereas amylose uses 1–4 linkages between alpha-glucose molecules). The condensation reaction that bonds one beta-glucose molecule to the next requires every second beta-glucose molecule to be oriented "upside down" compared to the glucose it is bonded to. This orients the hydroxyl group of carbon #1 with the hydroxyl group of carbon #4.

When beta-glucose molecules form their 1–4 linkages, the result is a very linear polymer with no branches. You can visualize the resulting molecule as similar to a very long thin fibre. Many of these "fibres" run parallel to each other and form a multitude of hydrogen bonds with the adjoining fibres (see Figure 8). This pattern continues as bundles of fibres made of beta-glucose form even larger bundles held together by the cross-linking attractions of hydrogen bonds. The result is a very stable molecule of cellulose.

B1.1 Figure 8 Glucose molecules form linear polymers as a result of 1–4 linkages with additional hydrogen bonds, forming a strong and stable macromolecule.

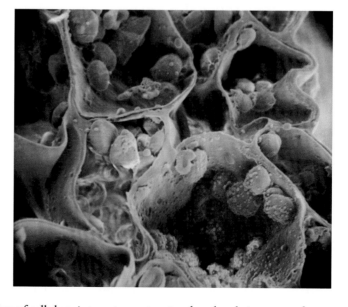

A scanning electron micrograph (SEM) of sliced open plant cells. The plant cell walls are composed mainly of cellulose and are clearly visible. In the interior of the cells are chloroplasts, which produce and store carbohydrates such as starch.

Research is underway to convert cellulose waste (leftover plant materials) into products that would replace items that currently are made of plastics. Plastic pollution is a long-term global problem.

The function of cellulose is to act as a structural molecule in nature, for example cellulose is found in plant cell walls. As well as being strong, cellulose is insoluble in water and the fibres allow water and other substances to pass freely into and out of plant cells. Very few organisms produce the enzyme (cellulase) necessary to digest cellulose, therefore it is not considered to be an energy storage molecule.

B1.1.7 – Conjugated carbon molecules

B1.1.7 – **Role of glycoproteins in cell–cell recognition**
Include ABO antigens as an example.

So far we have only described the structure and function of a few biochemically important molecules. Each molecule has been a carbohydrate, lipid, protein or nucleic acid. Sometimes two or more of these categories of molecules are bonded together in order to accomplish a specific function. Table 3 shows three general examples.

Carbon components	Type of conjugated molecule
Lipid + protein	Lipoprotein
Carbohydrate + lipid	Glycolipid
Carbohydrate + protein	Glycoprotein

B1.1 Table 3 The composition of some common conjugated molecules

Glycoproteins and cell–cell recognition

Various kinds of proteins are associated with cell membranes (see Chapter B2.1). Membrane proteins may or may not be conjugated. Membrane proteins are responsible for a variety of functions, including:

- cell to cell chemical communication (cell signalling)
- transport of molecules in and out of a cell
- cell to cell adhesion
- catalysis as a result of enzymes adhering to the inside or outside of the cell membrane
- recognition of body cells (self) versus non-body cells (not-self) for immune system functions.

Glycoproteins and ABO blood types

Glycoproteins on the surface of red blood cells determine a person's ABO blood type. Red blood cells can have two possible types of glycoprotein on their plasma membranes. The two proteins, A and B, are called **antigens**, because their presence can trigger the immune system. A major component of the immune system is the ability of some white blood cells to recognize "self" from "not-self".

People that have blood type AB have both the A and B type antigens, and thus their immune system will not be triggered by the presence of either. People with blood type O have neither the A or B protein, and their immune system will be triggered by the presence of either of the A and B antigens. The immune system of people who inherit only A or only B antigen is triggered by the presence of the protein that they do not have.

This is important when a blood transfusion is given. People with blood type O are considered to be universal donors because they can give blood to others with type O and those with types A, B or AB. People with blood type AB are universal recipients because they can receive blood from types AB, A, B and O.

An easy way to understand appropriate blood types for transfusions is to remember that a person cannot receive an A or B antigen unless they have genetically inherited that glycoprotein.

ABO blood type	Glycoprotein found on red blood cell plasma membrane	Can receive a transfusion from	Can give blood to
A	A	A and O	A and AB
B	B	B and O	B and AB
AB	A and B	AB, A, B, O	AB only
O	Neither A or B	O only	AB, A, B, O

The four ABO blood types and their associated glycoproteins and transfusion possibilities

B1.1.8 – Lipid solubility

B1.1.8 – Hydrophobic properties of lipids

Lipids are substances in living organisms that dissolve in non-polar solvents but are only sparingly soluble in aqueous solvents. Lipids include fats, oils, waxes and steroids.

Lipids can be categorized as fats, oils, waxes and steroids. Lipid molecules contain many areas of hydrocarbons, meaning areas containing just hydrogen and carbon. The covalent bond between carbon and hydrogen is a **non-polar covalent bond**. This means that lipids dissolve quite well in non-polar solvents but do not dissolve well in water. Organisms have evolved to take advantage of this very limited solubility, and have come up with some unique solutions to the problem of lipid insolubility when needed. One of those solutions is to conjugate the lipid with another molecule. Examples include glycolipids and lipoproteins.

B1.1.9 – Triglycerides and phospholipids

B1.1.9 – Formation of triglycerides and phospholipids by condensation reactions

One glycerol molecule can link three fatty acid molecules or two fatty acid molecules and one phosphate group.

Lipids are macromolecules composed of subunits. Lipids known as **triglycerides** contain one glycerol molecule and three fatty acid molecules. Lipids are formed from condensation reactions:

$$\text{1 glycerol + 3 fatty acids} \rightarrow \text{1 triglyceride + 3 water molecules}$$

Molecules known as **phospholipids** are formed if an inorganic phosphate group replaces one of the three fatty acids. The reaction would be:

1 glycerol + 2 fatty acids + 1 inorganic phosphate → 1 phospholipid + 3 water molecules

Glycerol is a three-carbon molecule, with each carbon bonding to one hydroxyl group initially. The fatty acids vary depending on the number of carbons in each and the possible presence of double bonds between one or more of the carbons. Each fatty acid always contains a terminal carboxyl group that is involved in the condensation reaction.

Formation of a phospholipid or triglyceride

Step 1
The OH group on the glycerol molecule aligns with OH group on the carboxyl group of the general fatty acid number one (R^1). A condensation reaction occurs with the release of one water molecule.

◀ **B1.1 Figure 9** Three condensation reactions with glycerol. As shown, the reaction could form a triglyceride or a phospholipid depending on the reactants. Each of the fatty acids is shown in an abbreviated form with the letter 'R' representing the long hydrocarbon chain.

Step 2
A new covalent bond is formed between fatty acid number one and the glycerol molecule. A second fatty acid (R^2) aligns with the second OH group and undergoes the second condensation reaction, with the release of another water molecule.

Step 3
A new covalent bond is formed between fatty acid number two and the glycerol molecule. A phosphate group aligns with the third OH group and undergoes another condensation reaction, with the release of one water molecule.

Step 4
A new covalent bond is formed between the phosphate group and the glycerol molecule. A phospholipid has been formed. A triglyceride would have been formed if a fatty acid was used in place of the phosphate group in this step. To fully appreciate the molecular product from these three condensation reactions, you should visualize R^1 and R^2 as long chains of carbons and hydrogens.

B1.1.10 – Properties of fatty acids

B1.1.10 – Difference between saturated, monounsaturated and polyunsaturated fatty acids

Include the number of double carbon (C=C) bonds and how this affects melting point. Relate this to the prevalence of different types of fatty acids in oils and fats used for energy storage in plants and endotherms, respectively.

Fatty acids can be divided into different groups depending on their structure. Three categories of fatty acids are found within lipids.

Saturated fatty acids are fatty acids that contain single bonds between the carbons. This means that all the other carbon bonds are to hydrogens (except for the carboxyl group). In other words, the molecule is "saturated" with hydrogens.

Saturated fatty acids have a relatively high melting point and are solid at typical room temperature. Triglycerides containing only saturated fatty acids are called "fats" and are often used by animals to store excess energy. Fats in animal meat and butter are examples, because they are solids at room temperature.

Monounsaturated fatty acids are fatty acids that have one double bond between two of the carbons in the hydrocarbon chain of the molecule. The location of the double bond can vary.

Triglycerides containing one or more monounsaturated fatty acids have a lower melting point than saturated fatty acids, and are liquid (oil) at typical room temperature. Some animals and many plants store energy in this form.

Polyunsaturated fatty acids are fatty acids that have more than one double bond in the hydrocarbon chain. The number and location of the double bonds can vary.

Triglycerides composed of polyunsaturated fatty acids also have a relatively low melting point and are liquid (oil) at room temperature. Many plants store energy in this form.

Two factors affect the melting point of lipids and the fatty acids they contain. One is the number of carbons in the fatty acids, and the other is the presence and number of double bonds. The highest melting points are found in lipids that contain the most carbons and the fewest number of double bonds.

▲ The structure of a triglyceride with two saturated fatty acids and one monounsaturated fatty acid. Saturated fatty acids are more linear in shape, while unsaturated fatty acids have bond angle changes because of the double bond(s). Note the three-carbon glycerol portion of the molecule on the left side.

B1.1.11 – Adipose tissue

> **B1.1.11 – Triglycerides in adipose tissues for energy storage and thermal insulation**
>
> Students should understand that the properties of triglycerides make them suited to long-term energy storage functions. Students should be able to relate the use of triglycerides as thermal insulators to body temperature and habitat.

Adipose tissue is composed of cells that store fat in the form of triglycerides. The quantity of triglycerides that is stored is determined by the organism's caloric intake compared to the calories burned. In Section B1.1.2 you learned about condensation reactions that form triglycerides. The reactions are most common when an organism eats foods that have more calories than the organism is using. Triglycerides can be used to supply energy when sufficient foods are not available for metabolic needs. In that circumstance, the stored triglycerides undergo hydrolysis reactions and the products (glycerol and fatty acids) are made available for energy in the process of cell respiration. Triglycerides are useful for long-term energy storage because they are insoluble in body fluids and thus will not move from their adipose storage sites. Per gram of substance, triglycerides provide approximately twice as much energy as that released by carbohydrates.

A depiction of an **adipocyte** (fat storage cell) growing larger as it accumulates triglycerides. The triglycerides are stored in one or more large vacuoles.

A thick layer of adipose tissue is typical of many animals that live in cold regions such as the arctic. Birds and mammals are **endotherms**, maintaining a steady internal temperature regardless of their environmental temperature. Seals, walruses and whales are all endotherms. Their thick adipose tissue is called blubber and is found between their skin and muscles. The blubber helps trap the heat generated by the inner metabolic activities of the animal.

The food intake of birds and mammals is high compared to other animals. This is because some of the food energy must be used to maintain a particular internal temperature as part of **homeostasis**.

B1.1.12 – Phospholipid bilayers

> **B1.1.12 – Formation of phospholipid bilayers as a consequence of the hydrophobic and hydrophilic regions**
>
> Students should use and understand the term "amphipathic".

Figure 9 on page 125 shows the stages in the formation of a phospholipid molecule. The presence of a phosphate group has consequences for the polarity of the molecule. Small molecules like water and even glucose are polar because of the polar covalent bonds that they contain. A phospholipid is a much larger molecule in comparison.

A drawing showing the amphipathic structure of a phospholipid. The highlighted spherical area is polar. The remaining portion of the molecule is non-polar.

A phospholipid has a polar end (the end with the phosphate group) and an even longer non-polar end (the two long hydrocarbon tails). Molecules like phospholipids, which have both hydrophilic and hydrophobic regions, are called **amphipathic** molecules.

Phospholipids in an aqueous solution solve the problem of having hydrophobic tails by forming a double layer or **bilayer**. In this bilayer, the hydrophobic fatty acid tails extend toward each other in order to keep away from the aqueous solutions inside and outside the cell. The polar phosphate groups are attracted to the aqueous solutions and so arrange themselves on the outside of the bilayer. This is the foundation of the **plasma membrane**. Many organelles within cells have one or more membranes that are used to separate the aqueous fluids within the organelle from the cytoplasm.

An artist's rendering of a phospholipid bilayer membrane. Notice that the hydrophilic heads of the bilayer are oriented outwards and the two sets of fatty acid tails are oriented inwards. This allows the membrane to interact with water on either side of the membrane and also to have an internal thickness that acts as a barrier to water and other polar molecules. The irregular shapes embedded into and extending through the membrane are a variety of proteins and carbohydrates. Many of these proteins and carbohydrates control the passage of polar molecules through the membrane.

B1.1.13 – Steroid hormones

B1.1.13 – Ability of non-polar steroids to pass through the phospholipid bilayer

Include oestradiol and testosterone as examples. Students should be able to identify compounds as steroids from molecular diagrams.

Hormones are chemical messenger molecules that are produced by a variety of glands in the body. After production, hormones are released into the bloodstream and have access to all body tissues. The body tissues that respond to any one hormone is called a target tissue of that hormone. One group of hormones, called **steroids**, are made from the lipid **cholesterol**. Cholesterol is primarily a hydrocarbon molecule. Steroids retain that hydrocarbon makeup and their fundamental structure is easy to identify.

The IB requires you to recognize the general structure of all steroids but not the atom by atom structure of any one steroid. Look for the four connected ring structures, as shown in Figure 10, to help you identify a molecule as a steroid.

testosterone   oestradiol

B1.1 Figure 10 The structure of two steroid hormones. Notice the similarity between the two molecules. Both are formed from the lipid cholesterol, which explains their hydrocarbon backbones. Each corner of each connected ring is a carbon atom with one or more implicit hydrogens to complete the four bonds per carbon. All other steroid hormones have a similar structure.

The production and action of **oestradiol** and **testosterone** are similar. Both are produced by **gonadal** tissue and are involved in the development of primary and secondary sex characteristics beginning at puberty. As lipid-based molecules, each of these two hydrophobic hormones is soluble through the lipid bilayer of cells, and directly enters both through the plasma membrane and nuclear membrane of their target tissue cells. Once inside the nucleus, the hormones direct the process of **transcription**, leading to the production of mRNA molecules.

Guiding Question revisited

In what ways do variations in form allow diversity of function in carbohydrates and lipids?

In this chapter we have learned that:

- carbohydrates exist in three forms, monosaccharides, disaccharides and polysaccharides
- monosaccharides are the smallest of the carbohydrate molecules and act as monomers of disaccharides and polysaccharides
- glucose is a monosaccharide and is often used in cell respiration as a direct source of cell energy
- glucose molecules can be joined in different ways to form both energy storage molecules (amylose, amylopectin and glycogen) and structural molecules (cellulose)
- lipids exist in many forms, such as triglycerides, phospholipids, cholesterol and steroid hormones
- lipids have low solubility in aqueous solutions
- amphipathic properties of phospholipids make them ideal for the formation of membrane bilayers
- the hydrophobic properties of steroid hormones enable the molecules to be soluble in the fatty acid layers of cell membranes
- conjugated forms of proteins, carbohydrates and lipids, such as glycolipids and glycoproteins, enable specialized functions in cells.

Guiding Question revisited

How do carbohydrates and lipids compare as energy storage compounds?

In this chapter we have discovered that:

- both carbohydrates and lipids are used in cells for chemical energy storage
- carbohydrate polysaccharide molecules like starch and glycogen are often used for short-term energy storage
- lipids in the form of triglycerides are often used for longer term energy storage
- carbohydrates hydrolyse into glucose, a soluble molecule that can be easily transported within cells and between cells
- one product of triglyceride hydrolysis is fatty acids, which have very low solubility in aqueous solutions and thus are not easily transported
- triglycerides can store approximately twice the chemical energy compared to the same mass of carbohydrates.

Exercises

ATP (adenosine triphosphate)

glycine

glucose

Q1. Study the figure above and answer the following questions.

    (a) Which functional group is found repeatedly within glucose?

    (b) Which two functional groups are found within the amino acid glycine?

    (c) Which three functional groups are found within ATP?

Q2. Predict the complete chemical formula for each of the following monosaccharides using the general formula of $C_nH_{2n}O_n$.

    (a) A triose.

    (b) A pentose.

    (c) A hexose.

Q3. A person has blood type A.

    (a) In a transfusion, what type(s) of blood can this person receive?

    (b) In a transfusion, to what blood type(s) can this person donate blood?

Q4. Energy storage polysaccharides like amylose, amylopectin and glycogen can add or remove glucose molecules as needed.

    (a) What type of reaction would add glucose to an already existing polysaccharide molecule?

    (b) Is water a reactant or a product of this reaction?

Q5. **(a)** What type of reaction would remove glucose from an already existing polysaccharide molecule?

    (b) Is water a reactant or a product of this reaction?

B1.2 Proteins

Guiding Questions

What is the relationship between amino acid sequence and the diversity in form and function of proteins?

How are protein molecules affected by their chemical and physical environments?

Cells use the naturally occurring 20 amino acids to synthesize polypeptides or proteins. Polypeptides are synthesized under the control of DNA, each polypeptide being coded for by a specific area of DNA molecule called a gene. It is the sequence of amino acids that determines the identity of a polypeptide. Proteins are incredibly diverse in their structure and function as a result of having 20 different building block units for their synthesis. The order of the amino acids in a protein determines its shape and therefore its function.

Protein structure can be changed by the chemical and physical environment. If the structure of the protein changes it may not be able to function correctly. The two most significant environmental factors are pH and temperature. If a protein is heated, or the pH around it changes, chemical bonds within the protein can be affected. The making or breaking of chemical bonds can cause the protein to become an alternative shape, one that may not then function as well.

B1.2.1 – The common structure of amino acids

B1.2.1 – Generalized structure of an amino acid

Students should be able to draw a diagram of a generalized amino acid showing the alpha carbon atom with amine, carboxyl, R-group and hydrogen attached.

In nature there are 20 different amino acids. Each of these is easy to identify as an amino acid because they all have a common structure. The IB requires you to be able to draw the structure that is common to all amino acids. The portion of the amino acid that is unique to each of the 20 is represented by the letter "R". The group of atoms making up the "R" portion of each amino acid can be called the variable group or the side chain (group).

where R = 1 of 20 variable groups

The molecular structure common to all 20 amino acids.

All amino acids are both an acid and a base. In an **aqueous solution** the carboxyl of each amino acid will act as an acid and donate a hydrogen ion, and the amine group will accept a hydrogen ion, therefore acting as a base. By doing this the amino acid becomes ionized.

Drawing the structure of a generalized amino acid is easier if you follow these steps.

1. Draw the single central (alpha) carbon atom and four covalent bonds.
2. Add an amine functional group to this central carbon.
3. Add a carboxyl functional group to the central carbon.
4. Add a hydrogen atom to the central carbon.
5. Finally, add the capital letter "R" to the only remaining covalent bond around the alpha carbon.

Usually the R-group is shown above the central carbon and the H group is shown below the central carbon, with the amine and carboxyl groups on either the left or right. But this is just a convention.

B1.2.2 – Condensation reactions bond amino acids together

B1.2.2 – Condensation reactions forming dipeptides and longer chains of amino acids

Students should be able to write the word equation for this reaction and draw a generalized dipeptide after modelling the reaction with molecular models.

B1.2 Figure 1 The condensation reaction that forms a covalent bond between two amino acids.

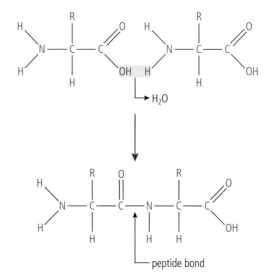

Practise drawing the reaction shown in Figure 1 from memory, using the general structure of amino acids. Once you feel confident, practise drawing only the dipeptide from memory without showing the condensation reaction.

The word equation for the reaction shown in Figure 1 would be:

amino acid 1 + amino acid 2 → dipeptide + water

In Figure 1, a molecule called a **dipeptide** is formed. The water molecule that forms comes from a hydroxyl group (−OH) from the carboxyl group of one amino acid, and a hydrogen ion (H^+) from the other amino acid. This frees up electrons to be shared between carbon and nitrogen atoms, so bonding the two amino acids together into a dipeptide.

Notice the carboxyl group on the right side of the dipeptide in Figure 1. This carboxyl group can be used for a condensation reaction with a third amino acid. The third amino acid will now have a carboxyl group that can be used to add a fourth amino acid, and so on. The new covalent bonds linking the amino acids together are called **peptide bonds**. This is where the term "**polypeptide**" comes from.

 SKILLS

Use model kits to practise modelling the reaction in Figure 1. Add as many amino acids to your polypeptide as possible. Each reaction will be identical to every other reaction, but remember the identity of the resulting dipeptide or polypeptide depends on the identity and sequence of amino acids used.

In this figure, five amino acids have been bonded together by four peptide bonds. The identity of the R (side chain) groups determine the identity and purpose of the protein. Notice that one amino group and one carboxyl group are left intact at the two ends of the protein.

B1.2.3 – Essential amino acids

B1.2.3 – Dietary requirements for amino acids

Essential amino acids cannot be synthesized and must be obtained from food. Non-essential amino acids can be made from other amino acids. Students are not required to give examples of essential and non-essential amino acids. Vegan diets require attention to ensure essential amino acids are consumed.

Our cells can synthesize 11 of the 20 amino acids from other amino acids, but nine have to come from our diet. Humans cannot remain healthy unless we eat foods that contain these **essential amino acids**.

A varied diet that includes different sources of protein should provide all the essential amino acids humans need. Meat is a good source of essential amino acids. Vegan diets or diets that rely heavily on just one protein source may have a limited amino acid content. For example, white rice is deficient in the amino acid lysine, so a white rice-rich diet may need to be supplemented with another source of lysine. Legumes, for example beans, contain low levels of the amino acid methionine. However, a diet that includes a daily mix of white rice and beans would offer all of the amino acids. You are not required to memorize the essential and non-essential amino acids.

B1.2.4 – The vast variety of polypeptides

B1.2.4 – Infinite variety of possible peptide chains

Include the ideas that 20 amino acids are coded for in the genetic code, that peptide chains can have any number of amino acids, from a few to thousands, and that amino acids can be in any order. Students should be familiar with examples of polypeptides.

The huge variety of polypeptides is possible because:

- DNA codes for the number and order of amino acids within polypeptides
- there are 20 different amino acids
- polypeptides can vary in length, from a few amino acids to thousands of amino acids
- some polypeptides are modified by cells after their initial synthesis
- amino acids can be arranged in any order.

Although each polypeptide synthesized by the same gene is identical, there is an immense number of gene and amino acid combinations. This almost infinite number of possible permutations means that the different polypeptides can also have specific functions.

TOK

Molecular models are just visual representations of structures that we cannot see. Models typically fit the best available data and allow our brain to view an image of what "could" be there. Technology enables us to make computer images of various types, and makes it relatively easy to manipulate the images. Are models misleading in some ways?

The number of peptide bonds within a polypeptide can be easily calculated by knowing the number of amino acids within the polypeptide. A dipeptide contains one peptide bond; a polypeptide with 42 amino acids would have 41 peptide bonds. The number of peptide bonds is always one less than the number of amino acids.

Some common polypeptides are:

- haemoglobin, an oxygen-carrying protein found in red blood cells
- keratin, found in hair, nails, claws and hooves
- lipase, a digestive enzyme that helps hydrolyse ingested lipids
- collagen, found in connective tissue in the body, including tendons and ligaments
- histones, proteins found in the nucleus of cells that help form chromatin and chromosomes
- insulin, a hormone that helps regulate blood sugar.

 There are numerous other examples of polypeptides that you will come across throughout this course. Look for examples as you work your way through the IB biology curriculum.

B1.2.5 – The effect of pH and temperature

B1.2.5 – Effect of pH and temperature on protein structure
Include the term "denaturation".

The function of a protein is very dependent on its structure. Some proteins are shaped like fibres while others are folded to form **globular proteins**. Proteins have a very precise three-dimensional shape resulting from intramolecular bonds between amino acids, for example **hydrogen bonds**.

How do abiotic factors influence the form of molecules?

The intramolecular bonds of proteins are susceptible to alterations at above normal temperature. When protein molecules are placed into an environment that is at a higher temperature than their physiological optimum, the increased molecular motion puts a great deal of stress on many of the relatively weak hydrogen bonds. Often the sequence of amino acids connected by peptide bonds remains intact, but the hydrogen bonds that help shape the protein cannot stay in place under the stress. The result is that the protein loses its shape and much or all of its function. This loss of shape and therefore function is called **denaturation**. As long as the bonds between the amino acids remain intact, the protein will return to its normal shape and function if it is returned to its optimal temperature.

The protein on the left has been denatured by either heat or abnormal pH. The overall shape of the protein is altered and it can no longer function. Upon return to a physiologically normal temperature or pH, the protein re-forms its original shape and its function is restored. However, damage caused by excessive heat or pH conditions may lead to covalent bonds like peptide bonds breaking. If this occurs the protein will not be able to reshape itself.

normal protein   denaturation ▶   ◀ renaturation   normal protein

denatured protein

A similar phenomenon occurs when a protein is placed in a pH environment that is not close to its optimum pH. A protein will lose its normal three-dimensional shape, and thus lose its functionality in these circumstances. When a fluid environment such as cytoplasm or blood plasma is flooded with either H^+ ions (an acid) or $-OH$ ions (a base), the extra charges can prevent normal hydrogen bonding. The protein will not take on its "normal" shape and will not function normally. This denaturation is usually reversible as long as the underlying polypeptide chain is not damaged.

Guiding Question revisited

What is the relationship between amino acid sequence and the diversity in form and function of proteins?

In this chapter you have learned that:

- DNA codes the amino acid sequence in a protein
- a polypeptide is created when condensation reactions occur between adjoining carboxyl groups and amine groups
- polypeptides vary in length from a few amino acids to thousands of amino acids
- the sequence of amino acids in a protein (primary structure) determines its molecular shape as well as its biological function
- the variety of polypeptides is nearly infinite.

Guiding Question revisited

How are protein molecules affected by their chemical and physical environments?

In this chapter you have learned that:

- environmental temperatures higher than an optimum affect the shape and thus the function of many proteins
- many of the bonds that create the shape of a protein are relatively weak hydrogen bonds between polar amino acids
- higher temperatures lead to increased molecular motion and place stress on hydrogen bonds
- breaking hydrogen bonds denatures a protein, leading to decreased activity
- an environment that is too acidic or basic will alter the positive and negative charges that are needed for hydrogen bonding
- protein in a non-optimal pH environment will change shape and its activity will decrease.

Exercises

Q1. What structural features do all amino acids have in common?

Q2. A given polypeptide contains 166 amino acids. How many peptide bonds does this polypeptide have?

Q3. Write the word equation for a condensation reaction between two generalized amino acids.

Q4. **(a)** An enzyme is heated in a laboratory to 8°C above its optimum temperature. When tested at this temperature it had lost its catalytic function. Explain why.

(b) Predict, with a reason, what would happen to the catalytic ability of this enzyme if it is returned to its optimum temperature.

(c) The same enzyme is heated to 50°C above its optimum temperature. When tested, it had lost its catalytic function. Predict, with a reason, what would happen to the catalytic ability of this enzyme if it is returned to its optimum temperature.

Q5. Identify the molecule that is not a protein.

A Haemoglobin

B Cellulose

C Lipase

D Keratin

Q6. Describe the health problem that could result if someone obtains nearly all of their protein intake from a single food source such as rice or beans.

Q7. **(a)** From memory, draw a dipeptide composed of two generalized amino acids.

(b) Label the peptide bond within this dipeptide.

B1 Practice questions

1. Draw the structure of a dipeptide. *(Total 3 marks)*

2. Outline how the structure of cellulose makes it a suitable component of cell walls. *(Total 2 marks)*

3. Explain the advantages of having both lipids and carbohydrates as energy stores in the human body. *(Total 2 marks)*

4. Label a peptide bond in the diagram of a polypeptide.

$$\text{H}_2\text{N}-\overset{\overset{\text{H}}{|}}{\underset{\underset{\text{H}}{|}}{\text{C}}}-\overset{\overset{\text{O}}{\|}}{\text{C}}-\overset{\overset{\text{H}}{|}}{\underset{\underset{\text{H}}{|}}{\text{N}}}-\overset{\overset{\text{H}}{|}}{\underset{\underset{\text{CH}_3}{|}}{\text{C}}}-\overset{\overset{\text{O}}{\|}}{\text{C}}-\overset{\overset{\text{H}}{|}}{\underset{\underset{\text{H}}{|}}{\text{N}}}-\overset{\overset{\text{H}}{|}}{\underset{\underset{\text{CH}_3}{|}}{\text{C}}}-\overset{\overset{\text{O}}{\|}}{\text{C}}-\overset{\overset{\text{H}}{|}}{\underset{\underset{\text{H}}{|}}{\text{N}}}-\overset{\overset{\text{H}}{|}}{\underset{\underset{\underset{\text{OH}}{|}}{\text{CH}_2}}{\text{C}}}-\overset{\overset{\text{O}}{\|}}{\text{C}}-\overset{\overset{\text{H}}{|}}{\underset{\underset{\text{H}}{|}}{\text{N}}}-\overset{\overset{\text{H}}{|}}{\underset{\underset{\text{H}}{|}}{\text{C}}}-\text{C}\overset{\nearrow O}{\searrow \text{OH}}$$

(Total 1 mark)

5. Sockeye salmon (*Onchorhynchus nerka*) spend the first years of their lives in the freshwater lakes of Alaska before migrating to marine waters. Their first months in marine waters are spent foraging and growing near the shore line. They then move to offshore regions of the North Pacific Ocean for 2 to 3 years.

Fork length

The graph shows fork length frequency of juvenile *O. nerka* caught during their first months in marine waters in autumn 2008 and ocean age *O. nerka* caught 15 months later during winter 2009 in the North Pacific Ocean.

Key: ■ autumn 2008 (juvenile *O. nerka*) ◇ winter 2009 (ocean age *O. nerka*)

(a) Identify the **total** number of *O. nerka* with fork length from 240 to 245 mm caught in autumn 2008 and winter 2009. (1)

(b) Compare the data in the graph for autumn 2008 and winter 2009. (3)

(c) Suggest **two** factors that could affect the distribution of *O. nerka* in the North Pacific Ocean. (2)

Lipid in *O. nerka* was measured to evaluate possible differences in energy status during their first 15 months at sea. The graph shows the relationship between fork length and lipid content for *O. nerka* caught during autumn 2008 and winter 2009.

Key: ■ autumn 2008 (juvenile *O. nerka*) ◇ winter 2009 (ocean age *O. nerka*)

(d) State the range of lipid content measured in *O. nerka* caught during autumn 2008. (1)

(e) Outline any correlation between total lipid content and fork length in autumn 2008 and in winter 2009. (2)

(Total 9 marks)

6. Outline the production of a dipeptide by a condensation reaction, showing the structure of a generalized dipeptide. *(Total 5 marks)*

THEME B Form and function

2 Cells

We look at a flower, praise its beauty, but rarely think about all the parts of the plant that played critical roles in its formation. Roots must take in water and nutrients, leaves convert sunlight into usable chemical compounds, and stems conduct essential materials up and down, night and day, before the structure that will become the flower has even formed. Each part of the plant has a structure that allows it to carry out an essential function so that the plant can remain alive. Cells are no different. Cells have many parts that work together to maintain the life of the cell. Each working part has a specific structure that enables it to carry out a function essential to the life of the cell. Some cells have unique tasks that they perform for the organism they are a part of. To accomplish these unique tasks, cells are diverse in size, shape and even organelle composition. We will examine some of these cell forms and functions in this chapter.

B2.1 Membranes and membrane transport

Guiding Questions

How do molecules of lipid and protein assemble into biological membranes?

What determines whether a substance can pass through a biological membrane?

Everywhere we look, we see protective exteriors around almost everything. There certainly are differences in these exteriors, but they all serve similar functions. They keep harmful factors out, they keep beneficial factors in, and they regulate what goes in and out of the structure. Cells are no different. They all have membranes around their exterior protecting them from what are often potentially damaging environments. These membranes are composed of lipids and proteins and are assembled in such a way that protection is provided to the cell interior. As well as providing protection, these lipid–protein complexes also control the movement of substances in and out of the cell.

These amazing membranes have receptors to monitor and respond to the surroundings, channels to allow specific molecules to be transported, carriers to maintain homeostatic conditions, and even structures to allow communication with other cells in the same organism or with other organisms in the same environment.

In this chapter, we will learn about the structure and function of the cell membrane.

B2.1.1 and B2.1.2 – Membrane structure

B2.1.1 – Lipid bilayers as the basis of cell membranes

Phospholipids and other amphipathic lipids naturally form continuous sheet-like bilayers in water.

B2.1.2 – Lipid bilayers as barriers

Students should understand that the hydrophobic hydrocarbon chains that form the core of a membrane have low permeability to large molecules and hydrophilic particles, including ions and polar molecules, so membranes function as effective barriers between aqueous solutions.

A study of beet cell membrane. Full details on how to carry out this activity with a worksheet are available in the eBook.

As early as 1915, scientists were aware that the structure of membranes isolated from cells included proteins and lipids. Further research established that the lipids were phospholipids. Early structural theories suggested that membranes were composed of phospholipids forming a bilayer, and on the inside and outside of this bilayer were thin layers of proteins.

In 1972, Seymour J. Singer and Garth L. Nicolson proposed that proteins are inserted into the phospholipid layer and do not form a layer on the phospholipid bilayer surfaces. They believed that the proteins formed a mosaic floating in a fluid bilayer of phospholipids.

Much of the evidence used to revise the model was obtained using an electron microscope. Another source of evidence was the study of cells and their actions in various environments and solutions. The ability to culture cells in the laboratory allowed many of these studies. Since 1972 further evidence has been gathered about the membranes, and only slight changes to the Singer–Nicolson model have been made.

Nature of Science

Using models is a way in which scientists can explain complex structures such as cellular membranes. Models are based on the knowledge available at the time a theory is suggested. Even though the early models of cell membranes were later proved wrong (because of new data), they helped in the development of the presently accepted model of cell membranes. Discuss why it is important to learn about theories that were later discredited.

The currently accepted model of the cellular membrane is known as the **fluid mosaic model**. The fluid mosaic model is discussed further in Section B2.1.10. All cellular membranes, whether plasma membranes or organelle membranes, have the same general structure. Membranes are flexible, supporting structures. They consist of several different types of molecules that allow them to function correctly.

Phospholipid structure

The "backbone" of the membrane is a bilayer produced from huge numbers of molecules called **phospholipids**. Each phospholipid is composed of a three-carbon compound called glycerol. Two of the glycerol carbons have fatty acids combined with them. The third carbon is attached to a highly polar organic alcohol that includes a bond to a phosphate group. Fatty acids are not water soluble because they are non-polar. However, because the organic alcohol with phosphate is highly polar, it is water soluble. This structure means that phospholipid molecules have two distinct areas when it comes to polarity and water solubility. One part of the molecule is water soluble and polar, and is referred to as **hydrophilic** (water-loving). This is the phosphorylated alcohol side. The other part is not water soluble and is non-polar. It is referred to as **hydrophobic** (water-fearing). Any molecules that have both a hydrophobic and a hydrophilic region are said to be **amphipathic**.

A model of a phospholipid ▶

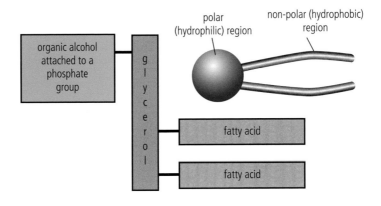

Phospholipid bilayer as a barrier

The hydrophobic and hydrophilic regions cause phospholipids to naturally align as a bilayer if there is water present. The hydrophobic regions are attracted to each other and the hydrophilic regions are attracted to the water in the cytoplasm or the extracellular fluid. Because the fatty acid "tails" do not attract each other strongly, the membrane tends to be fluid or flexible. This allows animal cells to have a variable shape and allows the process of **endocytosis**. What maintains the overall structure of the membrane is the relationship between its chemical makeup and the chemical properties of water.

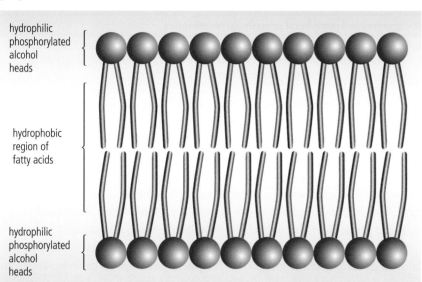

hydrophilic phosphorylated alcohol heads

hydrophobic region of fatty acids

hydrophilic phosphorylated alcohol heads

▲ This model of a phospholipid bilayer shows how phospholipid molecules behave in two layers. Both layers have the phosphorylated alcohol end of the molecules towards the outside and the fatty acid tails oriented towards each other in the middle.

Once the bilayer has formed, large molecules cannot pass through it easily because the molecules are tightly packed. Hydrophilic molecules, such as ions, might be smaller but they also find it difficult to move through the membrane because of the hydrophobic region in the middle of the bilayer. The bilayer therefore forms an effective barrier between the inside and outside of the cell. Because the cell does need large and polar molecules to pass into and out of it for certain functions, the structure of the bilayer allows the cell to control what passes through the membrane.

Because phospholipids naturally form continuous sheet-like bilayers in water, they act as a barrier between the inside and outside of the cell. They have low permeability to large molecules and hydrophilic particles. Hydrophilic particles such as ions and polar molecules do not pass easily through the bilayer. This allows the bilayer to function as an effective barrier between aqueous solutions.

B2.1.3 – Diffusion across cellular membranes

B2.1.3 – Simple diffusion across membranes

Use movement of oxygen and carbon dioxide molecules between phospholipids as an example of simple diffusion across membranes.

One type of transport that can take place through the membrane is diffusion. In diffusion, particles move from a region of higher concentration to a region of lower concentration. In a living system, diffusion often involves crossing a membrane. For example, oxygen is used by cells in respiration. There is therefore a lower oxygen concentration inside the cell compared to outside the cell. Oxygen diffuses into the cell as a result. Carbon dioxide diffuses in the opposite direction because carbon dioxide is produced by mitochondrial respiration inside the cell and is present in higher concentrations inside the cell compared to outside the cell. Both CO_2 and O_2 are small, uncharged molecules. They can move between the phospholipid molecules of the membrane, so their diffusion can occur easily.

B2.1.4 – Membrane proteins

B2.1.4 – Integral and peripheral proteins in membranes

Emphasize that membrane proteins have diverse structures, locations and functions. Integral proteins are embedded in one or both of the lipid layers of a membrane. Peripheral proteins are attached to one or other surface of the bilayer.

Another major component of cellular membranes is the proteins. It is these proteins that create the extreme diversity in membrane function. Proteins of various types are embedded in the fluid matrix of the phospholipid bilayer. This creates the mosaic or tile-like effect characteristic of cellular membranes. There are usually two major types of proteins. One type is referred to as **integral** proteins and the other type is referred to as **peripheral** proteins. Integral proteins show an **amphipathic character,** with both hydrophobic and hydrophilic regions within the same protein. These proteins will have their hydrophobic region in the mid-section of the phospholipid backbone. Their hydrophilic region will be exposed to the water molecules on either side of the membrane. Peripheral proteins, on the other hand, do not protrude into the middle hydrophobic region, but remain bound to the surface of the membrane. Peripheral proteins can be found on the surface of both the inner and outer sides of the membrane. Often these peripheral proteins are anchored to an integral protein.

Peripheral and integral proteins of a cell membrane. The peripheral proteins do not extend into the lipid bilayer(s) of the cell membrane like the integral proteins do. Peripheral proteins are attached to one of the two surfaces of the membrane.

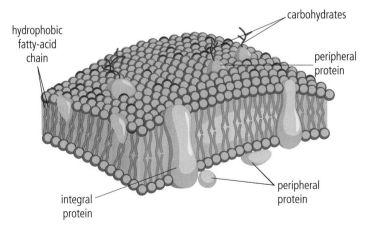

Although the protein components of cell membranes differ depending on the cell type and its particular function at a given time, several types of proteins are usually present. Examples are listed in Table 1.

Protein type	Description
Hormone-binding	These proteins have specific shapes exposed to the exterior that fit the shape of specific hormones. The attachment between the protein and the hormone causes a change in the shape of the protein, which results in a message being relayed to the interior of the cell.
Enzymatic	These proteins occur on either the interior or the exterior membrane surface. They are often grouped together so that a sequence of metabolic reactions, called a metabolic pathway, is catalysed.
Cell adhesion	This protein type allows temporary or permanent connections called junctions between cells. There are two types of junctions, gap junctions and tight junctions.
Cell-to-cell communication	Most of these proteins have carbohydrate molecules attached. They provide an identification label so that organisms can distinguish between self and non-self material.
Channel forming	Some proteins span the membrane, providing passageways for substances to be transported through.
Pumps for active transport	In active transport, proteins shuttle a substance from one side of the membrane to another by changing shape. This process requires the expenditure of energy in the form of adenosine triphosphate (ATP).

B2.1 Table 1 Types of protein usually present in a cell membrane

What are the roles of cell membranes in the interaction of a cell with its environment?

B2.1.5 and B2.1.6 – Membrane transport

B2.1.5 – Movement of water molecules across membranes by osmosis and role of aquaporins

Include an explanation in terms of random movement of particles, impermeability of membranes to solutes and differences in solute concentration.

B2.1.6 – Channel proteins for facilitated diffusion

Students should understand how the structure of channel proteins makes membranes selectively permeable by allowing specific ions to diffuse through when channels are open but not when they are closed.

There are two general types of cellular transport:

- passive transport
- active transport.

Passive transport does not require cellular energy (in the form of adenosine triphosphate, ATP) but active transport does. Passive transport takes place when a substance moves from an area of high concentration to an area of lower concentration. Movement is said to occur along a concentration gradient. The source of energy for this movement comes from the kinetic energy of the molecules. If left undisturbed, this directional movement will continue until equal concentrations of the substance are found in both areas and equilibrium is attained.

When active transport occurs, the substance is usually moved against a concentration gradient, so energy expenditure must occur. Equilibrium is not reached with active transport.

Examine Figure 1 illustrating chemical diffusion, an example of passive transport.

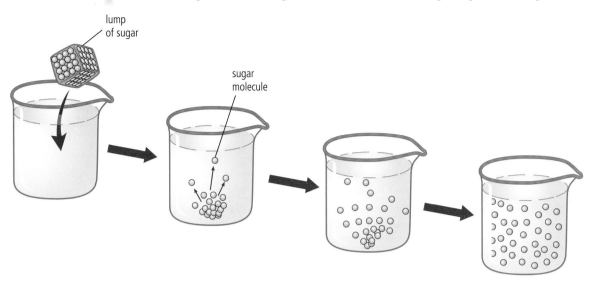

lump
of sugar

sugar
molecule

▲

B2.1 Figure 1 Chemical diffusion: note how the sugar molecules move from the area of higher concentration to the area of lower concentration. The particles in the liquid are present in a constant, random motion. When a high concentration of a molecule is present, there are more collisions between the molecules. This creates a net movement of particles into areas that are less concentrated.

B2.1 Figure 2 The cell wall of a plant makes it difficult to see the many changes that occur inside as a result of water movement. The rigid cell wall resists changes in shape. However, the cell membrane and cell contents are affected by water moving into and out of the cell.

Movement of water molecules

Osmosis is another type of passive transport: movement occurs along a concentration gradient. However, osmosis involves only the passive movement of *water* across a **partially permeable membrane**. A partially permeable membrane (also known as a **selectively permeable membrane**) is one that only allows certain substances to pass through (a permeable membrane would allow everything through). The concentration gradient of water that allows the movement to occur is the result of a difference in solute concentrations on either side of the membrane. A **hypertonic** solution has a higher concentration of solutes than a **hypotonic** solution (see Chapter D2.3). Water therefore moves from a hypotonic solution to a hypertonic solution across a partially permeable membrane (see Figure 2). If isotonic solutions occur on either side of a partially permeable membrane, no net movement of water is evident because equilibrium has been achieved.

Osmosis in a plant cell

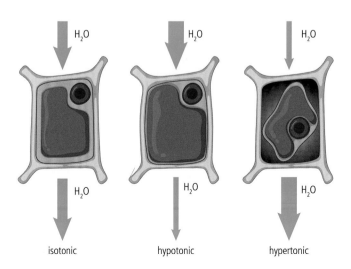

H_2O H_2O H_2O

H_2O H_2O H_2O

isotonic   hypotonic   hypertonic

The cell membrane is impermeable to many solute molecules. Therefore, in osmosis, only water moves across the cell membrane. This water moves with more ease than expected of a polar molecule. Usually, polar molecules cannot pass quickly through the cell membrane because of the hydrophobic properties of the middle membrane region. However, most cell membranes have protein channels called **aquaporins**, which allow water molecules to pass through them. Water molecules move randomly (like all other molecules that are diffusing) but if there is a higher concentration of water molecules in one area then there will be more molecules moving randomly towards the area of lower concentration and there will be a net movement towards the area with the lower concentration.

Aquaporins allow water to flow through cell membranes. There are different types of aquaporins depending on the organism and the cell types they are a part of. All aquaporins are embedded in the cell membrane as integral proteins and consist of amino acids producing repeating proteins. These proteins have non-polar areas on their exterior that allow the embedding of the aquaporin in the membrane. They also have polar areas internally associated with the channel to allow water to pass through.

Carrier and channel proteins

Facilitated diffusion is a particular type of diffusion that involves two types of integral proteins: carrier proteins and channel proteins.

Carrier proteins change shape in order to carry a specific substance (usually an ion) from one side of the membrane to the other. If a carrier protein is not working, no transport will occur. Carrier proteins can carry substances along a concentration gradient (in the case of facilitated diffusion) or against a concentration gradient (as in active transport). Carrier proteins can carry both water-soluble and insoluble molecules.

To remember the difference between diffusion and osmosis, think of "H₂Osmosis", linking the water to osmosis.

Channel proteins are different from carrier proteins in that they have pores through which molecules of appropriate size and charge can pass. Most channel proteins have "gates" that open and close in response to chemical or mechanical signals. Channel proteins do not change shape in the way that carrier proteins do: they just open and close a channel through which molecules can diffuse. Channel proteins only carry water-soluble molecules, and are specific for the ion that they carry. The presence of channel proteins and carrier proteins makes cell membranes selectively permeable: specific ions are allowed through the membrane at certain times. The rate of facilitated diffusion depends on several factors, including the concentration difference that exists across the membrane and the number of carrier proteins actively involved in transport and/or the number of channel proteins open.

An example of a disease involving facilitated diffusion is **cystinuria**. This occurs when the protein that carries the amino acid cysteine is absent from kidney cells. The consequence is a build-up of amino acids in the kidney, resulting in very painful kidney stones.

B2.1.7 – Active transport and pump proteins

B2.1.7 – Pump proteins for active transport

Students should appreciate that pumps use energy from adenosine triphosphate (ATP) to transfer specific particles across membranes and therefore that they can move particles against a concentration gradient.

Active transport requires work to be performed. This means energy must be used, so ATP is required. Active transport often involves the movement of substances against a concentration gradient. This process allows the cell to maintain interior concentrations of molecules that are different from exterior concentrations. Active transport can take place because of highly selective proteins in the membrane that bind with the substance to be transported. Different protein carriers involved in active transport differ in the way that they work. Look at Figure 3.

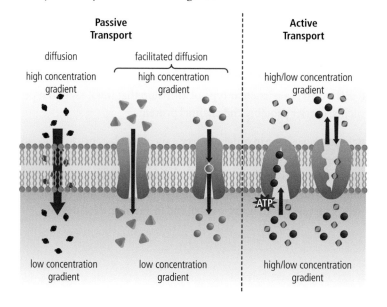

B2.1 Figure 3 A comparison of active and passive transport. Notice the descriptions of the concentration gradients for each type of transport. The active transport example is a protein pump requiring the use of ATP, such as in the sodium–potassium pump.

The sodium–potassium pump

The **sodium–potassium pump** is an extremely important example of active transport. It uses ATP to move ions directly against a concentration gradient. This is especially important in nerve cells, also called **neurons**, so that animals can respond appropriately to environmental stimuli.

What processes depend on active transport in biological systems?

B2.1.8 – Membrane permeability

B2.1.8 – Selectivity in membrane permeability

Facilitated diffusion and active transport allow selective permeability in membranes. Permeability by simple diffusion is not selective and depends only on the size and hydrophilic or hydrophobic properties of particles.

How easily a substance can move passively across a membrane depends on two major factors: size and charge. Substances that are small and non-polar will move across a membrane with ease. Substances that are polar or large, or both, do not cross membranes easily. Examples of small, non-polar substances are gases such as oxygen, carbon dioxide and nitrogen. Ions such as chloride ions, potassium ions and sodium ions have a great deal of difficulty crossing membranes passively, as do large molecules such as glucose and sucrose. Diffusion of small, simple molecules is therefore not selective. In contrast, the cell can be selectively permeable to large, charged molecules because they must travel through integral proteins.

Facilitated diffusion and active transport enable selective permeability in membranes through the control of the protein channels involved in the two processes. However, permeability by simple diffusion is not selective and depends on the size and chemical properties of the particles involved.

The size and polarity of molecules determine the ease with which various substances can cross membranes. These characteristics and the ability of molecules to cross membranes are arranged along a continuum:

small and non-polar molecules cross membranes easily   large and polar molecules cross membranes with difficulty

Challenge yourself

This challenge requires knowledge and understanding of cellular transport. Completing it will serve as a review of the cellular transport concepts we have considered so far.

1. A practical example of diffusion and osmosis is kidney dialysis. The kidneys are responsible for removing urea from the blood and also regulate the level of solutes in the body. If the kidneys are not functioning correctly this can be life-threatening. A process called **haemodialysis** can be used to remove urea artificially and restore the correct balance of solutes in the body.

   During haemodialysis, blood is passed through a system of tubes composed of selectively permeable membranes. These tubes are surrounded by a solution that is called the dialysate. The dialysate contains key solutes at levels close to the patient's normal blood levels. Wastes are kept at a low level in the dialysate. As blood moves through the tubes, the dialysate is constantly replaced to maintain ideal levels.

   Using your knowledge of osmosis, diffusion and membrane transport, suggest how haemodialysis works and why the dialysate must be constantly changed.

B2.1.9 – Glycoproteins and glycolipids

B2.1.9 – Structure and function of glycoproteins and glycolipids

Limit to carbohydrate structures linked to proteins or lipids in membranes, location of carbohydrates on the extracellular side of membranes, and roles in cell adhesion and cell recognition.

When cell membrane phospholipids have carbohydrate chains attached to them, they are known as **glycolipids**. **Glycoproteins** are cell membrane proteins that have chains of carbohydrates attached to them. Carbohydrate chains are only found on the exterior, extracellular side, of the cell membrane. These chains are quite diverse based on their sequences of sugar types and branching structures. Glycoproteins and glycolipids are important for cell identification and cell adhesion (cells sticking to each other).

In this diagram of a cell membrane, notice the structures labelled glycolipids and glycoproteins. These structures have carbohydrate chains attached. They occur on the cell membrane outer surface. From our earlier discussion of cell membranes, we know the outermost region and the innermost region of the cell membrane bilayer are hydrophilic. The middle portion of the lipid bilayer is hydrophobic. The relationship of the membrane regions to water allows maintenance of its structure as a rather stable bilayer.

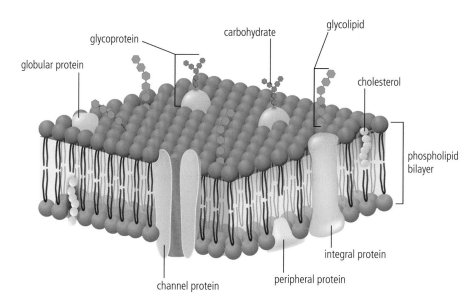

The characteristics of human blood types A, B and O are the result of carbohydrate chains. Carbohydrate chains allow the body to work out which cells belong to it (self) and which cells are from outside the body (non-self). This is especially important in procedures involving transplants. If the carbohydrate chains of the transplanted tissue or organ are not compatible, rejection will occur. Rejection means the receiving patient's immune system attacks the foreign cells, resulting in the possible failure of the transplant.

The **glycocalyx** is a thin sugar layer made up of carbohydrate chains attached to proteins that can cover a cell. It is common in animal cells. This animal cell "sugar coat" has many functions, including cell to cell adhesion, cell to cell recognition and reception of various signalling chemicals. The glycocalyx is also present on the surface of many bacterial and fungal cells, where it may have both adhesion and protective functions. When a glycocalyx occurs in plant cells, it often appears to help anchor the plant cell membrane to the cell wall.

B2.1.10 – The fluid mosaic model

B2.1.10 – Fluid mosaic model of membrane structure
Students should be able to draw a two-dimensional representation of the model and include peripheral and integral proteins, glycoproteins, phospholipids and cholesterol. Indicate hydrophobic and hydrophilic regions.

Figure 4 shows the fluid mosaic model of the cell membrane. You should be familiar with all the parts of a membrane by now, and how they work together to form a selectively permeable barrier.

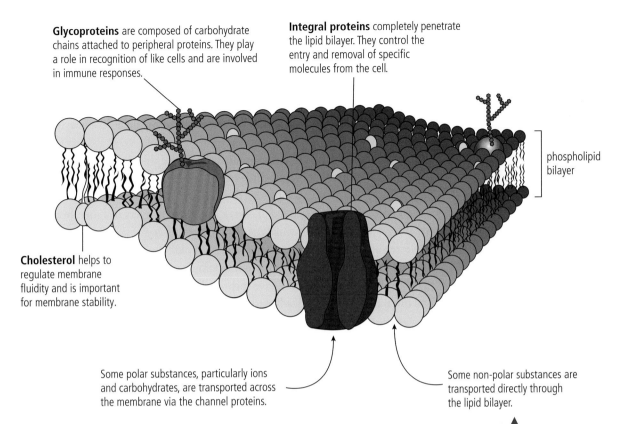

Glycoproteins are composed of carbohydrate chains attached to peripheral proteins. They play a role in recognition of like cells and are involved in immune responses.

Integral proteins completely penetrate the lipid bilayer. They control the entry and removal of specific molecules from the cell.

phospholipid bilayer

Cholesterol helps to regulate membrane fluidity and is important for membrane stability.

Some polar substances, particularly ions and carbohydrates, are transported across the membrane via the channel proteins.

Some non-polar substances are transported directly through the lipid bilayer.

▲ **B2.1 Figure 4** In the fluid mosaic model of a cell membrane, there is double layer of lipids (fats) arranged with their tails facing inwards. Proteins are thought to "float" in the lipid bilayer.

Cholesterol

Membranes must be fluid to function properly. They are a bit like olive oil in their consistency. Cholesterol molecules can be found at various locations in the hydrophobic region (fatty acid tails) of animal cells. These molecules have a role in determining membrane fluidity, which changes with temperature. The cholesterol molecules allow membranes to function effectively at a wider range of temperatures than if they were not present. They do this by interacting with the tails of the phospholipid bilayer. Plant cells do not have cholesterol molecules; they depend on saturated or unsaturated fatty acids to maintain proper membrane fluidity.

Study the fluid mosaic model in Figure 4. Practise drawing it with the following structures correctly labelled: phospholipids, integral proteins, peripheral proteins, glycolipids, glycoproteins and cholesterol. All the structures must be properly positioned within the drawing to earn marks in an exam. You should show the structure of a phospholipid as a circle with two parallel tails attached. You should also be able to identify the hydrophobic and hydrophilic regions.

cholesterol

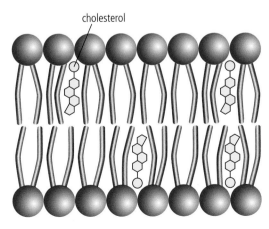

▲ Notice the position of the cholesterol molecules. They are closely associated with the phospholipid tails in animal membranes.

Guiding Question revisited

How do molecules of lipid and protein assemble into biological membranes?

In this chapter we have examined how:

- amphipathic lipids and phospholipids form continuous sheet-like bilayers when in water
- the bilayer forms because the hydrophilic portion of the phospholipid molecules is attracted to water, while the hydrophobic portion of the molecules faces inwards away from contact with water
- integral proteins are embedded in one or both lipid bilayers of a membrane
- peripheral proteins are attached to one or other surface of the lipid bilayer
- cholesterol is often present near the phospholipid tails of the cell membrane and has a role in the control of membrane fluidity
- glycoproteins and glycolipids have carbohydrate structures attached to them and often have roles in cell adhesion and recognition.

Guiding Question revisited

What determines whether a substance can pass through a biological membrane?

In this chapter we have discussed how:

- cell membranes have low permeability to large molecules and hydrophilic molecules
- diffusion and osmosis are examples of passive transport
- protein pumps allow the movement of materials across the cell membrane that otherwise would not be able to pass because of their chemical properties
- aquaporins are important in allowing polar water molecules to pass through cell membranes.

Exercises

Q1. Which **one** of the following is an example of active transport?

   A Facilitated diffusion.

   B Osmosis.

   C Movement of water through aquaporins.

   D Sodium–potassium pump at work.

Q2. Which type of compound occurs only on the surface of the cell membrane?

   A Integral proteins.

   B Carbohydrates.

   C Cholesterol.

   D Phospholipids.

Q3. Explain the orientation of the bilayer of phospholipid molecules in the plasma membrane using the terms hydrophobic and hydrophilic.

Q4. Why does a diet high in plants and plant products have relatively low cholesterol levels compared to a diet involving high amounts of animal products?

Q5. Which of the following is *not* a function of the cell membrane?
   A Cell adhesion.
   B Enzyme synthesis.
   C Active transport of specific substances.
   D Pump materials against a concentration gradient.

Q6. Name the structures found in the cell membrane that are involved with the transport of water.

Q7. What is the connection between carrier proteins and active transport?

B2.2 Organelles and compartmentalization

Guiding Questions

How are organelles in cells adapted to their functions?

What are the advantages of compartmentalization in cells?

Cells are the building blocks of all life forms. They come in a myriad of sizes and shapes. All are surrounded and protected by a multi-functional membrane, discussed in Chapter B2.1. Within most cells are highly specialized structures carrying out functions essential to the cell and/or organism. These specialized structures are called organelles. Organelles are adapted to their function. For example, mitochondria have infoldings of the inner membrane so that they have a larger internal surface area. This allows the reactions responsible for respiration to take place more efficiently. Some organelles are thought to have originated outside the cells they exist in today. A cell's function is reflected in the types and numbers of organelles present, leading to cell specialization, which is discussed in Chapter B2.3.

Most organelles are membrane bound, allowing compartmentalization within the cell. Compartmentalization allows unique processes to proceed without interference from chemicals or reactions occurring nearby in the cell. We will discuss these discrete cellular compartments in this chapter.

B2.2.1 – Cell compartmentalization

B2.2.1 – Organelles as discrete subunits of cells that are adapted to perform specific functions

Students should understand that the cell wall, cytoskeleton and cytoplasm are not considered organelles, and that nuclei, vesicles, ribosomes and the plasma membrane are.

NOS: Students should recognize that progress in science often follows development of new techniques. For example, study of the function of individual organelles became possible when ultracentrifuges had been invented and methods of using them for cell fractionation had been developed.

Cell compartment-
alization refers to
the division of a
cell into regions
or compartments
with single or
double membranes
between them.

The cell is a small but very busy unit, with different reactions occurring in close proximity within it. Selectively permeable membranes play an important role in allowing these reactions/functions to occur without interfering with one another. This isolation of reactions is referred to as **cell compartmentalization**, the result of which is that cells work much more efficiently than if all the reactions were mixed up together. Much of our research concerning the cell focuses on how cells work at the molecular level. The best way to do this is to reduce the cell to its component parts and study each part individually. This approach is known as **reductionism**. By studying localized parts and reactions, we can develop an understanding of the overall complex reactions of the cell.

Tools for cell research

The development of imaging in cell research is discussed in Chapter A2.2. Our understanding of the cell has advanced tremendously with improvements in light microscopes and electron microscopy and refinements in preparation techniques.

Another tool used in the study of the cell involves a process called **biochemical fractionation**. Fractionation refers to the separation and isolation of specific chemicals and/or structures so that detailed research can be carried out. Several techniques have been developed for cell research, each allowing the separation of different parts of the cell.

Centrifugation or **cell fractionation** allows the extraction of organelles from cells. Ultracentrifuges are often used for this process. Cells are first mixed in a tube with substances that break down the cell membranes. The sample is then spun at high speeds to isolate the different components by size and shape. Larger and heavier cell components can be separated off at lower speeds. Once separated, larger and heavier organelles are found at the bottom of the tube.

▲
A centrifuge separates components of a sample by spinning it at high speeds. Separation occurs because of the different densities of the component parts.

Chromatography is very effective at isolating pure substances such as amino acids, proteins, carbohydrates and plant pigments. A mixture of molecules is placed in a separating medium. The molecules separate out depending on their size and the speed with which they travel through the medium. There are several different types of chromatography, including gel and ion exchange chromatography.

◄ An example of column chromatography. Notice the colours in the column, which indicate different substances isolated from the original compound in the test tubes.

Gel electrophoresis separates molecules of different types by passing them through a gel using an electrical charge. The molecules are separated based on properties such as size and charge. This technique is commonly used in studies involving nucleic acids.

◀ An example of the apparatus used in gel electrophoresis to isolate molecules by size and charge.

migrated samples on the gel

gel under UV light

photo of the gel

Fluorescent dyes are often used in cell fractionation techniques to allow better viewing of the components. These dyes absorb light energy at a given wavelength and then re-emit that light at a longer wavelength allowing for enhanced viewing. Fluorescent dyes are highly specific and will only provide visibility when attached to certain molecules such as amino acids, peptides, antibodies or nucleic acids. Ethidium bromide is a fluorescent dye often used to observe DNA fragments produced by gel electrophoresis.

DNA fingerprinting using gel electrophoresis techniques is used around the world to convict criminals and identify victims. Specific segments of DNA are examined to determine relationships and identities. It is even possible to determine an individual's ancestral history using DNA analysis.

What separation techniques are used by biologists?

▲ This image of DNA fingerprints shows the high visibility of fragments that can be achieved when using the fluorescent dye known as ethidium bromide in gel electrophoresis. Ultraviolet light is provided to achieve the visibility.

Nature of Science

Science has progressed and continues to progress with the development of new study techniques. Not only has the microscope increased our knowledge of the cell, but ultracentrifuges and fractionation of cells have also greatly enhanced our understanding of the cell and its organelles.

Organelles: the compartments of the cell

Chapter A2.2 discusses the organelles of the cell. They are separate structures within the cell that carry out specialized functions. To carry out these particular functions, each type of organelle has a unique structure. Organelles are separated from the rest of the cell by a protective barrier, sometimes involving two membranes. This barrier is important because it allows the chemical reactions that take place in an organelle to happen without interference from the rest of the cell. However, not all parts of a cell are considered organelles.

Study Table 1, which summarizes the function of different cell organelles. See Chapter A2.2 for more details about the organelles.

Component	Organelle	General function
Cell wall	No	Encloses and protects plant cells
Cytoskeleton	No	Maintains cell shape, anchors organelles, facilitates cell movement
Cytoplasm	No	The region where most of the metabolic reactions in the cell occur
Nucleus	Yes	Genetic control
Vesicles	Yes	Storage and transport
Ribosomes	Yes	Protein synthesis
Plasma membrane	Yes	Regulates movement in and out of cell, transports materials to maintain the internal cell environment, cell recognition and communication
Cilia/flagella	Yes	Movement
Golgi apparatus	Yes	Modifies and stores endoplasmic reticulum products, forms lysosomes and transport vesicles
Mitochondria	Yes	Cellular energy (ATP) production
Chloroplasts	Yes	Conversion of light energy into chemical energy
Lysosomes	Yes	Digest worn out organelles and debris, digest materials brought into the cell by endocytosis

B2.2 Table 1 Different components of a cell

B2.2.2 – The nucleus and cytoplasm

B2.2.2 – Advantage of the separation of the nucleus and cytoplasm into various compartments

Limit to separation of the activities of gene transcription and translation – post-transcriptional modification of mRNA can happen before the mRNA meets ribosomes in the cytoplasm. In prokaryotes this is not possible – mRNA may immediately meet ribosomes.

The development of the nucleus in eukaryotic cells was a huge advantage compared to prokaryotic cells because it allowed some of the important cell processes to take place more efficiently.

What are examples of structure–function correlation at each level of biological organization?

SKILLS

Organelles and compartmentalization. Full details on how to carry out this activity with a worksheet are available in the eBook.

Transcription and translation are cell processes responsible for the production of proteins. In transcription, a DNA strand serves as a template or copy strand for the formation of messenger RNA (mRNA). Translation occurs when ribosomes use the code carried by mRNA to produce a polypeptide/protein. Transcription happens in the nucleus of eukaryotic cells, while translation is carried out in the cytoplasm. The separation of these two important cellular processes allows post-transcriptional modification of mRNA to occur in the nucleus before translation happens in the cytoplasm. In prokaryotic cells there is no isolation of these two processes, and mRNA can immediately come into contact with ribosomes and initiate translation without any modification occurring. (Transcription and translation are described in detail in Chapters D1.1 and D1.2. It is important to realize that this additional step of modification of the mRNA in eukaryotic cells decreases the chances of errors happening in the production of polypeptides.) It is the compartmentalization of the cell that allows this greater cell efficiency.

B2.2.3 – Compartmentalization of the cytoplasm

B2.2.3 – Advantages of compartmentalization in the cytoplasm of cells

Include concentration of metabolites and enzymes and the separation of incompatible biochemical processes. Include lysosomes and phagocytic vacuoles as examples.

All eukaryotic cells possess compartments or organelles that are involved with:

- energy production
- metabolism
- biosynthesis
- degradation.

However, it is important to note that the number and size of these compartments and organelles vary depending on the overall function of the cell in which they occur. For example, certain types of pancreatic cells called **acinar cells** specialize in the secretion of digestive enzymes. These cells are essential to life in humans and have a greatly enlarged endoplasmic reticulum (ER), Golgi apparatus and granule storage compartments.

Compartmentalization has allowed a division of labour within the cell, with specific tasks carried out by a single organelle or organelle-like structure. Enzymes can be kept in the areas where they will be most effective. Often, reaction pathways in cells rely on a series of enzyme-controlled reactions. Keeping reactions separate in different parts of the cell means that the metabolites and enzymes for each particular process can be concentrated in a particular area. This ensures that pathways run smoothly, can be easily controlled and do not interfere with each other.

Lysosomes participate in the breakdown of wastes and cellular components that need to be replaced. This breakdown requires some potentially destructive enzymes that could cause severe damage to the cell if they were not isolated by a membrane.

When endocytosis occurs, the result is often a phagocytic vacuole. This vacuole is a means of protecting the cellular contents from potential damage when phagocytosis occurs.

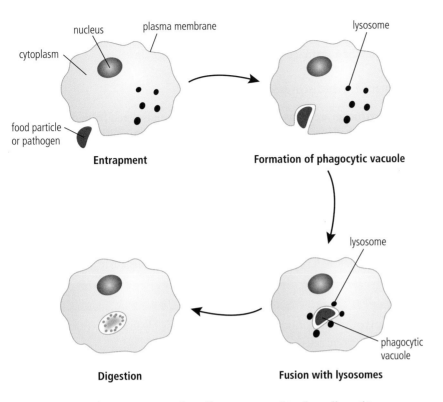

nucleus
plasma membrane
cytoplasm
food particle
or pathogen

Entrapment

lysosome

Formation of phagocytic vacuole

lysosome

phagocytic
vacuole

Fusion with lysosomes

Digestion

**A pathogen is a
disease-causing
organism. Grave
harm can be done
to a cell if pathogens
are not controlled.**

Once formed, the phagocytic vacuole will move around in the cell until it contacts a lysosome. The vacuole then fuses with the acidic lysosome, allowing inactivation and digestion of the threat. Phagocytosis plays a key role in defending cells against invading pathogens.

Compartmentalization does present challenges, however. The very fact that each area or organelle carries out one specific function means that the cell must develop a means of integrating all the separate functions. To accomplish this, some organelles are connected in a functional series, allowing the chemical pathways important to the cell to take place. Membrane pumps and carriers have evolved so that the products of one organelle can enter another, and important cell reactions can occur. By overcoming such challenges, compartmentalization has greatly enhanced the successful existence of the cell.

Guiding Question revisited

How are organelles in cells adapted to their functions?

In this chapter we have discovered that:

- there is a structure–function correlation with all cell organelles
- each organelle of the cell has a unique structure that allows that organelle to perform its function, for example lysosomes are packages of digestive enzymes surrounded by a double membrane
- the cell membrane around the lysosomes prevents the digestive enzymes from damaging healthy structures in the cell
- the lysosomes can move anywhere in the cell and have a flexible membrane that can fuse with vesicles when necessary.

Guiding Question revisited

What are the advantages of compartmentalization in cells?

In this chapter we have discovered that:

- compartments are areas of the cell that have been isolated from other parts of the cell so that specialized functions can be carried out within them
- splitting the cell up into compartments allows the concentration of metabolites and enzymes in a particular area to be controlled
- this makes cells processes more efficient and stops the enzymes from one cell process interfering with another process, for example lysosomes contain enzymes used to break down substances ingested into the cell and worn out cell components, and by packaging them in an organelle the digestive enzymes are contained and controlled
- separating the nucleus from the rest of the cell means that mRNA can be processed before it reaches the ribosomes which translate it.

Exercises

Q1. Why is post-transcriptional modification possible in eukaryotic cells but not in prokaryotic cells?

Q2. Which cell fractionation process uses an electrical current to separate molecules of different size and electrical properties? What groups of cell molecules are often separated using this technique?

Q3. What are the advantages of compartmentalization to the cell?

Q4. Which cellular organelle possesses destructive enzymes necessary for the breakdown of wastes and damaged cell organelles?

    A Ribosome.

    B Lysosome.

    C Chloroplast.

    D Nucleus.

Q5. Explain how reductionism is used in the research involving the cell.

Q6. Which of the following is not regarded as an organelle?

    A Nucleus.

    B Chloroplast.

    C Cell membrane.

B2.3 Cell specialization

Guiding Questions

What are the roles of stem cells in multicellular organisms?

How are differentiated cells adapted to their specialized functions?

Cells have an amazing capability to specialize. They carry a genetic code that allows the development of specific traits and functions that contribute to the survival of the organism. In humans we see specialization in cells such as those in muscles, the lungs, eggs and sperm. Plants also show specialization, with the development of unique characteristics portrayed by roots, stem, leaves and flowers. Each cell develops in a specific way so that it is best suited to its function. For example, root cells have a large surface area to help absorb water and nutrients from the soil. They also have many mitochondria so that they can carry out the active transport of ions into cells.

Stem cells have captured the interest of many people because they maintain at least some degree of versatility. Some types of stem cell formed early in an organism's development can differentiate into every type of cell that can exist in the adult form of the organism. Some cells remain unspecialized within the organism in order to be able to develop into new cells. For example, in plants meristematic tissue is found in buds and stems. This tissue can differentiate into any of the many types of tissue that plants need to grow. In this chapter, we will examine stem cells and some of the factors involved in cell specialization within a multicellular organism.

B2.3.1 – Cell reproduction and organism development

B2.3.1 – Production of unspecialized cells following fertilization and their development into specialized cells by differentiation

Students should understand the impact of gradients on gene expression within an early-stage embryo.

Many cells have the ability to reproduce themselves. In multicellular organisms this allows growth to happen. It also means damaged or dead cells can be replaced.

Multicellular organisms usually start their existence as a single cell called a **zygote**. The zygote is formed as a result of fertilization, which is part of sexual reproduction. The two cells that fuse in sexual reproduction are called **gametes**. Each gamete has one-half the genetic material of a zygote. The following shows the development progression for humans:

gametes → zygote → embryo → foetus → infant

From the formation of the single-celled zygote until the body structures begin to appear in approximately the ninth week of gestation, a developing human is called an embryo. Once the body structures appear, the embryo is then called a foetus, from the ninth week until birth.

The single-celled zygote can divide at a very rapid rate. Initially the cells produced are unspecialized. However, the cells of the zygote rapidly start to differentiate, a process that results in the formation of specialized cells. The number of different cell types that arise from the one original cell can be staggering. This differentiation process is the result of the expression of some genes but not others. Each body cell contains all the genetic information needed to produce the complete organism. However, each cell will develop in a very specific manner depending on which genes become active. What causes some genes to become active depends on the signals that the cell receives.

Human embryonic development

Zygote
(fertilized egg)

2-cell stage
(24 hours)

4-cell stage
(48 hours)

16-cell stage
(72 hours)

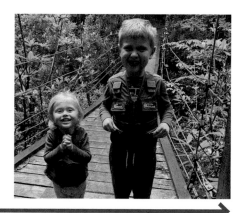

The protein known as bicoid is a morphogen that determines the anterior, head, end of a fruit fly. It is produced through specific directions from the *bicoid* gene of the fruit fly. A fruit fly embryo with defective *bicoid* genes will have posterior structures at both ends.

Cell signalling is the process by which information is transferred from the cell surface to the nucleus of a cell. This signalling process is essential in controlling gene expression and therefore differentiation. **Morphogens** are signal molecules that control cell differentiation. These signal molecules occur in gradients (areas of concentration differences) in different regions of the early embryo. The concentration of the signal molecules controls the regional development of the first cells into head and tail structures. The gradient of the signalling molecule results in different genes being expressed in different parts of the embryo, with the result that different parts of the embryo develop different features. As the embryo develops, other signalling molecules become factors in differentiation.

Cancer cells are examples of cells that undergo extremely rapid reproduction with very little or improper differentiation. The result is a mass of cells (tumour) with no useful function to the organism.

Some cells have a greatly diminished ability to reproduce once they become specialized, or lose the ability altogether. Nerve and muscle cells are good examples of this type of cell. Other cells, including epithelial cells such as skin, retain the ability to reproduce rapidly throughout their life. The cells formed from these rapidly reproducing cells will be the same cell type as the original cell.

How do cells become differentiated?

B2.3.2 – Stem cells

B2.3.2 – Properties of stem cells

Limit to the capacity of cells to divide endlessly and differentiate along different pathways.

There are populations of cells within organisms that retain their ability to divide and differentiate into various cell types. These cells are called **stem cells**. Stem cells retain the ability to divide indefinitely and can differentiate along different pathways, resulting in all the cell types an organism possesses.

Plants contain such cells in regions of **meristematic tissue**. Meristematic tissues are found near root and stem tips. The tissues are composed of rapidly reproducing cells that can become various types of tissue within the root or stem. Gardeners take advantage of these cells when they take cuttings from stems or roots and use them to grow new plants.

B2.3.3 – Stem cell niches

B2.3.3 – Location and function of stem cell niches in adult humans
Limit to two example locations and the understanding that the stem cell niche can maintain the cells or promote their proliferation and differentiation. Bone marrow and hair follicles are suitable examples.

When stem cells divide to form a specific type of tissue, they also produce some daughter cells that remain as stem cells. Figure 1 illustrates a common method stem cells employ when they divide. Note that this method allows continual production of a particular type of tissue while also providing for the continuation of stem cells, a process called **self-renewal**.

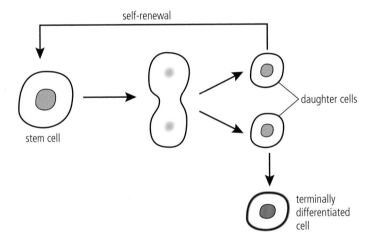

B2.3 Figure 1 Stem cell division

Stem cells have two unique properties.

1. They can self-renew. This is shown in Figure 1. When a stem cell divides, there are several possible outcomes: both daughter cells may remain as stem cells, or a stem cell and a differentiated cell may be formed, or both cells may be differentiated. Whatever the outcome, stem cells are maintained.

2. They can recreate functional tissues. Cells become differentiated when cell signalling ensures that specific genes are expressed as the cell develops.

For stem cell research, scientists examine certain locations or **stem cell niches** in humans. In a stem cell niche the stem cells are present in high numbers as a result of regular proliferation, but they also demonstrate differentiation. Bone marrow and hair follicles are both stem cell niches in humans.

In the bone marrow, the stem cells that produce blood cells are found alongside self-renewing stem cells. As the blood cells are produced, the differentiated cells are transported away via a large array of supporting blood vessels. The renewal process ensures a constant supply of stem cells to continue differentiation.

Hair follicles exist in the skin, and large numbers of epithelial stem cells are found in the bottom, rounded area of a hair follicle. These stem cells are **multipotent**. They are involved with hair growth, skin and hair follicle regeneration, and the production of sebaceous (oil-producing) glands associated with hair follicles.

Stem cell niches in humans have also been studied in the central nervous system, the intestinal system and in muscle fibre bundles. One feature that all these niche areas have in common is the presence of signalling factors that bring about both self-renewal and cell differentiation.

B2.3.4 – Types of stem cell

B2.3.4 – Differences between totipotent, pluripotent and multipotent stem cells

Students should appreciate that cells in early-stage animal embryos are totipotent but soon become pluripotent, whereas stem cells in adult tissue such as bone marrow are multipotent.

In the early 1980s, scientists found and described embryonic stem cells in mice. Similar stem cells have since been discovered in the embryo stage of many organisms. There are different types of stem cells.

Name of stem cell	Major characteristics
Totipotent	Capable of continued division and possesses the ability to produce any tissue in the organism. Very low numbers of cells are totipotent. Only exist in the very early stages of embryo development. They may form a complete organism.
Pluripotent	Arise from totipotent cells and only exist in the early embryonic stage. They can mature into almost all the different cell types that exist in an organism. Unlike totipotent cells, they cannot produce a complete organism.
Multipotent	Only forms a limited number of cell types. Bone marrow tissue that produces different types of blood cell is multipotent. They occur later in the development of the embryo and are present during the remainder of an organism's life.
Unipotent	Only forms a single cell type, such as sperm cells in mammals. They usually form late in the embryonic stage and exist in the functioning organism.

▲

Different types of stem cell

Because of their unique properties, medical experts saw the possibilities of using stem cells to treat certain human diseases. However, one problem discovered early on in stem cell research was that stem cells cannot be distinguished by their appearance. They can only be isolated from other cells based on their behaviour.

Recently some very promising research has been directed towards growing large numbers of embryonic stem cells in culture so that they can be used to replace differentiated cells lost because of injury and disease. This involves **therapeutic**

In the same way that niche environments provide all the support needed for normal stem cell function, malignancies or cancer may also occur as a result of changes in a particular niche. Tumour growth is often associated with blood vessel development. The added vasculature provides nutrients and transport that may increase the severity of the cancer. One approach to anticancer therapy includes attempting to limit this blood vessel support to tumours and cancer regions.

At about 4–6 days into development, totipotent cells specialize and become pluripotent stem cells. In adult tissue, specialization has progressed to the point that multipotent stem cells such as in bone marrow are now present rather than pluripotent cells. Even though unipotent stem cells may form only a single cell type, they retain the property of self-renewal that distinguishes them from non-stem cells.

There has been much sharing of data involving stem cell research. However, many nations have banned or restricted research in this area because of local cultural and religious traditions.

cloning. Parkinson's and Alzheimer's diseases are caused by the loss of proper functioning brain cells, and it is hoped that implanted stem cells could replace many of these lost or defective brain cells, thus relieving the symptoms of the disease. With some forms of diabetes, the pancreas is depleted of essential cells, and it is hoped that a stem cell implant in this organ could have positive effects. However, currently most of the research on stem cells is being carried out in mice: it will probably be some time before this approach to treatment becomes widespread in humans.

Stem cells are being utilized in several ways by scientists around the world. One area of research involves using human embryonic stem cells to understand human development better. This research involves studying cell division and differentiation. Other scientists are using stem cells to test the safety and effects of experimental therapeutic drugs. Information in this area is essential to our understanding of how these drugs might affect differentiating cells in existing organisms. Another interesting area of study involves cell-based therapies, especially as they may have a positive influence on the treatment of diseases and traumas such as Alzheimer's disease, spinal cord injuries, heart disease, diabetes, burns and strokes.

There are ethical issues involved in stem cell research. The use of pluripotent stem cells is particularly controversial. These cells are obtained from embryos, largely obtained from laboratories carrying out **in vitro fertilization (IVF)**. Harvesting these cells involves the death of an embryo, and some people argue that this is the taking of a human life. Others argue that this research could result in a significant reduction in human suffering and is, therefore, acceptable.

Where do you stand in the debate about the nature of stem cell research? How do you feel about the sources of pluripotent stem cells? Should scientific research be subject to ethical constraints or is the pursuit of all scientific knowledge intrinsically worthwhile?

B2.3.5 – Cell size and specialization

B2.3.5 – Cell size as an aspect of specialization

Consider the range of cell size in humans including male and female gametes, red and white blood cells, neurons and striated muscle fibres.

The size of cells and organelles is discussed in Chapter A2.2. The microscope is an essential tool for the study of cells because of their small size. The function of a cell determines how large the cell must be.

What are the advantages of small size and large size in biological systems?

Cell type	Size
Sperm cell	3 μm in diameter, 50 μm in length
Egg cell	120 μm
Fat cell	50–150 μm
Red blood cell	7.5 μm
White blood cell	12–15 μm
Skeletal muscle cell	10–50 μm in width, 40 mm in length
Neuron (nerve cell)	350 μm in length

Sizes of various human cell types

The male and female gametes in humans, the egg and sperm cells, form a zygote during fertilization. This zygote may then go through development, as discussed earlier, to become an embryo, then a foetus, and finally an infant. This development involves a tremendous increase in the number of cells present as well as the differentiation of cells in various regions to become specialized tissues. The sperm cells are relatively small because they only carry out the function of transporting genetic material so that a viable zygote can be formed.

Red blood cells carry oxygen through the organism and have several important adaptations:

- they contain haemoglobin that can combine with and release oxygen
- they have a biconcave disc shape that allows more surface area for oxygen absorption
- they lack mitochondria as well as a nucleus
- they are flexible and size limited because they need to move through narrow blood capillaries.

White blood cells are larger than red blood cells. Their main function is defence against infections. They retain their nucleus throughout their lifetime. There are several distinct types of white blood cells and each type has a specific function. Blood cells are discussed fully in Chapter B3.2. Many possess vesicles with enzymes that can kill microorganisms. The enzymes present are also used in the breakdown of harmful cellular debris brought into the cell by phagocytosis. Their increased size is because of the necessary presence of the nucleus, granules and organelles such as mitochondria.

▲
A human blood smear under a light microscope. The larger cells are the white blood cells. The smaller cells are red blood cells. There are normally 4.2–6.2 million red blood cells per cubic millimetre of blood. In the same cubic millimetre, there are normally 5,000–10,000 white blood cells.

There are several types of neurons (nerve) cells. The distinct types of neuron exhibit adaptations for specialized roles. Motor neurons carry impulses from the brain or spinal cord that allow muscles to respond appropriately to stimuli. This nerve cell type has long fibres called axons that can carry impulses up and down the body over long distances.

A false-coloured micrograph of a motor neuron. The extensions leaving the main cell body allow it to carry impulses in multiple directions. The nucleus is obvious in the cell bodies shown.

The axons of the motor neuron can extend up to 1 m in the human body. This allows efficient and rapid transmission of nerve impulses from the brain and spinal cord to muscles in the limbs to produce movement.

Striated muscle fibres are also specialized cells found in skeletal muscle. Each muscle fibre is a single muscle cell. The fibres are cylindrical and surrounded by membranes capable of impulse propagation. Striated muscle fibres can be up to 12 cm long and are longer than the muscle fibres found in smooth or cardiac muscle. The bands visible in the micrograph (Figure 2) represent units of contraction within the muscle fibre. This muscle type can only produce movement by contraction or shortening. Because the fibre is relatively long, all the units contracting together produce a significant movement.

B2.3 Figure 2 Bands of striated skeletal muscle fibres are shown in this light microscope micrograph. Notice the bands, position of nuclei and lack of plasma membranes in the picture.

As seen with these examples, different cell types not only show unique adaptations involving shape and structure but also in their size. Cell size is largely dictated by two major factors.

1. Basic processes of cell physiology, such as the need for materials to move in and out of the cell. This usually involves the surface area-to-volume ratio, which is discussed in B2.3.6.

2. Cell division apparatus. If cells are too large or too small the mitotic spindle will not function properly.

Cell size is set as the cell goes through its differentiation process to become a particular type of cell within an organism. All the adaptations come together to produce the most efficient cell possible for the specific function it has.

B2.3.6 – Constraints on cell size

Surface area-to-volume ratio. Full details on how to carry out this activity with a worksheet are available in the eBook.

> **B2.3.6 – Surface area-to-volume ratios and constraints on cell size**
>
> Students should understand the mathematical ratio between volume and surface area and that exchange of materials across a cell surface depends on its area whereas the need for exchange depends on cell volume.
>
> **NOS:** Students should recognize that models are simplified versions of complex systems. In this case, surface area-to-volume relationship can be modelled using cubes of different side lengths. Although the cubes have a simpler shape than real organisms, scale factors operate in the same way.

The cell is obviously a small object. You may wonder why cells do not grow to larger sizes. The **surface area-to-volume ratio** of a cell limits the size a cell can reach. In a cell, the rate of heat and waste production, and rate of resource consumption, are functions of (depend on) the volume. Most of the chemical reactions of life occur inside a cell, and the size of the cell affects the rate of those reactions. The surface of the cell, the membrane, controls what materials move in and out of the cell. A cell with more surface area per unit volume can move more materials in and out of the cell, for each unit volume of the cell.

As the width of an object such as a cell increases, the surface area also increases, but at a much slower rate than the volume. This is shown in Table 1: the volume increases by a factor calculated by cubing the radius; at the same time, the surface area increases by a factor calculated by squaring the radius.

B2.3 Table 1 Surface area-to-volume ratio

Factor	Measurement		
Cell radius (r)	0.25 units	0.50 units	1.25 units
Surface area ($4\pi r^2$)	0.79 units	3.14 units	19.63 units
Volume ($4/3\pi r^3$)	0.06 units	0.52 units	8.18 units
Surface area : volume	13.17	6.04 : 1	2.40 : 1

Sphere formulae:

surface area = $4\pi r^2$
volume = $4/3\pi r^3$

This means that a large cell, compared to a small cell, has less surface area to bring in materials that are needed and to get rid of waste. Because of this, cells are limited in the size they can reach and still be able to carry out the functions of life. This means that large animals do not have larger cells; instead, they have more cells.

Nature of Science

Models are useful to help us understand complex systems. For example, you may have looked at surface area-to-volume ratios using cubes that have different lengths of side. Clearly most cells are not regular cubes but this is still a good model because the scale factors work in the same was as they do in cells. The surface area-to-volume ratio relationship can also be modelled using spheres of different radii.

Guiding Question revisited

What are the roles of stem cells in multicellular organisms?

In this chapter we have described how:

- stem cells can reproduce throughout their life
- stem cells can differentiate along different pathways
- totipotent stem cells are the rarest in number and they can differentiate into any type of cell in the organism
- stem cells divide to form a specific type of cell, and they also produce some daughter cells that remain as stem cells
- stem cell niches exist in areas of an organism, and these areas have signaling factors that bring about self-renewal and differentiation of the stem cells present.

Guiding Question revisited

How are differentiated cells adapted to their specialized functions?

In this chapter we have learned how:

- cell signalling is the process that controls gene expression at various stages of embryo and organism development
- morphogens function in the embryo to control regional development of specific cell types
- cell function has a great influence on the size of a cell
- cells involved in secretion are small so that their surface area-to-volume ratio is larger
- many different cells exist in an organism, and all have a size and shape that means they can carry out their essential function efficiently.

Exercises

Q1. The very first embryonic cells would be which of the following kind of stem cell?

A Multipotent.

B Totipotent.

C Differentiated.

D Pluripotent.

Q2. Which of the following cells lacks a nucleus in its differentiated stage?

A Neuron.

B Gamete.

C Muscle cell.

D Red blood cell.

Q3. What normally happens to a spherically shaped cell as it grows larger?

A Surface area-to-volume ratio stays the same.

B Surface area-to-volume ratio decreases.

C Surface area-to-volume ratio increases.

D Its ability to bring in adequate nutrients increases.

Q4. What characteristic do nerve cells possess that allow them to carry out their functions?

Q5. What are morphogens and why are they important?

B2 Practice questions

1. Describe how the properties of the molecules that make up a cell membrane help to maintain the cell membrane structure.

   (Total 3 marks)

2. Describe the process of active transport.

   (Total 3 marks)

3. Explain how a channel protein allows a cell membrane to be selectively permeable.

   (Total 4 marks)

4. Describe the advantages of compartmentalization in the cytoplasm of cells.

   (Total 4 marks)

5. Explain how the surface area-to-volume ratio influences cell sizes.

   (Total 3 marks)

6. Compare and contrast totipotent, pluripotent and multipotent stem cells.

   (Total 3 marks)

THEME

B Form and function

3 Organisms

◀ Blood vessels form a characteristic "tree" pattern as they branch out into a network whilst feeding tissues with oxygen and nutrients. One of the fundamental challenges that organisms faced in their evolutionary development from single celled to multicellular creatures was the distribution of nutrients and subsequent removal of waste products. In large multicellular organisms, such as humans, each cell must have access to molecules that are only available in their environment. Many of those cells are in the interior of the organism and thus cannot rely on direct molecular transfer from and to the outside environment. Both animals and plants have evolved adaptations that allow them to take in molecules from the environment, circulate the molecules to the interior cells within branching vessels, and then use the branching vessels to take waste products to locations that facilitate removal from the organism.

B3.1 Gas exchange

Guiding Questions

How are multicellular organisms adapted to carry out gas exchange?

What are the similarities and differences in gas exchange between a flowering plant and a mammal?

Multicellular organisms have the problem of getting the air molecules that they need from their environment to cells that may be deep within them. Diffusion alone cannot solve this bioengineering problem. Thus, many multicellular life forms have adaptions that combine a respiratory gas exchange system with a fluid transport system. Lungs for gas exchange coupled with a circulatory system for transport is a common example.

Plants, like animals and other life forms, require an exchange of gases with the atmosphere. Photosynthesis typically comes to mind when people think about the life processes associated with plants. Plants are also aerobic organisms and require oxygen for their cells just like animals do. Plants have adaptations that allow the atmospheric gases of both cell respiration and photosynthesis to be exchanged with their environment.

B3.1.1 – The exchange of gases between organisms and their environment

B3.1.1 – Gas exchange as a vital function in all organisms

Students should appreciate that the challenges become greater as organisms increase in size because surface area-to-volume ratio decreases with increasing size, and the distance from the centre of an organism to its exterior increases.

Most organisms are aerobic, meaning that they require oxygen to metabolize energy from organic substances such as glucose. In addition, organisms need to remove metabolic waste products such as carbon dioxide. A few organisms, including many single-celled life forms, can exchange oxygen and carbon dioxide directly with the atmosphere through their plasma membranes. However, that is not an option for larger multicellular organisms where metabolically active tissues may lie deep within the organism and far away from their environment. These organisms have evolved complex adaptations to exchange respiratory gases between the atmosphere or water habitat and their tissues.

The problem of getting gases directly to and from an organism's interior cells is compounded by the **surface area-to-volume ratio**, which changes as an organism gets larger. Surface area is a squared function of its dimensions and that is why we give surface area a square unit (such as cm² or m²). Volume is a cubed function (such as cm³ or m³). Another way of expressing this idea is that the surface area-to-volume ratio decreases with increasing size.

How do multicellular organisms solve the problem of access to materials for all their cells?

Surface area alone cannot be considered an important factor for solving biological problems like providing respiratory gas exchanges. It is the ratio of surface area to volume of the organism that is the important factor. A microscopic protist has a lower surface area than an elephant but the surface area-to-volume ratio of a protist is much higher than that of an elephant.

The volume of an organism is a reflection of its metabolic need to exchange respiratory gases. An organism's ability to take in and release substances is limited by its outer layer surface area. Only the smallest organisms can rely on direct exchange of respiratory gases with their environment, all others must have anatomical and physiological adaptations to get oxygen to internal tissues and take carbon dioxide away.

B3.1.2 – Gas exchange surfaces

B3.1.2 – Properties of gas-exchange surfaces

Include permeability, thin tissue layer, moisture and large surface area.

Organisms that have evolved adaptations for gas exchange must have specialized tissues designed for the molecular exchanges. The specialized tissues are found in the skin of some small organisms, gills of many aquatic organisms, and the lungs of some larger terrestrial organisms. The exchange of gases sometimes occurs between the air and the living tissue (lungs) or between water and the living tissue (gills). In many organisms the gases are immediately exchanged to blood vessels to be circulated to body tissues.

Gas exchange surfaces are characterized by:

- being thin (often only one cell layer), to keep diffusion distances short
- being moist, to encourage gas diffusion
- having a large surface area, for maximum diffusion
- being permeable to respiratory gases (oxygen and carbon dioxide).

These properties allow the maximum volume of gases to be exchanged across the surface in the smallest amount of time.

A freshwater salamander known as an axolotl (*Ambystoma mexicanum*). The axolotl only lives in one small area in Mexico and is critically endangered. This amphibian, even as an adult, has six external gills for respiratory gas exchange with water.

B3.1.3 – Concentration gradients at exchange surfaces in animals

B3.1.3 – Maintenance of concentration gradients at exchange surfaces in animals

Include dense networks of blood vessels, continuous blood flow, and ventilation with air for lungs and with water for gills.

Oxygen and carbon dioxide are exchanged by diffusion. This means that **concentration gradients** must be maintained for oxygen to diffuse into the blood and carbon dioxide out of the blood.

In Figure 1 there are two fluids to take into account for respiratory gas concentrations. One fluid is the environmental water passing over the gill tissue. The other is the blood within the capillaries of the gills. The concentration of respiratory gases in the environmental water does not change as long as the body of water maintains good ecological health and the water is not stagnant around the gills. The concentrations of oxygen and carbon dioxide do change within the blood of the organism, however.

When the blood is first circulated to the gills, it has recently been within capillaries of the muscles and other body tissues. The body cells are continuously respiring, which utilizes oxygen and produces carbon dioxide. The blood that leaves body tissues contains a higher concentration of carbon dioxide and a lower concentration of oxygen compared to levels before the blood reached the active body tissues. The blood will then be transported to the gill tissue and the exchanges occur once again.

Diffusion gradients also explain the gas diffusion that takes place in animals with lungs. Within the lungs are numerous dense capillaries that contain blood that has recently come from respiring body tissues. The concentration of oxygen in the lung capillaries is lower than that of air inspired into the lungs. In addition, the concentration of carbon dioxide in the lung capillaries is higher than that in the air inspired.

Two events must occur to keep concentration gradients in place:

- water must be continuously passed over the gills/air must be continuously refreshed (ventilated) in the lungs
- there must be a continuous blood flow to the dense network of blood vessels in both the body tissues and the tissues of the gills or lungs.

i All animals that use gills are **exothermic** (cold blooded). One of the reasons for this is the relatively high metabolic rates necessary to be **endothermic** (warm blooded). The low oxygen levels available in bodies of water would not support the metabolic rate needed for a constant internal body temperature.

i The only blood vessels that permit the exchange of substances are capillaries. Capillaries are only one cell thick.

B3.1.4 – Gas exchange in mammalian lungs

B3.1.4 – Adaptations of mammalian lungs for gas exchange

Limit to the alveolar lungs of a mammal. Adaptations should include the presence of surfactant, a branched network of bronchioles, extensive capillary beds and a high surface area.

A colorized X-ray showing the anatomy of the respiratory system. The large central air tube is the trachea, which branches into the right and left bronchi leading to each lung. Within the lungs further branching occurs repeatedly, into small bronchioles. At the end of each tiny branch is an air sac called an alveolus. Each lung contains about 300 million alveoli.

B3.1 Figure 2 On the left is a cluster of alveoli connected to the smallest bronchi of a bronchiole tree. Much of the air in each alveolus is refreshed every time you inspire and expire. One alveoli group is shown with a dense surrounding capillary bed. On the right side is a sectioned view of a single alveolus and capillary.

Our lungs have an amazing capacity to expose life-giving air to an incredibly large surface area of gas exchange tissue. The lungs do this by subdividing their volume into microscopic spheres called alveoli. Each **alveolus** is at a terminal end of one of the branches of tubes that started as the trachea. Every time you breathe in (**inspire**) and breathe out (**expire**) you replace most of the air in millions of alveoli.

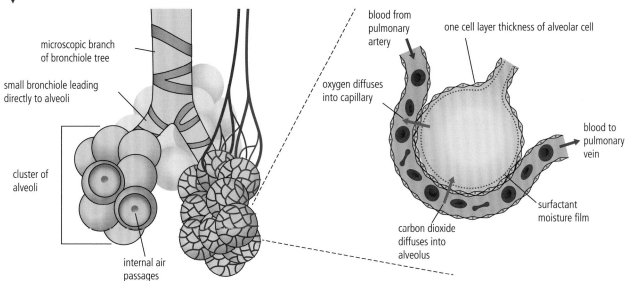

microscopic branch of bronchiole tree

small bronchiole leading directly to alveoli

cluster of alveoli

internal air passages

blood from pulmonary artery

one cell layer thickness of alveolar cell

oxygen diffuses into capillary

blood to pulmonary vein

carbon dioxide diffuses into alveolus

surfactant moisture film

The inner surface of each alveolus is lined by a thin phospholipid and protein film called a **surfactant**. Specific alveolar cells secrete the surfactant, which coats the inside of each alveolus. The surfactant acts to reduce the surface tension of the moist inner surface and helps prevent each alveolus from collapsing each time air is expired.

The inside of each lung is subdivided into several lobes, which are in turn subdivided into the millions of spherical alveoli all connected by small tubes called **bronchioles**. All of the bronchioles are ultimately connected into the trachea for access to inspired and expired air. The spherical shape of the alveoli provides a vast surface area for the diffusion of oxygen and carbon dioxide.

The diffusion of respiratory gases is also helped by the dense network of capillaries surrounding the alveoli. Each alveolus has close access to a capillary (see Figure 2).

Air inspired into the alveoli has a higher concentration of oxygen and lower concentration of carbon dioxide compared to the blood in a nearby capillary. The two types of gas diffuse according to their concentration gradient (see Figure 2). Because capillaries are just one cell thick and each alveolus is just one cell thick, the respiratory gases only need to diffuse through two cells to enter or exit the blood stream. The oxygen-rich blood is now ready to return to the heart to be pumped out into actively respiring tissues. The entire process is ongoing as long as the heart continues to send blood to the capillaries within the lungs, and air continus to be refreshed within the alveoli.

B3.1.5 – Lung ventilation

B3.1.5 – Ventilation of the lungs
Students should understand the role of the diaphragm, intercostal muscles, abdominal muscles and ribs.

We breathe in and out continuously all our lives. The mechanism is a series of events that thankfully usually happens without our conscious thought. The tissue that makes up our lungs is passive and not muscular, therefore the lungs themselves are incapable of purposeful movement. However, there are muscles surrounding the lungs, including the diaphragm, muscles of the abdomen, and the external and internal intercostal muscles (surrounding your ribs). All of these muscles work collectively to either increase or decrease the volume of the thoracic cavity, leading to pressure changes in the lungs.

The mechanism of breathing is based on the inverse relationship between pressure and volume. **Boyle's law** states that an increase in volume will lead to a decrease in pressure, and vice versa. Your lungs are located within your thoracic cavity (or **thorax**). The thoracic cavity is closed to the outside air. Inside the thoracic cavity are the lungs, which only have one opening, through your trachea (via your mouth and nasal passages). The diaphragm is a large, dome-shaped muscle that forms the "floor" of the thoracic cavity. When it contracts it flattens the dome shape and increases the volume of the thoracic cavity. See Figure 3.

B3.1 Figure 3 On the left, contraction of the diaphragm and external intercostal muscles raising the rib cage results in an increased volume of the thoracic cavity. This allows the lungs to expand, lowering their internal pressure. Air enters the nose or mouth and continues down to each of the alveoli. On the right, relaxation of diaphragm and contraction of the internal intercostal results in an increased pressure on the lung. This leads to an expiration of air.

 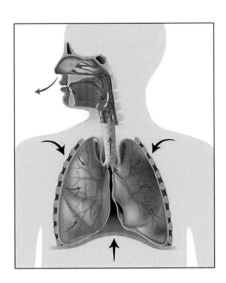

Contraction of the diaphragm increases the volume of the thoracic cavity by flattening the curvature of the muscle.

Following very heavy exercise, the pain you may feel in your midsection may be because you have overused the various muscles involved in frequent and deep breathing. These muscles include the diaphragm, abdominal muscles and intercostal muscles.

Inspiration (breathing in)

Inspiration occurs in a series of steps.

1. The **diaphragm** contracts, increasing the volume of the thoracic cavity.

2. At the same time, the **external intercostal muscles** and one set of abdominal muscles both contract to help raise the rib cage. These actions also help increase the volume of the thoracic cavity.

3. Because the thoracic cavity has increased in volume, the pressure inside the thoracic cavity decreases. This leads to less pressure "pushing on" the passive lung tissue.

4. The lung tissue responds to the lower pressure by increasing its volume.

5. This leads to a decrease in pressure inside the lungs, also known as a **partial vacuum**. Air comes in through the open mouth or nasal passages to counter the partial vacuum within the lungs, and fills the alveoli.

All the steps become more frequent and exaggerated when you are exercising and thus breathing deeply. For example, the abdominal muscles and intercostal muscles achieve a greater initial thoracic volume. This leads to deeper breathing and more air moving into the lungs.

Challenge yourself

1. The process leading to an expiration (breathing out) is the result of the opposite action of muscles compared to inspiration. Using the five steps above, list the steps of an expiration. The layer of muscles that contract and move the rib cage are the **internal intercostal muscles**. The first step is given to you:

   Step 1 The diaphragm relaxes, decreasing the volume of the thoracic cavity.

B3.1.6 – Lung volume

B3.1.6 – **Measurement of lung volumes**
Application of skills: Students should make measurements to determine tidal volume, vital capacity, and inspiratory and expiratory reserves.

A device known as a **spirometer** is used to measure lung volume. A range of air volumes can be measured, including the following.

- Tidal volume – the volume of air that is breathed in or out during a typical cycle when a person is at rest. The term "tidal" volume comes from the idea of an ocean tide coming in and out.
- Inspiratory reserve volume – the maximum volume of air that a person can breathe in (measured from the maximum point of the tidal volume).
- Expiratory reserve volume – the maximum volume of air that a person can breathe out (measured from the minimum point of the tidal volume).
- Vital capacity – the sum of the inspiratory reserve volume, the tidal volume and the expiratory reserve volume.

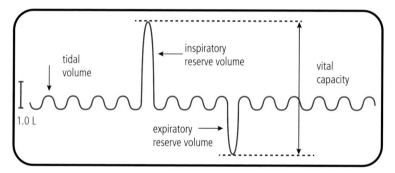

Figure 4 A graphic representation of the various lung volumes that can be measured or calculated with the help of a spirometer.

SKILLS

The IB asks you to take the measurements shown in Figure 4 using an inexpensive spirometer.

Study your spirometer and take note of the units and graduations that you will be using. Most spirometers do not hold a position on these markings, so you need to be able to read numbers from the device very quickly.

1. First, use the spirometer to measure a person's tidal volume. Ask your subject to stand and breathe in and out using the device for about 15 seconds. When you think their breathing pattern is stable, measure the low and high volume readings for each breath. The difference between low and high readings is their tidal volume.

2. Give your subject a rest period of at least one minute. Ask your subject to breathe in and out normally again (recording the tidal volume), then at some point, when they seem comfortable, ask them to take in the maximum volume of air they can. Record this maximum value. Calculate their inspiratory reserve volume by subtracting the maximum reading of their tidal volume from this maximum inspiration volume.

3. Once again give your subject a rest period, so that they can relax. This time ask them to breathe in and out normally (tidal volume) and at some point, when they are comfortable, ask them to release the maximum volume of air out they can. Calculate their expiratory reserve volume by subtracting the minimum reading of their tidal volume from this maximum expiration volume. (Note that not all spirometers can measure the volume of air taken in.)

4. Calculate your subject's vital capacity by summing the inspiratory reserve volume, tidal volume and expiratory reserve volume.

i According to the Institute for Health Metrics and Evaluation (IHME), 13% of deaths worldwide are the direct result of smoking tobacco products, with an additional 2% of deaths the result of second-hand smoking. Lung cancer and chronic obstructive pulmonary disease (COPD) are two of the principal diseases leading to these fatalities.

B3.1.7 – Gas exchange in leaves

B3.1.7 – Adaptations for gas exchange in leaves

Leaf structure adaptations should include the waxy cuticle, epidermis, air spaces, spongy mesophyll, stomatal guard cells and veins.

Plants are adapted to exchange respiratory gases with the atmosphere. A typical leaf is thin, comprising only a few cell layers, so that the diffusion of gases can be quick and efficient. This also permits a relatively large surface area-to-volume ratio for efficient diffusion. There are two primary energy-related processes within plants, cell respiration and photosynthesis. The cells of a plant are always using aerobic cell respiration to synthesize adenosine triphosphate (ATP) molecules for energy-requiring reactions. In addition, when light is available, plants are using photosynthesis to make sugars as fuel for cell respiration. The summary reactions for cell respiration and photosynthesis are the opposite of each other.

Cell respiration: glucose + oxygen $\rightarrow$ carbon dioxide + water

Photosynthesis: carbon dioxide + water $\rightarrow$ glucose + oxygen

The rates of these two sets of reactions are not equal, however. In a plant the rate of cell respiration is fairly constant, while the rate of photosynthesis is heavily dependent on light availability. When conditions are optimal for photosynthesis its rate is far greater compared to cell respiration.

Look at the summary reactions again. During the day, when photosynthesis can occur at a high rate, a plant requires carbon dioxide and releases oxygen. A leaf has various structural adaptions to help facilitate this exchange of gases. Refer to Figure 5 in the next section as you read about these adaptations.

- A **waxy cuticle**: a wax lipid layer that covers the surface of leaves and prevents uncontrolled and excessive leaf water loss by evaporation.
- An upper epidermis: small cells on the upper surface of leaves that secrete a waxy cuticle.
- **Palisade mesophyll**: a densely packed region of cylindrical cells in the upper portion of the leaf. These cells contain numerous chloroplasts and are located to receive maximum sunlight for photosynthesis.
- **Spongy mesophyll**: these loosely packed cells are located below the palisade layer and just above the stomata. They have few chloroplasts and many air spaces, providing a large surface area for gas exchange.
- Veins: these structures enclose the fluid transport tubes called **xylem** and **phloem**. Water moves up from the root system to the leaves in the xylem. Water and dissolved sugars are distributed to other parts of the plant in the phloem. Veins are located centrally within a leaf, to provide access to all the cell layers.
- A lower epidermis: small cells on the lower surface of leaves that secrete a waxy cuticle. Guard cells forming stomata are embedded in this layer.
- **Stomata** (singular stoma): numerous microscopic openings on the lower surface of leaves. Each stoma is composed of two **guard cells**. A pair of guard cells can create an opening or close it, as needed. When open, stomata permit carbon dioxide to enter the leaf and at the same time water vapour and oxygen to exit the leaf. These three gases move by diffusion as a result of their concentration

gradients. At night many plants close their stomata. Their location on the lower surface of leaves limits water loss as a result of transpiration, because the lower surface of leaves experiences lower temperatures compared to the upper surface.

B3.1.8 – Leaf tissue distribution

B3.1.8 – Distribution of tissues in a leaf

Students should be able to draw and label a plan diagram to show the distribution of tissues in a transverse section of a dicotyledonous leaf.

Figure 5 shows how the different tissues are distributed within a leaf.

The IB requires you to be able to draw and label a sectioned plan diagram of a dicotyledonous leaf, similar to Figure 5. However, the internal detail of the cell types shown here is not required for a plan diagram, and a two-dimensional diagram is acceptable.

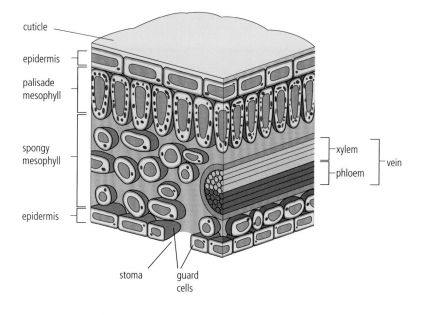

B3.1 Figure 5 A transverse section of a dicotyledonous leaf

B3.1.9 – Transpiration

B3.1.9 – Transpiration as a consequence of gas exchange in a leaf

Students should be aware of the factors affecting the rate of transpiration.

The evaporation of water through open stomata is called **transpiration**. Transpiration is a natural consequence of a leaf's function to accomplish photosynthesis. A leaf must open its stomata to allow carbon dioxide to enter as a reactant of photosynthesis. Excess oxygen is then diffused out while the stomata are opened. The mesophyll area of the leaf is very humid and water will also evaporate through any open stomata. The leaf can open or close its stomata but it cannot filter which gases pass through the openings. The water evaporated can be traced back to the water that entered the roots and has now reached the upper sections of the plant. Transpiration can amount to a significant volume of water when conditions are optimal. The factors that influence the rate of transpiration are described in Table 1.

B3.1 Table 1 Factors affecting the rate of transpiration ▶

Environmental factor	Effect on rate	Explanation
Increased light (see note below)	Increases	Light stimulates guard cells to open stomata. Increased light also stimulates the rate of photosynthesis to increase. Open stomata permit diffusion of carbon dioxide in and oxygen out
Increased temperature	Increases	Increased molecular movement, including increased evaporation of water
Increased wind speed	Increases	Wind removes water vapour at the entrance of stomata, thereby increasing the water concentration gradient between the inside and outside of the leaf
Increased humidity	Decreases	Increased humidity lessens the water concentration gradient between the inside and outside of the leaf

Note that if a lack of light results in stomata being closed, the rate of transpiration will be zero. In that situation changing the other three factors will have no effect.

B3.1.10 – Stomata

B3.1.10 – Stomatal density

Application of skills: Students should use micrographs or perform leaf casts to determine stomatal density.

NOS: Reliability of quantitative data is increased by repeating measurements. In this case, repeated counts of the number of stomata visible in the field of view at high power illustrate the variability of biological material and the need for replicate trials.

Studies have shown that the density of stomata varies between species of plants and even varies within a single species based on long-term environmental factors. In order to study any factor that may be correlated with stomata density, you need to be able to view the stomata and measure the area you are viewing. Stomata density can be expressed as number of stomata mm^{-2} or number of stomata μm^{-2}.

B3.1 Figure 6 A light micrograph of an unknown species of leaf showing paired guard cells forming stomata. The scale bar in the lower right corner allows you to calculate the area of the photograph. ▶

Challenge yourself

2. Calculate the density of stomata (stomata mm⁻²) in Figure 6. You will need a metric ruler. Measure the scale bar showing 100 µm using the metric ruler, and then measure the length and width of the photograph. Use the proportions of the measured photo to the measured scale bar to calculate the actual length and width. Express your stomata count to the nearest whole number of stomata; count any stoma where at least one half of the stoma can be seen. Express your stomata density to the nearest whole number of stomata mm⁻².

Nature of Science

When working with living material, it is normal for a great deal of variation to exist in factors such as stomatal density. Replicated counts can minimize any variation arising from human error.

SKILLS

The IB requires you to use microscopic leaf sections to determine stomata density. This can be accomplished in a variety of ways, and some suggestions for viewing stomata and taking the measurements necessary for calculating the density are given here. These suggestions are not intended to provide step by step directions, but instead are hints that will help you in an investigation of your own design.

- If available, learn to use a micrometre slide to help calculate areas. This is a microscope slide that has very precise length measurements etched into it.
- Alternatively, access online sources for the diameter of field of view at various powers of magnification. Apply the formula for the area of a circle once you know the diameter.
- A cast of a stomata can be made using clear fingernail polish and clear sticky tape. Paint the underside of a leaf with clear fingernail polish and let it dry for 8–10 minutes. Stick clear tape over the dried polish area, gently smooth it to get a good adhesion, and then peel back the tape. Transfer the tape to a clean microscope slide and view it under a light microscope (use clear tape not "magic" tape).
- If available, use a set of micrographs of stomatal tissue that have scale bars included. The scale bar will allow you to calculate the total area of the photograph, as you did in the Challenge yourself.

Guiding Question revisited

How are multicellular organisms adapted to carrying out gas exchange?

In this chapter we have described how:

- multicellular organisms must have adaptations that solve the problem of a decreased surface area-to-volume ratio as organisms gain size
- animals use lungs and gills to exchange respiratory gases between the exterior air and interior tissues
- transport vessels within animals move respiratory gases from the lungs or gills to tissues deep within the organism
- gas exchange surfaces are adapted for efficient diffusion, with continuous blood flow and thin tissue layers
- air within lungs is constantly renewed, with changing volume and pressures within the thoracic cavity.

Guiding Question revisited

What are the similarities and differences in gas exchange between a flowering plant and a mammal?

In this chapter we have discussed how:

- gas exchange of oxygen and carbon dioxide occurs in both flowering plants and mammals
- the exchanges occur in both as a result of diffusion along concentration gradients
- the exchanges occur between thin, moist tissues with minimal cell layers
- mammalian gas exchange occurs between alveoli within the lungs and adjacent blood capillaries
- exchanges in flowering plants occur through multiple small leaf openings called stomata
- flowering plants can stop gas exchanges for periods of time by closing stomata, whereas mammals must have a constant exchange
- during daylight hours, the diffusion of gas exchange in flowering plants shows a net diffusion of oxygen out and carbon dioxide in as a result of photosynthesis
- the diffusion of gas exchange in mammals shows a net diffusion of oxygen in and carbon dioxide out as a result of aerobic cell respiration.

Exercises

Q1. When living organisms are relatively large, the distance from their exterior to the organism's tissues at its centre increases, creating a problem for respiratory gases to be exchanged. A second problem is the small tissue surface area exposed to the air (or water) in relation to the organism's large volume. In your own words, explain this second problem using a comparison of a single-celled amoeba to a multicellular rabbit.

Q2. In terms of diffusion, what happens in a mammal's lungs if its heart stops beating and the blood is no longer transported to and from the capillaries surrounding the alveoli?

Q3. Specific cells inside alveoli secrete a substance called a surfactant. What is the function of this secretion and what could happen if the surfactant was not coating the inside of the alveoli?

Q4. You are looking for a good visual representation of the interior of a small portion of a mammal's lungs. Which of these would be the best physical representation?

    A A balloon blown up with air.

    B Two balloons blown up.

    C A large cluster of grapes.

    D A collection of small boxes arranged to look like a lung.

Q5. What is a spirometer used for?

Q6. Leaves are usually broad, flat and thin. What prevents a leaf from desiccating in the hot sun?

Q7. Assume a leaf is in bright sun and its stomata are open. Identify three other environmental factors that would lead to a relatively high rate of transpiration.

B3.2 Transport

Guiding Questions

What adaptations facilitate transport of fluids in animals and plants?

What are the differences and similarities between transport in animals and plants?

Chapter B3.1 discusses the adaptations of organisms for efficient gas transport within multicellular organisms. In this chapter the focus will be the transport of fluids in both animals and plants.

In most animals there are blood vessels called arteries and veins that circulate fluids. These two types of blood vessels are connected by capillaries, thin-walled microscopic vessels that permit chemical exchanges with the cellular tissues. The heart pumps blood within these vessels.

In plants transport of fluids is accomplished within vessels known as xylem and phloem. The fluid is water with dissolved substances unique to each type of vessel. The two types of vessels use different mechanisms to ensure water movement.

B3.2.1 – Capillaries and chemical exchange

B3.2.1 – **Adaptations of capillaries for exchange of materials between blood and the internal or external environment**
Adaptations should include a large surface area due to branching and narrow diameters, thin walls, and fenestrations in some capillaries where exchange needs to be particularly rapid.

Capillaries receive their blood from the smallest of arteries called **arterioles**. Within body tissues an arteriole branches into what is called a **capillary bed**. This is a network of capillaries that all receive blood from the same arteriole. There are millions of arterioles and capillary beds in your body. A single capillary bed will drain its blood into the smallest of veins called a **venule**.

An illustration of an arteriole feeding blood into a capillary bed that drains blood into a venule. All chemical exchanges in the lungs and body tissues happen within these capillary beds.

When blood enters a capillary bed much of the pressure and velocity of the fluid is lost. Blood cells line up in single file because the **lumen** (inside diameter) of each capillary is only large enough to accommodate one cell at a time. Each capillary is a small tube composed of a single-cell thickness of inner tissue and a single-cell thickness of outer tissue. Both of these cell layers are very permeable to many substances, either through the membranes or between the membranes forming the tube. The total surface area and extensive branching of capillary beds is very high, so no cell in the body is far from a capillary. Some metabolically active tissues in the body are especially enriched with capillary beds. This is referred to as **highly vascular tissue**. Most capillaries exchange molecules within the tissues of an organism, although the capillaries within lungs and gills exchange molecules between the blood and the external environment.

Some tissues have capillary beds that are designed to be even more permeable to substances than a typical capillary. These capillaries are said to be **fenestrated**. The fenestrations are small slits or openings that allow relatively large molecules to exit or enter the blood and allow increased movement of all molecules in a given period of time. Examples of fenestrated capillaries include the numerous small capillaries of the kidneys and areas of the intestine where movement of molecules needs to be rapid.

Capillaries are adapted to their function by:

- having a small inside diameter
- being thin walled
- being permeable
- having a large surface area
- having fenestrations (in some).

B3.2.2 – Arteries and veins

B3.2.2 – Structure of arteries and veins

Application of skills: Students should be able to distinguish arteries and veins in micrographs from the structure of a vessel wall and its thickness relative to the diameter of the lumen.

Arteries take blood away from the heart and veins take blood back to the heart. The identification is not based on the level of blood oxygenation.

Arteries and veins are identified according to whether the vessel receives blood from the heart and takes that blood to a capillary bed (artery) or receives blood from a capillary bed and takes that blood back to the heart (vein).

Because arteries receive blood directly from the heart and the blood is under relatively high pressure, they are lined with a thick layer of smooth muscle and elastic fibres. The lumen of arteries is relatively small compared to veins. Veins receive low pressure blood from capillary beds. They are relatively thin walled with a large lumen to carry the slow-moving blood.

SKILLS You need to be able to identify an artery and a vein from a micrograph similar to Figure 1. The thicknesses of the walls and the lumen size will help you identify the two types of vessels.

B3.2.3 – Adaptations of arteries

B3.2.3 – Adaptations of arteries for the transport of blood away from the heart

Students should understand how the layers of muscle and elastic tissue in the walls of arteries help them to withstand and maintain high blood pressures.

Arteries are adapted to transport high pressure blood away from the heart. The heart contracts and relaxes on a rhythmic schedule. When the heart contracts a surge of blood enters an artery and its branches. Each artery has a relatively thick layer of **smooth muscle** controlled by the **autonomic nervous system** (ANS). The ANS controls those functions in your body that are necessary but not controlled consciously. The smooth muscle changes the lumen diameter of arteries to help regulate blood pressure.

In addition to smooth muscle, the wall of each artery contains the proteins elastin and collagen. The muscular and elastic tissues permit arteries to withstand the high pressure of each blood surge and keep blood continuously moving. When blood is pumped into an artery, the elastin and collagen fibres are stretched and allow the blood vessel to accommodate the increased pressure. Once the blood surge has passed the elastic fibres recoil and provide further pressure, propelling the blood forwards within the artery. In this way the blood in arteries maintains a high pressure between pump cycles of the heart.

pressure stretches elastic fibres

recoil of elastic fibres

▲ High pressure caused by each contraction of the heart pushes outwards on the elastic wall of each artery. In between contractions the stretched area recoils and helps maintain the high pressure characteristic of arteries.

B3.2.4 – Measuring the pulse rate

B3.2.4 – Measurement of pulse rates

Application of skills: Students should be able to determine heart rate by feeling the carotid or radial pulse with fingertips. Traditional methods could be compared with digital ones.

The pulse rate or heart rate is a measurement of the number of times your heart beats in a minute. Each time the heart contracts and sends blood directly into arteries, the "pulse" of pressure described in Section B3.2.3 can be felt in an artery.

You can take your own pulse rate by feeling for the pulse using your index and middle fingers at two possible locations.

- The carotid artery – feel for this artery on either side of your trachea (windpipe) in your neck.
- The radial artery – feel for this artery on your wrist with the palm of your hand facing upwards. You should feel the pulse 2 cm from the base of your thumb.

SKILLS

It may take a little practice, but you can learn to identify the feeling of the arteries as they "pulse" with each heartbeat. Once you are confident in identifying the pulse, use a clock or timer to determine the pulse rate. You can count for the full 60 seconds or alternatively count for 30 seconds and then multiply your count by two. Remember that a data set requires repetition to minimize uncertainty in measurement. Once you have a good data set using the traditional method of feeling for the pulse, compare that data set with one or more generated using digital methods.

B3.2.5 – Adaptations of veins

B3.2.5 – Adaptations of veins for the return of blood to the heart

Include valves to prevent backflow and the flexibility of the wall to allow it to be compressed by muscle action.

Veins are blood vessels that return blood back to the heart after the blood has passed through a capillary bed. Blood loses a great deal of pressure and velocity in capillary beds. To account for this, veins have thin walls and a larger internal diameter. The unidirectional flow of the relatively slow-moving blood in veins is aided by internal valves that help prevent backflow of the blood. In addition, the thin walls of veins are easily compressed by surrounding muscles. One of the many reasons to stay active!

Calf muscle acts as pump for deep leg veins

Valves prevent backflow

Blood flow caused by muscle contraction

Valve open

Valve closed

Calf muscle relaxed

Calf muscle contracted

◀ Contraction of skeletal muscle stimulates blood flow by squeezing veins. Internal valves ensure a one-way flow.

B3.2.6 – Occlusion of coronary arteries

B3.2.6 – Causes and consequences of occlusion of the coronary arteries

Application of skills: Students should be able to evaluate epidemiological data relating to the incidence of coronary heart disease.

NOS: Students should understand that correlation coefficients quantify correlations between variables and allow the strength of the relationship to be assessed. Low correlation coefficients or lack of any correlation could provide evidence against a hypothesis, but even strong correlations such as that between saturated fat intake and coronary heart disease do not prove a causal link.

The heart is a very active and thick muscle. Like any muscle it requires oxygen and nutrients to stay active and healthy. The arteries that supply blood to cardiac muscle are called **coronary arteries**.

◀ A depiction of the coronary arteries (in red) that feed oxygen and nutrients directly into the muscle tissue of the heart.

Coronary heart disease is a term often used to describe the narrowing of coronary arteries by plaque.

Over time a person may develop a build-up of cholesterol and other substances in the lumen of arteries. This build-up is called **plaque** and the restriction in blood flow it causes is called an **occlusion**. As you can see in Figure 2, plaque build-up is progressive and can severely decrease the artery's blood flow. If the occluded artery is a coronary artery, it may result in a heart attack because the cardiac muscle in one or more areas of the heart will be deprived of oxygen.

B3.2 Figure 2 Progressive stages in the build-up of plaque, consisting of cholesterol and other lipids. If blood flow is restricted within one or more of the coronary arteries, an occlusion can lead to a heart attack.

SKILLS

You need to be able to evaluate **epidemiological** data relating to the incidence of coronary heart disease (CHD). Epidemiological studies deal with the incidence, distribution and control of conditions such as heart disease. Many factors have been correlated with CHD, including sex, age, family history, diet, diabetes, hypertension, high cholesterol, weight and smoking. It is very difficult to measure the effects of any one factor and its impact on the incidence of CHD. Almost all factors have an impact on one or more other factors. One of the most important considerations in evaluating data with any of these risk factors is to remember that you are considering a correlation and not a "cause and effect" relationship.

Challenge yourself

Answer the following questions relating to the figure.

1. When comparing sex and smoking only, which factor appears to have a higher correlation coefficient with the predicted 10-year risk of developing coronary heart disease?

2. At what age do the correlations appear to increase most significantly?

3. At what age is the predicted 10-year risk of coronary heart disease 10% for (a) a male non-smoker and (b) a male smoker?

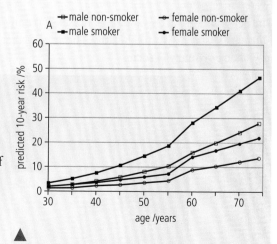

Predicted correlation of age, sex and smoking status with a future 10-year risk of developing coronary heart disease.

4. One of the methods used for collecting the data was a questionnaire. Evaluate the use of this data collection method.

5. Does the data show that smoking causes an increase in coronary heart disease?

Nature of Science

Correlation coefficients quantify correlations between variables and allow the strength of the relationship to be assessed. Low correlation coefficients or a lack of any correlation can provide evidence against a hypothesis, but even strong correlations such as that between saturated fat intake and coronary heart disease do not prove a causal link.

B3.2.7 – Water transport from roots to leaves

B3.2.7 – Transport of water from roots to leaves during transpiration

Students should understand that loss of water by transpiration from cell walls in leaf cells causes water to be drawn out of xylem vessels and through walls by capillary action, generating tension (negative pressure potentials). It is this tension that draws water up in the xylem. Cohesion ensures a continuous column of water.

Unlike many animals, plants do not have a heart to pump fluids for distribution to various tissues. In order to bring water and dissolved minerals up from the roots, a plant relies on a tension force generated by transpiration. Chapter B3.1 describes how transpiration is the evaporation of water from leaves through open stomata. The water is located in the air spaces created by the spongy mesophyll layer of the leaf. The loss of water by transpiration causes water to be pulled through the cell walls of nearby xylem tissue by capillary action. This creates tension (a negative pressure) at the upper end of each xylem tube. The tension results in the movement of water up the xylem, and the entire column of water moves up because of cohesion. This upwards movement of water with dissolved minerals is called the **cohesion-tension theory**. Chapter A1.1 provides the background to this topic.

 In a typical plant more than 90% of the water taken in by the roots is lost by transpiration. A mountain range in the USA is called the Great Smokey Mountains because of the continual haze above it created by transpiration from the abundant trees present.

 In southwest Utah in the United States there is a quaking aspen tree with so many "tree shoots" that it covers more than 43 ha (108 acres). Each shoot appears to be a separate tree, but DNA analysis has shown that all of the "trees" are in fact a single organism sharing a common root system.

The moisture haze above a portion of the Great Smoky Mountains, USA.

B3.2.8 – Adaptations of xylem vessels

B3.2.8 – Adaptations of xylem vessels for transport of water

Include the lack of cell contents and incomplete or absent end walls for unimpeded flow, lignified walls to withstand tensions, and pits for entry and exit of water.

Plants that form wood have many concentric rings of once active xylem tissue. As the plant grows by increasing its girth, the xylem near the outside (under the bark) still conducts water but the interior xylem does not. The interior xylem provides excellent support, allowing the growth of massive trees.

Imagine many cylinder-shaped plant cells stacked up on each other to make a long tube. When alive these cells would have had complete cell walls, plasma membranes and typical plant cell organelles. Now imagine that all of these cells die leaving behind only their thick cylinder-shaped cell walls. Even the end walls where the cells were joined to each other in the tube completely or partially degenerate. This describes the formation of xylem tubes.

The dead xylem tubes have cell walls fortified with **lignin** for strength. The lignin provides resistance to collapse of the tubes because of the tension created by transpiration. The partial or total lack of cell walls between adjoining cells of the xylem tube allows unobstructed water flow upwards. Xylem also has small pits (microscopic holes) in its sidewalls that allow the easy flow of water in and out as needed.

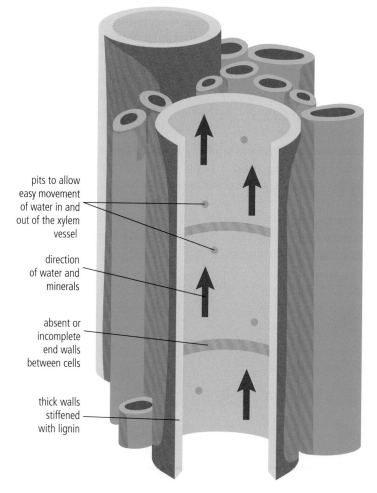

pits to allow easy movement of water in and out of the xylem vessel

direction of water and minerals

absent or incomplete end walls between cells

thick walls stiffened with lignin

A xylem vessel shown in section. Each vessel extends from a root to the upper parts of the plant where the leaves are located.

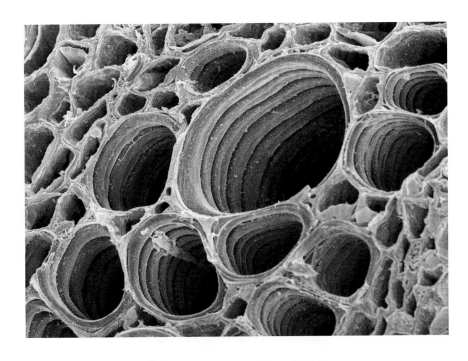

A scanning electron micrograph (SEM) of xylem tissue. You are looking at the dead remains of once living cells that now form tubes. The cell walls are composed of cellulose with a high content of lignin for strength.

B3.2.9 – Tissues in a dicotyledonous stem

B3.2.9 – Distribution of tissues in a transverse section of the stem of a dicotyledonous plant

Application of skills: Students should be able to draw plan diagrams from micrographs to identify the relative positions of vascular bundles, xylem, phloem, cortex and epidermis. Students should annotate the diagram with the main functions of these structures.

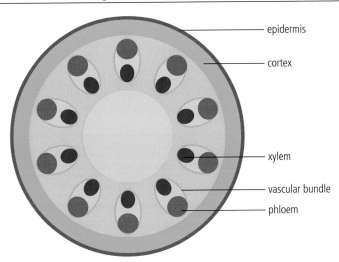

epidermis

cortex

xylem

vascular bundle

phloem

B3.2 Figure 3 A transverse section of a typical dicotyledonous stem. Dicotyledons are one of two categories of flowering plants; monocotyledons are the other category. Only dicotyledons have the arrangement of tissues shown here.

SKILLS

You will be required to draw, label and annotate a transverse section plan diagram of a dicotyledonous stem by looking at microscopic sections of prepared slides. A plan diagram does not show individual cells. Your plan diagram should look similar to the one shown in Figure 3, but keep in mind that actual tissue is likely to be more complex. The annotations should provide brief functions for all the labelled structures. Table 1 outlines the functions of the different tissues.

B3.2 Table 1 A summary of the functions of five dicotyledonous stem tissues

Xylem is the dead leftover walls of cells, while phloem is a living tissue.

Tissue	Function
Epidermis	Prevents water loss and provides protection from microorganisms
Cortex	An unspecialized cell layer that sometimes stores food reserves
Xylem	Transport tubes that bring water up from the roots
Phloem	Transports carbohydrates, usually from leaves to other parts of the plant
Vascular bundle	Contains multiple vessels of both xylem and phloem

B3.2.10 – Tissues in a dicotyledonous root

B3.2.10 – Distribution of tissues in a transverse section of a dicotyledonous root

Application of skills: Students should be able to construct diagrams from micrographs to identify vascular bundles, xylem and phloem, cortex and epidermis.

B3.2 Figure 4 A transverse section of a typical dicotyledonous root.

SKILLS

You will be required to draw, label and annotate a transverse section plan diagram of a dicotyledonous root by looking at microscopic sections of prepared slides. A plan diagram does not show individual cells. Your plan diagram should look similar to the one shown in Figure 4, but keep in mind that actual tissue is likely to be more complex. The annotations should provide brief functions for all the labelled structures. Table 2 outlines the functions of the different tissues.

B3.2 Table 2 Tissues and their functions within a dicotyledonous root

You may be presented with plan diagrams of a dicotyledonous stem and/or root and asked to provide labels and annotations. Practise for this possibility.

Tissue	Function
Epidermis	Grows root hairs that increase the surface area for water uptake
Cortex	An unspecialized cell layer that stores food reserves
Xylem	Transport tubes for water and minerals, starting in the roots
Phloem	Transport tubes that receive sugars from leaves
Vascular bundle	The area in the centre of the root containing xylem and phloem

Guiding Question revisited

What adaptations facilitate transport of fluids in animals and plants?

In this chapter we have described the following in relation to both animals and plants:

- the use of water as a solvent to carry dissolved substances to tissues
- the use of tube-like structures (arteries, veins, xylem and phloem) for fluid transport

- multiple branches of tube structures leading to an increase in surface area in tissues
- the use of fluid pressure to provide the force necessary to move liquids.

Guiding Question revisited

What are the differences and similarities between transport in animals and plants?

In this chapter we have described how:

- both have fluid moving in tube-like structures
- both have mechanisms for creating a pressure differential to move fluid
- both allow the chemical exchanges necessary for life processes
- both increase the surface area by having multiple vessel branches
- animals use blood as a transport medium, while plants use water with dissolved substances.

Exercises

Q1. State whether each of these descriptions applies to an artery or a vein.
- **(a)** Thin muscular wall.
- **(b)** Capable of changing internal diameter (lumen).
- **(c)** Carries blood away from the heart.
- **(d)** Has internal valves to ensure one-way flow.
- **(e)** Carries high pressure blood.

Q2. Data has been collected correlating a high fat diet with eventual occlusions of coronary arteries. Why is it incorrect to say that high fat diets cause heart attacks?

Q3. Why is it important for a plant to have a **continuous** water column all the way from the roots to the leaves?

Q4. Why is it vital for capillaries to have thin walled structures?

Q5. Occlusion of arteries by plaque can lead to a highly diminished blood flow through any artery affected. Which arteries are occluded when someone has a heart attack?

Q6. Blood flow in veins is aided by skeletal muscle contraction of those muscles surrounding a vein. Describe structures in veins that promote a one-way flow of blood back towards the heart.

Q7. Samples of fluids within vascular bundles of a dicot plant were taken and tested for chemical composition. Identify whether the fluid sampled was taken from xylem or phloem if the content of the sample was found to be:
- **(a)** sucrose dissolved in water
- **(b)** a variety of minerals dissolved in water.

Q8. Why is it important for xylem vessels to be fortified with lignin?

B2 Practice questions

1. Outline the process of inspiration in humans.

   (Total 4 marks)

2. Coronary heart disease (CHD) is common in some families, with men being more susceptible to the disease than women. Researchers in Finland carried out an investigation to determine whether the pattern within families was the same for women as for men. The graph shows how the risk of developing CHD in men and women of certain ages depends on whether they had a brother or sister with the disease.

key:
- ■ men whose brothers had CHD
- ● women whose brothers had CHD
- □ men whose sisters had CHD
- ○ women whose sisters had CHD
- ✳ male control
- ✱ female control

   (a) State the risk of a man developing CHD between the ages of 55–59 if his brother had CHD. (1)

   (b) Calculate the increase in risk over the control group for a woman of 60–64 of developing CHD if her sister had the disease. (1)

   (c) Compare the results for the men and the women. (3)

   (d) Suggest **two** reasons why a man is more likely to develop CHD if his brother had the disease. (2)

   (Total 7 marks)

3. (a) Xylem and phloem contain structures that are adapted for transport. Outline the differences between these structures in xylem and phloem. (2)

(b) Explain how the properties of water allow it to move through xylem vessels. (2)

(Total 4 marks)

4. Explain the relationship between the structure and functions of arteries, capillaries and veins.

(Total 9 marks)

5. Explain the need for, and the mechanism of, ventilation of the lungs in humans.

(Total 8 marks)

6. The leaves of plants are adapted to absorb light and use it in photosynthesis. Draw a labelled diagram to show the arrangement of tissues in a leaf.

(Total 6 marks)

THEME

B Form and function

4 Ecosystems

A reticulated giraffe (*Giraffa reticulata*) in its native habitat in Kenya. This species of giraffe and all of the other organisms that live in this ecosystem are well adapted for their environment. With the exception of a few invasive species, organisms have lived in specific ecosystems for many thousands of generations. Each generation is a genetic "package" that provides efficient adaptations for that environment. Evolution provides very small changes to these genetic packages in keeping with the changes that occur to the environment. The living organisms of an ecosystem are highly dependent on each other for survival. Sometimes this can be as simple as one type of organism providing nutrition for another. Other times it involves much more complex interactions that we must study carefully to truly understand the nuances of the interspecies dependencies.

B4.1 Adaptation to environment

Guiding Questions

How are the adaptations and habitats of species related?

What causes the similarities between ecosystems within a terrestrial biome?

Organisms have complex lives. Few people think about the conditions necessary for earthworms, gopher tortoises, giraffes and other species to stay alive, but the environmental conditions they need are numerous and quite specific. A habitat must provide an organism with the basic requirements to stay alive. Organisms develop adaptations over time that allow them to be successful in their environment. The mechanisms for developing those adaptations explain evolution. Any genetic variation that permits a greater survivability in a given habitat will be passed on by the process of reproduction. Habitats change over time, and the adaptations of the organisms that live within those habitats must also undergo change.

A combination of mean annual precipitation and air temperature creates predictable terrestrial land areas called biomes. These biomes are found in various locations on Earth, but the ecosystems each one supports are also predictable, with similar characteristics. Convergent evolution leads to organisms living in each type of biome solving physiological challenges in similar ways.

B4.1.1 – What is a habitat?

B4.1.1 – Habitat as the place in which a community, species, population or organism lives

A description of the habitat of a species can include both geographical and physical locations, and the type of ecosystem.

A **habitat** is a place where organisms live. If more than one species have similar requirements, then a habitat can be a place where a **community** of multiple species lives. Habitats provide the organisms that live there with the basic requirements they need to stay alive long-term. Organisms need shelter, food, water, oxygen and often light.

Habitats can be described by their geographical or physical location *and* by the type of ecosystem they exemplify. Imagine you are visiting the Everglades National Park in Florida, USA, and use your phone or other global positioning system (GPS) device to find your location. You would be working out your geographical location. Even though that GPS location might be accurate, it would give a very incomplete description of your surroundings. More useful information can be provided by describing your surroundings, for example as containing shallow water, large areas of sawgrass plants, alligators, and numerous bird species. Such a description would give others a much better idea of the type of ecosystem you are visiting rather than just your location.

Living organisms do not live in isolation, instead they share habitats with each other. Each living organism has an impact on the other living organisms with which it shares a home.

B4.1.2 – Adaptation to the abiotic environment

B4.1.2 – Adaptations of organisms to the abiotic environment of their habitat

Include a grass species adapted to sand dunes and a tree species adapted to mangrove swamps.

We are going to consider two examples of how organisms have adapted to a relatively harsh abiotic environment.

Sand dune grass species

The sea oat (*Uniola paniculata*) is a species of grass that lives on and creates sand dunes along the eastern seaboard of the United States.

Sea oats (*Uniola paniculata*) helping to form a sand dune. Blowing sand accumulates around the base of the plants. As the sand gets higher, the plants grow higher, to make sure that the seeds stay above the level of the sand. Look for the seed heads on the upper portions of the plants.

Sea oats are drought resistant and, like other dune grasses, have a large shallow root system. They also have narrow leaves, to help reduce transpiration. Sea oats will close their stomata if soil/sand conditions around the roots remain dry for an extended period of time. The sandy "soil" they live in does not hold water for very long, so dense interwoven roots are needed to maximize the take-up of water during the short period of time it is available after rain. This intricate root system is also important because the massive intertwined roots help to hold the sand in place and prevent beach erosion. The reason that sand dunes grow taller is that blowing sand accumulates and is held by the root system of sea oats.

Sea oats thrive in full sun and easily tolerate salt spray; they can even survive complete immersion in saltwater for a short period of time. Sea oat plants produce **nodes** and **rhizomes** near their base, above the sand line. When covered by blowing sand, these asexual growth shoots are stimulated and produce shoots above the newly accumulated sand. Sexual reproduction is accomplished with the production of seed heads that resemble those of a true oat plant.

Mangrove tree species

Red mangrove (*Rhizophora mangle*) is a tropical and subtropical tree that grows along the saltwater tidal zone in Bermuda, Florida, the West Indies and other areas of tropical America. The prop roots of this tree extend above the water line, forming a "spider-like" support system. The roots above the water line also absorb air. The air is used to oxygenate the root tissues, which are below the water line and buried in mud. The roots below the water line filter salt out of the water, so that the tree has access to fresh water. Red mangroves are adapted to the changing water levels characteristic of saltwater tides. The tangled root growth under the trees provides a protective habitat for many fish and other marine animals. Marine animals often use this habitat as a nursery for their young.

Their adaptations for growing in a salty environment allow sea oats and mangroves to live in an environment that is inhospitable to most other plant species.

Red mangroves (*Rhizophora mangle*) are adapted for their saltwater tidal habitat.

Red mangroves produce an unusual fruit, containing a seed that germinates and begins to grow before falling from the parent plant. The young plant is called a **propagule**. The propagule eventually falls from the tree and floats in the water below.

After absorbing water, the propagule orientates itself in shallow water, with its roots downwards (the same orientation as shown in Figure 1), and begins its early root growth. A shoot with early leaves grows from the opposite end. This is an adaptation for plant dispersal in a marine environment.

Mangrove species are now legally protected in most of the areas where they are found. The roots of mangroves prevent erosion and encourage a build-up of sediments. The thickets of roots absorb and dissipate the energy of major storms such as hurricanes, helping to minimize coastal erosion.

B4.1 Figure 1 A propagule of a red mangrove tree. What appears to be a seed pod is actually a young growing tree waiting to drop and begin its life away from the parent tree.

B4.1.3 – Abiotic variables

B4.1.3 – Abiotic variables affecting species distribution

Include examples of abiotic variables for both plants and animals. Students should understand that the adaptations of a species give it a range of tolerance.

Abiotic factors are the non-living components of an ecosystem.

Common abiotic factors include:

- water availability
- temperature range
- light intensity and duration
- soil composition
- pH range
- salinity.

Because of the complexity of habitats, the distribution of living organisms is dependent on many abiotic factors. Any one of those abiotic factors can act as a **limiting factor** if that factor is outside the tolerance zone of an organism. Organisms do not need an abiotic factor to be held at a constant level, but instead adapt to tolerate an acceptable range of values. Figure 2 shows the predicted population size of a species along a gradient for a particular environmental factor.

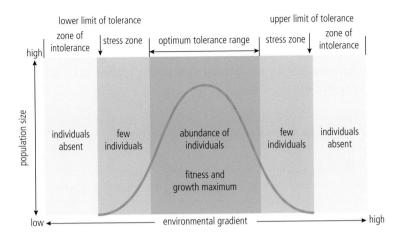

B4.1 Figure 2 The environmental gradient on the *x*-axis could be one of any number of abiotic factors, such as water availability or temperature.

As you can see from Figure 2, abiotic factors can and do affect population sizes, but often the abiotic factor can be far from the optimum before an organism is excluded from an area. Some organisms have developed special adaptations that extend their tolerance range within their habitat. Many species of catfish can take in oxygen through their skin, which permits them to live in oxygen-poor habitats. There are even species of catfish that can burrow into wet mud in order to survive a drought.

Being able to tolerate high or low values for certain abiotic factors can provide habitat opportunities for some organisms. Here are some examples of organisms that tolerate unusual habitats:

- red mangroves – high salinity shorelines
- sea oats – sandy soil along beaches
- polar bear – low air temperatures in arctic regions
- thermophilic bacteria – natural water sources at temperatures of 60–80°C.

Each of these organisms not only has a wide range of tolerance for the abiotic factor listed, but its optimum value is also unusually high or low compared to many other similar organisms. This allows the organism to experience less competition within a given habitat. For example, red mangrove trees have little competition for the saltwater shorelines that they inhabit, as few other plants can live in a high saline water habitat.

What are the properties of the components of biological systems?

B4.1.4 – Limiting factors

B4.1.4 – Range of tolerance of a limiting factor

Application of skills: Students should use transect data to correlate the distribution of plant or animal species with an abiotic variable. Students should collect this data themselves from a natural or semi-natural habitat. Semi-natural habitats have been influenced by humans but are dominated by wild rather than cultivated species. Sensors could be used to measure abiotic variables such as temperature, light intensity and soil pH.

A limiting factor is an abiotic (or biotic) factor that limits the population size or even presence of a particular species in a habitat. It is possible to work out the point at which a factor starts to limit the abundance of a species by carrying out practical experiments. The data collected needs to include a measure of the abundance of the species being studied and the level of the abiotic factor. For example, you could study the presence of a particular type of woodland plant in different light conditions.

SKILLS

You should be able to design and carry out a study where an abiotic limiting factor is correlated to the **distribution** of an animal or plant species. This is best accomplished as a small group or class project.

You can collect this data from a natural or semi-natural habitat. A semi-natural habitat is one that may have been influenced by humans but is still dominated by wild, rather than cultivated, species. The design should be based on counting population numbers along a **transect**. A transect is a scaled line (such as a long tape measure) that is laid along the entire length of the area you plan to investigate. The organism of interest is counted at specific intervals along the transect.

There are several types of transects. Two that you may be interested in using are the **line transect** and **belt transect**. A line transect is usually used to simply determine whether an organism is present or not at set intervals. When using a belt transect, a quadrat is placed at regular intervals along the transect and the number of individuals within each area counted.

Before you start you will need to decide:

- which abiotic factor you will measure (ideally you will choose one that is variable along the transect line)
- which organism will be counted
- where exactly you will set the transect
- how long the transect will be
- what intervals you will use
- what type of transect you will use (if you use a belt transect, how wide the belt will be).

Many of these preparatory steps are interrelated, such as the choice of area and the organism to be counted. For example, a transect running from the edge of a lake to higher ground might be used to measure soil water content and the presence of a native plant species. In an area where there is light and shade, you might want to measure light levels and the abundance of an invasive plant species.

If you carry out this study as a small group or entire class, a discussion of preparatory steps will be helpful. Remember to measure the abiotic factor at each interval, as well as the abundance of the organism. This should allow you to see whether there is a correlation between them, and at which point the abiotic factor becomes a limiting factor for the organism.

A quadrat counting grid placed along a transect line in a coral reef environment. A photograph is being taken of the randomly selected grids, which will be used to count organisms at a later date.

Nature of Science

Measurements can be taken using sensors for data logging. The measurements can be taken rapidly and/or automatically over longer periods of time. Measurements taken this way are accurate and reliable. Sensors are available for light, temperature, pH, carbon dioxide levels, and many other abiotic factors.

B4.1.5 – Coral reef formation

B4.1.5 – Conditions required for coral reef formation

Coral reefs are used here as an example of a marine ecosystem. Factors should include water depth, pH, salinity, clarity and temperature.

Coral reefs are found in less than 1% of the ocean's surface area yet, amazingly, an estimated 25% of all marine species live in and around coral reefs. Corals are the result of a **symbiotic** relationship between coral polyps and a microscopic algae called zooxanthellae. Both organisms in this mutualistic relationship require suitable growing conditions. The small size of the ocean surface area populated by coral reefs is an indication that the combination of all the right abiotic factors for these symbiotic species is rare.

Abiotic factor	Limiting effect
Water depth	Light only penetrates to relatively shallow depths. Zooxanthellae are photosynthetic and require adequate light levels. Most of the ocean floor is too deep to allow enough light to support coral reef growth.
Water temperature	Corals only survive in a narrow range of water temperatures (between 20°C and 28°C). Global warming is resulting in temperatures that are too warm for corals to tolerate. When the water becomes too warm, corals become stressed, and they expel the symbiotic zooxanthellae living in their tissues. Bleached coral is the result.
Salinity	Corals need the correct amount of salt in the water around them. Areas with freshwater run-off may not be of the correct salinity.
Water clarity	Water needs to be clear for light to pass through it. If there is a lot of sediment or pollution in the water, the clarity decreases and the zooxanthellae may not receive enough light.
Water pH	Increased carbon dioxide from fossil fuel emissions is being absorbed into ocean water, resulting in a lowered pH that is detrimental to coral growth. A lower pH (**acidification**) results in less calcium carbonate compounds being available in the water for corals to use when building reefs.

▲ Abiotic factors affecting the growth and health of coral reefs. Reefs have a range of tolerance for each of these factors, but human activities are resulting in the upper or lower tolerance limits for many of the factors being exceeded.

Atmospheric and ocean carbon dioxide content plotted with ocean pH over a selected period of time. Carbon dioxide is given in parts per million (ppm). As carbon dioxide is in solution in the ocean, the value given is the partial pressure of carbon dioxide (partial pressure is the pressure of one gas within a mixture of gases).

Challenge yourself

Use the figure below to answer the following questions.

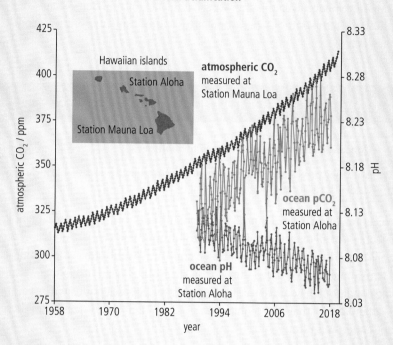

Ocean acidification

1. What is the apparent correlation between carbon dioxide in the atmosphere and the partial pressure of carbon dioxide in the Pacific Ocean around the Hawaiian islands?

2. Why do you think a mountain top in the Hawaiian islands was chosen as a long-term monitoring station for carbon dioxide?

3. Atmospheric measurements show regular cycles, with a minimum and maximum carbon dioxide level each year. Suggest a reason why these cycles are so consistent.

4. Does this data show a causal link between atmospheric carbon dioxide levels and ocean pH?

5. Some people suggest that oceans are a good "sink" or reservoir for excess carbon dioxide in the atmosphere. Describe one potential issue with increased carbon dioxide levels in oceans.

B4.1.6 – Terrestrial biomes

B4.1.6 – Abiotic factors as the determinants of terrestrial biome distribution

Students should understand that, for any given temperature and rainfall pattern, one natural ecosystem type is likely to develop. Illustrate this using a graph showing the distribution of biomes with these two climatic variables on the horizontal and vertical axes.

A **biome** is a large geographical area that contains communities of plants and animals that are adapted to living in that environment. Biomes are often named after the dominant vegetation type that is found within the biome. For example, a grassland biome contains many different grasses. The desert and tundra biomes are exceptions, but biologists can still predict what plant and animal species will be present in those biomes. Biomes of any one type can be found in various locations on Earth, because they are characterized by specific temperatures and rainfall levels, which are not restricted to one geographic location. Deserts, for example, are found in Africa, Asia, America and Australia. Biomes can be subdivided based on other environmental conditions. For example, there are hot deserts and cold deserts. All tropical forest biomes have plentiful rain, but some have more than three times the rainfall that others receive. For any given temperature and rainfall pattern, one natural ecosystem type called a biome is likely to develop.

Because biomes are created by varying conditions of precipitation and temperature, they can be plotted on a graph using the two environmental conditions as the horizontal and vertical axes.

B4.1 Figure 3 Biomes identified by mean annual precipitation and mean annual temperature. Taiga is also known as boreal or coniferous forest. Notice that some environmental conditions lead to an overlap of biomes.

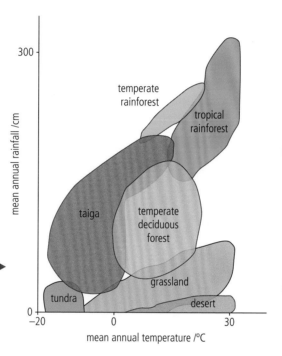

Spend some time making sure that you understand Figure 3, including what is shown on each axis. Note that the average annual temperature starts at −20°C.

Average annual temperatures and average annual rainfall determine global biomes.

B4.1.7 – Biomes, ecosystems and communities

B4.1.7 – Biomes as groups of ecosystems with similar communities due to similar abiotic conditions and convergent evolution

Students should be familiar with the climate conditions that characterize the tropical forest, temperate forest, taiga, grassland, tundra and hot desert biomes.

Biomes have identifiable abiotic characteristics but any one type of biome can be scattered across many different locations on Earth. These locations do not usually share geographic borders. Biomes can contain many ecosystems. An ecosystem is made up of the physical environment and the plants and animals that live there and interact with each other. Biomes are often much too large for each set of shared interactions.

The plants and animals found in similar biomes that are geographically separated will have different genetic backgrounds. If you visit a desert community anywhere in the world you will probably find similar organisms in each. However, while their **morphology** and **physiology** will be quite similar these organisms will usually have little genetic similarity.

Left: A sundew plant (*Drosera capensis*) secretes a sticky digestive juice on the ends of filaments to entrap and digest a fly. The fly will be used as a source of nitrogen.

Right: Pitcher plants (*Nepenthes spp.*) have evolved a jug-like shaped container that holds digestive secretions in its base. The plants produce a slippery substance around the lip of their container. Insects slide from the lip into the container, where they are digested. This provides a source of nitrogen for the plants.

▼

The reason for this similarity is a type of evolution called **convergent evolution**. Convergent evolution occurs when two or more organisms solve an environmental problem by independent (unshared) genetic adaptations. Similar species that live within the same ecosystem are often genetically related to each other as often they are a result of **adaptive radiation.** Thus, they have a fairly recent common ancestor and a very similar set of genes. In contrast, species that live in the same type of biome but in different parts of the world may solve challenges in a similar way but have distinct genetic differences. Species evolve in order to adapt to challenges, and can do so by different sets of "trial and error" adaptations that lead to a very similar solution. The adaptations that work best will be similar in different locations because the abiotic conditions they are responding to are similar.

Carnivorous plants of different species have independently solved the challenge of living in poor soils by developing adaptations to capture and digest insects as a source of nitrogen. These different adaptations for the same purpose are examples of convergent evolution.

Convergent evolution will never result in two or more species becoming one.

How much has evolutionary theory advanced since molecular biology has been used to determine the relatedness of organisms? Common structural features of organisms were originally the only evidence that indicated organisms were related. Convergent evolution often gives a false representation of relatedness between organisms. DNA similarities can not only show relatedness, but also provide evidence about how long ago two organisms diverged in their evolutionary path.

Biome	Climatic conditions	Communities include
Hot desert	Very low annual rainfall (less than 300 mm per year), hot temperatures during the daytime but cold at night.	Sparse vegetation, often with spines for leaves, burrowing animals only active during the cooler night time.
Grassland	Semi-arid climate with somewhere between 500 mm and 950 mm rainfall per year. Temperatures vary depending on latitude. The annual range can be between −20°C and 30°C. Grasslands can have seasons (i.e. a wet season and a dry season).	Vegetation dominated by grass species. Little significant tree growth because of the lack of water. Animal species dominated by grazers and few predators.
Tundra	Cold temperatures (between −40°C and 18°C). Low precipitation (150 mm to 250 mm per year). In the winter it is dark for long periods.	No trees because of the lack of water and short growing season. The soil is frozen for most of the year. Animals have adapted to hibernate for long periods of time or to migrate when the conditions on the tundra become too difficult.
Taiga or conifer forest	Very cold winters and relatively high precipitation in the form of snow. Temperatures can range from −40°C to 20°C. Usually between 300 and 900 mm of rain per year.	Evergreen forests dominated by conifer trees. Animals must have adaptations for a very cold climate. The largest terrestrial biome on Earth by landmass.
Temperate forest	Four seasons with no extremes of temperatures, abundant year-round precipitation (somewhere between 750 and 1500 mm of rain). Soil enriched by leaf drop each year.	Area dominated by **deciduous** broad-leafed trees. Rich variety of animal species.
Tropical forest	High annual rainfall (from 2,000 mm to 10,000 mm per year). Warm temperatures (around 20°C to 25°C). Nutrient-poor soil as plants are rarely deciduous.	Very high plant and animal biodiversity.

▲
Climatic conditions and community types typical of biomes

B4.1.8 – Hot deserts and tropical rainforests

B4.1.8 – Adaptations to life in hot deserts and tropical rainforest
Include examples of adaptations in named species of plants and animals.

We are going to consider examples of how organisms have adapted to hot deserts and tropical rainforests.

Hot desert biome adaptations

The saguaro cactus

The saguaro cactus (*Carnegiea gigantea*) is native to the Sonoran desert in southwestern United States. As a desert plant, most of its adaptations are related to water gathering and retention. The thick waxy skin is completely waterproof and is covered by bristles as a defence against grazers. The saguaro cactus has a single long taproot that it sends down to retrieve deep water when it is available, but it also has a massive shallow root system to absorb occasional rainwater. After a rare rain shower, the water taken up is stored in sponge-like tissue. This stored water maintains the low water needs of the plant until the next rainfall.

The growth of the saguaro cactus is very slow. At ten years of age, its height is only about 2 cm. The cactus will not reach its full height of about 14 m until it is about 200 years old.

207

The saguaro cactus (*Carnegiea gigantea*), sometimes called the organ pipe cactus. The larger cacti shown in this photograph will be more than 100 years old.

The fennec fox

The fennec fox (*Vulpes zerda*) is a small fox native to the desert areas of the Sahara in North Africa. It has many adaptations for desert life, the most notable being its very large ears. The ears are highly **vascular** and help dissipate heat. In addition, the large ears help the animal locate small prey animals moving underground. Like most desert animals, fennec foxes are nocturnal hunters. They spend their daylight hours in large underground dens shaded from the Sun. They obtain their water primarily from their food, although they will drink from a water source, if available. Fennec foxes have kidneys adapted to reabsorb most of the water that passes through them, and they only rarely urinate.

A fennec fox (*Vulpes zerda*) showing its characteristic large, highly vascular ears.

Sometimes a species is introduced into a new habitat that is very well suited to that habitat, but is also very damaging to that habitat. Europe alone is estimated to have over 100 terrestrial vertebrates, 600 terrestrial invertebrates, and 300 aquatic species that are invasive.

Tropical rainforest biome adaptations

The kapok tree

A tropical rainforest is characterized by high temperature and abundant rainfall. In tropical latitudes there is also plentiful sunlight. Collectively, those abiotic factors lead to abundant and varied plant growth. The abundant growth creates competition for available sunlight, and those species that can grow the tallest will have access to more sunlight. One of the species that can grow very tall is the kapok tree (*Ceiba pentandra*). This tree forms part of the upper canopy layer of rainforests in Costa Rica and the Amazon. In order to support rapid growth and a very tall trunk in a relatively shallow soil, a kapok tree makes a strong foundation from **buttress roots** that extend above ground.

Is light essential for life?

Indigenous people in the Amazon use kapok fibre from the seed coats to wrap around their blowgun darts. The fibres create a seal that allows pressure to build-up before the dart is forcefully expelled through the blowgun tube.

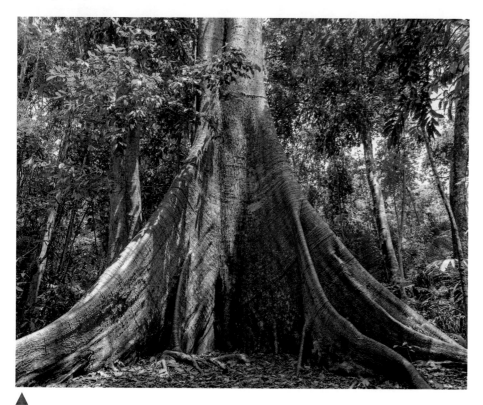

The buttress roots of a kapok tree (*Ceiba pentandra*) are an adaptation that creates a strong foundation for the trees to grow very large in shallow rainforest soil.

Poison-dart frogs

Poison-dart frogs are well adapted to their tropical rainforest environments. As amphibians, they must reproduce by laying their eggs in water. The rainforest provides small pools of water inside the many bromeliad plants that are found in the canopy. Predators such as snakes and lizards are numerous, but these small frogs have developed highly toxic chemicals in their skin as a result of their diet of poisonous insects. They have evolved to have very bright colours and body patterns as a warning to predators. This is known as warning coloration because the predators have coevolved an instinct to avoid brightly coloured frogs.

A poison-dart frog species (*Dendrobates tinctorius azureus*) showing bright coloration and distinctive body pattern. Predators evolve instincts that help them avoid prey animals with this type of bright coloration.

The toxic chemicals produced by poison-dart frogs are used by Indigenous peoples to coat the darts they use in blowguns for hunting.

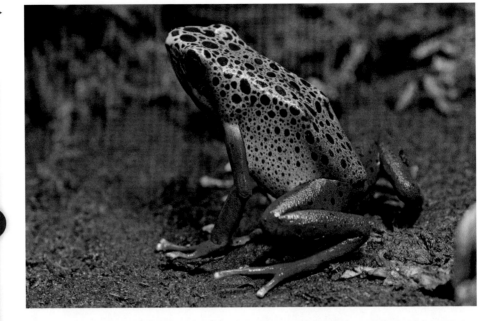

Guiding Question revisited

How are the adaptations and habitats of species related?

In this chapter you have learned that:

- different habitats offer various ranges of abiotic factors
- organisms often have a range of tolerance for any one abiotic factor
- any one abiotic factor can exclude a species from a habitat if the factor is outside the range of tolerance of those organisms
- adaptations help to extend the range of tolerance for organisms, for example
  - a kapok tree can grow to great heights because of its buttress roots
  - a fennec fox can tolerate a hot desert environment because of its large and highly vascular ears
  - a saguaro cactus can survive drought periods in the desert because of its adaptations to store water
  - carnivorous plants can survive nitrogen-deficient soils because they can capture and digest insects.

Guiding Question revisited

What causes the similarities between ecosystems within a terrestrial biome?

In this chapter you have learned that:

- a biome is the largest geographic biotic unit
- biomes are often named after their dominant vegetation type
- biomes are created by predictable rainfall levels and temperature ranges
- the organisms making up the ecosystems within any type of biome (e.g. tropical rainforest) are likely to have many similar adaptations even if they have limited genetic similarity

- this is because the organisms are all adapting to the same environmental conditions
- convergent evolution is the driving force for these similar adaptations
- convergent evolution leads to similar solutions to challenges within similar ecosystems.

Exercises

Q1. Identify three adaptations of sea oats that make this species well suited to growing along beach shorelines.

Q2. Describe what is meant by the optimum tolerance range as related to an organism and an abiotic factor?

Q3. Why are there desert biomes in many different locations around the globe? Choose one of the following answers.

    A Deserts are often surrounded by mountainous regions.

    B Deserts are found in any area of land with very low annual rainfall.

    C Deserts are inhabited by organisms with water conservation adaptations.

    D Deserts are found in any area of land that has very high temperatures.

Q4. Explain the relationship between increased atmospheric carbon dioxide levels and the pH of the oceans.

Q5. Outline why water clarity is important to coral reefs.

Q6. Some species of tropical rainforest frogs have bright colours and obvious body markings even though they are not poisonous to predators. Suggest a reason why these adaptations are still an effective defence against predators?

B4.2 Ecological niches

If all organisms ate the same types of food, there would be considerable competition for the food. Specializing in one category of food or method of feeding ensures that the number of competitors for a species is reduced. Making your own food, as photosynthetic organisms do, allows organisms to be somewhat more independent. However, we will see that even these organisms need help getting certain nutrients such as nitrogen. If an organism can become well adapted to what and how it feeds, it can occupy a niche that no other organism occupies, and hopefully flourish as a consequence.

In order to occupy a specific niche in an ecosystem, organisms have to adapt. Physical adaptations, and sometimes behavioural adaptations, are crucial. The morphology of an organism's body and teeth, or stems, roots and leaves, allows it to obtain certain resources from its environment. Physical and behavioural adaptations help predators find, pursue and kill prey, but also help the prey to hide, deter predators or escape. In this way, each species plays a particular role within an ecosystem.

B4.2.1 – Species and ecosystems

B4.2.1 – Ecological niche as the role of a species in an ecosystem

Include the biotic and abiotic interactions that influence growth, survival and reproduction, including how a species obtains food.

Each species plays a unique role within a community

The unique role that a species plays in the community is called its **niche**. The concept of niche includes where the organism lives (its spatial habitat), and what its role is in nature: what and how it eats (its feeding activities), and its interactions with other species. The ecologist Eugene Odum once said "If an organism's habitat is its address, the niche is the habitat plus its occupation".

Spatial habitat

Every type of organism has a unique space in an ecosystem. The physical area inhabited by any particular organism is its **spatial habitat**. For example, leopard frogs (*Rana pipiens*) live in the ponds in dunes in Indiana, USA. They burrow in their spatial habitat, which consists of mud in between the grasses on the edge of the ponds. Some of the aspects of an organism's habitat are **abiotic factors**, meaning they are made up of non-living things. Sand and water are two important components of the mud where leopard frogs live, and are both abiotic. Abiotic factors also include sunlight, soil type, pH and temperature.

The leopard frog (*Rana pipiens*)

There are also **biotic factors** in an ecosystem, meaning other living organisms. Biotic interactions can include feeding relationships, the provision of shelter such as nest sites, or the presence of parasites within the environment. Feeding relationships can be complex and involve many other species in the ecosystem. For example, an organism may be in competition with another organism for the food supply. It may itself be the prey for a larger predator. It may harbour parasites in its intestines. These interactions are complex and difficult to piece together, but they indicate the important role of the organism in the ecosystem. Interactions between the abiotic factors and biotic factors greatly influence the health, growth, survival rate and reproduction of an organism. A change in acidity of rainwater, for example, could greatly affect the ecosystem where the leopard frog lives, and if one organism is affected by this change it can have an effect on others, because the biotic and abiotic factors are all connected.

B4.2.2 – Obligate anaerobes, facultative anaerobes and obligate aerobes

B4.2.2 – Differences between organisms that are obligate anaerobes, facultative anaerobes and obligate aerobes
Limit to the tolerance of these groups of organisms to the presence or absence of oxygen gas in their environment.

Many species on Earth not only survive perfectly well without oxygen gas, some are poisoned by it. How well a species reacts to the presence of something in its environment is called **tolerance**. An organism with low tolerance to oxygen gas means it does not survive well when the gas is present. No tolerance at all would lead to death in the presence of oxygen.

The chemical transformation of food nutrients into energy that requires oxygen is referred to as **aerobic respiration**. The chemical transformation of food into energy that does not require oxygen is called **anaerobic respiration**. Some organisms are adapted to do one or the other, and some can do both.

Obligate anaerobes

Obligate anaerobes are single-celled organisms that have no tolerance to the presence of oxygen and are poisoned by it. The prokaryotes present on Earth for the first billion years of life were intolerant of oxygen. Initially that was not a problem, because it was not until photosynthesis evolved that oxygen started to collect in the atmosphere and in the water. Today, to escape from Earth's atmosphere, obligate anaerobes live in places where the air cannot reach them, such as in soil, deep water or the intestines of animals, including humans.

Facultative anaerobes

Facultative anaerobes are capable of carrying out both anaerobic and aerobic respiration. Baker's yeast (*Saccharomyces cerevisiae*) is a single-celled fungus that can use oxygen to convert sugar to energy when oxygen is available. When oxygen is not present, the yeast cells can switch to anaerobic respiration. They are neither hurt by nor killed by the presence of oxygen.

Sourdough bread is made using a starter, visible in the glass container, that contains yeast as well as food for the yeast, such as wheat flour. The bubbles in the starter as well as the bread are filled with carbon dioxide gas that the yeast cells have produced.

One example of a sudden drop in oxygen levels that can be detrimental to aerobic organisms is when a power station releases warm water into a river. Gases dissolve better in cold water, so if the water is suddenly warmed, the oxygen level will drop.

Obligate aerobes

Obligate aerobes require oxygen and cannot convert food nutrients into energy without it. If oxygen in their environment is greatly reduced (**hypoxia**) or absent (**anoxia**), these organisms die. Fish, such as trout, that live in freshwater streams, survive well when dissolved oxygen levels in the water are between 7 and $12\,\mathrm{mg\,L^{-1}}$. But if the dissolved oxygen levels drop below $3\,\mathrm{mg\,L^{-1}}$, the fish could die. Unlike yeast cells, fish cells in hypoxic environments cannot switch to anaerobic respiration.

B4.2.3 – Photosynthesis

B4.2.3 – Photosynthesis as the mode of nutrition in plants, algae and several groups of photosynthetic prokaryotes

Details of different types of photosynthesis in prokaryotes are not required.

Roughly a thousand million years after life first evolved on Earth, photosynthesis began. Bacteria living 3,500,000,000 years ago developed the ability to convert carbon dioxide and water into sugar using energy from sunlight. The waste product of photosynthesis is oxygen gas, O_2, and it is thanks to photosynthetic organisms that we have oxygen in the atmosphere today. The process of photosynthesis provides a remarkable bridge between non-living matter and organic matter as it transforms air and water into food.

For about 3 thousand million years, the only organisms photosynthesizing were single-celled organisms. We can still find various forms of photosynthetic bacteria today. Examples include not only cyanobacteria but green sulfur bacteria such as those in the genus *Chlorobium*, or purple bacteria such as those in the genus *Rhodospirillum*.

Microscopic view of filaments of cyanobacteria that can photosynthesize.

In the last 400 million years or so, other organisms, such as algae and, of course, aquatic and terrestrial plants, have adopted this type of metabolism. Algae are eukaryotes and can be single-celled, such as *Chlorella vulgaris*, or they can be multicellular such as kelp (e.g. *Ecklonia maxima*), a type of seaweed. In these organisms, the green pigment **chlorophyll** is used for the process of photosynthesis (see Chapter C1.3).

Chlorophyll is also found in aquatic plants and terrestrial plants and it is what gives leaves their green colour. The vast majority of ecosystems in the world get their initial energy from sunlight, which is used to convert inorganic matter into food using photosynthesis. If you greatly reduced or turned off sunlight from our planet, most of life as we know it would die. Almost all living things rely either directly or indirectly

In 1815, a volcano in Indonesia sent up so much volcanic dust into the atmosphere that a haze persisted in the sky for over a year, causing temperatures to drop worldwide. 1816 was declared the "year without a summer", and a massive reduction in solar energy reaching the surface meant reduced photosynthesis. This caused crop failures and led to famine in many countries.

on energy from the Sun. Organisms that can make their own food from inorganic substances using techniques such as photosynthesis are called **autotrophs**. Because they not only produce food for their own growth but can also be eaten by other organisms, they are often referred to as **producers**.

B4.2.4 – Holozoic nutrition

B4.2.4 – Holozoic nutrition in animals

Students should understand that all animals are heterotrophic. In holozoic nutrition food is ingested, digested internally, absorbed and assimilated.

Organisms need to either make their own food or get it from other organisms. In Section B4.2.3 we saw that photosynthesis means that certain organisms are able to make their own food. Organisms that cannot make their own food but rely on eating other organisms are called **heterotrophs**.

Examples of heterotrophs include:

- zooplankton
- sheep
- fish
- birds.

There are different types of heterotrophs, depending on how they get their nutrients. **Holozoic nutrition** refers to a way of getting nutrients by ingesting all or part of an organism. The eaten organism's parts are ingested and broken down into nutrients (digested) that can then be absorbed into the bloodstream (absorption) and used within the body (assimilation). Humans use holozoic nutrition, as do our pet cats and dogs. Organisms that obtain their food in this way are called **consumers**.

B4.2.5 – Mixotrophic nutrition

B4.2.5 – Mixotrophic nutrition in some protists

Euglena is a well-known freshwater example of a protist that is both autotrophic and heterotrophic, but many other mixotrophic species are part of oceanic plankton. Students should understand that some mixotrophs are obligate and others are facultative.

Organisms that are both autotrophic and heterotrophic are capable of making their own food *and* ingesting nutrients from other organisms. Such a mode of nutrition is called **mixotrophic nutrition**. This is useful if levels of sunlight are too low at times to support the organism through photosynthesis alone, and there is not enough food in the environment for a heterotrophic existence. The genus *Euglena* is made up of species that are single-celled protists that have photosynthetic pigments but also can ingest food from the water around it. This is an example of mixotrophic nutrition.

Euglena spirogyra is a eukaryote and has membrane-bound chloroplasts filled with green chlorophyll.

Obligate mixotrophs need both systems to grow and thrive. **Facultative mixotrophs** can survive on one system but use the other as a supplement.

 A lot of different terminology is used in this chapter. Make sure you know what unfamiliar terms mean. As we saw with both anaerobes and mixotrophs, when the term obligate is used, it means that the organism must use that mode and no other. When the term facultative is used, it means that the organism can sometimes use that mode but is capable of using another.

 What are the relative advantages of specificity and versatility?

 Humans can get all the vitamin D they need from a healthy, well-balanced diet. But, facultatively, we can obtain vitamin D by exposing our skin to ultraviolet rays.

B4.2.6 – Saprotrophic nutrition

B4.2.6 – Saprotrophic nutrition in some fungi and bacteria

Fungi and bacteria with this mode of heterotrophic nutrition can be referred to as decomposers.

Organisms called **saprotrophs** live on or in non-living organic matter, secreting digestive enzymes and then absorbing the products of digestion. Saprotrophs play an important role in the decay of dead organic materials. The fungi and bacteria that are saprotrophs are also called **decomposers**, because their role is to break down waste material. A mushroom that is growing on a fallen tree is secreting enzymes into the dead tissue of the tree trunk, in order to break down the complex molecules within the tree tissue. The mushroom then absorbs the simpler energy-rich carbon compounds that are released by the action of the enzymes. Slowly, over time, the tree trunk decomposes as the molecules inside the wood are digested and reused.

Fungi such as mushrooms on the forest floor are saprotrophs, helping to decompose material such as this fallen tree.

 Holozoic consumers ingest part or all of an organism by swallowing and extracting nutrients using their gut. Saprotrophic decomposers release enzymes onto their food and the digestion happens outside their bodies. They then absorb the digested nutrients.

B4.2.7 – Diversity of nutrition in archaea

B4.2.7 – Diversity of nutrition in archaea

Students should understand that archaea are one of the three domains of life and appreciate that they are metabolically very diverse. Archaea species use either light, oxidation of inorganic chemicals or oxidation of carbon compounds to provide energy for ATP production. Students are not required to name examples.

Living things can be classified into three domains: Bacteria, Archaea and Eukarya. Organisms in the domain Archaea show remarkable diversity in the methods they use to obtain nutrients and energy, including:

- photosynthesis – generating cellular energy with the help of sunlight
- chemosynthesis – generating cellular energy from reactions involving inorganic molecules (without the help of sunlight)
- heterotrophic nutrition – obtaining nutrition by eating other organisms.

The archaea in the genus *Halobacterium* are able to perform a type of photosynthesis that is very different from that of organisms that use chlorophyll. Another pigment, bacteriorhodopsin, is used in these archaea to generate cellular energy, i.e. adenosine triphosphate (ATP). The "halo" part of their name refers to salt, because these microbes like to live in very salty environments such as the Great Salt Lake or the Dead Sea. These archaea are not considered autotrophs in the way plants are because, although they can generate ATP from sunlight, they get the carbon they need from other organisms.

When an organism is capable of producing its own food using chemical reactions, without the need for sunlight, it is called a **chemoautotroph**. This way of generating energy is called **chemosynthesis**. The archaeon *Ferroplasma acidiphilum*, for example, lives in very acidic environments (e.g. lower than pH 2) and gets its energy by oxidizing ferrous iron. It can be found living in wastewater runoff from iron mines. Archaea in the oceans and in soils use ammonia, NH_3, as a source of energy, and allow bacteria to convert the nitrogen compounds they generate into forms that can be used by plants.

Some archaea rely on organic food sources for their energy needs. These heterotrophs include members of the genus *Pyrococcus*, which can use amino acids, starch or maltose for food.

> Bioremediation is the concept of using microbes to clean up a toxic environment, such as those around mines and certain industrial sites. Because some archaea can survive in extreme conditions and have such diverse ways of using substances in their environment for food, they can be used to convert toxic compounds into safer ones.

B4.2.8 – The relationship between dentition and diet

B4.2.8 – Relationship between dentition and the diet of omnivorous and herbivorous representative members of the family Hominidae

Application of skills: Students should examine models or digital collections of skulls to infer diet from the anatomical features. Examples may include *Homo sapiens* (humans), *Homo floresiensis* and *Paranthropus robustus*.

NOS: Deductions can be made from theories. In this example, observation of living mammals led to theories relating dentition to herbivorous or carnivorous diets. These theories allowed the diet of extinct organisms to be deduced.

We are primates. We belong to the family **Hominidae**, the great apes, which include the following genera (singular genus):

- *Pongo*, orangutans, of which there are three extant (living) species
- *Gorilla*, of which there are two extant species
- *Pan*, chimpanzees, of which there are two extant species
- *Homo*, of which there is one extant species, modern humans (*Homo sapiens*).

Dozens of species in the family Hominidae are now extinct, so the only evidence we have of their presence on Earth is their fossil remains, such as bones, skulls and teeth, and even, in some cases, fossil DNA. Occasionally we may also find evidence of the tools that they used. Examples of extinct species of Hominidae include:

- *Australopithecus africanus*, which was present on Earth about 3 million years ago
- *Homo erectus*, which was present on Earth about 2 million years ago
- *Paranthropus robustus*, which was present on Earth about 1 million years ago
- *Homo floresiensis*, which was present on Earth about 100,000 years ago and, because of the first specimen's small stature (with a height just over 1 m), has been nicknamed "the Hobbit"
- *Homo neanderthalensis*, which also was present on Earth about 100,000 years ago, and encountered (and interbred with) modern humans.

What an organism eats tells us a lot about its place in an ecosystem. Knowing what eats what can help us determine which species occupy which niches, for example. Working out what extinct species ate can therefore help us work out what niches they may have occupied. Palaeontologists often look at teeth and jawbones, which are much better preserved in the fossil record than, for example, soft digestive organs.

A fossil jawbone of *Homo floresiensis*. The small canines and flat, broad molars suggest a plant-based diet, although tools and other archaeological evidence indicates that they were hunters, so they probably consumed some meat as well.

Paranthropus robustus also has the dentition of a herbivore.

Tooth anatomy

Incisors, your front teeth, are found at the front of your mouth; your **canines** are to the side of the incisors; then your **premolars** are in the middle of your mouth, followed finally by the **molars** at the back. Incisors are used for cutting off bite-sized pieces of food. Next time you eat an apple or a sandwich, notice how the incisors act as scissors. Primates that eat mostly plant material such as leaves (leaf eaters are called **folivores**) and fruit (fruit eaters are called **frugivores**) tend to have large incisors. Canines are sharper and are used for ripping and tearing tougher materials such as meat. Premolars are for crushing or slicing up food, and molars are for grinding food and reducing it to a paste before swallowing. Generally speaking, the narrower and more serrated (pointed) the crowns (tops) of the premolars and molars are, the better adapted they are for eating meat, whereas the more rounded or blunt (not pointed) they are, the better adapted they are for eating plant material. Herbivores tend to have bigger incisors than carnivores.

Each type of tooth has a name and a specific function, such as chopping, tearing off or grinding up food. Human teeth are shown here, but the same names are used in other mammals.

incisor   canine   premolar   molar

Diets of the great apes

If we look at the diets of extant species (those that are not extinct, at least not at the time of writing), we see that orangutans eat mostly fruit, which explains why they live in trees and occasionally also eat leaves. Some orangutans supplement their diet with insects, eggs or honey, so they can be considered omnivores, but they are essentially frugivores. Gorillas eat almost exclusively plant material: leaves, roots and stems. Some occasionally eat ants or termites, but gorillas are herbivores and more specifically folivores. Chimpanzees are omnivores, preferring fruit but also eating leaves and stems from plants, as well as meat. Chimpanzees consume invertebrates such as ants, termites and bees (and their honey), but also vertebrates such as monkeys, birds (and their eggs), antelope and warthogs, especially the young, which are easier to catch. Humans are also omnivores, eating fruit and grains, but consume vertebrates such as birds, fish, pigs and cattle as well.

Connecting dentition with diet

If we look at herbivores in general, they tend to have large incisors and wide premolars and molars that have rounded peaks and valleys for shearing and crushing plant material. Carnivores tend to have sharp, pointy teeth, not just their incisors and canines, but even their premolars and molars can be serrated and narrow rather than wide and rounded. Omnivores' teeth are somewhere in between. Their canines are not as long and pointed as carnivores. Their molars are of an intermediate width, not as wide as herbivores but not as narrow as carnivores. Their premolars and molars are usually rounded rather than serrated.

Looking at the teeth of chimpanzees, it appears that their dentition matches their diet. They have small incisors and long pointed canines for eating meat, for example. However, if we look at human teeth, even though meat plays a big role in many people's diet, our canines, premolars and molars are not shaped like carnivore teeth. Orangutans have long pointed canines and yet they do not eat meat. The complication is that teeth are not only used for eating. Some animals use sharp teeth to intimidate rivals or fend off intruders. Gorillas are herbivores but have very intimidating canines. Male chimpanzees tend to have much more prominent canines than females. Such evidence should be kept in mind when trying to determine diet solely from dentition. While there are general theories about the morphology of teeth and the diets of the animals, there are also exceptions.

One other aspect that experts look at in both extant and extinct species is **microwear**, small abrasions or removal of a tooth's surface, made as organisms chew, which can reveal the type of food they were eating. Softer foods will leave different marks compared to harder foods, and foods that have grit in them from soil will scratch teeth in a particular way that can be seen and analysed under a microscope.

TOK

When it is claimed that the dentition of an extinct hominid skull indicates that it was a herbivore, how much evidence is necessary before we decide that the claim can be considered valid? Do three or more experts have to agree? Or is it enough that one eminent, highly respected expert declares it? Is it important that their results are published in peer-reviewed journals, or is it sufficient that they made their observations and declarations without publishing? Is it necessary to compare their evidence to evidence from other fields, such as archaeology, to corroborate the findings, or are the results of a single specialty enough? How can we judge when the evidence provided is adequate?

X-ray images of the skulls of a gorilla (*Gorilla gorilla*), a male chimpanzee (*Pan troglodytes*) and a human (*Homo sapiens*).

Nature of Science

Deductions can be made from theories. For example, observation of living mammals has led to theories relating dentition to herbivorous and carnivorous diets. These theories allowed the diet of extinct organisms to be deduced. But nature is full of diversity, and theories have to be modified as new evidence arises. As we have seen, teeth can be used for self-defence, not just for eating, and certain hominids that lack classic carnivore teeth can still be meat eaters, by using tools instead of teeth to cut and tear flesh. The ability to use fire to cook meat also reduces the need for specialized carnivore teeth, as cooked meat is much more tender and easier to chew.

B4.2.9 – Adaptations of herbivores and plants

B4.2.9 – Adaptations of herbivores for feeding on plants and of plants for resisting herbivory

For herbivore adaptations, include piercing and chewing mouthparts of leaf-eating insects. Plants resist herbivory using thorns and other physical structures. Plants also produce toxic secondary compounds in seeds and leaves. Some animals have metabolic adaptations for detoxifying these toxins.

How herbivores are adapted to eating plant material

Plants are not always easy to eat. Their leaves tend to be protected by thick layers of cells with semi-rigid cell walls, and not many organisms possess the enzymes necessary to break down cellulose, the chains of carbohydrates that make up plant fibre. Some insects solve this problem by piercing the plant and drinking the sugar dissolved in the sap inside. This is the case for aphids, small insects that you might find on a rosebush, which use modified mouthparts called **stylets**. Other insects, such as grasshoppers and caterpillars, use their sharp pinching **mandibles** to cut into grass blades and leaves, to help ingest them. Such insects are considered pests by farmers who do not want to see their crops damaged.

Herbivorous vertebrates such as cows and sheep have specialized back teeth that are broad and flat for grinding plant matter, and their digestive systems are adapted for digesting it. Cows are **ruminants**, which means that they swallow grass or hay before fully chewing it, then regurgitate it later when they are resting in order to chew it some more, a process called **chewing the cud**. Cows and many other herbivores harbour bacteria and archaea in their digestive systems that help them break down the cellulose. Giraffes' long legs and necks allow them to access leaves from their favourite tree, the acacia, and their tough tongues can resist the pointy thorns.

How plants protect themselves from herbivores

Herbivory means to feed on plants. Because plants cannot run away from the animals that want to eat them, they have other adaptations for defending themselves. Thick bark is difficult for many insects and some animals to penetrate, and therefore protects the plant against animals like aphids. Thorns and spikes are useful for deterring herbivores.

The common nettle (*Urtica dioica*) has tiny hairs of silica on its stem and on the underside of leaves, which are filled with chemical irritants. When animals approach the plant and rub against them, the silica breaks, scratching the skin, and the irritant inside is released into the damaged skin. This generates an unpleasant stinging sensation like a burn, as well as swelling of the skin, which the animal is likely to remember next time it considers eating the plant.

▲
The stinging nettle (*Urtica dioica*): notice the pointy hairs on the underside of the leaf.

Plant poisons, called **phytotoxins,** are made from secondary compounds (see the Global context box) and can cause nausea, cardiac problems or hallucinations when ingested. Foxgloves, in the genus *Digitalis*, for example, produce toxins that will make many types of mammals very sick, including humans. The castor bean (*Ricinus communis*) produces seeds that are rich in nutritious oil, making them tempting to eat. Humans make castor oil from these seeds and use it for food as well as industrial purposes. To protect the seeds from animals that want to eat them, the plant produces a phytotoxin called ricin, which is highly toxic. Fortunately, in the process of manufacturing castor oil, this toxin is removed.

As plants evolve chemical deterrents such as alkaloids and tannins, animals evolve ways of neutralizing the toxins so that they are not poisoned. In ruminants and insects that rely on microbes for digestion, the microbes that live in the gut can detoxify many plant poisons. If a ruminant eats a small quantity of a toxic plant, colonies of microbes that can degrade the poison will start to grow in its gut. As it eats more of the plant, it can cope with more toxins as the colonies proliferate. However, it would eventually be poisoned if it continued to eat the toxic plant. Animals seeking out new food sources use a technique called cautious sampling, in which they do not eat too much of a plant the first time it is encountered. In browsing herbivores such as the moose (*Alces alces*), proteins in their saliva have evolved to neutralize tannins. If a mammal is not killed by a toxin, the toxin will travel through the blood to the liver, where it will be neutralized.

Secondary compounds, also called secondary metabolites, are molecules that are not necessary for the normal growth or reproduction of the organism, but which can be used by the organism as a toxin for defence. Humans sometimes use secondary compounds as medicines or stimulants. Examples include quinine, penicillin and caffeine. Foxgloves have long been used as medicinal plants, and a heart medication has been developed from them.

B4.2.10 – Adaptations of predators and prey

> **B4.2.10 – Adaptations of predators for finding, catching and killing prey and of prey animals for resisting predation**
>
> Students should be aware of chemical, physical and behavioural adaptations in predators and prey.

How predators find, catch and kill prey

Chemical adaptations

When we think of predators, we often picture a cheetah running after a gazelle, but not all predators rely on speed and physical strength to catch their prey. Some use surprise tactics, or inject chemicals into their prey, while others can use chemical compounds to lure their prey by trickery.

The black mamba (*Dendroaspis polylepis*) is a venomous snake that lives in southern and eastern parts of Africa. The venom in its bite contains neurotoxins that paralyse its prey. After biting and injecting the venom, the black mamba waits until its victim, such as a small bird or rodent, is no longer moving, then it will ingest it whole.

Pheromones are organic molecules used to send messages through the air, and some of them are intended to attract mates. Certain species of orb-weaver spiders are capable of producing chemicals that mimic the sex pheromones of moths. They release the pheromones and wait for their prey to arrive. After following the scent, the moth finds that it has been invited to dinner not by a mate but by a predator.

Physical adaptations

To find and chase down prey, predators need to be able to detect their prey using senses such as sight, smell or even electrolocation. Birds of prey such as hawks and eagles have excellent eyesight for detecting prey. Owls have eyes that are well adapted for seeing in low light at night. Bats and dolphins use echolocation, which involves sending out ultrasonic vibrations, and their brains process how the waves bounce off objects (including prey) in the environment, in order to perceive their environment with sound. Sharks have specially adapted organs in their heads called ampullae of Lorenzini. These sensing organs detect changes in electromagnetic fields, allowing them to detect prey. As a fish or seal swims, its nervous system releases small discharges of electricity that can be detected by sharks. An acute sense of smell helps birds like vultures find rotting flesh, or fish find prey in low-light conditions.

The Malaysian orchid mantis is well adapted for attracting prey by mimicking a flower that has delicious nectar, but it is also well adapted for catching insects (using appendages that grasp) as well as eating insects (with sharp mouthparts to chew through tough exoskeletons).

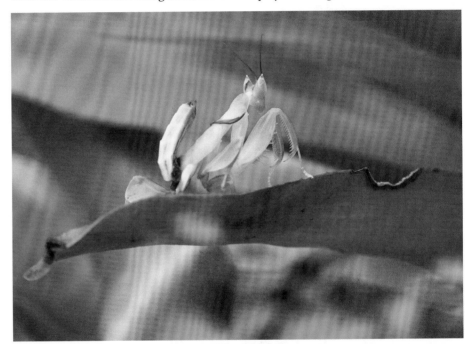

But finding prey is only part of the story. Catching the prey and then eating it also present challenges. The ability to fly, run or swim not only rapidly but also with precision is key in chasing down prey. Then claws, beaks, teeth and a well-adapted digestive system are needed to kill the prey and extract nutrients. Predators also need a brain that can quickly assess rapidly changing circumstances and make complex decisions involving the time, energy and risk involved in pursuing prey. If the risk of exhaustion or injury is too great, the predator must know when to give up and try again another time.

Behavioural adaptations

Some predators are **ambush predators**. They hide and wait for prey to come near and then pounce on them. This is true for many spiders, notably those that build a web, when they hide at one end of the web and wait. Anglerfish such as frogfish hide on the ocean floor and use a lure called an **illicium** (a long thin appendage protruding from their head) to attract prey. They then open their mouths in a fraction of a second to engulf the prey, before the prey is even aware that they are there.

Teamwork is sometimes a successful adaptation. **Pack hunting** is common in wolf species, and the chances of bringing down a large animal are greater if several wolves work together than if one wolf hunts on its own. For this to work, there must be an established relationship of trust between individuals. The group knows which animal is the leader of the pack, and which animals are subservient. Other social animals that engage in group hunting, or even what could be considered warfare, are the hymenoptera: ants, termites, bees and wasps. Soldier ants, for example, participate in raiding parties and use their large mandibles to kill and dismember prey, then bring the dead prey back to the colony for a feast.

A type of predator you have probably seen on nature documentaries is the **pursuit predator**, which relies on speed to outrun its prey. Cheetahs are the fastest land mammal, at least over short distances, and they are well-adapted to chasing down gazelles, especially those that are very young, very old or unhealthy. But speed is not the only strategy: endurance can work sometimes, too. The idea then is to keep pursuing the prey for many hours until it drops from fatigue. This is known as **persistence hunting**. Humans living as hunter gathers use persistence hunting.

How animals that are preyed upon resist predation

Chemical adaptations

Organisms that are preyed upon can try to run away, but they can also produce chemicals to dissuade or fool predators. One adaptation is to produce chemicals that taste bad or that poison the predator. This is the strategy used by poison dart frogs. One highly poisonous dart frog, *Phyllobates terribilis*, produces an alkaloid on its skin that can interfere with muscle function, including heart muscles, causing death. First Nations people of Colombia have successfully used this poison for hunting by applying it to the tips of the darts they use in their blowguns.

Poison dart frogs like this one secrete toxins on their skin that can paralyse or kill animals that try to catch it. The bright pigments in their skin act as a warning sign.

▼

The harlequin filefish (*Oxymonacanthus longirostris*) has an interesting adaptation: it picks up the characteristic smell of the coral reefs it feeds on, so that it smells like coral. Predators such as cod fish are not attracted by the smell of coral, so will search elsewhere for prey.

Physical adaptations

One of the best ways to not get eaten is to not be detected by a predator. **Camouflage**, the ability of an organism to take on the appearance of its surroundings, can work well against predators that rely on vision. Some organisms only have one fixed adaptation, such as the coloration patterns on the wings of moths, whereas others can adapt to their environments. Certain species of octopus can not only modify cells in their skin to adapt to the colours of their background, but they can generate bumpy or smooth textures on their skin to mimic the surfaces they hide on.

Poison dart frogs use dramatically bright and unusual colours, such as yellow, blue and red, to inform potential predators that they are poisonous. This technique is called **aposematism**. Some other prey animals mimic these warnings. Non-venomous kingsnakes look like coral snakes, which are poisonous. Warning vocalizations can deter predators and warn fellow prey that there is danger. Many birds such as jays and blackbirds have specific warning calls, and they are not alone. Non-human primates such as monkeys can use different types of calls, depending on the type of threat.

Growing a protective shell can be another way of deterring some predators. Invertebrates such as grasshoppers and lobsters have exoskeletons, and clams and mussels have hard shells for protection. Among vertebrates, turtles and tortoises have shells to reduce the chances that a predator might consider them as prey. An alternative way of dissuading a predator is to have sharp spines, like the porcupine.

Behavioural adaptations

Many animal behaviours are instinctive and encoded in their DNA. These can include fleeing at the sight of a predator, hiding, forming groups or using certain types of dissuasive behaviour to ward off predators.

The expression "there is safety in numbers" is not just a figure of speech. It has been tested using both field observations and controlled experiments. A large group attacked by a predator will suffer fewer kills than solitary individuals or small groups that are attacked. For example, when threatened, a herd of elephants will group together, with the largest adults placing themselves facing outwards and the young positioned in the centre. Predators will often be dissuaded from attacking a group because the risk of getting injured is greater than the chance of successfully taking down a vulnerable, solitary, juvenile, for example.

Wildebeests know instinctively that there can be safety in numbers. When bison or wildebeest stick together in a group on the prairie, there are more eyes watching out for predators in all directions, so the chances of being surprised by an ambush are lessened.

B4.2.11 – Harvesting light

B4.2.11 – Adaptations of plant form for harvesting light

Include examples from forest ecosystems to illustrate how plants in forests use different strategies to reach light sources, including trees that reach the canopy, lianas, epiphytes growing on branches of trees, strangler epiphytes, shade-tolerant shrubs and herbs growing on the forest floor.

In order for leaves to photosynthesize at their optimal rate, they need to catch as much sunlight as possible, using the best angle possible for the longest possible period of the day. Leaves tend to be flat and angled towards the Sun. Chloroplasts are concentrated on the top surface to catch as much light as possible and convert it to food.

Trees

One adaptation to maximize access to sunlight is to position leaves far above the ground, so that they are above their competitors. This is the case for the tallest tree species, whose sturdy woody trunks allow them to dominate the **canopy**. The canopy is the upper layer of a forest where the crowns (tops) of trees are found; zones below the canopy are called the **understorey**, where shorter trees can be found. The **shrub layer** contains the shortest trees and shrubs, while the **forest floor** is home to smaller, non-woody plants. In a dense forest, every square centimetre of the forest floor is almost always in shade, because leaves from plants at every level above it absorb the sunlight and shield the zones below.

But there is a price to pay for reaching high above the ground: building a sturdy trunk and strong supporting branches requires a sizeable investment in energy and nutrients.

Lianas

Plants that cannot build trunks big enough to reach the canopy can use another adaptation: borrowing support from a nearby tree. Lianas are vines that take root on the forest floor and use trees as a scaffold, allowing them to grow into the canopy to obtain more light. When seeds germinate, most seedlings seek out light and bend towards it. Liana seedlings do the opposite: they grow towards shade, which means they grow towards tree trunks, and can then start to climb.

It should be obvious that liana vines are direct competitors for trees, not only for sunlight but also for minerals in the soil and space on the forest floor for germinating seeds. The bigger and more entangled the lianas get, the more harm they do to the trees, causing the trees to grow less well. Lianas can eventually kill a tree.

Epiphytes

Similar to lianas, epiphytes take advantage of the height and strength of trees to get up into the understorey or canopy to access sunlight. The difference is that their roots are not in the soil on the forest floor. Have you ever seen moss growing on a tree branch? It is getting all the moisture it needs from water trickling along the branch when it rains, or from humidity in the air. The orchids we can buy at a florist's are epiphytes. In nature, they attach their roots to tree trunks and are well adapted to survive on very little water. One of the best ways to kill an orchid that you have adopted as a houseplant is to overwater it.

▲ Even with some trees cleared to make a path through this forest, most of the sunlight is blocked by leaves, and only a few small patches of light reach the forest floor.

For each form of nutrition, what are the unique inputs, processes and outputs?

Orchids are epiphytes for their entire lives: their roots do not need to be in soil down on the forest floor, they can survive on very little water while living on tree trunks or branches.

The strangler fig growing around this tree is an example of a hemi-epiphyte, because it can take root in the soil at the base of the tree.

Strangler figs are examples of **hemi-epiphytes**. They spend the early part of their life in a tree without any roots in the soil. In addition to pushing their stems upwards to get more sunlight, they then push some of their stems downwards to reach the ground and start growing roots. The intertwining of the strangler fig's stems and branches can completely encircle the host tree's trunk.

Growing in the shade

Shade-tolerant shrubs grow on the forest floor between trees and are well adapted to absorbing the wavelengths of diffuse sunlight that remain after passing through other leaves, notably the longer wavelengths in the red part of the spectrum.

Some well-known and popular foods such as bananas and ginger are from herbaceous plants that grow in the understorey of forests in the tropics, and are well adapted to growing in the shade. **Herbaceous plants**, otherwise called **herbs**, are those that do not produce a woody stem with bark the way trees do. Banana plants are herbs because the part that looks like a woody stem is, in fact, not made of wood but of rigid layers of the bases of the leaves. Other examples of herbaceous plants that can grow on the forest floor are wildflowers and berries such as strawberries.

B4.2.12 – Ecological niches

B4.2.12 – Fundamental and realized niches

Students should appreciate that fundamental niche is the potential of a species based on adaptations and tolerance limits and that realized niche is the actual extent of a species when in competition with other species.

The **fundamental niche** of a species is the potential niche that it could inhabit, given the adaptations of the species and its tolerance limits. The **realized niche** of a species is the actual niche that it inhabits. The realized niche can be different to the fundamental niche because of competition with other species.

The habitat of the red fox (*Vulpes vulpes*) in the USA is the forest edge. Its food consists of small mammals, amphibians and insects. It interacts with other species, such as

TOK

When someone declares, "I don't need to know what relationships there are between organisms in a distant tropical rainforest or at the bottom of the sea", how would an expert in ecosystems respond? Probably by saying that ignorance of these systems means that if the conditions in those places are modified over time by processes such as deforestation and human-induced climate change, we won't know what existential threats there are for other parts of the world or other species, including our own. What are the implications of not having knowledge?

the mosquitoes that suck its blood and scavengers that eat its leftovers. Its physical characteristics and behaviour allow it to survive in all seasons, including cold, snowy winters. The forest edge is the fundamental niche of the red fox.

What has happened to the red fox's fundamental niche in recent decades? The forest edge has been turned into farmland in many places. Some of the species eaten by the red fox no longer live there. For example, amphibians are particularly sensitive to changes in their environment and to the pesticides that farmers use. The red fox has less physical space and there is less food availability. In addition, there is direct competition from the coyote (*Canis latrans*), whose own niche has also been modified by changes in the environment cause by human activity. This new and narrower niche is called the fox's realized niche.

An activity on fundamental and realized niches in two species of *Paramecium* can be found on this page of your eBook.

B4.2.13 – Competitive exclusion

B4.2.13 – Competitive exclusion and the uniqueness of ecological niches

Include elimination of one of the competing species or the restriction of both to a part of their fundamental niche as possible outcomes of competition between two species.

The **principle of competitive exclusion** states that no two species in a community can occupy the same niche. If they do coexist for a certain period of time, as is currently happening with the fox and the coyote mentioned in Section B4.2.12, the numbers of both populations will tend to decrease. In the long run, it is often the case that one species will replace the other. This is easier to see in microbial populations, which reproduce at a very fast rate.

Do not confuse the Russian ecologist Gause with the German mathematician Gauss, who lived a hundred years earlier and is notable for many reasons, among them the idea of a normal distribution around a mean represented by a bell-shaped curve, also called a Gaussian curve or Gaussian distribution.

The competitive exclusion principle was demonstrated in 1934 by a Russian ecologist, G. F. Gause. He performed a laboratory experiment with two different species of *Paramecium*: *P. aurelia* and *P. caudatum* (see Figure 1). His experiments showed the effects of **interspecific competition** between two closely related organisms. Interspecific means between two or more different species. When each species was grown in a separate culture, with the addition of bacteria for food, they did equally well. When the two were cultured together, with a constant food supply, *P. caudatum* died out while *P. aurelia* survived. *P. aurelia* out-competed *P. caudatum*. The experiment supported Gause's hypothesis of competitive exclusion. When two species have a similar need for the same resources in the same space at the same time, one will be excluded. One species will die out in that ecosystem and the other will survive. *Paramecium aurelia* must have had a slight advantage that allowed it to out-compete *P. caudatum*.

B4.2 Figure 1 The results of Gause's experiment to demonstrate the competitive exclusion principle.

It is not easy to observe this phenomenon happening in ecosystems, because it has usually already had its effect: when we look at a niche, we find only one species occupying it within an ecosystem. If we base our observations on what happens when an introduced or invasive species takes over, it is hard to know if it is only competitive exclusion that is happening or if there are other factors too. Currently in Britain the population of eastern grey squirrels (*Sciurus carolinensis*) is growing (the species was introduced from North America), while the population of red squirrels (*Sciurus vulgaris*; the native species) is falling to such low levels that there is a worry this species will be permanently driven out of some areas. It is difficult to know if this change is caused by the inability to share a niche, or if it is the impact of, for example, human activity or disease.

Guiding Question revisited

What are the advantages of specialized modes of nutrition to living organisms?

In this chapter you have learned that:

- some organisms use a particular form of nutrition so that they do not compete with other organisms for the same food
- some organisms have adaptations for eating a diverse range of foods
- autotrophs make their own food, whereas consumers such as heterotrophs and saprotrophs need to eat other organisms to get their nutrition
- some organisms are mixotrophic, and can use autotrophic nutrition at certain times and heterotrophic nutrition at others
- each organism is adapted to obtain nutrition for itself but avoid being eaten, and those adaptations can be seen in morphological features such as dentition, production of secondary metabolites and position of leaves.

Guiding Question revisited

How are the adaptations of a species related to its niche in an ecosystem?

In this chapter we have discussed how:

- an organism's niche is where it is best adapted to survive
- normally a particular niche can only be occupied by one species
- in order to continue to occupy a specific niche in an ecosystem, organisms have to adapt.

Exercises

Q1. Which of the following terms describes the place where an organism lives and the role it plays in nature?

   A Niche.   **B** Habitat.

   C Ecosystem.   **D** Community.

Q2. Which of the following primates does not belong to the great apes?

   A Orangutan.   **B** Gorilla.

   C Lemur.   **D** Human.

Q3. The Venus fly trap plant can photosynthesize but it needs to take in nitrogen by trapping and digesting insects. Which of the following terms describe it?

   I. Obligate mixotroph.   II. Obligate aerobe.   III. Decomposer.

   A I and II only.   **B** I and III only.

   C II and III only.   **D** I, II and III.

Q4. Outline two behavioural adaptations found in predators to help them catch their prey.

Q5. Using a named example, outline the advantages of mixotrophic nutrition.

Q6. Compare and contrast the concept of an organism's fundamental niche with its realized niche.

B4 Practice questions

1. In South Korea, flocks of birds of the tit family (Paridae) forage together on trees for food. Researchers observed four species of Paridae to determine whether they shared the same habitat in the trees and whether their position on the tree depended on their size. The leafy part of the tree (crown) was divided into nine sections, three according to height from the ground and three according to the distance from the tree trunk. Observations were also made of birds foraging in the bushes surrounding the trunk and on the ground below the tree. The chart shows the relative use of each section of the habitat by the birds.

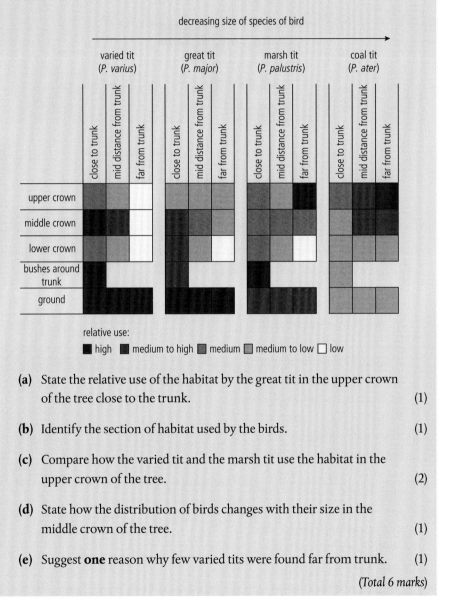

    (a) State the relative use of the habitat by the great tit in the upper crown of the tree close to the trunk. (1)

    (b) Identify the section of habitat used by the birds. (1)

    (c) Compare how the varied tit and the marsh tit use the habitat in the upper crown of the tree. (2)

    (d) State how the distribution of birds changes with their size in the middle crown of the tree. (1)

    (e) Suggest **one** reason why few varied tits were found far from trunk. (1)

    (Total 6 marks)

2. During the 1980s, a tiny invasive crustacean, *Bythotrephes cederstroemii*, entered the eastern Great Lakes from Europe (probably via freshwater or mud in the ballast water of merchant ships) and eventually colonized Lake Michigan. *Bythotrephes* reproduces very quickly and eats common zooplankton, disrupting the food web by directly competing with small juvenile resident fish. *Bythotrephes* avoids predation by larger fish through the timing of its activities, which have been investigated in offshore waters of Lake Michigan at various depths during the day and night.

Bythotrephes / individuals m^{-3} water

(a) State the depth range showing the most *Bythotrephes* during the night. (1)

(b) Describe the distribution of *Bythotrephes* during the day. (2)

(c) Deduce the responses of *Bythotrephes* to temperature and light. (2)

(d) Explain the change in distribution of *Bythotrephes* between day and night in terms of its position in the lake food chain. (2)

(Total 7 marks)

3. An experiment was set up to investigate how populations change over time. Rotifers and water fleas are small protists that are found in freshwater plankton. Two flasks, X and Y, were set up where each could grow alone and they each thrived when fed a constant supply of food. A third flask, Z, was set up in the same conditions into which both species were introduced. The graph shows the results from flasks X and Z.

(a) Describe the changes in the experiments in flasks X and Z over time. (2)

(b) Explain the differences in population changes in flasks X and Z. (3)

(Total 5 marks)

4. Ecologists studied the distribution of five species of insectivorous wood warblers of the genus *Dendorica* living on different parts of coniferous trees in mature forests.

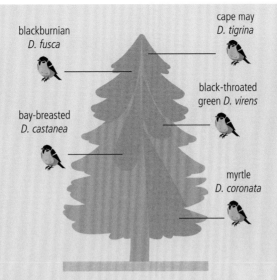

blackburnian *D. fusca*

cape may *D. tigrina*

black-throated green *D. virens*

bay-breasted *D. castanea*

myrtle *D. coronata*

(a) Distinguish between the distribution of *D. tigrina* and that of *D. coronata*. (1)

(b) Outline the principle of competitive exclusion. (2)

(c) Other than position in the tree, suggest two ways in which the niches of the warblers in the ecosystem may differ. (2)

(d) The diagram shows the realized niches of the five species of warbler. Suggest how the fundamental niche of *D. castanea* might differ from its realized niche. (2)

(Total 7 marks)

5. Explain the importance of saprotrophic nutrition in a woodland.

(Total 3 marks)

6. Using a named example, outline how a plant can protect itself from herbivory by producing certain chemicals

(Total 3 marks)

THEME

C Interaction and interdependence

1 Molecules

◀ The butterfly *Anartia jatrophae* utilizing the leaves of a plant for stability and rest. The interdependence of plants and animals is evident all around us every day.

Molecules are a connecting thread for all life on Earth. Waste products from one organism become essential building blocks for other life forms. For example, respiration requires oxygen, which plants release into the atmosphere as a waste product from photosynthesis. In turn, photosynthesis needs carbon dioxide, which is one of the products formed during respiration. Chemical reactions occur continuously in all living organisms.

Chemical reactions are essential to life and can occur on their own. However, it is essential that these reactions take place at a rate that is favourable to maintaining life. As you will see, enzymes control the rate of all chemical reactions in organisms, including cellular respiration and photosynthesis.

C1.1 Enzymes and metabolism

Guiding Questions

In what ways do enzymes interact with other molecules?

What are the interdependent components of metabolism?

Enzymes are catalysts. They do not work in isolation, however; they interact with substrates and other molecules. These interactions can control the rate of reactions.

Living organisms are dependent on a relatively small number of different atoms, mainly carbon (C), hydrogen (H) and oxygen (O) (with others in smaller amounts). These molecules are taken in from the surrounding environment and then rearranged to make all the molecules that the organism needs. All components of metabolism are interdependent. Living organisms depend on complex sequences of reactions.

Adenosine triphosphate (ATP) provides energy for many of the cellular activities essential for life. The production of ATP is complex and involves many pathways of sequential chemical reactions. The control of these reactions is the focus of this chapter.

C1.1.1 – Enzymes as catalysts

C1.1.1 – Enzymes as catalysts
Students should understand the benefit of increasing rates of reaction in cells.

Most reactions within a cell proceed too slowly on their own to sustain the life processes. However, in the presence of **catalysts** these reactions occur much faster, so that essential life functions can be maintained. In organisms, organic catalysts are known as **enzymes**.

C1.1.2 – Metabolism

C1.1.2 – Role of enzymes in metabolism
Students should understand that metabolism is the complex network of interdependent and interacting chemical reactions occurring in living organisms. Because of enzyme specificity, many different enzymes are required by living organisms, and control over metabolism can be exerted through these enzymes.

Metabolism includes all the chemical reactions that occur in an organism. These chemical reactions may be independent of one another, or they may interact with other reactions. Each chemical reaction is controlled by a specific enzyme. Because of this specificity, there are many, many enzymes in each organism. All chemical reactions involve **reactants** and **products**. Reactants are the substances that participate in a reaction, while products are the substances that are formed.

$$A + B \quad \rightarrow \quad C + D$$
$$\text{reactants} \qquad\qquad \text{products}$$

C1.1.3 – Anabolism and catabolism

C1.1.3 – Anabolic and catabolic reactions
Examples of anabolism should include the formation of macromolecules from monomers by condensation reactions including protein synthesis, glycogen formation and photosynthesis. Examples of catabolism should include hydrolysis of macromolecules into monomers in digestion and oxidation of substrates in respiration.

Some metabolic reactions use energy to build complex organic molecules from simpler organic molecules. These reactions are said to be **anabolic**, and the process is called **anabolism**. The metabolic reactions that break down complex organic molecules, with the release of energy, are called **catabolic** reactions and the process is called **catabolism**. Table 1 summarizes anabolic and catabolic reactions.

C1.1 Table 1 Anabolic and catabolic reactions ▶

Anabolic reactions	Catabolic reactions
Build macromolecules (and release water) from monomers by condensation reactions	Break down macromolecules into monomers by hydrolysis (the splitting of molecules by adding water)
Require energy input to occur	Release energy as they occur
Examples include photosynthesis, protein synthesis and glycogen formation	Examples include digestion and the oxidation of substrates in respiration

Chapter B1.1 discusses how condensation and hydrolysis reactions are common in living systems, allowing all the functions of life to be maintained. As you work through this chapter, you should keep in mind the characteristics of these two general metabolic reactions.

Energy and life

All organisms maintain their structure and function through chemical energy. In general, energy is the capacity to cause change, to do work.

Table 2 lists several forms of energy particularly important to organisms.

Form of energy	Description
Kinetic energy	Energy of motion, including movement of molecules within objects
Potential energy	Stored energy or energy in a form that is not being used at a point of time
Chemical energy	A form of potential energy that is available for release when a chemical reaction occurs
Thermal energy	A form of kinetic energy stored within objects. Capable of being transferred from one object to another as heat

There are also other forms of energy, including mechanical, sound, radioactivity and electric current. These different forms of energy can require different methods of measurement, but one that is often used is heat. The unit used to measure heat in biology is the **kilocalorie** (kcal).

Adenosine triphosphate (ATP) is the energy currency of a cell. Study the diagram of ATP and the similar molecule adenosine diphosphate (ADP) (Figure 1).

ATP has many functions in an organism, including:

- supplying the energy needed to synthesize large molecules called macromolecules
- supplying the energy necessary for mechanical work, such as muscle action, chromosome movement and cilia or flagellum motion
- providing energy to move substances across the cell membrane, such as the sodium–potassium pump.

◀ **C1.1 Table 2** Different forms of energy

A kilocalorie is 1000 calories (cal). A calorie is the amount of heat necessary to raise the temperature of 1 gram of water by 1 degree Celsius (°C). Food labels use the symbol C for calories, which is the same as a kilocalorie. The joule is another unit of energy. One joule equals 0.239 cal.

◀ **C1.1 Figure 1** This diagram shows the structure of both ADP and ATP. Adenine and ribose combine to form the molecule adenosine. ATP has three phosphate groups attached to adenosine, while ADP has only two. Note the locations of high-energy bonds represented by wavy lines. The high-energy bonds, especially the one located between the second and third phosphate in ATP, are the source of energy for chemical reactions within a cell.

What are examples of structure–function relationships in biological macromolecules?

C1.1.4 – Globular proteins and active sites

C1.1.4 – Enzymes as globular proteins with an active site for catalysis

Include that the active site is composed of a few amino acids only, but interactions between amino acids within the overall three-dimensional structure of the enzyme ensure that the active site has the necessary properties for catalysis.

Almost all enzymes are proteins. Protein enzymes are long chains of amino acids that have taken on a very specific three-dimensional shape. Think of a flexible metal wire that can be bent many times into what is called a **globular** shape. This shape is complex and at first glance appears to be random, but in enzymes (and other **globular proteins**) the complex shape is not random: it is very specific. Somewhere in the three-dimensional shape of the enzyme is an area that matches the shape of that enzyme's substrate. This area of the enzyme is called the **active site**. The shape of the active site closely matches the shape of one particular substrate. It is important to note that the active site is composed of only a few amino acids. It is the interaction between the amino acids in the overall three-dimensional enzyme shape that provides the active site with the properties necessary to carry out catalysis. This three-dimensional shape is essential to the action of the enzyme. If it is changed in any way, the enzyme is said to be **denatured**, and it will no longer function as a catalyst. Several factors can cause denaturation by affecting the chemical bonds amongst the amino acids present. These factors will be discussed later in this chapter.

A ribozyme is a ribonucleic acid (RNA) molecule that acts as an organic catalyst. Ribozymes are often able to catalyse their own assembly. Some believe that life on Earth began with a simple RNA molecule capable of self-assembly.

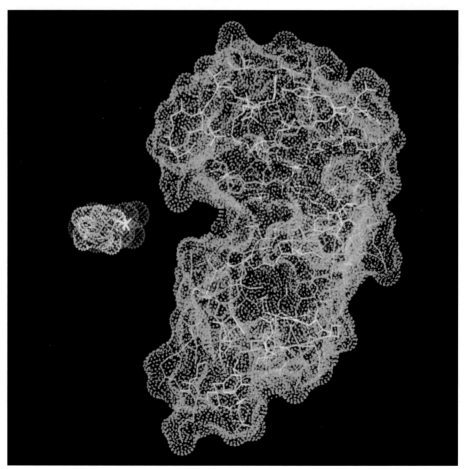

A computer graphic showing an enzyme (the larger molecule on the right) and its substrate. Notice the active site on the left-hand side of the enzyme.

C1.1.5 and C1.1.10 – Enzyme activation

C1.1.5 – Interactions between substrate and active site to allow induced-fit binding

Students should recognize that both substrate and enzymes change shape when binding occurs.

C1.1.10 – Effect of enzymes on activation energy

Application of skills: Students should appreciate that energy is required to break bonds within the substrate and that there is an energy yield when bonds are made to form the products of an enzyme-catalysed reaction. Students should be able to interpret graphs showing this effect.

The induced-fit model

In the 1890s, Emil Fischer proposed the **lock-and-key** model for enzyme action. In this model, the lock represents the enzyme's active site, and the key represents the substrate. Because the three-dimensional shape of the internal portion of the lock is complex and specific, only one key will fit it. At the time this model provided a good explanation of the specificity of enzyme action. However, as knowledge about enzyme action has increased, Fischer's model has been modified into what is now known as the **induced-fit model** of enzyme action.

Nature of Science

Models often change over time as more evidence is gathered, resulting in changing hypotheses, theories and predictions. Peer review amongst researchers is essential to develop and verify the most accurate models possible.

Research has shown that many enzymes undergo significant changes in their conformation (shape) when substrates combine with their active site. A good way to visualize this model of enzyme action is to think of a hand and glove, the hand being the substrate and the glove being the enzyme. The glove looks a bit like the hand. However, when the hand is placed in the glove, there is an interaction that results in shape changes of both the hand and the glove, thus providing an induced fit. The changes in the shape of the substrate (the hand in this analogy) causes stresses upon its chemical bonds. The bonds become destabilized, which favours reactions and increases reaction rates.

Activation energy

It is not enough for an enzyme's substrate(s) to enter an active site. The substrate(s) must enter with a minimum rate of motion, kinetic energy, that will provide the energy necessary for the reaction to occur. Enzymes do not provide this energy; they simply lower the energy minimum that is required. The energy being referred to is called the **activation energy** of the reaction. Thus, enzymes lower the activation energy of reactions. Enzymes are not considered to be reactants and are not used up in the reaction. An enzyme can function as a catalyst many, many times.

Activation energy is the energy necessary to destabilize the existing bonds in a substrate so that a reaction can proceed. Reactions that require larger amounts of activation energy tend to proceed more slowly than those requiring smaller amounts.

TOK The work by Emil Fischer was carried out before details of enzyme structure were known. What is the role of imagination and intuition in the creation of hypotheses in the natural sciences?

Catalysts lower the activation energy needed for a reaction to proceed. Chemical reactions are reversible, which means they can occur in both directions. By reducing activation energy, catalysts increase the rate of a chemical reaction in both the forward and reverse directions. You can think of a catalyst as lowering the energy barrier that is preventing or hindering a reaction from occurring. In living systems enzymes act as catalysts.

Exergonic and **exothermic** reactions release energy when they occur. The products of an exergonic reaction have less energy than the reactants had because of this released energy. **Endergonic** and **endothermic** reactions result in products that have a higher energy level than the reactants.

This is because there are fewer molecules colliding with sufficient energy to overcome the initial energy requirement. There are two ways of overcoming the energy barrier and increasing the rate of these chemical reactions.

1. Increasing the energy of the reacting molecules and thus increasing the rate of collisions, usually by the addition of heat.
2. Lowering the activation energy that is required to stress particular chemical bonds in the reactants so that the bonds can be broken more easily.

Because living systems are vulnerable to higher temperatures, most chemical reaction rates in organisms are increased by the action of enzymes.

In addition, an enzyme cannot force a reaction to occur that would not otherwise happen without the enzyme. However, the reaction will be much more likely to occur with an enzyme present because the input of energy (activation energy) required will be lower.

 Enzymes accelerate exothermic reactions by lowering the activation energy required. The activation energy is needed to destabilize the chemical bonds in the reactant. The upper curve shows the activation energy when no enzyme is involved. The lower curve shows the activation energy required when an enzyme is present to catalyse the reaction.

Reactions proceed until there is equilibrium between the relative amounts of reactants and products. It is important to note that, even though enzymes lower the activation energy of a particular reaction, they do not alter the proportion of reactants to products at equilibrium. Some reversible chemical reactions require a different enzyme to lower the activation energy in the reverse direction.

 Exergonic and endergonic reactions are **bioenergetic reactions**. Molecules are rearranged and energy can be used to do work. Exothermic and endothermic reactions are primarily **thermodynamic reactions.** The energy is given out or taken in in the form of heat. In chemistry it is more usual to have exothermic and endothermic reactions. In biology, exergonic and endergonic are more common.

SKILLS

Breaking chemical bonds requires energy, and when chemical bonds form they release energy. Look at the figure below. In A, more energy is released when the chemical bonds form in the products than is needed to break the bonds in the reactants. Overall, the reaction releases energy (it is exergonic). In B, more energy is needed to break the bonds in the reactants than is released when the products form. Overall, the reaction takes in energy (it is endergonic).

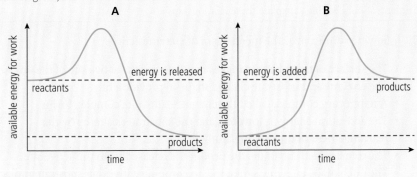

1. What is the initial energy called that starts the reaction in both graphs?
2. What type of reaction results in products that have a higher energy level than the substrates?
3. Photosynthesis is an endergonic reaction. Explain why an endergonic reaction is useful to cells.
4. What would be the effect of adding an appropriate enzyme to each reaction type?
5. Some reactions will occur spontaneously after a level of activation energy is provided. This means no additional energy will be required for the reaction to proceed. Which of the above reactions represents a spontaneous reaction?
6. An additional supply of energy beyond activation energy is required for the reaction shown in B. Where is that energy stored in the product produced?

C1.1.6 – Molecular motion

C1.1.6 – Role of molecular motion and substrate-active site collisions in enzyme catalysis

Movement is needed for a substrate molecule and an active site to come together. Sometimes large substrate molecules are immobilized while sometimes enzymes can be immobilized by being embedded in membranes.

Active sites and movement are both important to enzyme action and control of chemical reactions. The active site must join with the substrate based on shape. However, movement is also important. In order for substrates to react, they need to collide and they need to find the enzyme's active site. Molecules therefore need enough energy to move and collide. Often, the substrate or enzyme is anchored or immobilized in a membrane, allowing a more efficient joining of substrate and active site. The processes of cellular respiration and photosynthesis utilize enzymes embedded in membranes to carry out essential reactions efficiently. This is discussed in Chapters C1.2 and C1.3. The evolution of life has progressed largely as the result of the development of more efficient chemical reactions.

Immobilized enzymes are utilized in many present-day industrial practices. Attached to stationary surfaces, they are used in the following processes:
- conversion of carbohydrates to ethanol-based fuels (biofuels)
- production of dairy products and several beverages
- diagnosis of various diseases
- manufacture of penicillin and other antibiotics.

C1.1.7 – Mechanism of enzyme action

C1.1.7 – Relationships between the structure of the active site, enzyme–substrate specificity and denaturation

Students should be able to explain these relationships.

The following summarizes the mechanism of enzyme action.

- The surface of the substrate makes contact with the active site of the enzyme.
- The enzyme and substrate change shape to provide a fit.
- A temporary complex called the enzyme–substrate complex forms.
- The activation energy is lowered, and the substrate is altered by the rearrangement of the existing atoms.
- The transformed substrate, the product, is released from the active site.
- The unchanged enzyme is then free to combine with other substrate molecules.

Enzyme action can also be summarized by the following equation:

$$E + S \leftrightarrow ES \leftrightarrow E + P$$

where E is the enzyme, S is the substrate, ES is the enzyme–substrate complex, and P is the product.

As you have seen, the structure of the enzyme is key to its function. The active site must be a specific shape to allow the formation the enzyme–substrate complex. Each enzyme is specific for its substrate and anything that changes the shape of the enzyme (including denaturization) will affect the rate at which the enzyme works.

C1.1.8 – Factors affecting enzyme-catalysed reactions

Collision theory states that reactants of a chemical reaction must collide with one another with sufficient energy to react. They must also collide in the correct orientation so that chemical bonds are affected, allowing the chemical reaction to proceed. Many factors play a role in these collisions, including the concentration of reactants, temperature, nature of the reactants and catalysts present.

C1.1.8 – Effects of temperature, pH and substrate concentration on the rate of enzyme activity

The effects should be explained with reference to collision theory and denaturation.

Application of skills: Students should be able to interpret graphs showing the effects.

NOS: Students should be able to describe the relationship between variables as shown in graphs. They should recognize that generalized sketches of relationships are examples of models in biology. Models in the form of sketch graphs can be evaluated using results from enzyme experiments.

When you are considering the various environmental factors that affect enzyme-catalysed reactions, you must first remember that all chemical reactions are fundamentally caused by molecules colliding. If the molecules that are colliding do so at a high enough speed, and the molecules have the capability of reacting with each other, then there is a chance that a reaction will occur. Enzymes cannot change those fundamentals.

Effect of temperature

Imagine an enzyme and its substrate floating freely in a fluid environment. Both the enzyme and substrate are in motion and the rate of that motion is dependent on the temperature of the fluid. Fluids with higher temperatures will have faster moving molecules (more kinetic energy). Reactions are dependent on molecular collisions and, as a rule, the faster molecules are moving, the more often they collide, and with greater energy. Reactions with or without enzymes will increase their reaction rate as the temperature (and thus molecular motion) increases. However, reactions that use enzymes do have an upper limit (see Figure 2). That limit is based on the temperature at which the enzyme (as a protein) begins to lose its three-dimensional shape because the intramolecular bonds are being stressed and broken. When an enzyme loses its shape, including the shape of the active site, it is said to be denatured. **Denaturation** can be temporary, as in many instances the intramolecular bonds will re-establish when the temperature returns to a suitable level. Denaturation will be permanent if the increasing temperature is such that it prevents a return to the native, biologically shaped molecule.

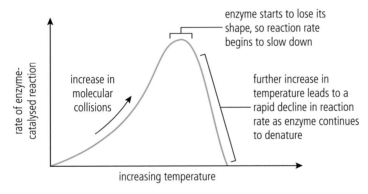

C1.1 Figure 2 The effect of increasing temperature on the rate of an enzyme-catalysed reaction.

Effect of pH

The active site of an enzyme typically includes a few amino acids. Some amino acids have areas that are charged either positively or negatively. The charged areas of the substrate must match the charged areas of the amino acids in the active site. When a solution has become acidic, the concentration of the hydrogen ions (H^+) rises. The hydrogen ions can bond with the negative charges of an enzyme or a substrate, and prevent matching of the charges between the two. A similar scenario occurs when a solution has become too basic: hydroxide ions (OH^-) can bond with the positive charges of a substrate or an enzyme, and once again prevent proper charge matching between the two. Either of these scenarios will result in an enzyme becoming less efficient. If the pH changes a great deal, bonds within the enzyme may start to break, causing the enzyme to lose its shape and thus become denatured.

Whether or not an enzyme is permanently destroyed by denaturation is largely dependent on whether covalent bonds (such as peptide bonds) have broken. Deoxyribonucleic acid (DNA) determines the order of amino acids, and they have no way of reassembling properly if they become detached from each other.

There is no one pH that is best for all enzymes (see Figure 3). Many of the enzymes operating in the human body are most active when in an environment that is near neutral. There are exceptions to this, however; for example, pepsin is an enzyme that is active in the stomach. The environment of the stomach is highly acidic, and pepsin is more active in an acidic pH.

C1.1 Figure 3 The effect of pH on the rate of an enzyme-catalysed reaction. This illustrates that there is no single pH that is best for all enzymes.

The pH scale is a logarithmic scale. This means that each whole number on the pH scale represents an increase or decrease by a power of 10. Thus, a solution with a pH of 4 has a 10 times greater concentration of hydrogen ions compared to a solution with a pH of 5. That same solution with a pH of 4 has a 100 times greater concentration of hydrogen ions compared to a solution with a pH of 6.

Effect of substrate concentration

If there is a constant amount of enzyme, as the concentration of a substrate increases, the rate of reaction will increase as well (see Figure 4). This is explained by the **collision theory**. If the concentration of reactant molecules increases, there are more molecules to react and collide with each other and the enzymes. There is a limit to this, however, because enzymes have a maximum rate at which they can work and only one active site. If every enzyme molecule has an active site that is occupied, adding more substrate to the solution will not increase the reaction rate further (see Figure 4).

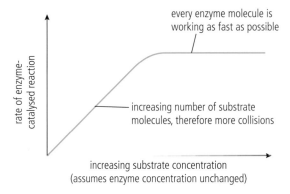

	pH
	1
strongly acidic	2
	3
	4
weakly acidic	5
	6
neutral	7
weakly alkaline	8
	9
	10
	11
strongly alkaline	12
	13
	14

The pH scale. Most fluids within the human body are close to neutral. The pH of blood plasma is typically 7.4, making it very slightly alkaline.

C1.1 Figure 4 The effect of increasing the substrate concentration on the rate of an enzyme-catalysed reaction.

What biological processes depend on differences or changes in concentration?

Nature of Science

Graphs are often used to show the relationship that exists between variables. Graphs such as the ones we have just examined concerning temperature, pH and substrate concentration on enzyme activity are examples of models in biology. These models in the form of sketch graphs can be evaluated using data from enzyme experiments.

An important skill to develop in science is the ability to interpret graphs. In this exercise you will examine and interpret two graphs involving enzyme activity. As you now know, temperature and pH all affect enzyme activity. The temperature at which a particular enzyme works most efficiently is known as the enzyme's **optimum temperature.** Similarly, the pH at which an enzyme works most efficiently is known as its **optimum pH**. Study the two graphs and answer the following questions.

1. Which of the two lines in the top graph represents an enzyme working within the human body?
2. The human stomach is strongly acidic while the human intestine is much less acidic. Which letter represents an enzyme functioning in the human intestine?
3. Pepsin is one of the enzymes that functions in the human stomach. What is likely to be the cause of its drastic decrease in rate of reaction as it moves into the small intestine?
4. In the top graph, the enzyme represented by letter B is placed in an environment with the temperature held at 40°C. It is then placed back into its optimum temperature environment. Will the enzyme work with a higher rate of activity at the higher temperature?

C1.1.9 – Measuring enzyme-catalysed reactions

C1.1.9 – Measurements in enzyme-catalysed reactions

Application of skills: Students should determine reaction rates through experimentation and using secondary data.

Measuring the rate of an enzyme-controlled reaction is similar to measuring the rate of any other reaction. It is possible to measure either the rate at which the substrate is used or the rate at which the product is produced. For example, in the investigation provided (see the link to eBook) the enzyme used is lactase, which is an enzyme that digests lactose sugar into glucose and galactose. The rate of the enzyme activity is measured by monitoring the rate at which glucose is produced.

See Section C1.1.5 for Section C1.1.10: they have been presented together.

Investigation of factors affecting enzyme activity. Full details of how to carry out this experiment with a worksheet are available in the eBook.

Guiding Question revisited

In what ways do enzymes interact with other molecules?

In this chapter we have looked at how:

- enzymes increase the rates of reactions in cells
- enzymes are globular proteins with an active site, and it is the active site that combines with a substrate to catalyse a reaction
- both enzymes and substrates change shape when binding occurs
- if the active site of an enzyme is changed, it will not be able to bind to the substrate and increase the rate of the reaction
- extremes of pH and temperature can change the shape of the enzyme and therefore the active site.

Guiding Question revisited

What are the interdependent components of metabolism?

In this chapter we have looked at how:

- metabolism often involves both cyclical and linear pathways
- the enzymes of metabolism and, therefore, the rates of metabolism are greatly affected by changes in pH, temperature and substrate concentration
- enzymes lower activation energy, allowing chemical reactions to occur faster
- feedback inhibition prevents the cell from wasting resources and energy by producing more products than are needed.

Exercises

Q1. Explain why enzymes only work with specific substrates.

Q2. What similar response in enzymes is caused by changes in both temperature and pH?

    A   Lowering of activation energy.

    B   Rate of movement of the enzyme.

    C   Rate of movement of the substrate.

    D   The three-dimensional shape of enzyme is altered.

Q3. What is activation energy?

    A   Energy released from the hydrolysis of a molecule.

    B   Energy required to initiate a chemical reaction.

    C   Energy of motion.

    D   Energy produced by a condensation reaction.

Q4. Compare the energy of the products in an endothermic and an exothermic reaction.

Q5. Describe the role of ATP in a cell in terms of exergonic and endergonic reactions.

Q6. Compare the types of chemical reactions associated with anabolism and catabolism.

C1.2 Cell respiration

Guiding Questions

What are the roles of hydrogen and oxygen in the release of energy in cells?

How is energy distributed and used inside cells?

All cells can convert organic molecules into the usable chemical energy known as adenosine triphosphate (ATP) through the process of cell respiration. This conversion involves a cascade of chemical reactions all with controlling mechanisms, so that efficiency and lack of damage to the cell are achieved. These chemical reactions occur in specialized cellular regions and involve molecules known as enzymes. Enzyme actions (discussed in Chapter C1.1) are essential to the process of cell respiration.

Hydrogen and oxygen play important roles in the production of ATP, the key energy-providing molecule of the cell. Energized hydrogen is an essential part of the ATP-generation process, while oxygen acts as the final electron acceptor in respiration to form water.

C1.2.1 – ATP structure and function

C1.2.1 – ATP as the molecule that distributes energy within cells

Include the full name of ATP (adenosine triphosphate) and that it is a nucleotide. Students should appreciate the properties of ATP that make it suitable for use as the energy currency within cells.

Organic molecules contain energy stored in their molecular structures. Each covalent bond in a molecule of glucose, an amino acid, or a fatty acid represents stored chemical energy.

Cells break down (or metabolize) their organic nutrients by slow oxidation. A molecule, such as glucose, is acted on by a series of enzymes. The function of these enzymes is to catalyse a series of reactions in which the covalent bonds are broken (oxidized) one at a time and new products are formed that have a lower energy. The goal of releasing energy in a controlled way is to store the released energy in the form of ATP molecules. If a cell does not have glucose available, other organic molecules may be substituted, such as fatty acids or amino acids.

The structure of ATP is discussed in Chapter C1.1. Nucleotides are compounds that consist of a 5-carbon sugar bonded to a nitrogenous base and a phosphate group. ATP is a nucleotide because it contains the 5-carbon sugar ribose, the nitrogenous base adenine, and three phosphate groups.

ATP has a specific chemical structure that allows it to function as the energy currency of the cell. The last two phosphate groups of ATP are attached to the main molecule by high-energy bonds.

Oxidation is a general type of chemical reaction resulting in products with lower potential energy than the reactant(s). When it occurs in cells, there is an increase in compounds with carbon to oxygen (C–O) bonds. The opposite reaction to oxidation is **reduction**. When reduction occurs, the products contain more potential energy than the reactants. Reduction in cells creates a higher number of compounds with carbon to hydrogen (C–H) bonds. Because oxidation and reduction reactions always occur together, these chemical reactions are known as **redox reactions**. Oxidation results in loss of electrons, while reduction involves a gain of electrons.

The two high-energy bonds (shown as wavy lines) attaching the final two phosphates to the molecule are important to its function. If the last phosphate is removed, the molecule becomes adenosine diphosphate (ADP). When two phosphates are missing, the molecule is known as adenosine monophosphate (AMP).

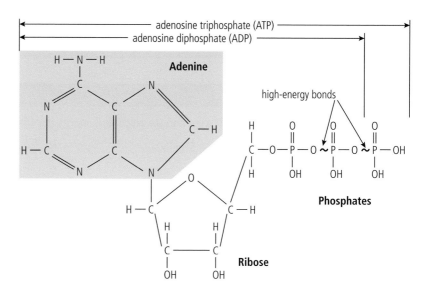

In what forms is energy stored in living organisms?

Because the phosphate groups are negatively charged, they repel one another, resulting in an unstable covalent bond between the two, referred to as a high-energy bond. These unstable bonds have a low activation energy and are easily broken by hydrolysis. This hydrolysis reaction is exergonic or energy releasing. The released energy is then free to perform cellular work.

C1.2.2 – Life processes within cells require ATP

C1.2.2 – Life processes within cells that ATP supplies with energy

Include active transport across membranes, synthesis of macromolecules (anabolism), movement of the whole cell or cell components such as chromosomes.

Cellular work carried out using the energy released from the high-energy bonds of ATP includes:

- active transport across cell membranes (discussed in Chapter B2.1)
- synthesis of macromolecules by anabolism (discussed in Chapter C1.1)
- movement of the whole cell by cilia or flagellum action
- movement within the cell of cell components, such as chromosome movement in mitosis or meiosis.

C1.2.3 – ATP and ADP

C1.2.3 – Energy transfers during interconversions between ATP and ADP

Students should know that energy is released by hydrolysis of ATP (adenosine triphosphate) to ADP (adenosine diphosphate) and phosphate, but energy is required to synthesize ATP from ADP and phosphate. Students are not required to know the quantity of energy in kilojoules, but students should appreciate that it is sufficient for many tasks in the cell.

Because ATP is needed for all cell activities, it must be continually produced. This production involves a cycle, as shown in Figure 1.

▲
C1.2 Figure 1 The ATP cycle. A cyclic pathway allows the formation of ATP from ADP and inorganic phosphate. This reaction requires energy. When hydrolysis of ATP occurs, the third phosphate group and energy are both released. This energy is then available for cellular work. A molecule of ADP is also formed.

In the ATP cycle it is important to note that energy is required to synthesize ATP from ADP and phosphate. This energy is then stored in the high-energy bond that exists between the second and third phosphate groups. Because of the input of energy and the higher potential energy of ATP than ADP, this reaction is an **endergonic** reaction. When ATP undergoes hydrolysis to form ADP plus a separated phosphate group, energy is released that makes this an **exergonic** reaction.

Nature of Science

Quantitative measurements are more objective in scientific research than qualitative measurements. Scientists use sensors to measure the amount of energy released when ATP undergoes hydrolysis. The sensors used are limited in precision and accuracy. However, with repeated procedures, measurements and peer reviews, more reliability in the measurements is possible.

i Most cells do not store large amounts of ATP. Cells can fulfil their continuous energy needs because the ATP cycle is constantly turning, as long as ADP, phosphate groups and energy are available. Because of this cycle, ATP is known as a renewable resource.

C1.2.4 and C1.2.5 – ATP and cell respiration

C1.2.4 – Cell respiration as a system for producing ATP within the cell using energy released from carbon compounds

Students should appreciate that glucose and fatty acids are the principal substrates for cell respiration but that a wide range of carbon/organic compounds can be used. Students should be able to distinguish between the processes of cell respiration and gas exchange.

C1.2.5 – Differences between anaerobic and aerobic cell respiration in humans

Include which respiratory substrates can be used, whether oxygen is required, relative yields of ATP, types of waste product and where the reactions occur in a cell. Students should be able to write simple word equations for both types of respiration, with glucose as the substrate. Students should appreciate that mitochondria are required for aerobic, but not anaerobic, respiration.

It is important to recognize that cellular respiration is different from gas exchange. Gas exchange is the process by which an organism obtains sufficient oxygen (O_2) and disposes of the waste gas carbon dioxide (CO_2). Cell respiration is the enzyme-controlled metabolic pathway that produces ATP.

Cell (cellular) respiration is the process by which most organisms on Earth synthesize ATP for cellular functions. It involves the release of energy from carbon compounds, especially glucose ($C_6H_{12}O_6$) and fatty acids. Carbohydrates (other than glucose), proteins and many other carbon-containing compounds can also be used in respiration.

We will focus on glucose as the carbon-containing compound because it is commonly catabolized in cellular respiration. It is important to keep the following equation in mind as we discuss this life-sustaining process for the cell:

The complete breakdown of glucose in cell respiration results in carbon dioxide, water and energy release. This catabolic reaction involves the removal of electrons from glucose (oxidation) and the acceptance of those electrons by oxygen (reduction).

Glucose is a high-energy molecule compared to carbon dioxide and water. Therefore, as this reaction proceeds, energy is released. The pathways of cellular respiration allow the slow release of energy from the glucose molecules so that ATP can be produced more efficiently.

Anaerobic cellular respiration

All cells begin the process of cell respiration in the same way. This initial stage is called **glycolysis**. Glucose enters a cell through the cell membrane and is found in the cytoplasm. Enzymes then catalyse reactions to ultimately cleave the 6-carbon glucose molecule into two 3-carbon molecules. Each of these 3-carbon molecules is called **pyruvate**. Some, but not all, of the covalent bonds in the glucose are broken during this series of reactions. Some of the energy that is released from the breaking of these bonds is used to form a small number of ATP molecules. Study Figure 2.

C1.2 Figure 2 A simplified version of the events of glycolysis

Notice in Figure 2 that two ATP molecules are needed to begin the process of glycolysis, and a total of four ATP molecules is formed. This is a net gain of two ATPs.

As we have said, glycolysis is the metabolic pathway that is common to most organisms on Earth. As can be seen in Figure 2, oxygen is not needed for glycolysis to proceed. Some organisms derive all their ATP without the use of oxygen. These organisms are said to carry out **anaerobic** cell respiration. The breakdown of organic molecules for anaerobic ATP production is called fermentation. There are two types of fermentation: **alcoholic fermentation** and **lactic acid fermentation**.

Our focus here is on the cell respiration that occurs in humans. If oxygen is not present after the initial stage of cell respiration, then in humans lactic acid fermentation commences. If your exercise rate exceeds your body's capacity to supply adequate oxygen, at least some of the glucose entering cell respiration will follow the anaerobic pathway called lactic acid fermentation. Study Figure 3.

C1.2 Figure 3 A simplified version of the events of lactic acid fermentation

The lack of adequate oxygen results in the conversion of each pyruvate produced by glycolysis into lactic acid molecules. Like pyruvate, lactic acid molecules are 3-carbon molecules. Lactic acid fermentation allows glycolysis to continue because there is not a build-up of pyruvate. However, only two ATP molecules are generated from anaerobic respiration.

Aerobic cellular respiration

Aerobic cell respiration produces many more ATP molecules than anaerobic cell respiration, and it occurs in the presence of oxygen. Like anaerobic respiration, aerobic respiration begins with glycolysis producing two molecules of pyruvate in the cytoplasm of the cell. The two pyruvates then enter the mitochondrion.

Once inside the mitochondrion, the pyruvate molecules are turned into a 2-carbon compound that enters the next stage of respiration, which is called the **Krebs cycle**. The preparatory reaction is known as the **link reaction** and takes place in the matrix of the mitochondria. The Krebs cycle also takes place in the matrix of the mitochondrion, and is a series of reactions that begins and ends with the same molecule. A net gain of two ATPs occurs in the Krebs cycle.

Alcoholic fermentation is a type of anaerobic respiration that occurs in yeast. Yeast cells take in glucose, and glycolysis occurs as shown in Figure 2. As there is no oxygen present, the pyruvate molecules are then converted into two molecules of ethanol. Ethanol is a 2-carbon compound. The third carbon of pyruvate is lost from the system, ultimately combining with oxygen to form carbon dioxide. The carbon dioxide is released into the cellular environment.

You may have experienced the muscle burn that occurs as a result of lactic acid accumulation during intense exercise. The burn goes away when adequate supplies of oxygen are provided to the muscle so that aerobic cell respiration can occur. The lactic acid is carried to the liver via the bloodstream, where it is converted to glucose, then glycogen.

This high-resolution, false-colour transmission electron micrograph (TEM) shows a single mitochondrion. It is in this organelle that the preparatory (link) reaction and the Krebs cycle occur.

The final stage of aerobic respiration is the **electron transport chain**, which occurs in the cristae of the mitochondrion. Most ATP molecules produced from the breakdown of glucose are made in the electron transport chain: 30–34 ATPs are produced in this stage. More details about the stages of aerobic respiration are given in later sections of this chapter.

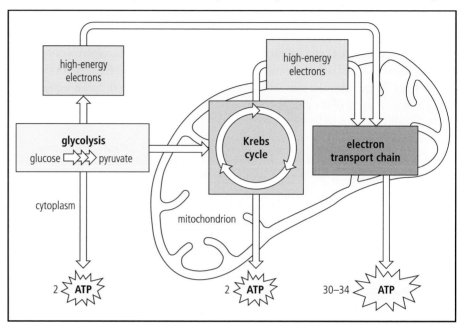

Research is continuing to determine the exact number of ATP molecules produced by aerobic cellular respiration. Roughly 30–40% of the available energy in a glucose molecule is transferred to ATP. The rest is transferred out of the cell in the form of heat.

An overview of aerobic cell respiration. There is a link or preparatory reaction that changes the pyruvate produced by glycolysis into a 2-carbon compound that enters the Krebs cycle. The link reaction and Krebs cycle occur in the matrix of the mitochondrion, while the electron transport chain occurs in the cristae

A summary of anaerobic and aerobic respiration

Examine the overview given in Table 1 of anaerobic and aerobic respiration in humans.

C1.2 Table 1 Anaerobic and aerobic cell respiration

Anaerobic cell respiration	Aerobic cell respiration
Does not require oxygen but does require glucose	Requires oxygen and glucose
Takes place in the cytoplasm of the cell	Begins in the cytoplasm
Glucose is split into two molecules of pyruvate	The product of the first part of respiration is two molecules of pyruvate made from glucose
If the oxygen supply is inadequate, in humans, pyruvate is made into lactic acid. This fermentation will occur in the cytoplasm	If oxygen supply is adequate, pyruvate will move into the mitochondria
No mitochondria are needed	Pyruvate is converted into a 2-carbon compound in the matrix of mitochondria
Net gain of two ATPs	The 2-carbon compound enters the Krebs cycle, also in the mitochondrial matrix
	Carbon dioxide is produced as a waste product of the Krebs cycle
	30–34 ATPs are produced in the cristae of the mitochondria

Key points to remember are:

- both types of cellular respiration initially take place in the cytoplasm
- in both cases glucose (a 6-carbon molecule) is broken down into two molecules of pyruvate (a 3-carbon molecule)
- the production of ATP is very low in anaerobic cellular respiration compared to aerobic cellular respiration
- anaerobic cell respiration occurs outside the mitochondria (in the cytoplasm) and does not require oxygen
- aerobic cell respiration starts in the cytoplasm but finishes within the mitochondria and requires oxygen
- the final products of anaerobic respiration in humans are lactic acid and ATP
- the final products of aerobic respiration are carbon dioxide, water and ATP.

C1.2.6 – The rate of cell respiration

C1.2.6 – **Variables affecting the rate of cell respiration**
Application of skills: Students should make measurements allowing for the determination of the rate of cell respiration. Students should also be able to calculate the rate of cellular respiration from raw data that they have generated experimentally or from secondary data.

Because cell respiration involves a series of chemical reactions, there are many factors that affect its overall rate. Some of these factors are listed below.

- Temperature: the optimum temperature for the rate of cell respiration is 20–30°C. Significantly higher and lower temperatures greatly decrease the rate.
- Carbon dioxide concentration: an increase in carbon dioxide concentration adversely affects the rate of cell respiration.
- Oxygen concentration: lower concentrations of oxygen lower the rate of cell respiration. The absence of oxygen results in anaerobic respiration.
- Glucose concentration: low levels of glucose in the cell will decrease the rate of cell respiration.
- Type of cell: some types of cells require more energy than others. Those that require more energy have higher cell respiration rates.

Factors that affect cell respiration can be determined experimentally by calculating the rate of cell respiration using raw or secondary data. **Respirometers** are often used to calculate the rate of cell respiration. The following Worked example demonstrates how the rate of cell respiration can be calculated using a respirometer.

Worked example

Respirometers are devices used to measure an organism's rate of respiration by measuring the oxygen rate of exchange. They are sealed units in which any carbon dioxide produced is absorbed by an alkali such as soda lime or potassium hydroxide. Absorbing the carbon dioxide allows an accurate measurement of oxygen exchange. These devices may work at a cellular level or at a whole-organism level. Look at the graph and answer the questions. The *y*-axis of the graph represents the relative amount of oxygen used.

Oxygen consumption by germinating and non-germinating pea seeds at 12°C and 22°C.

1. In the germinating pea seeds, what type of respiration is occurring? What is the evidence for this answer?

2. Why is the oxygen consumption of non-germinating pea seeds very low?

3. Why would the germinating seeds show a greater oxygen consumption at 22°C than at 12°C?

4. Predict how the graph would look for non-germinating seeds at 12°C.

Solution

1. Aerobic. There is a significant amount of oxygen consumption occurring.

2. They are not carrying out respiration and have a low metabolic rate.

3. At 22°C the rate of respiration is faster than at 12°C. Therefore, there is a greater oxygen consumption at the higher temperature.

4. The line of the graph would be almost right on the non-germinating (22°C) line that exists now. A prediction that it would be just slightly lower is best.

Nature of Science

The use of animals in such experiments has ethical implications. It is essential to refer to the IB animal experimental policy before carrying out any procedures on animals. Scientists have obligations to assess the risks associated with their work. Most aim to do no harm. Ethical and environmental consequences must be constantly considered when carrying out research.

Experiment to determine the effect of environmental factors on anaerobic respiration rates. Full details of how to carry out this experiment with a worksheet are available in the eBook.

Guiding Question revisited

What are the roles of hydrogen and oxygen in the release of energy in cells?

In this chapter we have described how:

- oxidation of carbon compounds occurs when hydrogen is removed from a molecule, each hydrogen atom is made of one electron and one proton
- oxygen is required in aerobic cell respiration, but not for anaerobic cell respiration
- glycolysis does not require oxygen to proceed
- the energy to produce ATP from ADP comes from the energy present in the electrons of hydrogen removed during oxidation of carbon compounds.

What are the consequences of respiration for ecosystems?

Guiding Question revisited

How is energy distributed and used inside cells?

In this chapter we have discussed how:

- ATP is known as the energy currency of the cell and is used to distribute energy within cells
- ATP is a nucleotide with a high-energy phosphate bond
- ATP is essential for active transport across cell membranes, the synthesis of macromolecules needed by the cell, and all types of movement involving the cell
- cell respiration is the metabolic pathway that produces ATP in the cell.

Exercises

Q1. Explain the properties of ATP that make it suitable for use as the energy currency of the cell.

Q2. Explain why very high temperatures dramatically decrease the overall rate of cell respiration.

Q3. State the final products of anaerobic respiration.

Q4. State the final products of aerobic respiration in humans.

Q5. Explain reasons for referring to ATP as the energy currency of the cell.

Q6. Explain the products and major locations within the cell where the anaerobic phases of cellular respiration occur.

C1.3 Photosynthesis

Guiding Questions

How is energy from sunlight absorbed and used in photosynthesis?

How do abiotic factors interact with photosynthesis?

Plants are an amazing group of organisms. Their ability to absorb light energy and convert it into chemically stored energy opened the world to the extraordinary development of all other organisms on our planet. Photosynthesis is the cornerstone of almost all life forms that exist today. We are continuously studying plants to better understand their ability to absorb and use sunlight. Artificial leaves and solar panels are just a couple of the spinoffs from this research.

Of course, plants need more than just sunlight to carry out photosynthesis. They need water, the correct temperature, carbon dioxide and a host of chemicals to convert sunlight into chemical energy efficiently. Many of the factors required are abiotic or non-living factors. Today we are able to manipulate some of these abiotic factors in greenhouses and other agricultural settings, so that crop yields are higher than ever before.

C1.3.1 – Light energy and life processes

C1.3.1 – Transformation of light energy to chemical energy when carbon compounds are produced in photosynthesis

This energy transformation supplies most of the chemical energy needed for life processes in ecosystems.

Life on the eastern Colorado plains (USA) and everywhere else in the world ultimately depends on the process of photosynthesis. Nearly every oxygen molecule we breathe in was once part of a water molecule that was turned into oxygen through photosynthesis. All the energy present on Earth has a connection to this process. An understanding of photosynthesis is essential to appreciate the intertwined ecosystems of our planet.

Plants and other photosynthetic organisms produce foods that start food chains. Most living things rely on the Sun to provide the energy needed for both warmth and food production. However, the sunlight that strikes Earth must be converted into a form of chemical energy if it is to be useful. It is this transformation of light energy into chemical energy that fuels the life processes in ecosystems. The chemical energy produced by photosynthesis can be in the form of glucose. If you recall, glucose is also the molecule that organisms use for the process of cell respiration.

Chapter C1.2 discusses respiration and considers how the cell breaks down chemical bonds in glucose to produce adenosine triphosphate (ATP). In this chapter we are going to find out how chemical bonds are made to produce carbon compounds. The raw materials of photosynthesis are carbon dioxide and water but the process also requires light. The overall word equation is:

$$\text{carbon dioxide} + \text{water} \xrightarrow{\text{light}} \text{glucose} + \text{oxygen}$$

C1.3.2 and C1.3.3 – The equation for photosynthesis

C1.3.2 – Conversion of carbon dioxide to glucose in photosynthesis using hydrogen obtained by splitting water

Students should be able to write a simple word equation for photosynthesis, with glucose as the product.

$$\text{carbon dioxide} + \text{water} \xrightarrow{\text{light}} \text{glucose} + \text{oxygen}$$

C1.3.3 – Oxygen as a by-product of photosynthesis in plants, algae and cyanobacteria

Students should know the simple word equation for photosynthesis. Students should know that the oxygen produced by photosynthesis comes from the splitting of water.

During photosynthesis water is split to give hydrogen and oxygen. Oxygen is a by-product of photosynthesis: it is not needed by the plant and is released into the atmosphere. The hydrogens released during the splitting of water reduce carbon dioxide to form glucose. The chemical equation for photosynthesis showing the accompanying redox reactions is shown in Figure 1.

As we proceed in our discussion of photosynthesis, we will refer to this equation often. Plants, algae and cyanobacteria all carry out photosynthesis, and as a result all release oxygen into the atmosphere as a by-product.

Organisms that use light to produce their own food are called **photoautotrophs**. *The United States Department of Agriculture estimates that up to 200 billion metric tons of sugar are produced each year by these organisms.* **Heterotrophs** *are organisms that cannot produce their own food.* **Autotrophs** *are organisms that can produce their own food. Not all autotrophs are photoautotrophs, some are* **chemoautotrophs** *(chemotrophs), which use chemical reactions to obtain their energy.*

SKILLS

What gas is produced during photosynthesis? Full details of how to carry out this experiment with a worksheet are available in the eBook.

C1.3 Figure 1 The redox reactions of photosynthesis

What are the consequences of photosynthesis for ecosystems?

C1.3.4 – Photosynthetic pigments and light absorption

C1.3.4 – Separation and identification of photosynthetic pigments by chromatography

Application of skills: Students should be able to calculate R_f values from the results of chromatographic separation of photosynthetic pigments and identify them by colour and by values. Thin-layer chromatography or paper chromatography can be used.

Most plant leaves appear green to our eyes. If you were able to zoom into leaf cells and look around, you would see that the only structures in a leaf that are actually green are the **chloroplasts**. Plants contain a variety of pigments in the chloroplasts. The photosynthetic pigment that dominates in most plant species is the molecule **chlorophyll**. There are actually several different types of chlorophyll, and these produce the characteristic green colour of most plants. Of the different types of chlorophyll, chlorophyll a and chlorophyll b are the most common. We can also find a group of pigments called carotenoids. The carotenoids usually include the specific pigments known as carotene and xanthophyll. Each of the pigments and their concentrations are unique to a plant species. The pigments that a particular plant contains can be separated by a process known as chromatography.

Inside each of these plant leaf cells are many green chloroplasts. Each chloroplast is loaded with light-absorbing pigments.

SKILLS

In Chapter B2.2, **chromatography** is presented as one of the techniques that allows the isolation or separation of pure substances from a complex material. Gel and ion-exchange chromatography are two types that are often used. Another very common type of chromatography is paper chromatography. This technique uses a paper sheet or strip through which a solution can pass to separate pure substances. The paper is referred to as the stationary phase because it does not move. The mobile phase is a solvent that moves up the stationary phase (paper) by capillary action. As the mobile phase moves through the stationary phase, it carries the components of the mixture being analysed with it. During this movement, pure substances separate from each other because they have different migration rates across the stationary phase. It is an inexpensive yet very powerful means of analysing the components of a solution. This technique is used often to separate the different pigments found in plants. By using this technique, a variety of pigments have been found in different plants. Examine the chromatogram on the right.

Pigment	R_f value
carotene	0.95
xanthophyll	0.71
chlorophyll *a*	0.65
chlorophyll *b*	0.45
leaf stain	

the pigments separate, as they move different distances up the stationary phase

▲ A paper chromatograph obtained from the leaf of a plant showing the various pigments that are necessary for photosynthesis. Each of the pigments migrated at a different rate to produce the separate bands, which are labelled with the name of the pigment. The R_f value is the ratio of the distance travelled by the pigment to the distance travelled by the solvent (mobile phase).

Worked example

The chromatogram on the right was obtained in the laboratory. Calculate the R_f value for each of the pigments by using the following formula:

$$R_f = \frac{\text{distance moved by substance}}{\text{distance moved by solvent}}$$

Once you have calculated the R_f value for each pigment, answer the questions below.

(a) Explain which of the pigments was most soluble in the solvent.

(b) These colours were not visible in the leaf of the plant. Explain why they are visible on the chromatograph.

(c) Suggest why the calculated R_f values are different to those in the chromatogram above.

Distance migrated

12.0 cm — solvent front
11.5 cm — carotene

— xanthophyll
4.5 cm — chlorophyll *a*
2.0 cm — chlorophyll *b*

Solution

(a) Carotene was most soluble in the solvent because it moved farther than the other pigments on the chromatogram.

(b) Pigments are often masked by other pigments in a plant because some colours are more intense. Also, the quantity of a pigment present in a plant may affect whether it is masked or not.

(c) There are different solvents that can be used to separate plant pigments. Each solvent interacts with the pigments present in different ways. For example, the solubilities of the pigments may be different in alternative solvents. This would cause variation in R_f values. The attraction of the pigment to the stationary phase may also change if different materials are used.

Separation of photosynthetic pigments by chromatography. Full details of how to carry out this experiment with a worksheet are available in the eBook.

Chromatography allows us to determine the pigments present in chloroplasts. Pigments with higher R_f values are more soluble in the solvent being used, and they are often smaller in size. Higher R_f value pigments have a lower affinity for the paper used in the chromatography activity.

Plants make use of the same part of the electromagnetic spectrum that our eyes can see. We call this the visible portion of the spectrum. Sunlight is a mixture of different colours of light. You can see these colours when you let sunlight pass through a prism.

The electromagnetic spectrum. Notice that the visible light portion of this spectrum has colours with wavelengths between 400 nm and 740 nm.

The visible light spectrum includes many colours but, for the purpose of considering how plant pigments absorbs light energy, we are going to consider three regions of the spectrum:

- the red end of the spectrum
- the green middle of the spectrum
- the blue end of the spectrum.

Substances can do one of only two things when they are struck by a particular wavelength (colour) of light. They can:

- absorb that wavelength (if so, energy is being absorbed and may be used)
- reflect that wavelength (if so, the energy is not being absorbed and you will see that colour).

Worked example

You are walking outside with a friend who is wearing a red and white shirt. Explain why the shirt appears to be red and white.

Solution

Sunlight is a mixture of all the wavelengths (colours) of visible light. When sunlight strikes the red pigments in the shirt, the blue and the green wavelengths of light are absorbed, but the red wavelengths are reflected. Thus, our eyes see the colour as red. When sunlight strikes the white areas of the shirt, all the wavelengths of light are reflected, and our brain interprets the mixture as white.

C1.3.5 and C1.3.6 – Absorption and action spectra

C1.3.5 – Absorption of specific wavelengths of light by photosynthetic pigments

Include excitation of electrons within a pigment molecule, transformation of light energy to chemical energy and the reason that only some wavelengths are absorbed. Students should be familiar with absorption spectra. Include both wavelengths and colours of light in the horizontal axis of absorption spectra.

C1.3.6 – Similarities and differences of absorption and action spectra

Application of skills: Students should be able to determine rates of photosynthesis from data for oxygen production and carbon dioxide consumption for varying wavelengths. They should also be able to plot this data to make an action spectrum.

The ability of photoautotrophs to absorb light energy is determined by the pigments present on the membranes of the chloroplasts. It is the pigments that absorb the light. The amount of light absorbed plotted against the wavelength of light produces the **absorption spectrum** for that pigment. The **action spectrum** indicates the rate of photosynthesis at different wavelengths of light. The rate of photosynthesis can be calculated from either the rate of oxygen produced, or the rate of carbon dioxide used up (see Section C1.3.7). Study the figure on the next page showing the absorption and action spectra of some common photosynthetic pigments.

When reading any graph it is essential you note what each axis represents. By looking at the two *y*-axes in the figure, it is clear which one represents the absorption spectrum and which represents the action spectrum. In the case of the *x*-axes, note that the blue light has a shorter wavelength than red light.

Determining the photosynthetic rate and production of an action potential

The rate of photosynthesis can be determined in a number of ways. Two of the most commonly used methods involve measuring oxygen production over time and carbon dioxide consumption over time. Once the photosynthesis rate has been calculated using either of these two methods, an action spectrum can be constructed, such as shown in the second graph in the figure below. The first graph is an example of an absorption spectrum, showing the absorption of light by various photosynthetic pigments of a plant.

The top graph represents the absorption spectrum of common photosynthetic pigments. Carotenoid pigments are yellow, orange and red pigments, and the absorption spectrum for them as a group is presented here. The bottom graph represents the action spectrum of photosynthesis, plotting the overall photosynthetic rate at different wavelengths (colours).

1. At wavelengths where there is a higher percentage of light absorption, the rate of photosynthesis is also high. Absorption is greatest for the blue (400–500 nm) and red (600–700 nm) wavelengths.

2. The least light absorption occurs at with the green (500–600 nm) wavelength. Plants reflect green light rather than absorb it: this is why they look green to us.

3. When absorption of light by a pigment is low, the contribution of that pigment to the rate of photosynthesis will be low. Green light is therefore the least effective type of light for photosynthesis.

A summary of the differences and similarities between absorption and action spectra is presented in Table 1.

Absorption spectrum	Action spectrum
Varies depending on type of photosynthetic pigment present	Varies depending on type of photosynthetic pigment present
Represents the amount of light energy being absorbed by the photosynthetic pigment	Represents the rate of the photosynthetic process being carried out by the pigment
For the plant, this spectrum represents the light absorbed by all the pigments present	For the plant, this spectrum represents the rate of photosynthesis as a result of all the pigments present
Chlorophylls *a* and *b* have a high absorption of light energy in the violet-blue and red light wavelengths	Chlorophylls *a* and *b* create a relatively high efficiency rate of photosynthesis
Pigments like carotenoids absorb light energy at different wavelengths compared to chlorophyll *a* and *b*	Pigments like carotenoids allow photosynthesis at different wavelengths
Other pigments are not as efficient at absorbing light energy as chlorophylls *a* and *b*	Other pigments are not as effective at achieving high rates of photosynthesis as chlorophylls *a* and *b*

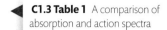 **C1.3 Table 1** A comparison of absorption and action spectra

Light energy behaves as if it exists in discrete packets called **photons**. Shorter wavelengths of light have greater energy within their photons than longer wavelengths. Photons can transfer their energy upon interaction with other particles. This transfer of energy occurs many times during photosynthesis.

When a pigment absorbs light, the energy is used to raise an electron in the pigment to a higher energy level. This is known as **excitation of electrons**. Raising an electron to a higher energy level requires a specific amount of energy (or specific photons of light). This explains why different pigments absorb different wavelengths of light: they each need a different wavelength to excite electrons. Once the electrons are excited to a higher energy level, this energy can then be used to make chemical bonds. The net result is that light energy is transformed into chemical energy.

What are the functions of pigments in living organisms?

C1.3.7 – Measuring the rate of photosynthesis

C1.3.7 – Techniques for varying concentrations of carbon dioxide, light intensity or temperature experimentally to investigate the effects of limiting factors on the rate of photosynthesis

Application of skills: Students should be able to suggest hypotheses for the effects of these limiting factors and to test these by experimentation.

NOS: Hypotheses are provisional explanations that require repeated testing. During scientific research, hypotheses can either be based on theories and then tested in an experiment or be based on evidence from an experiment already carried out. Students can decide in this case whether to suggest hypotheses for the effects of limiting factors on photosynthesis before or after performing their experiments. Students should be able to identify the dependent and independent variable in an experiment.

Look again at the summary reaction for photosynthesis:

$$6CO_2 + 6H_2O \rightarrow C_6H_{12}O_6 + 6O_2$$

This balanced equation shows us that carbon dioxide molecules are reactants and oxygen molecules are products of photosynthesis. If you recall some of the information you have learned about cell respiration, you will see that the reverse is true for that process. In other words, in cell respiration oxygen is a reactant and carbon dioxide is a product.

At any given time of year, any one plant has a fairly consistent rate of cell respiration. Not only is this rate consistent throughout the day and night, but it is also at a relatively low level. Plants need ATP for various biochemical processes, but the levels they need are typically far lower than animals need.

The same consistency is not true regarding the rate of photosynthesis. The photosynthetic rate is highly dependent on many environmental factors, including the intensity of light and air temperature. During the daytime, especially on a warm, sunny day, the rate of photosynthesis may be very high for a particular plant. If so, the rate of carbon dioxide taken in by the plant and the rate of oxygen released will also be very high. Because the plant is also carrying out cell respiration, a correction needs to be made to the carbon dioxide and oxygen levels. At night, the rate of photosynthesis may drop to zero. At that time, a particular plant may be releasing carbon dioxide and taking in oxygen to maintain its relatively low but consistent rate of cell respiration.

▲ A diagram showing the elements of a photosynthometer, used to measure the rate of photosynthesis with various independent variables.

Measuring the rate of oxygen production or carbon dioxide intake allows direct measurement of photosynthetic rate, if a correction is made for cell respiration.

This student is measuring oxygen production by an aquatic plant. Oxygen is collected and the volume measured using a **photosynthometer** centre. With this apparatus, you can determine the effect of various environmental conditions, such as varying light intensities or temperature, on the rate of photosynthesis.

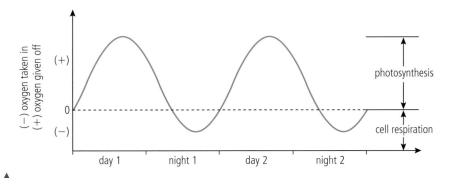

A graph showing the oxygen given off and taken in by a hypothetical plant over a 48-hour period. When the line intersects 0, the oxygen generated by photosynthesis is equal to the oxygen needed for cell respiration.

Another common method for measuring photosynthesis is to track changes in the biomass of experimental plants. However, the mass of plants is an indirect reflection of photosynthetic rate, as an increase or decrease in biomass can be caused by a variety of factors as well as the photosynthetic rate.

SKILLS

Experimental methods for measuring the rate of photosynthesis

Many techniques can be used to determine the rate of photosynthesis, including the following.

- Counting bubbles given off by aquatic plants. The bubbles are of oxygen and their number per unit of time can be recorded. The volume of the oxygen produced can also be measured.
- Solutions such as hydrogen carbonate indicators or universal indicators can be used with aquatic plants to measure pH changes. Carbon dioxide in water produces carbonic acid. As carbon dioxide is consumed, the pH of the solution becomes less acidic, with a corresponding colour change in the indicator. Colour changes can be recorded over time.
- Electronic sensors can also be used to measure the amount of carbon dioxide and oxygen, in both aquatic and carefully controlled atmosphere settings.

The effects of temperature, light wavelength, light intensity and other independent variables can all be determined using the above techniques. When looking at wavelengths of light, the data can then be used to produce an action spectrum.

Limiting factors

In 1905, British plant physiologist Frederick Frost Blackman proposed the **law of limiting factors**. It stated that a process that depends on multiple factors will have a rate limited by the factor at its least favourable (lowest) value. In many cases, this is the factor that is in "shortest" supply. Photosynthesis as a metabolic process has many potential limiting factors, including the amount of water, sunlight, temperature, carbon dioxide, chloroplasts and chlorophyll. Scarcity of any one of these factors influences the rate of photosynthesis. Study Figures 2–4, illustrating how light intensity, temperature and carbon dioxide concentration can affect the rate of photosynthesis. After becoming familiar with the graphs, complete the Challenge yourself activity.

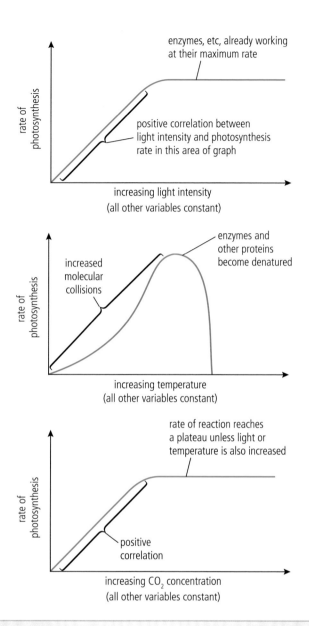

C1.3 Figure 2 The effect of increasing light intensity on the rate of photosynthesis.

C1.3 Figure 3 The effect of increasing temperature on the rate of photosynthesis.

C1.3 Figure 4 The effect of increasing carbon dioxide concentration on the rate of photosynthesis.

Challenge yourself

Use Figures 2–4 above to answer the following questions about photosynthesis and limiting factors.

1. Look at Figure 2. Explain why the early part of the graph is labelled as a positive correlation.

2. Look at Figure 3. Why does the denaturing of enzymes and other proteins at higher temperatures dramatically lower the rate of photosynthesis?

3. For Figure 4, suggest what could cause a change from the plateau to an increasing rate, besides increasing CO_2 concentration.

4. Design a procedure to investigate the effect of one of the limiting factors mentioned above on the rate of photosynthesis. Note that water for photosynthesis experiments can be made to be free of dissolved carbon dioxide by boiling and then cooling it.

5. Suggest a hypothesis that can be tested using your procedure. Remember that your hypothesis for one of these experiments should be based on your knowledge and understanding of photosynthesis.

Nature of Science

When designing your procedure to investigate the effect of a limiting factor on the rate of photosynthesis, you may have had an idea about what would happen. A hypothesis is a provisional explanation that requires repeated testing to verify it, and is based on knowledge. During scientific research, hypotheses can be based on theories and then tested in an experiment, or they may be based on evidence from an experiment that has already been carried out. It is acceptable to suggest hypotheses for the effects of limiting factors on photosynthesis before or after performing experiments. You should be able to identify the dependent and independent variables in an experiment. **Dependent** variables are the outcome of the procedure or the value that you are measuring, in this case the rate at which carbon dioxide is used or oxygen is produced. The **independent** variable is the variable that is being changed in the experiment, for example the temperature, or the wavelength of the light.

The ability to ask a meaningful question or create a hypothesis is often quite challenging to the scientist. What is the role of imagination and intuition in the creation of hypotheses in the natural sciences?

C1.3.8 – Carbon dioxide levels and future rates of photosynthesis

C1.3.8 – Carbon dioxide enrichment experiments as a means of predicting future rates of photosynthesis and plant growth

Include enclosed greenhouse experiments and free-air carbon dioxide enrichment experiments (FACE).

NOS: Finding methods for careful control of variables is part of experimental design. This may be easier in the laboratory but some experiments can only be done in the field. Field experiments include those performed in natural ecosystems. Students should be able to identify a controlled variable in an experiment.

Many experiments have been carried out or are ongoing to determine the effects of increasing carbon dioxide levels in our atmosphere on plants and photosynthetic rates. From our discussion of carbon dioxide as a limiting factor, you might predict that increasing carbon dioxide levels would increase plant productivity. This has been verified. However, it is interesting to note that the growth rate of weeds increases more than the growth rate of crop plants and trees.

A study recently conducted by Peter Wayne and colleagues in controlled greenhouses involved ragweed (*Ambrosia artemisiifolia*) plants, which produce pollen that causes allergy symptoms in many people. The results of the study showed that a doubling of carbon dioxide levels stimulated ragweed pollen production by 61%. To a person with ragweed pollen allergy, this is not good news.

As well as many greenhouse-based experiments, studies are being conducted in natural settings. Natural settings often provide conditions that cannot be controlled in the same way as laboratory-based procedures can. These studies are known as free-air

Studies are being carried out all over the world to better understand the present and future effects of an ever-increasing amount of carbon dioxide in our atmosphere. Early hypotheses included the possible beneficial effects of higher carbon dioxide levels. However, at present relatively few plants are able to increase photosynthetic rates in a beneficial way to our planet. Most studies have found serious problems with increased levels of carbon dioxide. As carbon dioxide has increased, so has temperature, resulting in reduced water and mineral availability and lowered photosynthesis rates. Also, with increased carbon dioxide levels, studies have shown that many food crops (such as wheat) have lower protein levels. Increased carbon dioxide levels also seem to lead to greater plant ingestion by animals. This may be because the plants are producing less defensive chemicals (toxins). Overall, the rate of photosynthesis by plants appears to be dropping.

carbon dioxide enrichment experiments (FACE). FACE allows us to examine the effects of increasing carbon dioxide levels on plants in natural and agricultural ecosystems. By artificially increasing the carbon dioxide levels in these natural systems, we gain a more reliable picture of what may happen in the real-world future of steadily increasing carbon dioxide levels.

Nature of Science

Control of variables is an essential part of experimental design. Controlling as many variables as possible increases our ability to arrive at valid and reliable conclusions after an experimental procedure. **Reliability** is the degree to which an assessment tool produces stable and consistent results. **Validity** refers to how well a test measures what it is intended to measure. Controlling variables is usually much easier in the laboratory, but some experiments can only be done in the field. Field experiments include those performed in natural ecosystems, such as described in FACE studies.

Guiding Question revisited

How is energy from sunlight absorbed and used in photosynthesis?

In this chapter we have discussed how:

- absorption of light for photosynthesis is essential to produce carbon compounds (chemical energy) and atmospheric oxygen
- oxygen is formed from the splitting of water during photolysis and is a waste product of photosynthesis
- photosynthetic pigments such as the carotenoids and chlorophylls absorb specific wavelengths of light
- the light-dependent reactions require the pigments of photosystems I and II to proceed efficiently.

Guiding Question revisited

How do abiotic factors interact with photosynthesis?

In this chapter we have discussed how:

- carbon dioxide and water are two essential reactants of photosynthesis
- carbon dioxide is converted to glucose after reduction during carbon fixation
- temperature, light intensity and carbon dioxide concentration are all potential abiotic limiting factors of photosynthesis.

Exercises

Q1. Compare and contrast action and absorption spectra.

Q2. List the reactants and the products of photosynthesis.

Q3. Explain two reasons why the splitting of water in the photosynthetic process is so important to ecosystems.

C1 Practice questions

1. Which of the following statements is **true** about enzymes?

   A   They are used up in the reactions they catalyse.

   B   Allosteric inhibitors bind to the active site.

   C   They lower the energy of activation for a reaction.

   D   They supply the energy of activation for a reaction.

   (Total 1 mark)

2. Explain why plant cells need mitochondria.   *(Total 3 marks)*

3. Explain the effect of varying temperature on the rate of reaction of an enzyme controlled reaction.   *(Total 3 marks)*

4. At the start of glycolysis, glucose is phosphorylated to produce glucose 6-phosphate, which is converted into fructose 6-phosphate. A second phosphorylation reaction is then carried out, in which fructose 6-phosphate is converted into fructose 1,6-bisphosphate. This reaction is catalysed by the enzyme phosphofructokinase. Biochemists measured the enzyme activity of phosphofructokinase (the rate at which it catalysed the reaction) at different concentrations of fructose 6-phosphate. The enzyme activity was measured with a low concentration of ATP and a high concentration of ATP in the reaction mixture. The graph below shows the results.

   (a) (i)   Using **only** the data in the above graph, outline the effect of increasing fructose 6-phosphate concentration on the activity of phosphofructokinase, at a low ATP concentration.   (2)

   (ii)   Explain how increases in fructose 6-phosphate concentration affect the activity of the enzyme.   (2)

   (b) (i)   Outline the effect of increasing the ATP concentration on the activity of phosphofructokinase.   (2)

   (ii)   Suggest an advantage to living organisms of the effect of ATP on phosphofructokinase.   (1)

   (Total 7 marks)

5. Biosphere 2, an enormous greenhouse built in the Arizona desert in the USA, has been used to study five different ecosystems. It is a closed system so measurements can be made under controlled conditions. The effects of different factors, including changes in carbon dioxide concentration in the greenhouse, were studied. The data shown below were collected over the course of one day in January 1996.

(a) (i) Identify the time of day when the Sun rose. (1)

    (ii) Identify the time of minimal CO_2 concentration. (1)

(b) Determine the maximum difference in the concentration of CO_2 over the 24-hour period. (1)

(c) Suggest reasons for changes in CO_2 concentration during the 24-hour period. (2)

(Total 5 marks)

6. Inflammation of human tissues often causes pain. Cyclooxygenases (COX) are a group of enzymes that play a role in causing inflammation. Analgesics are drugs that can reduce pain. The graph below shows how increasing concentrations of the analgesic drug dipyrone affects the activity of three different cyclooxygenases, COX-1, COX-2 and COX-3.

(a) Outline the relationship between dipyrone concentration and COX-3 activity. (2)

(b) Deduce whether dipyrone is an inhibitor of COX-2. (2)

(c) Evaluate the potential of dipyrone as an analgesic using the data in the graph. (2)

(Total 6 marks)

◀ The Hawaiian bobtail squid (*Euprymna scolopes*) has a symbiotic relationship with the bacterium *Vibrio fischeri*. The bacteria reside in the light organ of the squid's mantle and produce light by bioluminescence. The light produced by the bacteria provides protection against predators for the squid. The *Vibrio* bacteria control the production of this light using a process of cell signalling called quorum sensing.

All living things receive and deliver signals of several different types. Two major types of signalling are chemical and neural. Chemical signals are produced by structures in specialized regions of the cell and/or organism. Once produced, they can be transported by many different means. Receptors are then capable of receiving these chemical signals and initiating a response within the same or different cell and/or organism. The *Vibrio* bacteria mentioned above produce such a chemical that initiates the result of bioluminescence within the squid.

Neural signalling involves electrical impulses and activity. Nerve and muscle cells within animals demonstrate neural signalling. This neural signalling is essential to maintaining proper and coordinated functioning within the cell and/or organism. In medicine, the electrical impulses of neural signalling are often utilized to study activity within various organs. Some common examples of medical applications of neural signalling within animals and humans include studies of the heart known as electrocardiograms (ECG) and studies of the brain known as electroencephalograms (EEG).

C2.1 is not included as it is for HL students only.

C2.2 Neural signalling

Guiding Questions

How are electrical signals generated and moved within neurons?

How can neurons interact with other cells?

Very few people think of the human body in terms of electrical pathways. However, life would not occur without them. Chemicals, especially charged atoms known as ions, are constantly on the move within our systems. The movement of charged particles into and out of neurons results in electrical impulses, and these impulses can then be propagated through the nerve cells.

To maintain the functions of life, our body must constantly monitor and evaluate changes in our environment, both internally and externally, so that chemical reactions and electrical pathways can persist. Once changes are detected, messages (impulses) must be conducted to the spinal cord and/or brain so that the appropriate responses can occur. A complex system of nerve cells known as neurons interacts with all areas of the body to allow this control and communication. Neurons connect with one another at junctions known as synapses so that electrical pathways can connect all cells. Neurotransmitters are signalling chemicals that allow impulse transmission across synapses. Neurotransmitters are also used at the junctions between nerves and muscles.

C2.2.1 – The role of neurons

C2.2.1 – Neurons as cells within the nervous system that carry electrical impulses

Students should understand that cytoplasm and a nucleus form the cell body of a neuron, with elongated nerve fibres of varying length projecting from it. An axon is a long single fibre. Dendrites are multiple shorter fibres. Electrical impulses are conducted along these fibres.

The brain and spinal cord make up the **central nervous system** (**CNS**). These two structures receive sensory information from various receptors around the body, and

273

then interpret and process that sensory information. If a response is needed, some portion of the brain or spinal cord initiates that response.

The cells that carry this information are called **neurons**. **Sensory neurons** carry information to the CNS, and **motor neurons** carry response information to muscles.

Together, the sensory neurons and motor neurons make up the **peripheral nerves**. A neuron is an individual cell that carries electrical impulses from one point in the body to another, and does so very quickly. When many individual neurons group together into a single structure, that structure is called a **nerve**. You can think of a nerve as being like a large cable made up of many individual, smaller cables. Each smaller cable within that large cable represents a single neuron. Study Figure 1.

C2.2 Figure 1 A light microscope photograph of a section of a nerve. The very large circle is the entire nerve, and each small circle within it is one of the axons of a neuron contained within that nerve.

central nervous system

peripheral nervous system

Neurons can be very long. In the human body, there are neurons that extend from the lower portion of the spinal cord all the way to the big toe: single cells that extend about 1 m! Of course, not all neurons are that long. In fact, some neurons are quite short.

Neuron structure

A single neuron is made up of **dendrites**, a **cell body** and an **axon**. Axons are long, single fibres; dendrites are multiple, shorter fibres. At the end of the axon are **synaptic terminal buttons**, which release chemicals called **neurotransmitters** that continue the impulse chemically to the next neuron(s) or a muscle. An impulse is always carried from the dendrite end of a neuron along the membrane of the cell body down the axon, and results in a release of a neurotransmitter.

The central nervous system (CNS) consists of the brain and spinal cord. The peripheral nervous system (PNS) is made up of the nerves and branches that enter and leave the spinal cord and brainstem.

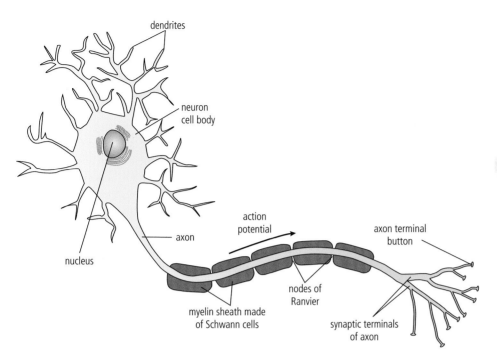

C2.2 Figure 2 The structure of an individual motor neuron. The function of the myelin sheath and nodes of Ranvier are discussed later in this chapter.

An axon is a relatively long single fibre. Dendrites are multiple shorter fibres. The cell body of the neuron connects the dendrites with the axon and contains the organelles to keep the neuron alive and functioning. Electrical impulses are conducted along these fibres.

How is the structure of specialized cells related to function?

Figure 2 shows an **action potential** moving from the cell body towards the axon terminal buttons. Action potential is another name for a nerve impulse, and it is electrical in nature because it involves the movement of positively charged ions.

The term "nerve impulse" is misleading because it is not the nerve that carries the impulse: the individual neurons within the nerve are each capable of carrying the impulse. Because axons of neurons are often relatively long, it is convenient to think of the axon as the conductor of a neuron impulse. The axons of neurons in some organisms, including humans, have surrounding membranes that are collectively called the **myelin sheath**. We will talk more about the myelin sheath in the following sections. However, to learn how a neuron conducts an impulse, we will first discuss an axon without a myelin sheath.

C2.2.2 and C2.2.3 – Generation and transmission of an impulse along a neuron

C2.2.2 – Generation of the resting potential by pumping to establish and maintain concentration gradients of sodium and potassium ions

Students should understand how energy from ATP drives the pumping of sodium and potassium ions in opposite directions across the plasma membrane of neurons. They should understand the concept of a membrane polarization and a membrane potential and also reasons that the resting potential is negative.

C2.2.3 – Nerve impulses as action potentials that are propagated along nerve fibres

Students should appreciate that a nerve impulse is electrical because it involves movement of positively charged ions.

An **action potential** is a sequence of events that allows an impulse to travel through a neuron. When a neuron is ready to send an impulse, it is **polarized** and has a

A neuron axon at resting potential. Think of the axon as a three-dimensional tube, and thus the ion movements shown are occurring all around the tube. Notice that there is a negative net charge inside the fibre relative to the outside because there are more positive sodium ions (Na⁺) outside than potassium ions (K⁺) inside. The diagram on the right is a model showing the actual number of K⁺ and Na⁺ ions inside and outside the neuron relative to one another.

▼

resting potential across the membrane. The resting potential is created by the **active transport** of sodium and potassium ions in two different directions across the cell membrane of the neuron. Most of the sodium ions are actively transported out of the axon's cytoplasm and into the intercellular fluid, and most of the potassium ions are transported into the cytoplasm from outside the cell. This transport of sodium and potassium in opposite directions is the result of an active transport mechanism called the **sodium–potassium pump** (described in Chapter B2.1). The sodium–potassium pump works by transporting three sodium ions out of the cell for every two potassium ions transported in. It is an example of active transport, which means that adenosine triphosphate (ATP) is required to provide the energy for this pumping action to occur. In addition, there are negatively charged organic ions permanently located in the cytoplasm of the axon. The net result is that the outside of the cell becomes positively charged in relation to the inside. This potential difference in charge across the cell membrane is called the **membrane potential**.

The nerve impulse is the action potential that is propagated through the neurons, immediately followed by the sodium–potassium pump to restore the resting potential, in a wave-like action. This allows the neuron to propagate another action potential straight away.

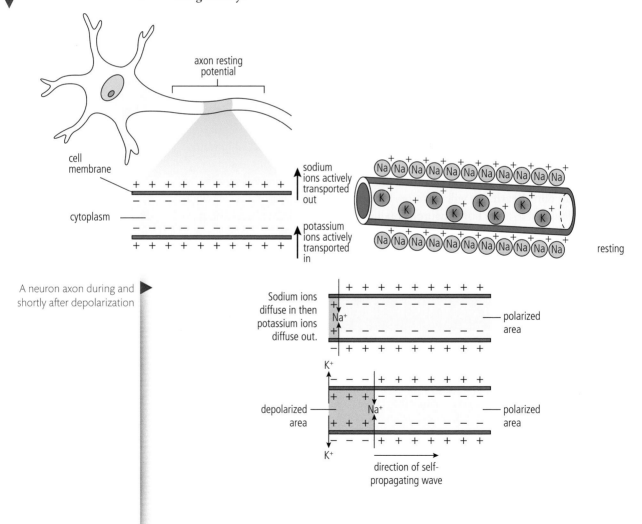

A neuron axon during and shortly after depolarization ▶

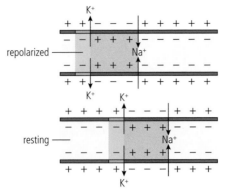

◄ Return to the resting potential

repolarized

After sodium ions and potassium ions diffuse, both are actively transported back to their resting potential locations.

resting

ATP drives the pumping of sodium and potassium ions in opposite directions across the plasma membrane of neurons. There is a net negative charge inside the neuron membrane in relation to outside the membrane. The membrane is therefore said to be polarised. The neuron is said to have a negative resting potential because there are to fewer positive ions and more negative ions inside the neuron than outside. When depolarization occurs, the inside of the membrane becomes positive due to more positive ions moving into the neuron through the plasma membrane. This depolarization results in the movement of positively charged ions and produces the nerve impulse.

C2.2.4 – The speed of nerve impulses

C2.2.4 – Variation in the speed of nerve impulses

Compare the speed of transmission in giant axons of squid and smaller non-myelinated nerve fibres. Also compare the speed in myelinated and non-myelinated fibres.

Application of skills: Students should be able to describe negative and positive correlations and apply correlation coefficients as a mathematical tool to determine the strength of these correlations. Students should also be able to apply the coefficient of determination (R^2) to evaluate the degree to which variation in the independent variable explains the variation in the dependent variable. For example, conduction speed of nerve impulses is negatively correlated with animal size, but positively correlated with axon diameter.

Several factors contribute to the speed at which an action potential moves along a nerve fibre. One factor is whether there is a myelin sheath surrounding the fibre or not. Figure 2 shows a neuron with a myelin sheath made of **Schwann cells**. Schwann cells wrap themselves around a nerve fibre (axon) and act as insulators: there is no ion movement in an axon covered by Schwann cells. Schwann cells are spaced evenly along an axon, with small gaps between them; these gaps are called **nodes of Ranvier**.

When myelin sheaths are present, action potentials skip from one node of Ranvier to the next as the impulse progresses along the axon towards the synaptic terminals. In this case, the action potential does not have to undergo the time-consuming and energy-expensive ion movements in the membrane underneath the myelin material. The myelin sheath prevents charge leakage through the membrane, thus acting as an insulator. The advantage of this type of impulse transmission is two-fold:

- the impulse travels much faster compared to an impulse in non-myelinated fibres, because the in/out movement of ions only occurs at the nodes of Ranvier
- less energy (in the form of ATP) is expended for the transmission of impulses, because the sodium–potassium pump is only working at the nodes.

Neurons with myelinated axons and nodes of Ranvier. Action potentials only occur at these nodes. The photomicrograph is a false-colour scanning electron micrograph (SEM) of a nerve (bundle of neurons) with myelin sheaths. The darker (blue) colour indicates the axons, and the surrounding paler (yellow) colour indicates the myelin sheath of each axon. ▶

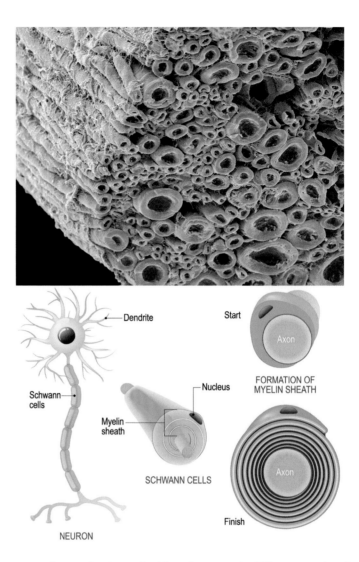

C2.2 Table 1 Axon conduction velocities ▼

Table 1 compares the conduction velocities of neurons in different species. Axons found in squid are particularly large and are unmyelinated.

Axon source	Axon diameter / μm	Myelin present?	Conduction velocity / m s⁻¹
Giant squid axon	500	No	25
Human leg axon	20	Yes	120
Human skin temperature receptor axon	5	Yes	20
Human internal organ axon	1	No	2

Two observations can be made from Table 1:

- myelinated axons conduct action potentials faster than non-myelinated axons
- axons with a greater diameter have a faster transmission velocity than those with a smaller diameter.

Observations are being made all the time in science. From Table 1 we can see that axon diameter is positively correlated with impulse conduction velocity. Correlation means there is a statistical association between two variables. Causation means that a change in one variable actually causes a change in another variable. Remember, just because there appears to be a correlation does not mean there is causation.

A scatter graph plots two continuous variables on opposite axes and can be used to look for a correlation. If one variable increases as the second increases, this is a positive correlation. If one variable increases while the second decreases, this is a negative correlation.

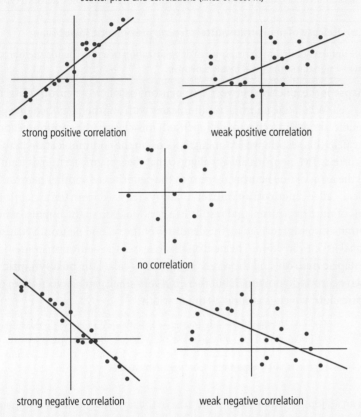

scatter plots and correlations (lines of best fit)

strong positive correlation

weak positive correlation

no correlation

strong negative correlation

weak negative correlation

Plotting a line of best fit helps visualize the correlation. Statistical tests can also be used to analyse correlation. A correlation coefficient quantifies the strength of the linear relationship between two variables, and can be denoted using the symbol R. For a perfect positive correlation, $R = 1$. If there is a weak correlation, R is closer to 0.

Look at the scatter graphs above. If a data point is a long way from the line of best fit, then there is a lot of variance in that data point. In a data set with lots of variance, the correlation will be low. Squaring the correlation coefficient (to get R^2) provides the coefficient of determination: this value gives you some idea of the shared variance in the data set as a whole. If there is a smaller amount of variance, all the data points are closer to the line of best fit. This means that changes in the independent variable can be used to predict changes in the dependent variable. For example, if we studied the relationship between the number of hours of biology revision a student did and their exam score, we might find a correlation of 0.9. This very high value suggests that spending time on revision helps you gain a good exam score. In this case the coefficient of determination is $R^2 = 0.81$. To look at it another way, 81% of the variance in exam scores can be explained by the amount of revision done.

In the case of impulses in axons, scientists have found that the conduction velocity of a nerve impulse is negatively correlated with animal size. They also have shown a strong positive correlation between conduction velocity of a nerve impulse and axon diameter.

C2.2.5 and C2.2.6 – Synapses, neurotransmitters, and their actions

C2.2.5 – Synapses as junctions between neurons and between neurons and effector cells
Limit to chemical synapses, not electrical, and these can simply be referred to as synapses. Students should understand that a signal can only pass in one direction across a typical synapse.

C2.2.6 – Release of neurotransmitters from a presynaptic membrane
Include uptake of calcium in response to depolarization of a presynaptic membrane and its action as a signalling chemical inside a neuron.

In what ways are biological systems regulated?

A **synapse** is the junction between two neurons or where a neuron contacts a muscle cell. When one neuron communicates with another, or when one neuron communicates with an effector cell such as a muscle cell, the communication is chemical. In this section we are going to focus on the communication between two neurons. Two neurons always align so that the synaptic terminals found at the end of the axon of one neuron are next to the dendrites of another neuron. The chemical, called a **neurotransmitter**, is always released from the synaptic terminal buttons of the first neuron, and results in a continuation of the impulse when the neurotransmitter is received by the dendrites of the second neuron. Synapses only pass the signal in one direction. The neuron that releases the neurotransmitter is called the **presynaptic neuron**, and the receiving neuron is called the **postsynaptic neuron**. The two neurons do not touch but there is a very small gap between them called the **synaptic cleft**, which is around 20 nm in width.

Approximately 50 different neurotransmitters have been identified as active in the human brain. An imbalance of just one can result in conditions such as schizophrenia or severe depression. Many pharmaceuticals have been developed to treat these conditions, based on knowledge of how synapses and neurotransmitters work.

When an action potential reaches the area of the terminal buttons, it initiates the following sequence of events.

1. The action potential arrives at the terminal button, which results in depolarization of the presynaptic membrane and the uptake of calcium ions (Ca^{2+}) into the terminal buttons.

2. The calcium ions act as a signalling chemical, activating a pathway that moves vesicles containing the neurotransmitter through the cell. The vesicles then fuse with the presynaptic membrane.

3. The neurotransmitter is released from the fused vesicles into the synaptic cleft.

4. The neurotransmitter binds with a protein receptor on the postsynaptic neuron membrane.

5. This binding results in an ion channel opening, and sodium ions diffusing in through this channel.

6. This initiates the action potential to begin moving down the postsynaptic neuron, because it is now depolarized.

7. Any neurotransmitter that is bound to the protein receptor is released back into the synaptic cleft. The neurotransmitter in the synaptic cleft is degraded (broken into two or more fragments) by enzymes.

8. The ion channel in the postsynaptic membrane closes to sodium ions.

9. Neurotransmitter fragments diffuse back across the synaptic cleft, to be reassembled in the terminal buttons of the presynaptic neuron.

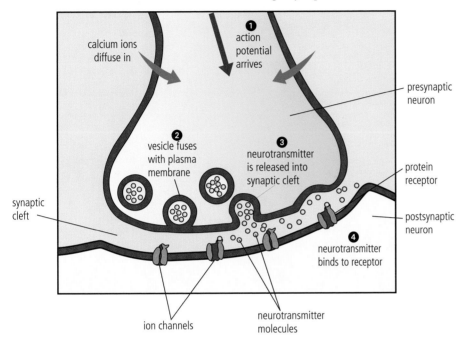

◀ The mechanism of synaptic transmission

C2.2.7 – Acetylcholine and the generation of a postsynaptic potential

C2.2.7 – Generation of an excitatory postsynaptic potential

Include diffusion of neurotransmitters across the synaptic cleft and binding to transmembrane receptors. Use acetylcholine as an example. Students should appreciate that this neurotransmitter exists in many types of synapse including neuromuscular junctions.

Acetylcholine is a very common neurotransmitter and is found at synapses between two neurons, and at synapses between a neuron and a muscle cell (a **neuromuscular junction**). When an impulse passes across a neuromuscular junction, the stimulus causes the muscle to contract. **Acetylcholinesterase** is an enzyme found in many synaptic clefts of muscles and nerves. It immediately breaks down the neurotransmitter acetylcholine into fragments, so the transmission of the action potential from the presynaptic membrane to the postsynaptic membrane occurs only once.

The freckled sea snail (*Aplysia parvula*) ▶

Nature of Science

The fields of psychology, chemistry, biology and medicine all combine to contribute to our knowledge of memory and learning. One of the many complications for research on memory and learning is the sheer complexity of the human brain. Often, complex biological systems are best studied by using simpler "models" that represent the more complex activity.

Biologists often use invertebrates, which have a simpler nervous system compared to humans and other vertebrates, as a model. One interesting invertebrate is a sea snail called *Aplysia parvula*. This marine snail can be stimulated to retract its siphon when it is touched, as part of its defence mechanism. The snail can learn from experience, and can keep its siphon protected for a longer period of time after being given a chance to learn. In addition, repeated touching of the siphon leads to a greater number of synapses between neurons in the very simple brain of *Aplysia*. This can be observed and documented because *Aplysia* has very few, but very large, neurons that can easily be seen.

Guiding Question revisited

How are electrical signals generated and moved within neurons?

In this chapter we have looked at how:

- neurons are specialized cells that carry electrical impulses
- as part of their specialization they have a long axon
- for an action potential to occur, depolarization must first occur within the neuron
- depolarization involves the opening of voltage-gated sodium and then potassium channels, with a more positive electrical charge resulting within the neuron
- after depolarization of a neuron region occurs, it immediately undergoes repolarization
- the sodium–potassium pump is required for repolarization to occur
- electrical impulses are propagated through neurons and only move in one direction, i.e. from the dendrites to the other end of the axon
- myelin is an insulating material that speeds up the movement of nerve impulses through neurons
- larger axons carry impulses at a faster rate than smaller axons.

Guiding Question revisited

How can neurons interact with other cells?

In this chapter we have learned how:

- electrical impulses are transmitted from one neuron to another at a synapse
- an impulse in a neuron can be passed to another neuron
- an impulse can also be passed via a synapse to a muscle, where the impulse causes the contraction of the muscle, or to another effector cell, causing a specific reaction
- neurotransmitters are chemicals that diffuse across the synaptic cleft
- impulses or signals can only pass in one direction across a synapse
- depolarization at the presynaptic membrane causes an influx of calcium ions, which in turn causes vesicles carrying neurotransmitters to release their contents into the synapse
- the neurotransmitter diffuses across the synapse and binds to the postsynaptic membrane, generating an excitatory postsynaptic potential
- neurotransmitters such as acetylcholine exist at many different types of synapse, including neuromuscular junctions.

Exercises

Q1. Explain how a resting potential is produced in a neuron.

Q2. Name two variables that affect the speed of action potentials in neurons.

Q3. Explain why saltatory conduction is a faster means of electrical impulse transmission in a neuron.

Q4. Which of the following best describes the resting neuron?

    A Inside of neuron is more positively charged than outside.

    B Inside of neuron is more negatively charged than outside.

    C Inside and outside of neuron have the same electrical charge.

    D Sodium–potassium pump is active to establish polarization.

Q5. The sodium–potassium pump is:

    A Active in depolarization.

    B Important only at the synapse.

    C Essential for repolarization of the neuron.

    D Not active in neurons, only muscle cells.

Q6. Name the neuron parts labelled on the diagram.

Q7. What is the role of calcium ions in the transmission of an action potential at a synapse?

Q8. Which letter of the following represents an axon that is demonstrating depolarization?

C2 Practice questions

1. Describe the major features of neurons that allow them to carry out their function.

   (Total 3 marks)

2. Explain the steps required to produce an excitatory postsynaptic potential.

   (Total 4 marks)

3. Which of the following is **not** a factor in the transmission of an action potential from a presynaptic neuron to a postsynaptic neuron?

   A   Neurotransmitter is reassembled in the presynaptic neuron.

   B   Calcium ions act as a signalling chemical.

   C   The neurotransmitter is degraded in the synapse.

   D   The sodium–potassium pump is active, to cause depolarization in the postsynaptic membrane.

   E   The length of the axons.

   (Total 1 mark)

4. Explain how nerve impulse conduction takes place with myelinated axons.

   (Total 3 marks)

5. Draw a graph which would represent a strong positive correlation between two variables.

   (Total 3 marks)

◀ The human nervous system is just one of the many systems in the body that enable communication and thus integration of body systems. We are aware of only some actions of the nervous system, as others are accomplished by the autonomic nervous system. In addition, chemical communication between tissues by hormones and the endocrine system, and cell-to-cell communication by the immune system, provide different means of transmission for important signals around the body.

C3.1 Integration of body systems

Guiding Questions

What are the roles of nerves and hormones in integration of body systems?

What are the roles of feedback mechanisms in regulation of body systems?

Multicellular organisms have specialized cells that form tissues. Tissues are responsible for specific functions within an organism and, as a consequence, make up organs. In order for the tissues within organs to work in concert with each other, there must be communication between the tissues. Without a means of communication, the emergent property represented by an entire organism would not be possible. A human being or a magnolia tree can only exist because their cells, tissues and organs have evolved ways of communicating with each other.

Two systems have evolved within animals specifically to enable communication between, and thus integration of, body systems.

- A nervous system, designed to receive sensory information through structures called receptors and send motor responses to muscles, resulting in movement.
- An endocrine system, consisting of glands that respond to chemical signals in the body with the production and release of a variety of hormones. Hormones affect the activity of specific cells known as the target tissue of a particular hormone.

Humans and many other organisms have evolved feedback mechanisms that inform the body of the need for action by one or more body tissues. Usually these feedback mechanisms work to keep a factor within a normal, homeostatic, range. Examples of factors kept within a normal range are body temperature and levels of glucose in the bloodstream. If a factor rises above a particular level, or set point, the body will initiate one set of actions to bring levels down. If a factor goes below the set point, another set of actions will be initiated to bring that factor back to the set point. This is called negative feedback control.

C3.1.1 – Coordinating systems

C3.1.1 – System integration

This is a necessary process in living systems. Coordination is needed for component parts of a system to collectively perform an overall function.

Living organisms are complex. That complexity is the result of millions of years of evolutionary adaptations that represent only the relatively few mutations that gave an organism a greater chance to survive and reproduce. All life started as single-celled organisms and has expanded into the rich variety of single-cell and multicellular life that exists today.

The Atlantic puffin (*Fratercula arctica*) and many other animals use the same specialized systems as humans. All of its cells, tissues, organs and body systems work collectively as a puffin.

Coordination is needed for the component parts of organisms to collectively perform complex functions.

In order for complex organisms to evolve to survive in their environments, it was necessary for cells to become specialized for certain functions. Groups of specialized cells became specialized **tissues**, and groups of specialized tissues became **organs**. Some organs have evolved to work collectively to accomplish certain functions, and have become **body systems**. Body systems are specialized for functions such as obtaining nutrients, discarding waste and reproduction. All of the body systems working in unison represent the entire organism.

Even though cells in a multicellular organism are specialized, they are all using the same DNA as their genetic code. A specialized cell, such as a muscle cell, uses some genes that other cell types do not.

Some organisms have evolved into organized collections of billions of cells. The cells in one part of the organism often need to communicate with other cells where cell-to-cell communication is impossible. Two systems of communication have evolved to enable communication within organisms and facilitate efficient processes. One is found in both plants and animals, and involves chemicals called **hormones** that are produced in one location and then carried within fluids to other locations in the body. The other communication system is specific to animals and involves **electrical signals** sent from one location to another by a nervous system.

C3.1.2 – Hierarchy of body subsystems

C3.1.2 – Cells, tissues, organs and body systems as a hierarchy of subsystems that are integrated in a multicellular living organism

Students should appreciate that this integration is responsible for emergent properties. For example, a cheetah becomes an effective predator by integration of its body systems.

Multicellular animals and plants have evolved a common hierarchy of organization that permits effective communication and functioning within their environments. It is because of this hierarchy of subsystems that living organisms are able to survive and interact with their surroundings (Table 1). Each level of organization allows greater efficiency and complexity. All living organisms continue to evolve to improve adaptations to the environment in which they exist at each level of cellular organization.

Subsystem	Animal example	Plant example
Cell	Smooth muscle cell	Guard cell
Tissue	Muscular wall	Stoma
Organ	Bladder	Leaf
Organ system	Urinary	Vascular
Organism	White-tailed deer (*Odocoileus virginianus*)	Magnolia tree (*Magnolia grandiflora*)

◀ **C3.1 Table 1** Levels of organization within an animal and a plant

Emergent properties are those that exist when the sum of all the parts creates features that do not exist within the individual components. This is the advantage of an organism level of complexity. A puffin does not exist as a puffin until it is a complete being with all its component parts working collectively. The organism level of organization results in a combination that is said to be greater than the sum of its parts.

C3.1.3 – Integration of organs in animals

C3.1.3 – Integration of organs in animal bodies by hormonal and nervous signalling and by transport of materials and energy

Distinguish between the roles of the nervous system and endocrine system in sending messages. Using examples, emphasize the role of the blood system in transporting materials between organs.

Organs in the body must work together in order to maintain body processes. Body processes include digestion, maintaining the heart rate, blood glucose levels, blood pressure, and many others. Sometimes the communication is by the **nervous system**, and often we have no idea that the communication is actually occurring. The part of your nervous system that communicates with your body tissues without your conscious knowledge is called your **autonomic nervous system** (ANS).

In addition, humans and many other animals use hormone production and secretion in an **endocrine system** to help communication between organs and to respond to special situations. The body tissue where a hormone exerts an effect is called the **target tissue** of the hormone. Because hormones are produced in endocrine glands, they must be transported to their target tissues by the bloodstream.

The nervous system and endocrine system often work together to integrate body processes. An example is the release of epinephrine (adrenaline) from the adrenal glands. Sensory organs transmit information to the nervous system that indicates epinephrine is needed as part of the fight-or-flight response. The autonomic nervous system then sends impulses to the adrenal glands to release epinephrine. This hormone leads to a variety of body responses, including increased heart rate and increased flow of blood to muscles, to prepare the body for immediate increased activity. This integration between body systems is thought to have evolved as a survival mechanism.

Even though the nervous and endocrine systems often work together to integrate the systems in the body, each has its own characteristics.

In the nervous system:

- electrical impulses are used to send messages
- cells called neurons are used to transmit and receive impulses
- portions of the system control voluntary actions while other portions control involuntary actions
- responses occur quickly but are short lived.

In the endocrine system:

- hormones are used to send messages
- hormones travel through the bloodstream
- only involuntary functions are controlled
- responses are typically slow but are long lasting.

Most multicellular organisms have become so large that it is impossible for nutrients and waste products to be efficiently and directly moved from cell to cell. Transport vessels and aqueous fluids have evolved to serve that purpose. Humans and many other animals use blood circulating in arteries and veins to transport a variety of substances throughout the body tissues. The oxygen needed by leg muscles will be supplied by blood that has received that oxygen from lung tissues a short time before it is used. Urea produced as a by-product of protein metabolism in the liver will be transported by blood to the kidneys to be filtered out and become part of urine.

C3.1.4 – The brain and information processing

C3.1.4 – The brain as a central information integration organ

Limit to the role of the brain in processing information combined from several inputs and in learning and memory. Students are not required to know details such as the role of slow-acting neurotransmitters.

The brain is the most complex organ in the body. The brain regulates and monitors unconscious body processes such as blood pressure, heart rate and breathing. It receives a flood of messages from the senses, and responds by controlling balance, muscle coordination and most voluntary movements. Some parts of the brain deal with speech, emotions and problem solving. The brain relies on a variety of receptors to receive information. Some of these receptors relay information that we process at the conscious level, such as:

- **photoreceptors**, located within the retina of the eyes for visual information
- **chemoreceptors**, many located within our tongue for tasting
- **thermoreceptors**, located in the skin to provide information on changes in temperature
- **mechanoreceptors**, located in inner ear and sensitive to sound vibrations.

Other receptors send information to the brain at the subconscious level and are important components of our autonomic nervous system. Examples include:

- **osmoreceptors**, located in carotid arteries and the hypothalamus of the brain, which sense solutes and the water content of blood

- **baroreceptors**, located in carotid arteries and the aorta, which sense blood pressure based on how much a blood vessel is being stretched by internal pressure
- **proprioceptors**, located in muscles and joints, which provide the brain with a sense of balance and coordination, especially when the body is moving.

While receiving sensory information from a variety of receptors, the brain continuously processes the information and generates responses as needed. The process involves filtering the sensory information that is important from the information that is unimportant. This is necessary because the electrical signals travelling from sense organs are continuous and vast.

The brain communicates with the body in two ways. One involves impulses sent in and out of the spinal cord by 31 paired nerves called **spinal nerves**, which emerge directly from the spinal cord. The other involves the brain's own nerves, called **cranial nerves**. This set of 12 paired nerves connects various body parts to the brain through the **brainstem**. An example of cranial nerves is the pair of optic nerves that carry impulses from the retinas of the eyes to the brain.

The brain is divided into three main areas, each of which is subdivided further based on location and function.

- **Cerebrum**: the cerebrum is divided into right and left sides called **cerebral hemispheres**. Each hemisphere consists of four lobes: frontal, temporal, parietal and occipital, as shown in Figure 1. The neural processing carried out by these four lobes dominates our conscious activities. Although the four lobes interact with each other continuously, learning and memory activities are largely coordinated by the frontal lobe.
- **Cerebellum**: the cerebellum coordinates voluntary movements, and controls balance and equilibrium.
- **Brainstem** – in addition to relaying impulses between the cerebrum, cerebellum and spinal cord, the brainstem is responsible for most functions associated with the autonomic nervous system. The functions controlled by the brainstem are necessary for life, yet occur at a subconscious level. The **medulla** (or medulla oblongata) is an important part of the brainstem because it regulates both breathing and heart rate.

All parts of the brain must interact with each other for proper neural functioning, but the frontal lobe of the cerebrum is the dominant structure for learning and memory.

Very specific neural functions have been mapped by location on and within the brain. Specific brain activities are also associated with specific neurotransmitters that carry signals from one brain neuron to another. Some examples of neurotransmitters used within the brain are serotonin, dopamine and endorphins. There are many others.

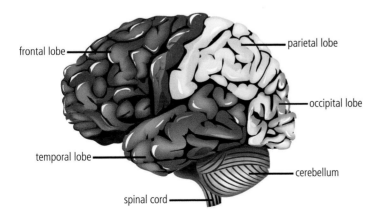

frontal lobe

parietal lobe

occipital lobe

temporal lobe

cerebellum

spinal cord

C3.1 Figure 1 The left cerebral hemisphere, cerebellum and spinal cord are shown in this illustration. The brainstem extends up from the spinal cord and is hidden in this view because it is covered by the temporal lobe of the cerebrum.

C3.1.5 – The spinal cord and unconscious processes

C3.1.5 – The spinal cord as an integrating centre for unconscious processes

Students should understand the difference between conscious and unconscious processes.

Together, the brain and spinal cord make up the **central nervous system (CNS)**. The spinal cord is a neural pathway between the body and the brain, but is capable of information processing on its own. The spinal cord controls some unconscious reflexes associated with balance and other skeletal muscle functions independently of the brain.

C3.1 Figure 2 A photograph of a section of the human spinal cord. In the centre, the darker "butterfly" shape is the grey matter of the spinal cord, capable of integration processing. Outside the grey matter is white matter, which carries impulses to and from the brain.

The 31 pairs of spinal nerves bring sensory information into the CNS from the body and allow motor (muscular) information to be sent out. Some of the information travelling in and out of the CNS is processed at the conscious level, meaning that the cerebrum of the brain is involved in the pathway.

The spinal cord has two types of neural tissue, **white matter** and **grey matter**. The white matter is composed primarily of axons of neurons and carries neural impulses to and from the brain. The grey matter of the spinal cord contains neurons and synapses involved in spinal cord integration processes. When sensory information enters the grey matter of the spinal cord and motor information is immediately sent back out, the pathway of the impulse is called a **reflex arc**. You will study a specific spinal reflex arc in Section C3.1.9.

C3.1.6 – Sensory neurons and conveying information

C3.1.6 – Input to the spinal cord and cerebral hemispheres through sensory neurons

Students should understand that sensory neurons convey messages from receptor cells to the central nervous system.

The beginning of a sensory neural pathway begins with a **receptor**. A receptor is a modified neuron that is capable of **transduction**. Transduction is the conversion of a physical stimulus into an electrical signal called an **action potential** that is carried along a neuron. Each type of receptor is specialized to transduce one specific type of physical stimulus. The retinal cells of the eyes transduce light, thermoreceptors in your

The quantity and variety of sensory information being sent to your brain at any given moment is vast. In fact, most sensory information has to be filtered out as unimportant by a process called **sensory gating**. At this moment try to make yourself aware of some of the sensory information that previously you were subconsciously filtering by sensory gating. This may include sights, sounds, things you are touching, temperature sensory information and a great deal more.

skin transduce heat or lack of heat, baroreceptors transduce pressure; there are many other receptor types. The neurons that carry impulses from receptors to the brain or spinal cord are called **sensory neurons**.

Example of a specific sensory path

The skin of your fingertips contains mechanoreceptors that sense pressure. When you touch your phone screen to type a message, the mechanoreceptors at the tips of your fingers and thumbs send action potentials up the small nerves in your hand and arm, to eventually join one of the 31 spinal nerves. Specifically, the action potentials travel to a spinal nerve that enters your spinal cord in the region of your chest. The neurons that carry this information enter the grey matter of the spinal cord but are directed out into the white matter (see Figure 2). The pathway of this spinal cord neuron carries the impulses up through your spinal cord, through the brainstem and into the parietal lobe of your cerebrum. This is where the sense of touch is interpreted or integrated. You are aware of this sensation, thus it is a conscious process, although it is only one sensation of many that you would be experiencing at that moment.

Each receptor that we perceive at the conscious level has a specific pathway to the CNS. Many receptors send their action potentials to a particular lobe of the cerebrum, for example:

- information from sound mechanoreceptors travels to the temporal lobe
- information from chemoreceptors for taste travels to the parietal lobe
- information from photoreceptors for visual information travels to the occipital lobe.

"cylinder" -shaped rod cell

"cone" or "vase" -shaped cone cell

A coloured scanning electron micrograph of a small section of the retina of an eye. The photoreceptors of the retina are shown. Rod cells that can sense black and white vision are shown in blue (they are rod-shaped). Cone cells responsible for colour vision are shown in green (they are cone-shaped).

C3.1.7 – Motor neurons and muscle stimulation

C3.1.7 – Output from the cerebral hemispheres of the brain to muscles through motor neurons

Students should understand that muscles are stimulated to contract.

The cerebrum uses the sensory information it receives to make decisions concerning movements. The action potentials are carried to muscle tissue by **motor neurons**. The portion of the cerebrum that sends the action potentials is called the **motor cortex** and is located in the most posterior portion of the frontal lobe of the cerebrum.

Motor neurons terminate in muscle tissue. Each axon of a motor neuron branches repeatedly and forms synapses with multiple fibres of a muscle.

A light micrograph showing a single motor neuron axon branching to form multiple synapses with fibres of skeletal muscle. Each synapse is called a motor end plate or neuromuscular junction.

Motor neurons form synapses with muscle fibres called **motor end plates** or **neuromuscular junctions**. When action potentials reach a motor end plate, a neurotransmitter called **acetylcholine** is released. This is the chemical signal that initiates contraction of the muscle.

C3.1.8 – Nerve fibres

C3.1.8 – Nerves as bundles of nerve fibres of both sensory and motor neurons
Use a transverse section of a nerve to show the protective sheath, and myelinated and unmyelinated nerve fibres.

The terms **nerve** and **neuron** are not interchangeable. A neuron is an individual cell of the nervous system, whereas a nerve is a collection of neurons surrounded by a protective sheath. Neurons may be sensory, motor or interneurons.

- **Sensory neurons** carry action potentials from receptors to the CNS.
- **Motor neurons** carry action potentials from the CNS to a muscle.
- **Interneurons** are located between sensory and motor neurons and are only found within the CNS.

Some of the 12 pairs of cranial nerves only contain sensory neurons, while others only contain motor neurons and others contain some of each type. The 31 pairs of spinal nerves contain both sensory and motor neurons, carrying action potentials in opposite directions. Nerves that contain both sensory and motor neurons are called **mixed nerves**.

Be careful with the use of the terms neuron and nerve. Nerve is the term used for a bundle of neurons surrounded by a protective sheath.

An individual neuron is either sensory or motor, depending on whether it is carrying action potentials towards or away from the CNS. An individual neuron cannot be mixed.

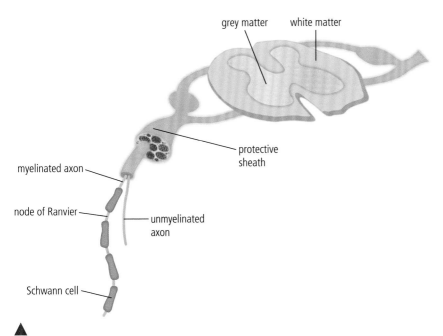

grey matter   white matter

protective sheath

myelinated axon

node of Ranvier

unmyelinated axon

Schwann cell

▲
A section view of an area of the spinal cord showing one of the 31 pairs of spinal nerves.

Neurons can be either **myelinated** or **unmyelinated**. Those that are myelinated have cells called **Schwann cells** wrapped around their axon, and intervening areas where there are no Schwann cells (also discussed in Chapter C2.2). The areas between Schwann cells are called **nodes of Ranvier**. The action potentials of myelinated axons are able to skip from one node of Ranvier to the next, making transmission of the action potential much faster compared to unmyelinated axons. Groupings of myelinated and unmyelinated axons are surrounded by protective sheaths.

C3.1.9 – Pain reflex arcs

C3.1.9 – Pain reflex arcs as an example of involuntary responses with skeletal muscle as the effector
Use the example of a reflex arc with a single interneuron in the grey matter of the spinal cord and a free sensory nerve ending in a sensory neuron as a pain receptor in the hand.

A **pain reflex arc** is an example of an involuntary response and involves only three neurons. The first of the neurons is a receptor neuron known as a **nocireceptor** or pain receptor. Imagine that you accidently hold a finger too close to an open flame. This results in nocireceptors located in the skin of your finger initiating **afferent** (sensory) action potentials. These action potentials travel through your hand and eventually join one of the spinal nerves. After entering the spinal cord, the afferent neuron synapses with a short **interneuron** (also called a relay neuron) located entirely within the grey matter of the spinal cord. The interneuron synapses with a motor neuron and the resulting action potentials go directly to arm muscles (the **effector**), which moves quickly to pull your finger away from the flame.

A schematic showing the three-neuron pathway of a pain reflex arc. ▶

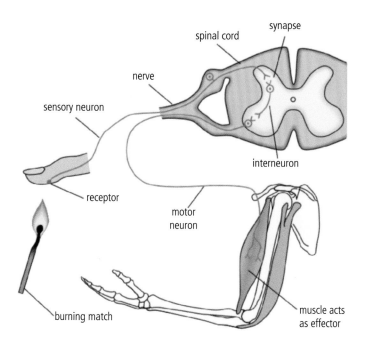

The action of pulling your finger away from the source of pain occurs much faster than truly sensing the pain. The reason for this is that the sensation of pain must travel to your cerebrum to be integrated by many neural synapses before a sensation is felt and a motor response formulated. The pain reflex arc has evolved to limit damage to body tissue by generating a quick reaction involving only three neurons. The unusual aspect of this is that the reflex arc uses skeletal muscle as the effector, tissue that is normally innervated by the frontal lobe of the cerebrum.

C3.1.10 – The cerebellum and skeletal muscle coordination

C3.1.10 – Role of the cerebellum in coordinating skeletal muscle contraction and balance
Limit to a general understanding of the role of the cerebellum in the overall control of movements of the body.

Although the **cerebellum** is a very important part of the brain associated with body movements, it does not initiate those movements. The initiation of muscle contractions and thus body movements is accomplished by the **motor cortex** of the cerebrum. As soon as a movement begins, the cerebellum receives feedback impulses from the area of the body that is moving and many sense organs. The cerebellum then sends out impulses to coordinate the movement. This results in smooth and balanced muscular activity, leading to coordinated movements. The cerebellum coordinates posture, balance, walking, hand and finger movements, eye movements, speech and much more. The term "muscle memory" often used by athletes is more to do with training coordinated movements by the cerebellum than actually training muscle.

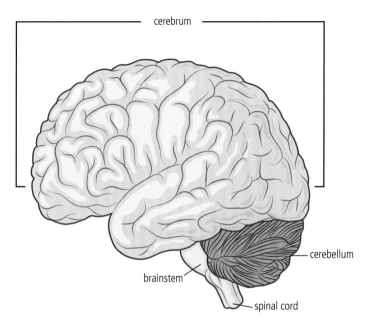

The three main portions of the human brain are shown, with the tissue of the cerebellum highlighted in a darker colour. Many body functions, including maintaining balance and coordinating movements, require interactions between all three of these parts plus the spinal cord, which extends down from the brainstem.

C3.1.11 – Melatonin secretion and sleep patterns

C3.1.11 – Modulation of sleep patterns by melatonin secretion as a part of circadian rhythms
Students should understand the diurnal pattern of melatonin secretion by the pineal gland and how it helps to establish a cycle of sleeping and waking.

A **circadian rhythm** is any pattern of behaviour or physiology that is based on a 24-hour cycle. The most obvious pattern of a circadian rhythm is our wake and sleep cycle. Many other organisms follow a circadian rhythm for sleep. Some animals, like ourselves, are **diurnal**, meaning that we are more active in daylight hours. Other organisms are **nocturnal** and are more active at night.

Evidence suggests that the circadian rhythm is largely controlled or modulated by a small endocrine gland called the **pineal gland**. This small gland is located near the centre of the brain between the cerebrum and brainstem. Its function is to produce a hormone called **melatonin.** This hormone regulates the sleep schedule. Studies have shown that melatonin levels are high during the night for diurnal animals and high during the day for nocturnal animals. Other studies have shown that light striking the retina of the eye inhibits melatonin production.

Nocturnal animals have specialized adaptations for sensory input during the night. Many nocturnal animals have a structure called the **tapetum lucidum** behind their retina. This is a layer that sends light back through the retina sensory receptors a second time by reflection. This doubles the intensity of light striking the sensory cells of the retina. You can tell when an animal has a tapetum lucidum because its eyes appear to glow when you shine a light at them.

A graph showing melatonin production by the human pineal gland over a time period of about 30 hours. Humans are diurnal and show increasing melatonin production soon after sunset. ▶

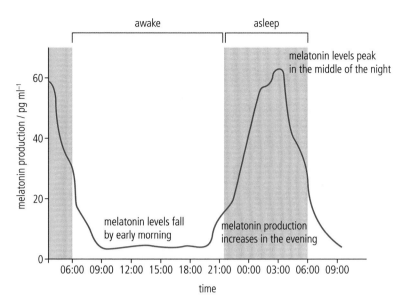

Over a prolonged period of time our bodies become naturally regulated to a circadian rhythm that is only interrupted by atypical events. One of those events is travelling through several time zones in a short period of time, sometimes known as "jet lag". Extended viewing of television, mobile phone and computer screens in the evening has also been shown to alter the natural circadian rhythm.

C3.1.12 – Epinephrine and vigorous activity

C3.1.12 – Epinephrine (adrenaline) secretion by the adrenal glands to prepare the body for vigorous activity

Consider the widespread effects of epinephrine in the body and how these effects facilitate intense muscle contraction.

When humans encounter a stressful situation, a hormone called **epinephrine** (also called adrenaline) is released from glands located on the upper or superior side of each kidney. These glands are called **adrenal glands**. Epinephrine release occurs as a result of a potentially harmful event, such as an animal attack, but also because some people choose recreational activities that stimulate release of the hormone. Like all hormones, epinephrine is released into the bloodstream and results in numerous responses by the body.

Epinephrine has widespread effects in the body, including:

- increasing the heart rate and blood pressure
- increasing the diameter (dilation) of air passages so more air can be received by the lungs
- dilation (increased size) of the pupils of the eyes
- increasing blood sugar levels by stimulating glycogen conversion to glucose in the liver
- increasing the blood supply to muscles.

Epinephrine prepares us for the fight-or-flight response, so called because the body's resources are called upon for immediate action in response to a threat or other stimuli that require a vigorous and immediate response. Intense muscle contractions are associated with epinephrine release.

C3.1.13 – The hypothalamus, pituitary gland and endocrine system

C3.1.13 – Control of the endocrine system by the hypothalamus and pituitary gland

Students should have a general understanding, but are not required to know differences between mechanisms used in the anterior and posterior pituitary.

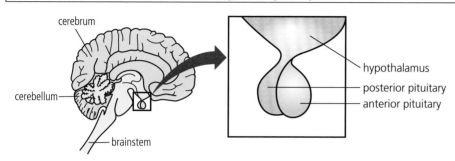

A section of the human brain showing the location of the hypothalamus and the two portions of the pituitary gland.

The **hypothalamus** is an area of the brain that acts as a link between the nervous system and the endocrine system. The hypothalamus contains some receptors that are associated with autonomic nervous system functions, and receives action potentials from other areas of the body that also contain this type of receptor. The hypothalamus is composed of both neurons and **glandular cells**. The glandular cells of the hypothalamus produce hormones that either stimulate hormone release by the **pituitary glands**, or inhibit their release.

Often the pituitary is referred to as a singular gland, but it is actually two glands that exist as different "lobes". The anterior and posterior lobes of the pituitary communicate with the hypothalamus in different ways, and each secretes its own hormones. The majority of these hormones are chemical signals released into the bloodstream that regulate the homeostasis of various physiological factors, such as metabolic rate, reproductive cell formation and water balance.

A good example is **antidiuretic hormone (ADH)** produced by the hypothalamus, which is sent to the posterior pituitary and when needed is secreted by the posterior pituitary. ADH helps control homeostatic levels of water in the body. The hypothalamus has specialized receptors called **osmoreceptors** that are capable of sensing the water content of blood as it passes through the hypothalamus. If the water content is relatively low, the hypothalamus will send action potentials to the cells in the posterior pituitary which then secrete ADH into the bloodstream. The target tissue of this hormone is the collecting tubules of nephrons in the kidneys. When the collecting tubules detect ADH, they reabsorb water that would have been released as part of urine. The target tissues of some other pituitary hormones controlled by the hypothalamus are shown in Figure 3.

Many hormones, such as ADH, work using a mechanism called **negative feedback**. The goal of negative feedback is to maintain homeostasis. ADH and kidney function maintain a homeostatic level of water in the body. If water in the body rises above the homeostatic level, more urine is produced. If the water level becomes too low, ADH is produced and water is reabsorbed before becoming part of urine.

Each hormone produced by an endocrine gland has one or more tissue types in the body that is the "target tissue" of that hormone. In many instances the target tissue is located far away from the endocrine gland. Thus endocrine glands secrete hormones into the blood for dispersal to all cells of the body, even though only the target tissue cells are affected by the hormone.

299

C3.1 Figure 3 The ▶
hypothalamus controls the
two portions of the pituitary
gland separately; both
the anterior and posterior
pituitary produce hormones
involved with homeostasis.
TSH = thyroid-stimulating
hormone; FSH = follicle-
stimulating hormone;
LH = luteinizing-hormone.

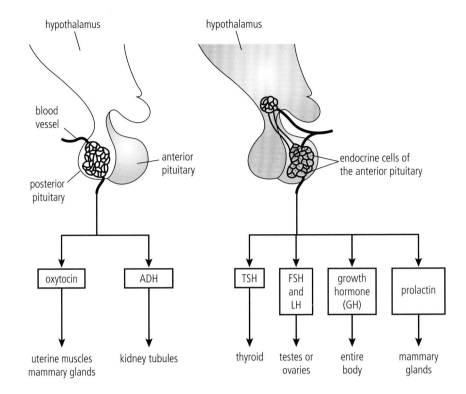

In Figure 3, note that
one of the hormones
secreted by the
posterior pituitary is
oxytocin. Oxytocin
is produced during
childbirth and is
controlled by a positive
feedback mechanism.
Oxytocin induces
uterine contractions,
and uterine
contractions stimulate
further secretion of
oxytocin.

C3.1.14 – Feedback control of heart rate

> **C3.1.14 – Feedback control of heart rate following sensory input from baroreceptors and chemoreceptors**
>
> Include the location of baroreceptors and chemoreceptors.
>
> Baroreceptors monitor blood pressure. Chemoreceptors monitor blood pH and concentrations of oxygen and carbon dioxide. Students should understand the role of the medulla in coordinating responses and sending nerve impulses to the heart to change the heart's stroke volume and heart rate.

Many physiological factors in the body change depending on environmental conditions and body activity. Changes to ventilation rate, body temperature and heart rate are brought back to set points by **feedback control** mechanisms. As an example, when the body is "at rest", the heart rate is under the control of the natural pacemaker within the heart known as the **sinoatrial (SA) node**. When you become active, muscle tissue requires additional oxygen and releases additional carbon dioxide as a result of the increased rate of cell respiration. An increase in heart rate and **stroke volume** is required in order to carry the additional respiratory gases to and from the lungs. Stroke volume is the volume of blood pumped out of the heart with each ventricular contraction.

Receptors known as **baroreceptors** and **chemoreceptors** are able to detect changes in the blood vessels and contents of the blood associated with an increase in the rate of cell respiration. Both baroreceptors and chemoreceptors are located in similar but not identical locations, as shown in Figure 4. The largest artery in the body is the **aorta**, and it forms an arch shape as it exits the left ventricle of the heart. One location for baroreceptors is on the arch of the aorta. Almost immediately, other major arteries

Check your
understanding of a
topic by explaining
diagrams to yourself
or a friend. If you can
verbalize a topic, you
will better understand it.

begin to branch from the aortic arch. Two of those major branches are the **carotid arteries**, which carry oxygenated blood to your head and brain. Just before the two carotid arteries branch, they form an enlargement called a **sinus**. Both of the carotid sinuses also have baroreceptors on the walls of the blood vessels.

Chemoreceptors are located in tissue near where the baroreceptors are located but outside the blood vessels. Each of the major arteries has small arteries that extend from the vessel and immediately branch into capillaries. The chemoreceptors monitor oxygen, carbon dioxide and pH levels in these capillaries. The chemoreceptor cells are capable of releasing a neurotransmitter that initiates action potentials that are carried to the **medulla**.

Baroreceptors

Baroreceptors are sensitive to pressure changes in arterial blood vessels. When blood pressure increases, the wall of an artery is distended or stretched outwards. This distention results in an increase in the rate of action potentials sent to the medulla. The medulla responds by sending impulses to the SA node to decrease the heart rate and decrease the force of contraction, leading to a lower stroke volume. When the blood pressure falls below normal, a decrease in action potentials sent to the medulla will lead to an increase in heart rate and stroke volume.

Chemoreceptors

Chemoreceptors are sensitive to levels of three different factors in the bloodstream that change as a result of an increase in cell respiration rate:

- oxygen levels, which decrease because oxygen is a reactant of cell respiration
- carbon dioxide levels, which increase because carbon dioxide is a product of cell respiration
- pH, which lowers as most carbon dioxide entering the blood combines with water and forms carbonic acid.

Each of these chemical changes has its own chemoreceptors that send an increased rate of action potentials to the medulla when there is an increase in rate of cell respiration. An increase in heart rate and stroke volume will result from the action potentials sent from the medulla to the SA node. The same chemoreceptors can sense the opposite physiological effects when exercise ceases, and action potentials will be sent to the heart to slow the heart rate and lower the stroke volume.

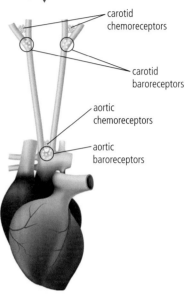

C3.1 Figure 4 The location of baroreceptors and chemoreceptors associated with changes in blood pressure, heart rate and stroke volume of the heart.

carotid chemoreceptors

carotid baroreceptors

aortic chemoreceptors

aortic baroreceptors

What are examples of branching (dendritic) and net-like (reticulate) patterns of organization?

Baroreceptors and chemoreceptors are neurons that are part of the **autonomic nervous system (ANS)**. Any responses arising from these receptors that decrease heart rate and stroke volume are associated with the **parasympathetic division** of the ANS, whereas responses arising from the same receptors that lead to an increase in heart rate and stroke volume are part of the **sympathetic division** of the ANS.

C3.1.15 – Feedback control of ventilation rate

C3.1.15 – Feedback control of ventilation rate following sensory input from chemoreceptors

Students should understand the causes of pH changes in the blood. These changes are monitored by chemoreceptors in the brainstem and lead to the control of ventilation rate using signals to the diaphragm and intercostal muscles.

When resting, the ventilation rate of the lungs is controlled by groups of cells called **respiratory centres** located in your medulla. When at rest, spontaneous action potentials are released by these cells, which travel to your diaphragm and intercostal muscles to maintain breathing at a relatively slow and controlled pace.

An illustration showing a portion of the brain and the 12 pairs of cranial nerves. The cranial nerves all connect into the brainstem and, among many other functions, carry both sensory and motor impulses related to heart and ventilation rates. If you look carefully, you can see branches of the cranial nerves forming the shape of the upper and lower jaws.

Chemoreceptors located in the medulla allow feedback control of the ventilation rate during and after exercise. Many chemoreceptors monitor the levels of carbon dioxide and pH in the blood passing through the medulla. The pH of blood typically falls within the small range of 7.35 to 7.45. In other words, blood is normally slightly alkaline. The response to exercise serves to keep the blood pH in this slightly alkaline range.

Body activity increases the rate of cell respiration and thus leads to an increase in carbon dioxide production. When carbon dioxide enters a red blood cell the following reaction, catalysed by carbonic anhydrase, occurs:

The production of carbonic acid and the resulting bicarbonate ion and hydrogen ion that occurs when carbon dioxide enters a red blood cell.

Strenuous exercise will lead to high levels of carbon dioxide being produced and a large number of hydrogen ions being produced. Chemoreceptors in the medulla sense this increase in hydrogen ions and send action potentials at a higher rate proportional to the number of hydrogen ions. All of the muscles associated with lung ventilation, especially the diaphragm and intercostal muscles, respond to these action potentials. Not only is the rate of ventilation increased but also the volume of air moving in and out is increased. When exercise decreases, hydrogen ion concentrations will also decrease and action potentials sent from the respiratory centres will decrease.

C3.1.16 – Control of peristalsis in the alimentary canal

C3.1.16 – Control of peristalsis in the digestive system by the central nervous system and enteric nervous system

Limit to initiation of swallowing of food and egestion of faeces being under voluntary control by the central nervous system (CNS) but peristalsis between these points in the digestive system being under involuntary control by the enteric nervous system (ENS). The action of the ENS ensures passage of material through the gut is coordinated.

Swallowing food is a voluntary action and is controlled by the **CNS**. Many hours later, after nutrients have been removed from the food, solid waste called **faeces** are egested, and that process is also under control of the CNS. A separate nervous system called the **enteric nervous system** (**ENS**) keeps the food (at various stages of digestion) moving along the **alimentary canal**. This movement of food is called **peristalsis** and is under **involuntary control**. The ENS is a web of sensory neurons, motor neurons and relay neurons embedded in the tissues of the alimentary canal, stretching from the lower portion of the oesophagus all the way to the rectum.

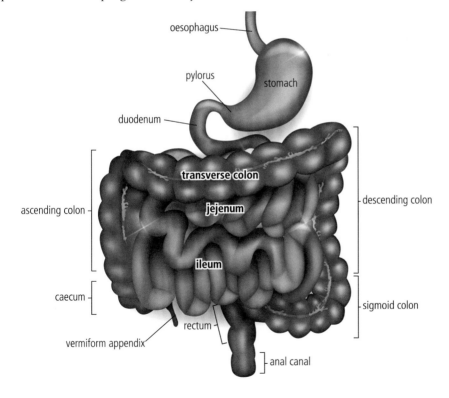

The alimentary canal is a long tube that begins at the mouth and finishes at the anus. Each part of the tube is innervated by neurons of the enteric nervous system (ENS). In addition, the tube is composed of smooth muscle (for peristalsis) and receives a rich blood supply for the absorption of nutrients from digested foods.

Peristaltic reflex

The **peristaltic reflex** is a series of smooth muscle contractions that occurs along the entire length of the alimentary canal to keep foods moving. Most smooth muscle in the body is controlled by areas of the brainstem. Peristalsis is controlled by the ENS because it has its own sensory, relay and motor neurons spread across the entire area of the alimentary canal.

There are many terms that end with "nervous system". Think about the meaning of each, as some are anatomical and some are functional in origin. Here are few used within this text:

- central nervous system (CNS), the term for the brain and spinal cord together
- peripheral nervous system (PNS), which comprises neurons and nerves outside the brain and spinal cord
- autonomic nervous system (ANS), which is any part of the nervous system that controls unconscious activities within the body
- enteric nervous system (ENS), which comprises those neurons that are located in, and control the unconscious peristaltic movement of food in, the alimentary canal.

A bolus of food keeps moving in a single direction within the alimentary canal because of the peristaltic reflex. Peristaltic movements are quite rapid in the oesophagus, creating a churning motion while food is in the stomach, and then slow down in the small and large intestine.

When food moves through the alimentary canal it forms into rounded masses, each called a **bolus**. It is these food masses that initiate the peristaltic reflex. Wherever a bolus is located in the tube of the alimentary canal, that area of the tube becomes distended, stimulating stretch receptors in the ENS. These (sensory) stretch receptors then synapse with nearby relay neurons. The relay neurons in turn synapse with two different types of motor neurons. One type of motor neuron releases an **excitatory neurotransmitter** to an area of smooth muscle "behind" the bolus of food. This stimulates that area of smooth muscle to contract, pushing the bolus along. At the same time, another type of motor neuron releases an **inhibitory neurotransmitter** "ahead" of the bolus. The smooth muscle ahead of the bolus relaxes in response, and opens the lumen (central space of the tube) for the food bolus to slide through. This reflex occurs many times along the gut, ensuring that the movement of the food material moves forwards in a coordinated manner.

Guiding Question revisited

What are the roles of nerves and hormones in integration of body systems?

Within this chapter you have learned:

- coordination and communication is necessary between cells, tissues and organs for them to work as an entire organism
- the nervous system and endocrine system are both important for information integration within many animals

- the brain is a central information integration organ in animals
- sensory information is carried to the brain and spinal cord through sensory neurons
- motor information is carried away from the brain and spinal cord to muscles through motor neurons
- the cerebellum coordinates impulses going to muscles to make movements smooth and efficient
- the hypothalamus and pituitary glands control many hormones secreted from endocrine glands.

Guiding Question revisited

What are the roles of feedback mechanisms in regulation of body systems?

Within this chapter you have learned:

- melatonin secretion by the pineal gland helps regulate the circadian rhythm of the sleep/wake cycle
- epinephrine secretion by the adrenal glands prepares the body for immediate rigorous activity
- baroreceptors and chemoreceptors send information to the medulla in order to regulate contraction rate and stroke volume of the heart
- the brainstem uses chemoreceptors to send information to regulate ventilation rate
- peristalsis is controlled by neurons within the alimentary canal that form the enteric nervous system
- most hormones help maintain homeostasis within the body and use negative feedback control to maintain a physiological variable within a narrow range.

Exercises

Q1. Identify the portion of the human brain most closely associated with the following functions.

   (a) Recalling memories from earlier in the day.

   (b) Coordinating motor action potentials for running.

   (c) Increasing heart rate during running.

   (d) Interpreting action potentials from the retinal cells of the eyes.

   (e) Learning new information.

Q2. What is the advantage of having one or more muscle movements occur through a pain reflex arc rather than initiated by the cerebrum?

Q3. Which of these is not a part of the enteric nervous system?

   A Stretch sensory neuron

   B Relay neuron

   C Interneuron within brainstem

   D Inhibitory neurotransmitter

Q4. Outline what happens to your blood pH and why when you exercise.

Q5. Which of these terms is least associated with a circadian rhythm?

A Pineal gland

B Cerebellum

C Melatonin

D Diurnal activity

Q6. Identify three specific muscles or muscle groups that receive a motor impulse when you increase the rate of ventilation during exercise.

Q7. Outline the difference between a "nerve" and a "neuron".

Q8. Nervous system receptors are neurons that are specialized to convert a specific physical stimulus to action potentials. What is the physical stimulus for each of these named receptor types?

(a) Photoreceptors

(b) Baroreceptors

(c) Chemoreceptors

(d) Nocireceptors

C3.2 Defence against disease

Pathogens are viruses, bacteria and other small organisms that can cause disease. Our first defence against pathogens is to prevent them from entering the body, our skin and mucous membranes acting as barriers. When a pathogen is able to enter the body, we have a two-layered immune system that responds to fight the infection. Our innate immune system consists of white blood cells called phagocytes that recognize pathogens as foreign and engulf them by endocytosis resulting in their digestion.

True immunity is built up over time by our adaptive immune system. This component of our immune system chemically recognizes the specific molecules that make up pathogens. These molecules are called antigens. A type of white blood cell called lymphocytes cooperates in the presence of specific antigens. This leads to cell cloning of specific lymphocytes to fight off the pathogens that carry the identified antigen. One important component of this response is the production of proteins called antibodies. Long-lived memory lymphocytes remain after an infection, providing long-term immunity.

There are many factors that influence the incidence of disease in populations. We only began to understand the causes of diseases about two hundred years ago. Since then, great advances have been made in disease prevention, especially in the way that human wastes are treated and sanitary conditions have been improved for water sources and preparing foods. Identification and treatments for viral diseases have been the most difficult to understand, with many viral diseases remaining and spreading in the population. Vaccines are our best protection against viruses, but there are challenges to their acceptance in some human populations. Bacterial diseases have been successfully treated by antibiotics, but their overuse is leading to the emergence of antibiotic-resistant strains of bacteria.

C3.2.1 – Infectious diseases are caused by pathogens

C3.2.1 – Pathogens as the cause of infectious diseases

Students should understand that a broad range of disease-causing organisms can infect humans. A disease-causing organism is known as a pathogen, although typically the term is reserved for viruses, bacteria, fungi and protists. Archaea are not known to cause any diseases in humans.

NOS: Students should be aware that careful observation can lead to important progress. For example, careful observations during 19th-century epidemics of childbed fever (due to an infection after childbirth) in Vienna and cholera in London led to breakthroughs in the control of infectious disease.

Pathogens are disease-causing organisms. Pathogens are any viruses, bacteria, fungi and protists that result in disease upon entry into the body. The vast majority of these

C INTERACTION AND INTERDEPENDENCE — Organisms

types of organisms are (thankfully) not pathogenic to humans. There are some that are only pathogenic to other animals, which explains why cats, dogs and farm animals are susceptible to their own diseases. Most of these diseases will not transmit to humans, although there are exceptions.

Nature of Science

Meticulous observations can lead to breakthroughs. In 1854, there was a major cholera outbreak in a suburb of London, UK. The *Vibrio cholerae* bacteria that causes cholera is found in the faeces of infected individuals. It was common practice in the 1800s for residents to empty human waste into areas in front of their homes. The bacteria moved down through the soil and infected a drinking well used by many in the community. A physician by the name of John Snow suspected cholera was being transmitted in water supplies. He created a map of all known infections and found one particular well had been used by all the infected people. The well was closed, and the number of cholera infections fell. The information and map created by Snow formed the basis of modern epidemiology studies that trace outbreaks of disease.

▲ This community water pump was responsible for 616 deaths in the 1854 cholera outbreak in London, UK.

Pathogenic organisms are not inherently "evil". They just happen to use human tissues as food and shelter. Their means of growth and secretions, however, can cause us harm. We have begun to learn a great deal about pathogenic organisms and have devised treatments for both before and after infection. Improvements in public health policies also help prevent the spread of pathogenic organisms. Throughout most of recorded history, humans had no real knowledge of the presence of pathogens and blamed infectious diseases on factors that now seem almost nonsensical to us. We must always remember that we live in an age where science is providing information not available to us even one or two generations ago.

Modern classification systems place all living organisms in one of three domains. One of these three domains is Archaea, comprising small single-celled prokaryotic cells that are fundamentally different from bacteria. No member of the Archaea domain is known to cause a human disease.

Nature of Science

In the mid-1800s, a physician called Ignaz Semmelweis, in Vienna, Austria, began to study a lethal disease commonly known as childbed fever. The disease had alarmingly high rates of infection and death in Vienna's maternity wards. Semmelweis began to note a difference in infection rates between two maternity wards. One ward was staffed by midwives and the other by physicians and medical students. The rate of infection and death on the physicians' ward was much higher than the ward staffed by midwives. Semmelweis began to narrow down and eliminate specific differences between the two wards, besides who staffed them. The one difference that proved to be consequential was that the physicians had often carried out autopsies before helping women in childbirth. Semmelweis postulated that the physicians were carrying small "particles" that caused disease from the autopsies to the pregnant women. He ordered the physicians to wash their hands with a chlorine solution before treating patients, and there was an immediate improvement in infection rate in the maternity ward.

C3.2.2 – Skin and mucous membranes as the first line of defence

C3.2.2 – Skin and mucous membranes as a primary defence
The skin acts as both a physical and chemical barrier to pathogens. Students are not required to draw or label diagrams of skin.

The best way to stay healthy is to prevent pathogens from having the opportunity to cause disease. One way to do this is to try to stay away from sources of infection. This is why it is still common to isolate (or quarantine) people who have highly transmittable diseases. Obviously, it is not possible to isolate yourself from every potential source of infection. Therefore, the human body has evolved some ingenious ways of making it difficult for pathogens to enter and start an infection. One of those ingenious ways is your skin.

Think of your skin as having two primary layers. The underneath layer is called the **dermis** and is very much alive. It contains sweat glands, capillaries, sensory receptors and dermal cells, which give structure and strength to the skin. The layer on top of this is called the **epidermis**. This epidermal layer is constantly being replaced as the underlying dermal cells die and are moved upwards. This layer of mainly dead cells forms a physical barrier against most pathogens because it is not truly alive. As long as our skin remains intact, we are protected from most pathogens that can enter living tissues. This is why it is important to clean and cover cuts and abrasions of the skin when they do occur.

A false-colour scanning electron micrograph (SEM) of the mucous membrane lining of the trachea. The large white cells are called goblet cells and they secrete mucus. Hair-like cilia (in pink) are also visible.

Pathogens can enter the body at the few locations that are not covered by skin. These entry points are lined with tissue cells that form a **mucous membrane**. The cells of mucous membranes produce and secrete a lining of sticky mucus. This mucus can trap incoming pathogens and so prevent them from reaching cells that they could infect. Some mucous membrane tissue is lined with **cilia**. Cilia are hair-like extensions capable of a wave-like movement. This movement carries trapped pathogens up and out of mucous-lined tissues such as your trachea. Table 1 shows some common areas that have a mucous membrane.

C3.2 Table 1 Areas of the body that have a mucous membrane ▶

Area with a mucous membrane	What it is and does
Trachea	The tube that carries air to and from the lungs
Nasal passages	Tubes that allow air to enter the nose and then the trachea
Urethra	A tube that carries urine from the bladder to the outside
Vagina	The reproductive tract leading from the uterus to the outside

According to an article published by the National Institutes of Health (NIH), bacteria outnumber their human hosts by about 10 to 1 cells. In a typical human adult, bacteria therefore account for about 2% of the human's body mass.

C3.2.3 – Blood clotting minimizes blood loss and infection

C3.2.3 – Sealing of cuts in skin by blood clotting

Include release of clotting factors from platelets and the subsequent cascade pathway that results in rapid conversion of fibrinogen to fibrin by thrombin and trapping of erythrocytes to form a clot. No further details are expected.

When small blood vessels such as capillaries, arterioles and venules are damaged, blood escapes from the closed circulatory system. Often the damaged blood vessels are in the skin, and so pathogens are then able to enter the body. Our bodies have evolved a set of responses to create a clot that "seals" the damaged blood vessels, so preventing excessive blood loss and helping prevent pathogens from entering the body.

Circulating in the blood plasma are a variety of molecules called **plasma proteins**. These proteins serve many purposes, including some that are involved in clotting. Two of the clotting proteins are **prothrombin** and **fibrinogen**. These two molecules are always present in blood plasma, but remain inactive until "called to action" by events associated with bleeding. Also circulating in the bloodstream are cell fragments known as **platelets**. Platelets form in the bone marrow, along with red blood cells (**erythrocytes**) and white blood cells (**leucocytes**), but do not remain as entire cells. Instead, one very large cell breaks down into many fragments, and each of the fragments becomes a platelet. Platelets do not have a nucleus and they have a relatively short cellular life span of about 8–10 days.

Consider what happens when a small blood vessel is damaged. The damaged cells of the blood vessel release chemicals that stimulate platelets to adhere to the damaged area, forming a "plug". The damaged tissue and platelets release chemicals called **clotting factors** that convert prothrombin to **thrombin**. Thrombin is an active enzyme that catalyses the conversion of soluble fibrinogen into the relatively insoluble **fibrin**. The appropriately named fibrin is a fibrous protein that forms a mesh-like network that helps to stabilize the platelet plug. More and more cellular debris becomes trapped in the fibrin mesh, and soon a stable clot has formed, preventing both further blood loss and the entry of pathogens.

C3.2.4 – A two-layered immune system: innate and adaptive

C3.2.4 – Differences between the innate immune system and the adaptive immune system

Include the idea that the innate system responds to broad categories of pathogen and does not change during an organism's life whereas the adaptive system responds in a specific way to particular pathogens and builds up a memory of pathogens encountered, so the immune response becomes more effective. Students are not required to know any components of the innate immune system other than phagocytes.

Humans are born with an immune system called the **innate immune system**. This first layer of the immune system responds to broad categories of pathogens and does not change during a person's lifetime. For example, the innate immune system would recognize any bacterium as a bacterium, rather than a specific species of bacterium.

The basis of the innate immune response is the ability to recognize those things that belong in the human body versus those that do not belong. In other words, it can recognize and respond to things that are "not-self". This includes bacteria, viruses, protists and fungi, and even things such as pollen and dust. The molecules of these foreign or not-self entities that can trigger an immune response are called **antigens**. The innate immune response involves activation of a group of leucocytes called **phagocytes**, which area capable of engulfing invading material by **endocytosis**.

A false-colour SEM of a large phagocyte (yellow) that has recognized a group of bacteria (pink rod shapes) as "not-self" and is in the initial stages of endocytosis.

The second layer of human immunity is called the **adaptive immune response**. This portion of our immune response develops over time and only after exposure to specific antigens of specific pathogens. The first exposure to a specific antigen leads to a series of cellular events culminating in molecules and cells that are long-lived and have the ability to defend the body against a specific pathogen. The specific long-lived white blood cells that are formed during the first exposure are called **memory cells**. Upon a second exposure to the same pathogen, these specific memory cells can be activated quickly. They can be so effective in fighting a pathogen that a person may not even realize that they were exposed a second time. The adaptive immune response becomes more effective with age, as a person becomes exposed to more pathogens.

C3.2.5 – The role of phagocytes

C3.2.5 – Infection control by phagocytes

Include amoeboid movement from blood to sites of infection, where phagocytes recognize pathogens, engulf them by endocytosis and digest them using enzymes from lysosomes.

Phagocytes are leucocytes (white blood cells) that are capable of an action called **amoeboid movement**. Cells capable of amoeboid movement can purposefully extend sections of their plasma membrane, followed by their cytoplasm and organelles. Phagocytes use this type of motion to squeeze their way through capillaries so that they can leave and enter the bloodstream in order to move through body tissues. When a phagocyte encounters something in body tissues that contains antigens and thus is not-self, it sends out plasma membrane extensions to engulf the foreign body in a process called **endocytosis**. The foreign body is brought inside the phagocyte, where the hydrolytic enzymes of **lysosomes** digest the potential invader. This response by phagocytes is non-specific and is part of the innate immune response.

A portion of the plasma membrane of a phagocyte engulfs a bacterium by endocytosis. Two stages are shown. The bacterium ends up being encased in a vesicle that is later digested by enzymes from one or more lysosomes.

C3.2.6 – The role of lymphocytes

C3.2.6 – Lymphocytes as cells in the adaptive immune system that cooperate to produce antibodies
Students should understand that lymphocytes circulate in the blood and are contained in lymph nodes. They should appreciate that an individual has a very large number of B-lymphocytes that each make a specific type of antibody.

There are many types of leucocyte that contribute to the human immune system. Two major types are called **B-lymphocytes** and **T-lymphocytes**. Sometimes their names are shortened to just B-cells and T-cells. Lymphocytes continuously circulate in the blood stream and are also contained within our lymphatic system, especially within lymph nodes. We are going to look at the function of B-lymphocytes first, and then explore the functions of T-lymphocytes.

A diagram showing the location of lymph nodes in the neck and head area. B-lymphocytes accumulate in lymph nodes. Each lymph node has lymph vessels bringing lymph fluid in, and other lymph vessels taking lymph fluid away.

The specific leucocytes called B-lymphocytes produce protein molecules called **antibodies** as part of the adaptive immune response. There are many types of B-lymphocytes, and each type is able to synthesize a specific antibody. Each specific antibody is able to recognize and bind to a specific antigen. If you had a measles infection, you would produce one type of antibody, and if you contract a virus that gives you influenza (flu), you would produce another type of antibody. Each type of antibody is different because each type has been produced in response to a different pathogen.

Each type of B-lymphocyte is a biological factory for synthesizing only one type of antibody.

Antibodies are a Y-shaped proteins. At the end of each of the branches of the Y is a **binding site**. The binding sites are where an antibody attaches itself to an antigen. Because the antigen is a protein on the surface of a pathogen (such as a bacterium), the antibody thus becomes attached to the pathogen. Each of us has many different types of antibody-producing B-lymphocyte cells and each can produce only one type of antibody.

The specificity that an antibody has for a certain antigen is not unlike an enzyme's specificity for a certain substrate or a hormone's specificity for a target protein found on only certain cells.

Computer artwork showing a single antibody (a blue, upside down Y shape in this picture) attached to an antigen on the surface of a pathogen.

Antibodies have specific mechanisms for fighting off infection by a pathogen. Any one antibody has two binding sites. If multiple antibodies bind to a cluster of pathogens, a clump is created because each antibody can potentially bind to two different pathogens. The antibody-bound cluster makes it easier for phagocytes to find and engulf the entire clump.

Many viruses attach to the plasma membrane of a body cell. The DNA or RNA of the virus is then injected into the cell. This cell then becomes a cellular factory to make more viruses. The protein coat of the virus, called a **capsid**, often remains on the outer plasma membrane of the cell and can be recognized by antibodies. Multiple antibodies of the same type use their binding sites to attach to proteins of the capsids. This is a way of marking the infected cell to be engulfed later by phagocytes.

C3.2.7 – Antigens trigger antibody production

C3.2.7 – Antigens as recognition molecules that trigger antibody production

Students should appreciate that most antigens are glycoproteins or other proteins and that they are usually located on the outer surfaces of pathogens. Antigens on the surface of erythrocytes may stimulate antibody production if transfused into a person with a different blood group.

An antigen is any substance that induces the immune system to produce antibodies. Most antigens are glycoproteins or other proteins and they are usually located on the outer covering of pathogens. These molecules, rather than the entire pathogen, are the molecular antigens that result in an immune response.

The adaptive immune response is based on many specifics:

- each type of B-lymphocyte makes a specific type of antibody
- each antibody is specific for one antigen
- each antigen is part of a specific group of molecules of a specific pathogen.

Antigens are usually proteins, and many are **glycoproteins** found embedded in the outer membrane of a pathogenic organism. This could be the plasma membrane of a bacterium or the outer cells of a protist or fungus. Viruses do not have a plasma membrane, but they do have a protein coat called a capsid and the capsid proteins act as antigens.

Other molecules, besides those in pathogens, can be recognized as antigens by our immune system. When organs are transplanted surgically, the organ or tissue transplanted must be "matched" very carefully by comparing the proteins of the donor and recipient. Transplanted hearts, kidneys and skin are examples of organs that can be transplanted. Many proteins must be taken into consideration, and rarely is there a perfect match. The exception is when identical twins are used as both donor and recipient.

Blood transfusions should only occur after the blood types of both donor and recipient have been tested and are known to be compatible. When a blood type is indicated by the notation AB⁺, for example, this notation is actually providing information about two blood types. One is called the ABO blood type, and the other is the Rh blood type. In this example, AB represents the ABO type, and + represents the Rh type. Blood typing is based on the presence or absence of three different antigens that are genetically inherited and found on the surface of erythrocytes. The three antigens are the A protein, B protein and Rh protein. The presence or absence of the three antigen proteins indicates a person's blood type.

Because pathogens have many antigens, a strong immune response may be a response to more than one of those antigens. In other words, more than one type of antibody can be produced in response to a single pathogen.

	Antigen found on the plasma membrane of erythrocytes
ABO blood type	
A	A protein
B	B protein
AB	A and B proteins
O	Neither A nor B protein
Rh blood type	
Positive	Rh protein
Negative	No Rh protein

The antigens associated with different blood types

For example:

- a person with blood type B⁻ has the B protein on their erythrocytes but does not have either the A protein or the Rh protein
- a person with blood type O⁺ has neither the A nor the B protein but does have the Rh protein.

For blood transfusions to be successful a person must not receive a protein that they do not already have, as determined by their own genetics. A person with blood type AB⁺ can receive blood from anyone because they already have all three antigens. A person with blood type O⁻ can only receive blood from someone who also has blood type O⁻, because they have none of the three antigens on their erythrocytes.

In a blood transfusion, a person cannot receive any of the three possible erythrocyte antigen proteins that they do not already have. This includes the A, B and Rh antigen proteins.

Challenge yourself

1. For each of the potential recipients shown, state all of the blood types that they could safely receive in a blood transfusion.

   (a) Recipient with blood type A⁺.
   (b) Recipient with blood type O⁺.
   (c) Recipient with blood type AB⁻.

The first blood transfusions were carried out before people had any knowledge of blood types. Transfusions were even carried out using farm animals as donors. As you can imagine, sometimes this accomplished more harm than good.

If someone receives a blood transfusion of an incompatible blood type, a transfusion reaction will occur. The reaction is an immune response to what the body identifies

315

as an antigen. If someone receives type B blood and they do not genetically produce the B antigen protein, antibodies will be produced that bind to the donated cells and **agglutination** (clumping) can occur. The resulting transfusion reaction may lead to minor effects but has been known to be fatal.

C3.2.8 – The role of helper T-lymphocytes

C3.2.8 – Activation of B-lymphocytes by helper T-lymphocytes

Students should understand that there are antigen-specific B-cells and helper T-cells. B-cells produce antibodies and become memory cells only when they have been activated. Activation requires both direct interaction with the specific antigen and contact with a helper T-cell that has also become activated by the same type of antigen.

Two important types of leucocytes that respond in the adaptive immune response are **helper T-lymphocytes** (T-cells) and **B-lymphocytes** (B-cells).

Helper T-lymphocytes chemically communicate with other leucocytes, including B-lymphocytes, to signal the presence of a specific antigen.

- There are many types of helper T-lymphocytes.
- Each type can only activate a specific B-lymphocyte.
- The same antigen that activates a specific B-lymphocyte will also activate a helper T-lymphocyte.
- Helper T-lymphocytes display antigens on their own plasma membrane.
- Helper T-lymphocytes release molecules called **cytokines** after finding a specific antigen to help activate a specific B-lymphocyte.
- Some helper T-lymphocytes are long-lived and are called **memory cells**.

B-lymphocytes produce a specific antibody that binds to a specific antigen.

- There are many types of B-lymphocytes.
- Each type produces an antibody specific to one antigen.
- Each type that produces a specific antibody must be activated before it can make antibodies.
- Activation requires exposure to an antigen of the pathogen and also exposure to an activated T-lymphocyte that is displaying the antigen and releasing chemicals called cytokines.
- Some B-lymphocytes are long-lived and are called memory cells.

C3.2 Figure 1 Activation of a B-lymphocyte by a helper T-lymphocyte. Both the B-lymphocyte and the helper T-lymphocyte have already encountered the pathogen. An antigen from the pathogen is displayed on the plasma membrane of the B-lymphocyte and on a receptor of the helper T-lymphocyte. A protein receptor on the B-lymphocyte must match a receptor on the helper T-lymphocyte. Cytokines from the helper T-lymphocyte are released and taken in by the B-lymphocyte. When all of these events have occurred, the B-lymphocyte is activated.

B-lymphocyte that has encountered its antigen

B-lymphocyte encounters activated helper T-lymphocyte that releases cytokines

C3.2.9 – Activation of a B-lymphocyte results in cloning

C3.2.9 – Multiplication of activated B-lymphocytes to form clones of antibody-secreting plasma cells

There are relatively small numbers of B-cells that respond to a specific antigen. To produce sufficient quantities of antibody, activated B-cells first divide by mitosis to produce large numbers of plasma B-cells that are capable of producing the same type of antibody.

The helper T-lymphocytes and B-lymphocytes described in Section C3.2.8 are antigen specific. The problem is that there is an incredible number of different antigens that may require a response. The immune system can only maintain a relatively low number of each type of cell that can respond to any one antigen. When specific B-lymphocytes are needed in an immune response, they first become activated and then undergo numerous mitotic cell divisions. In effect, they create **clones** of cells that have the genetic instruction to synthesize mass quantities of the antibodies that can bind to the antigens of a pathogen.

The basis of the adaptive immune response is that a few specific lymphocytes of each type are present in the body at all times. When a specific cell type is needed for an immune response, activation of that cell type leads to cloning to make many copies of that type of cell.

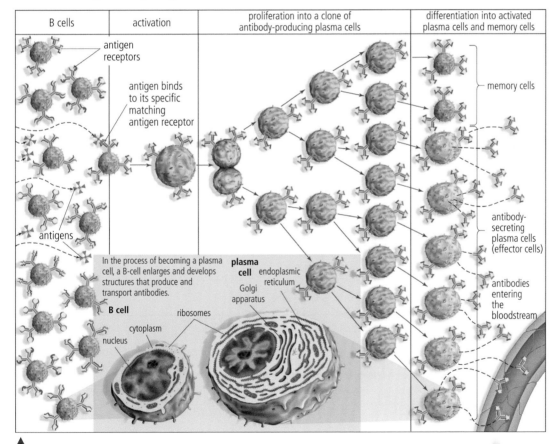

On the left is a representation of many types of B-lymphocytes. Only one type is activated by an antigen (and helper T-lymphocyte, not shown). The activated B-lymphocyte undergoes repeated mitosis to create a small "army" of the same type of B-lymphocyte. A few of these, shown on the upper right, are memory cells and will not produce antibodies during the current infection. The rest form antibody-secreting plasma cells that produce antibodies that can be useful in body tissues or be circulated in the bloodstream. Activated B-lymphocytes become larger, and develop many ribosomes, endoplasmic reticulum and Golgi bodies, all used for antibody production and secretion.

C3.2.10 – The role of memory cells

C3.2.10 – Immunity as a consequence of retaining memory cells
Students should understand that immunity is the ability to eliminate an infectious disease from the body. It is due to the long-term survival of lymphocytes that are capable of making the specific antibodies needed to fight the infection. These are memory cells.

The adaptive immune response requires an initial exposure to the antigen(s) of a particular pathogen. That first response is relatively long and is called the **primary immune response**. During this first exposure there are no memory cells, as the few lymphocytes that can respond to this pathogen have not yet been activated or cloned. The length of time it takes for a primary immune response to occur varies depending on the pathogen, but there is almost always sufficient time for symptoms of disease to develop. For example, if it is the first time a person is exposed to a specific cold virus, they will have symptoms of the cold. The primary immune response is taking place while the person is experiencing the symptoms, and will eventually result in the symptoms disappearing as the pathogen is eliminated from the body.

The second or any subsequent exposure to that same cold virus will trigger a **secondary immune response**. The memory cells that were produced during the primary infection continue to circulate in the bloodstream. These very long-lived cells, now in relatively large numbers, are capable of responding to the same pathogen very quickly. It is usually so quick that symptoms of the disease do not present, or are quite minor. The secondary immune response not only occurs faster it also produces many more antibodies than the first exposure.

Antibody production by the primary and secondary immune responses. The second exposure to the same antigen may be months or years after the first exposure. The production of antibodies is quicker after a second infection, and the number of antibodies is greater.

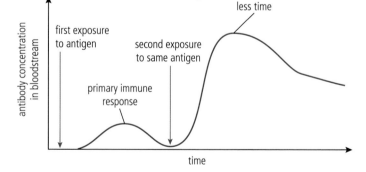

We have no true immunity to a pathogen during the first infection, but we do have an immune system that is usually able to eliminate a new pathogen. True immunity begins with a subsequent infection as a result of the activity of memory cells.

As shown by the recent **pandemic** caused by a coronavirus (SARS-CoV-2), pathogens are easily spread across the globe. SARS-CoV-2 is not the first pathogen to have moved from country to country despite interventions to stop its spread. Some previous pandemics have been referred to as plagues.

Sometimes the term immunity refers to the body's ability to eliminate an infectious disease during a primary immune response, when symptoms are likely to occur. This is because our immune system has both primary and secondary immune responses.

How do animals protect themselves from threats?

C3.2.11 – HIV transmission

C3.2.11 – **Transmission of HIV in body fluids**
Include examples of the means and implications of HIV (human immunodeficiency virus) transmission.

HIV is the abbreviation for a virus called **human immunodeficiency virus**. Just like any virus, HIV is very specific about which organisms and which cell types in an organism it infects. Unfortunately, the (host) cells it infects in humans is one of the key lymphocyte cell types involved in the human immune response.

HIV does not survive outside the body and is not transmitted by saliva, tears or sweat. It is also not transmitted by insects such as via mosquito bites. The fluids that can transmit HIV are blood, semen, rectal fluids, vaginal fluids and breastmilk.

The two most common ways that HIV is spread from person to person is by having unprotected sex with an infected person, and by using a hypodermic needle that has previously been used by someone who is infected. In addition, it is possible for an HIV-positive mother to infect her child during pregnancy, labour, delivery or breastfeeding. In some countries, receiving a blood transfusion can spread HIV, but this is no longer a risk in countries where blood and blood products are routinely tested for contamination. Some medical treatments, such as injections for treating haemophilia, have been known to spread HIV when the injected material was purified from human blood. In many areas of the world, these products are now produced by genetically engineered bacteria and there is no risk of transmitting HIV.

C3.2.12 – The result of HIV infection

C3.2.12 – **Infection of lymphocytes by HIV with AIDS as a consequence**
Students should understand that only certain types of lymphocyte are infected and killed, but that a reduction in these lymphocytes limits the ability to produce antibodies and fight opportunistic infections.

HIV is very specific about "choosing" which cell to infect. The host cells of HIV are known as helper T-lymphocytes or **CD4 T-lymphocytes**. CD4 is the name of the glycoproteins that are found on the plasma membrane of helper T-lymphocytes and are used by HIV in its mechanism for entering a cell.

Unfortunately for anyone infected with HIV, the helper T-lymphocytes will eventually be killed by the virus, but they are the same cells used to activate B-lymphocytes and some cells involved in an immune response (see Figure 1 on page 316). A person who has a very low helper T-lymphocyte count in their blood stream will not have a strong immune response to pathogens. This is the disease called **AIDS** or **acquired immune deficiency syndrome**. People that are HIV-positive and have progressed to AIDS are susceptible to **opportunistic infections**. These are infections that occur more often or with more severity in people with weakened immune systems. Opportunistic infections include tuberculosis, salmonella, pneumonia and several others.

Deaths from HIV have greatly decreased in countries with advanced medical care where HIV-positive patients take daily doses of medicines that reduce the potential for damage to the immune system.

The prevalence of people infected with HIV is very unevenly distributed around the world. Many sub-Saharan African countries continue to show a high percentage of people infected. In this graphic, published in the peer-reviewed journal *Nature*, the data points are shown as much smaller, more accurate clusters than ever shown before. This can be used to help distribute resources to fight AIDS.

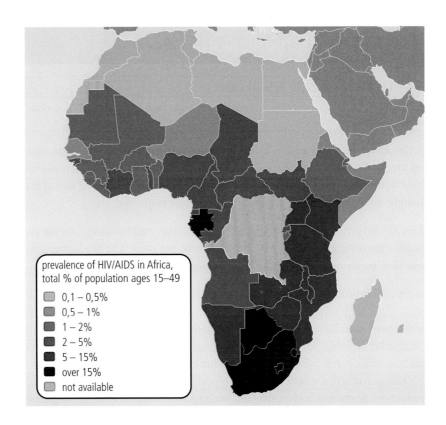

prevalence of HIV/AIDS in Africa, total % of population ages 15–49

- 0,1 – 0,5%
- 0,5 – 1%
- 1 – 2%
- 2 – 5%
- 5 – 15%
- over 15%
- not available

C3.2.13 – Antibiotics against bacterial infections

C3.2.13 – Antibiotics as chemicals that block processes occurring in bacteria but not in eukaryotic cells

Include reasons that antibiotics fail to control infection with viruses.

Bacteria are **prokaryotic** cells. Humans and other animals are composed of **eukaryotic** cells (see Chapter A2.2). There are many structural and biochemical differences between prokaryotic and eukaryotic cells. For example, while protein synthesis occurs in both types of cells, the processes are different. Also, bacteria have a cell wall, a structure that is not characteristic of eukaryotic animal cells.

Antibiotics are chemicals that take advantage of the differences between prokaryotic and eukaryotic cells: they selectively block some of the biochemical pathways needed by bacteria while having no effect on human or other animal cells. There are many categories of antibiotics, depending on the biochemical pathway that is being targeted. One type of antibiotic selectively blocks protein synthesis in bacteria, but has no effect on eukaryotic cells' ability to manufacture proteins. Another type of antibiotic inhibits the production of a new cell wall by bacteria, thus blocking their ability to grow and divide.

There are chemicals that have been developed as **antiviral medications**. These chemicals suppress a virus's ability to infect and multiply in the host's cells. Antiviral medications have become the standard of care for people with HIV and hepatitis C infections.

Viruses have no metabolism, which explains why antibiotics have no effect on them. Viruses make use of our own body cells' metabolism to create new viruses. Any chemical that could inhibit viral metabolic activity would also be damaging to our own body cells. Antibiotics should not be routinely prescribed for viral diseases.

C3.2.14 – Pathogenic resistance to antibiotics

C3.2.14 – Evolution of resistance to several antibiotics in strains of pathogenic bacteria
Students should understand that careful use of antibiotics is necessary to slow the emergence of multiresistant bacteria.
NOS: Students should recognize that the development of new techniques can lead to new avenues of research; for example, the recent technique of searching chemical libraries is yielding new antibiotics.

Bacterial resistance to antibiotics is a serious problem around the world. For too long, antibiotics have been used improperly and too frequently. The fundamental principles of evolution explain how pathogenic bacteria become resistant to any one antibiotic. When bacteria find their way into living tissues, frequently the environment is nearly perfect for their growth and cell division. Tissues are moist, full of nutrients, and relatively warm. In this type of growing environment some pathogenic bacteria can grow exponentially and double their numbers in as small a time period as 20 minutes.

Every cell division requires the DNA of a bacterial cell to replicate. Mutations occur spontaneously when DNA replicates. Most of the mutations are of no consequence, but when the DNA replication rate is very high, one or more mutations is likely to occur that is consequential. One of those random, but consequential, mutations may give a bacterial cell protection from the biochemical action of a particular antibiotic. One mutation is all that is required, as that bacterial cell will undergo binary fission repeatedly to grow into many cells. All bacterial cells that arise from the mutated cell will have the resistance to the antibiotic. In addition, the mutated strain may now cause infections in other people.

Bacterial species grown in a Petri dish showing resistance to antibiotics. The dull cloudy area is where bacteria are growing. The small white discs are impregnated with antibiotics. The clear area around many of the discs indicates that the antibiotics are preventing bacterial growth. The clear area is called a **zone of inhibition**. The presence of bacterial growth around three of the discs indicates those antibiotics that have little to no ability to prevent growth.

Nature of Science

A great deal of scientific research precedes the release of a new antibiotic before it can be prescribed. Antibiotics always come with directions for use that are based on this research. These directions include how long you should take the medicine for. If someone stops taking an antibiotic early, for example because their symptoms have improved, only the bacteria that are most sensitive to the antibiotic have been killed. A few resistant cells may not have been killed and can grow into a new resistant strain. You should always take an antibiotic for the full prescribed duration.

A few strains of pathogenic bacteria have emerged over the years that are resistant to several antibiotics. One example is methicillin-resistant *Staphylococcus aureus* (MRSA). This bacterial strain causes **staphylococcal infections** that are very difficult to treat. Responsible use of existing antibiotics is needed to prevent the emergence of more multiresistant bacteria. Responsible use includes:

- only prescribing an antibiotic when necessary
- taking the full course of an antibiotic, and not stopping when symptoms first subside
- reducing the spread of bacterial diseases by vaccination, hand-washing and proper food hygiene
- reducing or stopping the practice of adding antibiotics to farm animal feed.

Methods using artificial intelligence (AI) are being used to screen for new antibiotics. This involves predicting the interaction of chemicals using AI and the known chemicals in chemical libraries. Halicin, a chemical screened by AI, is a new and promising antibiotic that is being prepared for clinical trials.

Nature of Science

The development of new techniques can lead to new lines of research. The use of chemical libraries is a new approach to antibiotic development. Chemical libraries store chemicals along with all the known information about those chemicals. Searching chemical libraries is yielding new antibiotic treatments because synergistic effects of antibiotic combinations can be explored at the molecular level.

C3.2.15 – Zoonotic diseases

C3.2.15 – Zoonoses as infectious diseases that can transfer from other species to humans

Illustrate the prevalence of zoonoses as infectious diseases in humans and their varied modes of infection with several examples including tuberculosis, rabies and Japanese encephalitis. Include COVID-19 infection as an infectious disease that has recently transferred from another species, with profound consequences for humans.

Many infectious diseases are species specific. Those that can cross species, specifically animal to human, are called **zoonotic diseases**. The pathogen may be a virus, bacterium, protist or fungus. Some examples are described below.

Rabies

Rabies is a disease caused by a virus. Most human cases of rabies are the result of dog bites, although the dog may have received the virus from a wild animal. Cases of rabies occur throughout the world, but are more common in Africa and Asia. The rabies virus causes a progressive and fatal inflammation of the human brain and spinal cord. By the time symptoms begin to show in an infected person, it is too late for treatment. The best defence against rabies in humans is preventative vaccination of dogs. Seeking medical treatment shortly after a bite from a rabid animal can prevent death if the treatment is received quickly.

Tuberculosis

Zoonotic tuberculosis is a bacterial disease caused by *Mycobacterium bovis*. Humans are exposed to this bacterium through cattle. Ingestion of unpasteurized milk and milk products and infected meat are the primary means of transmission. Airborne transmissions are also possible, especially for those that work with cattle. The main symptom is damaged lung tissue, but other human tissues are also affected. The name tuberculosis comes from growths called tubercles that occur in the lymph nodes of an infected person.

Japanese encephalitis

Japanese encephalitis is caused by a virus that is transmitted through the bite of a species of *Culex* mosquito. The mosquito receives the virus from either a pig or wading bird. Most cases occur in southeast Asia and are quite mild, although a few cases have progressed to more serious symptoms, including coma and eventually death. There is a vaccine that prevents symptoms but it is not widely used in the rural areas where the disease is typically transmitted.

COVID-19

COVID-19 is a disease caused by a coronavirus known as **SARS-CoV-2**. It is almost certainly zoonotic, although no specific species has yet been identified as the first to infect humans. All other known coronaviruses can be zoonotic. SARS-CoV-2 has been shown to transfer easily from humans to other animals, such as dogs, cats and deer, and many other animals have tested positive for the virus. Most researchers classify COVID-19 as an "emerging infectious disease of probable animal origin", until more is learned about its origins. The virus quickly caused a global pandemic by spreading from person to person in 2019–2020. COVID-19 symptoms vary from **asymptomatic** (no symptoms) to severe and fatal respiratory damage. Variants of the virus are continuing to emerge, resulting in increased transmission rates. Vaccines have been developed but are not readily available in all areas of the world.

Current practices of keeping very dense populations of domesticated animals can lead to both increased animal-to-animal and animal-to-human zoonotic disease transmission.

C3.2.16 – Vaccines and immunity

C3.2.16 – Vaccines and immunization
Students should understand that vaccines contain antigens, or nucleic acids (DNA or RNA) with sequences that code for antigens, and that they stimulate the development of immunity to a specific pathogen without causing the disease.

In Section C3.2.10 you learned about the primary and secondary immune responses that result in the production and use of memory leucocytes. For many diseases, vaccines have been developed that act as the first exposure to a pathogen. A vaccine is typically composed of the chemical components of a pathogen after eliminating the disease-causing abilities of the pathogen. In traditional vaccine production, the pathogenic virus or bacterium is inactivated so that it cannot cause the disease. The pathogen or selected antigens from the pathogen are then injected into a person, resulting in the same immune response as if the pathogen had entered the host's body. This injection results in a primary immune response that then leaves behind memory cells that can be quickly triggered into action upon reinfection by the pathogen.

Recent advances in vaccine research and technology have led to a new approach. Instead of injecting an inactivated pathogen or antigen, the DNA or RNA molecules that code for the synthesis of specific protein antigens are injected. Body cells take in the nucleic acid and use their normal cell protein synthesis organelles and enzymes to produce antigens. These antigens, although produced by body cells, are recognized as foreign and stimulate a primary immune response without exposure to the pathogen. As with traditional vaccines, memory cells are produced to provide immunity.

A depiction of SARS-CoV-2 sitting in a protein receptor of a cell. Proteins that make the "spike" structures of the virus are synthesized by human cells after injection with one of the RNA vaccines.

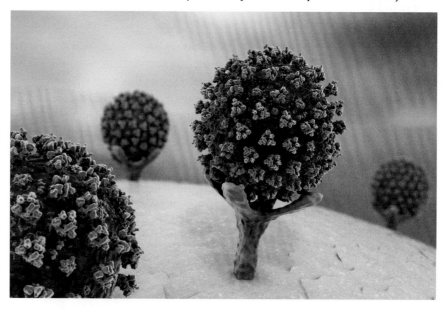

TOK

Social media has become a major news source for many people. What is the responsibility of social media platforms to ensure that shared information is fact based?

As part of the COVID-19 response, RNA vaccines were rapidly developed and manufactured. The protection provided by these vaccines has been excellent, especially for reducing the more serious symptoms that can result in hospitalizations. However, acceptance of the vaccines as safe and effective tools against COVID-19 has not been universal in some countries. Widespread misinformation concerning the safety and value of the vaccines has contributed to limited acceptance.

C3.2.17 – The role of herd immunity

C3.2.17 – **Herd immunity and the prevention of epidemics**
Students should understand how members of a population are interdependent in building herd immunity. If a sufficient percentage of a population is immune to a disease, transmission is greatly impeded. **NOS:** Scientists publish their research so that other scientists can evaluate it. The media often report on the research while evaluation is still happening, and consumers need to be aware of this. Vaccines are tested rigorously and the risks of side effects are minimal but not nil. The distinction between pragmatic truths and certainty is poorly understood.

Many pathogenic diseases spread as a result of person-to-person contact. When a large percentage of people in a given area (a herd) achieve immunity to a disease, there is a far reduced chance of the disease spreading. Even someone with no immunity is far less likely to get the disease when **herd immunity** has been achieved. The percentage of immune people that is needed to achieve herd immunity differs depending on the disease. Generally, the more contagious a disease, the higher the percentage needs to be. Measles, a highly contagious disease, requires 92–94% of the population to be immune. It is not yet certain what percentage of people need to be immune to COVID-19 to achieve herd immunity, or even if herd immunity is achievable. COVID-19 continues to produce new variants that may or may not be recognized by the immune response from a previous variant.

Nature of Science

Scientists publish their research so that other scientists can evaluate it. The media often report on the research while evaluation is still happening, and consumers need to be aware of this. Vaccines are tested rigorously and the risks of side effects are minimal but not nil. The distinction between pragmatic truths and certainty is poorly understood. A pragmatist can consider something to be true without needing to confirm that it is universally true.

C3.2.18 – Evaluating COVID-19 data

C3.2.18 – **Evaluation of data related to the COVID-19 pandemic**
Application of skills: Students should have the opportunity to calculate both percentage difference and percentage change.

There are many tools available to scientists for evaluating data. Two of the more common tools are calculating the **percentage difference** and **percentage change.** These very different calculations are often confused.

Calculating the percentage difference is useful when you are comparing two values that mean the same thing at the same time, for example if you are comparing the height of two people on the same day.

The percentage difference is the difference between two values divided by the average of the two values expressed as a percentage.

Worked example

Noah has a height of 176 cm and Mithun has a height of 184 cm. What is the percentage difference in their height?

Solution

Percentage difference = (184 cm − 176 cm) / ((184 cm + 176 cm) / 2) × 100

= 8 / 180 × 100 = 4.4%

There is a 4.4% difference between their heights.

Calculating the percentage change is useful when you are comparing two values that are separated by time.

Percentage change is the difference between the new and old values divided by the old value expressed as a percentage.

Worked example

Two years ago, Noah had a height of 160 cm. He now has a height of 176 cm. What is Noah's percentage change in height over the two years?

Solution

Percentage change = (176 cm − 160 cm) / 160 cm × 100

= 16 cm / 160 cm × 100 = 10%

There has been a 10% increase in Noah's height.

Challenge yourself

Herd immunity is difficult to achieve when a disease is highly contagious. How contagious a disease is represented by a value denoted as R_0, pronounced "R nought" or "R zero". The R_0 value is the estimated number of people that will be infected by a single infected person if everyone they contact is susceptible to the disease. The higher the R_0 number, the more contagious the disease is.

Use the data in the following table to answer the following questions.

2. What is the correlation between the R_0 value and the threshold percentage necessary for herd immunity?

3. Young people no longer receive a vaccine for smallpox. Suggest a reason why they do not need one.

4. In 1955, one company that produced a polio vaccine released some batches that contained active polio virus. Over 250 people contracted polio, with some resulting in paralysis. Discuss why information like this should be publicly available.

5. Calculate the percentage difference between the R_0 values of H1N1 and SARS-CoV-2 given in the table.

6. The virus causing COVID-19 has become more contagious over time as a result of mutations. The R_0 value of the original SARS-CoV-2 that emerged in 2019 was calculated to be 2.8 by the National Institutes of Health. Use the more recent R_0 value for the SARS-CoV-2 variant given in the table to calculate the percentage change in R_0 value.

Infectious diseases	R_0 value	Herd immunity threshold
Smallpox	5–7	80–85%
Mumps	4–7	75–86%
Measles	12–18	92–94%
Diphtheria	6–7	85%
Pertussis	12–17	92–94%
Polio	4–13	75–92%
Rubella	6–7	83–85%
H1N1 (2009 Pandemic)	1.6	40%
SARS	2–4	50–75%
SARS-CoV-2 (COVID-19)	5.7	82.5%

The concept of herd immunity based on how contagious a disease is. The R_0 and herd immunity values for SARS-CoV-2 are based on 2022 estimations.

The World Health Organization (WHO) is the United Nations agency that promotes good health practices and care throughout the world. Figure 2 shows the data collected by WHO, as of June 2022, on the number of SARS-CoV-2 cases in different areas of the world. Note, however, that testing protocols and reliability differed between the regions.

Total number of SARS-CoV-2 cases in June 2022

Europe
222,715,473
confirmed

Americas
158,984,066
confirmed

Western Pacific
62,063,771
confirmed

South-East Asia
58,251,326
confirmed

Eastern Mediterranean
21,828,710
confirmed

Africa
9,043,241
confirmed

C3.2 Figure 2 The total number of SARS-CoV-2 cases in June 2022 since virus transmission began in six WHO regions of the world.

How can false-positive and false-negative results be avoided in diagnostic tests?

Guiding Question revisited

How do body systems recognize pathogens and fight infections?

In this chapter you have learned:

- humans and many other animals have both innate and adaptive immune systems
- the innate immune response attempts to remove anything recognized as foreign or "not-self" without identifying it
- the adaptive immune response recognizes specific foreign entities and the response leaves behind immunity to that entity in the form of memory cells
- adaptive immunity involves recognition of molecules, called antigens, that make up pathogens
- specific helper T-lymphocytes are needed to recognize an antigen, and clone themselves to create higher numbers of that cell type
- specific B-lymphocytes are activated by helper T-lymphocytes, and produce antibodies that bind to antigens making up the pathogen
- long-lived cells of both types of lymphocyte act as memory cells to provide long-term immunity to the antigen
- vaccines are created by using inactive pathogens or nucleic acids injected into the body, which leads to the production of memory cells.

Guiding Question revisited

What factors influence the incidence of disease in populations?

In this chapter you have learned:

- skin and mucus membranes can often prevent entry of a pathogen into the body
- HIV/AIDS is a viral disease that can greatly lower the ability of a person's body to mount an effective immune response
- antibiotics are chemicals that selectively target the growth processes of bacteria and have been instrumental in treating bacterial infections
- antibiotic misuse and overuse have led to some pathogenic bacteria becoming resistant to one or more antibiotics
- vaccines are effective in creating long-term immunity in a population
- some pathogens are known to pass from species to species and are known as zoonoses
- herd immunity is achieved for a specific pathogen when a high percentage of the population has achieved immunity, making it unlikely that a pathogen would infect an unprotected person.

Exercises

Q1. Briefly state the function of each of the following during the process of blood clotting.

(a) Platelets

(b) Fibrinogen

(c) Thrombin

Q2. Which one of these is an unsafe transfusion of blood?

A O^+ to A^+

B A^+ to AB^+

C O^+ to O^-

D AB^- to AB^+

Q3. B-lymphocytes must be exposed to two things before becoming activated. What are those two things?

Q4. What is the cellular process that produces clones of selected lymphocytes?

Q5. Antibiotics will not help control a viral infection. Why?

Q6. Why do symptoms develop when the body undergoes a primary immune response?

Q7. A single pathogen can result in multiple adaptive immune responses in a person. Which answer best explains this.

A A pathogen may be related to a previous pathogen.

B A pathogen may contain many antigens recognizable by the immune system.

C A pathogen will always mutate inside the body.

D A pathogen is altered by the adaptive immune response.

C3 Practice questions

1. Annotate the diagram of the reflex arc to show the name and function of the structures labelled I and II.

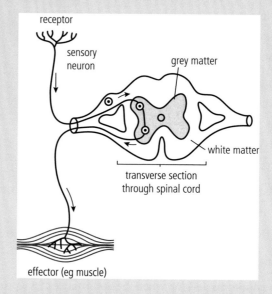

(*Total 2 marks*)

2. The heart responds quickly to physical activity. Describe how heart rate is controlled to meet increased circulatory demands.

(*Total 2 marks*)

3. Bacteria are prokaryotes that sometimes act as pathogens. Describe how the body can defend itself against pathogens.

(*Total 7 marks*)

4. Explain the evolution of antibiotic resistance in bacteria.

(*Total 6 marks*)

5. Ebola virus disease (EVD) is the disease in humans and other primates that is caused by the Ebola virus. Fruit bats are the reservoir for the virus and are able to spread the disease without being affected. Humans can become infected by contact with fruit bats or with people infected by the virus, their body fluids or equipment used to treat them.

The stacked bar graph shows the epidemiological data for EVD cases in Conakry, the capital city of Guinea, the surrounding suburbs, and rural areas in Guinea, from the beginning of January 2014 to the end of March 2015.

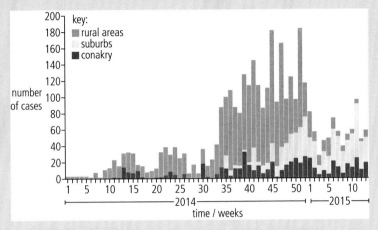

(a) Identify the week and year in which the first cases were recorded in the suburbs. (1)

(b) Based on the graph, compare and contrast the progress of the epidemic in the suburbs and rural areas. (3)

(c) Suggest **two** reasons for the overall decline in the epidemic after week 51. (2)

(Total 6 marks)

THEME

C Interaction and interdependence

4 Ecosystems

◀ Ecosystems are made up of complex interactions between organisms and their environment. In the Sonoran Desert, USA, plants and animals compete and cooperate in different ways.

Notice the word "system" in ecosystem: systems are made up of components that interact and are dependent on each other. Put many individuals together and you have a population; put many populations together and you have a community. New properties arise as the number of individuals and the number of species increase. These can include cooperation and competition, or herbivory and predation. Communities interact with their environment and usually depend on sunlight, temperature and water supplies to thrive.

A population can change over time. Researchers can use various sampling techniques to find out whether a population is growing or shrinking. Because organisms are connected in food chains and food webs, if one species can no longer survive in an ecosystem because of climate change or human activity, other organisms are also affected.

C4.1 Populations and communities

Guiding Questions

How do interactions between organisms regulate sizes of populations in a community?

What interactions within a community make its populations interdependent?

In this chapter we will see that nature has multiple ways of ensuring that no single species takes over an ecosystem and that each population has the potential to reach its maximum, but that many factors can limit populations or decrease their numbers. Such factors include the availability of food and water, presence of predators and the introduction of diseases. Some organisms engage in chemical competition by secreting molecules into their environment to reduce the chances of their competitors being successful.

No species lives in an isolated bubble. Plant eaters need plant material to survive, carnivores need other animals to eat, and some plants rely on microorganisms to provide nutrients such as nitrogen compounds to help them grow. This last example shows that sometimes two very different species can cooperate for mutual benefit.

C4.1.1 – Populations

C4.1.1 – Populations as interacting groups of organisms of the same species living in an area

Students should understand that members of a population normally breed and that reproductive isolation is used to distinguish one population of a species from another.

In biology, a **population** is defined as a group of individuals of the same species living in the same geographical area at the same time and able to interbreed. Here are a few examples of populations:

- emperor penguins (*Aptenodytes forsteri*) in Amundsen Bay, Antarctica
- bush-crickets (*Phaneroptera falcata*) living in a grassland in southern Europe
- rainbow trout (*Oncorhynchus mykiss*) along the northwest coast of the United States
- dandelions (*Taraxacum officinale*) growing in the same grassland as the bush-crickets.

Consider the emperor penguins. There is more than one population of them in Antarctica. How do we know where to draw the line between one population and the next? We can ask how probable it is that the individuals in population A might interbreed with individuals in population B. If penguin population A is separated by hundreds of kilometres of ice and open sea from population B, it is much more likely that penguins will stay and interbreed in their own population. We would be less sure of this if the populations were geographically closer to each other or if penguins could fly.

A population of emperor penguins (*Aptenodytes forsteri*)

When studying ecology, we can measure many things about a population. The most obvious is population size: the number of individuals. But populations are not fixed, they are dynamic, so we can also measure the change in population size over time, which could go up or down as a result of factors such as immigration and death. We will see that estimating the size of a population is not an easy task.

Other characteristics that can be used to describe a population are population density, its geographical distribution, and the maximum number of individuals that can be supported by the resources available. Once we understand a population, we can start to look at how it interacts with its environment and other species.

C4.1.2 – Estimating population size

C4.1.2 – Estimation of population size by random sampling

Students should understand reasons for estimating population size, rather than counting every individual, and the need for randomness in sampling procedures.

NOS: Students should be aware that random sampling, instead of measuring an entire population, inevitably results in sampling error. In this case the difference between the estimate of population size and the true size of the whole population is the sampling error.

Ideally, to know the size of the population we are studying, we would count every last individual. For emperor penguins, this would involve going to Amundsen Bay in Antarctica and counting each one. What would that look like? Once you started counting, how could you be sure you had not already counted a particular penguin? Some members of the population will be chicks keeping warm near their parents, and it will be hard to see all of them. Some individuals may be out in the water getting food. These are just a few examples of how, most of the time, it is simply impossible to count every last individual in a population. Instead, we need to rely on estimates. We count a sample and use that sample to estimate the overall population size.

There are two types of sampling: systematic and random. **Systematic sampling** is when a line or grid is set up and measurements or counting are carried out only at specified, regular intervals. For example, a 50 m measuring tape can be laid across a rocky shore and seaweed or snails can be counted in a 1 m² area every 5 m.

Random sampling is when arbitrarily chosen zones of the population's geographic distribution are sampled. Random directions and random distances between samples are used to try to overcome any bias that the investigators might have that would favour a particular area. Random sampling using quadrats and a method of mark and recapture are discussed in Sections C4.1.3 and C4.1.4.

A biology student using a quadrat on a rocky shore.

Nature of Science

Random sampling inevitably results in sampling error. The difference between an estimated population size and the true size of the whole population is the sampling error. If ecologists randomly take 200 samples from a prairie that are 1 m² each, and determine that the average number of ferns in each square metre is 1.3, they can deduce that, if the prairie is 10,000 m², the total population should be 13,000 ferns. But this assumes that the ferns are evenly distributed across the prairie. If there were actually only 8,000 ferns in the area studied, the sampling error would be quite large: 5,000 extra ferns were incorrectly estimated. This is one of the many reasons why, when given a number, scientists often ask how it was determined and what the degree of precision is. They want to know how much error might be involved and how close the estimate is to the true value.

Here is an allegory to illustrate sampling error. A man gets in a boat and rows out into the ocean. He fills a bucket with sea water and looks in the bucket to count the number of fish he sees. He sees none, and declares, "There are no fish in this ocean!"

C4.1.3 – Sampling sessile organisms

C4.1.3 – Random quadrat sampling to estimate population size for sessile organisms

Both sessile animals and plants, where the numbers of individuals can be counted, are suitable.

Application of skills: Students should understand what is indicated by the standard deviation of a mean. Students do not need to memorize the formula used to calculate this. In this example, the standard deviation of the mean number of individuals per quadrat could be determined using a calculator to give a measure of the variation and how evenly the population is spread.

Standard deviation is a quantity that measures the average difference between the measured values and the mean of those values. It describes how spread out the data are in relation to the mean.

Follow the downloadable activity in the eBook to use random quadrat sampling to estimate the population size of a local species of plant. Determine the **standard deviation** of the mean number of individuals per quadrat to give a measure of the variation and how evenly the population is spread.

Random sampling can be used to estimate population size for organisms that stay in one place, for example plants, lichens and corals. Such organisms are referred to as **sessile**; for much of their lives they do not change location. The example in the eBook shows how to estimate the population size of a sessile organism using a **quadrat** (see the photo on the previous page). A quadrat is a square of a particular dimension that can be made of a rigid material such as metal, plastic or wood. Using a quadrat means the surface area of the sample size is the same for each count you take.

C4.1.4 – Sampling motile organisms

C4.1.4 – Capture–mark–release–recapture and the Lincoln index to estimate population size for motile organisms

Application of skills: Students should use the Lincoln index to estimate population size. Population size estimate = $M \times N/R$, where M is the number of individuals caught and marked initially, N is the total number of individuals recaptured and R is the number of marked individuals recaptured. Students should understand the assumptions made when using this method.

The **capture–mark–release–recapture method** is a sampling technique that enables you to estimate the number of animals in an ecosystem. It is used instead of quadrats for **motile** organisms, those that move around. The technique involves catching some of a population and marking them. The marked animals are released back into the ecosystem and given a suitable period of time to remix with others in their population. A second sample of the population is then captured. Some in the second sample will be marked and some will be unmarked. The proportion of marked to unmarked individuals in the second sample is assumed to be the same as the proportion of the originally marked individuals to the whole population. The **Lincoln index** is then used to estimate the number of individuals in the population. Here is the formula:

$$\text{total population} = \frac{\text{number of individuals}}{\text{caught and marked initially}} \times \frac{\text{number of all individuals recaptured}}{\text{number of marked individuals recaptured}}$$

The Lincoln index can be written using variables instead of words:

$$\text{total population} = M \times \frac{N}{R}$$

where M is the number of individuals originally caught and marked, N is the total number of individuals recaptured, and R is the number of marked individuals recaptured.

You can try this method at home, in your classroom or on a field trip using the activity in your eBook.

The capture–mark–release–recapture method has some limitations:

- capturing and marking the animals may injure them
- the mark may make an animal more visible to predators and, if the marked animals are eaten, the second sample will not be reliable
- it assumes that the population is closed, with no immigration or emigration.

The last point is a rather big assumption, because very few populations are closed. Other assumptions are that all the individuals in a given area have an equal chance of being captured, that marked individuals will be randomly distributed after release, and that marking individuals will not make them more or less likely to be recaptured.

The mark and recapture technique can be used by experts to estimate the size of a bird population. Small metal bands with identification codes can be attached to a bird's leg to tag it, and when that same bird is recaptured, population ecologists know where and when it was originally captured.

C4.1.5 – Carrying capacity

C4.1.5 – Carrying capacity and competition for limited resources

A simple definition of carrying capacity is sufficient, with some examples of resources that may limit carrying capacity.

No habitat can accommodate an unlimited number of organisms: populations often grow, but they cannot continue to grow forever. There comes a time in the growth of a population when its numbers stabilize. This number, the maximum number of individuals that a particular habitat can support, is called the **carrying capacity**, and it is represented by the letter **K**.

Consider, for example, a given zone in a forest. There is a maximum number of trees that can grow there. This number is attained when enough trees are present to catch all the sunlight, leaving every square metre of the forest floor in shade. New tree seedlings trying to grow under the adult trees will have difficulty getting any sunlight.

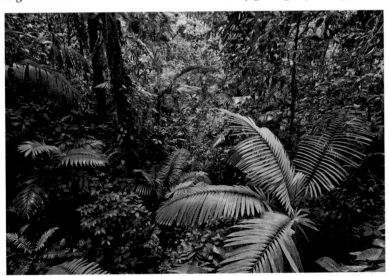

In tropical rainforests such as this one in Costa Rica, very little sunlight reaches the forest floor because a maximum number of trees and plants have spread their leaves to catch all the sunlight they can.

Young trees can store energy for years with very little vertical growth, waiting for a larger tree to die, leaving a hole in the canopy. The young trees then compete to use the opening to take the old tree's place. Those that lose this competition usually die. For a young tree to join the mature population, an old tree must die. Hence, there is no net increase in the tree population, at least at the canopy level.

A **limiting factor** is something that can prevent a population from getting bigger or reduce a population's size. Limiting factors that define the carrying capacity of a habitat include:

- the availability of resources, such as water, food, sunlight, shelter, space and oxygen (the latter notably in aquatic habitats)
- the build-up of waste, such as excrement and excess carbon dioxide
- predation
- disease.

What factors can limit capacity in biological systems?

C4.1.6 – Negative feedback

C4.1.6 – Negative feedback control of population size by density-dependent factors

Numbers of individuals in a population may fluctuate due to density-independent factors, but density-dependent factors tend to push the population back towards the carrying capacity. In addition to competition for limited resources, include the increased risk of predation and the transfer of pathogens or pests in dense populations.

As a population grows, its population density increases; there will be more individuals per unit of habitat, such as more plants per square metre of prairie or more plankton per litre of seawater. Some factors that change the size of a population are dependent on the density of the population, such as the spread of disease. These are called **density-dependent** factors. They tend to keep a densely populated area at or below its carrying capacity. Other factors are **density-independent**, meaning that it does not matter if there is a sparse population or an overcrowded population. Examples of density-independent factors include climate change, or a nearby forest fire or volcanic eruption. Populations both big and small will be affected, and the change in numbers can be an increase or a decrease, depending on whether the factor helps or hinders a population.

Complex systems require feedback mechanisms to keep parameters from reaching dangerously high or low levels. **Positive feedback** in a population is when something increases the population and the system encourages more of the same. For example, in a breeding population that is initially small, the more individuals that are produced by successful breeding, the more individuals there are to produce the next generation. The higher the population density, the higher the chances of a male encountering a female to produce even more individuals in the population. Positive feedback will help a population grow, but it can get out of control.

Negative feedback prevents a system from going too far in one direction. If a population continued to grow uncontrollably, for example, individuals would start to run out of resources such as space, food or water. The chances of diseases spreading also increase when population sizes are large and density is high. The chances of attracting predators to an area increase as the number of prey individuals increases. Negative feedback in the form of competition for resources and the spread of disease works to control the size of the population so that it cannot go above its carrying capacity. When it comes to population dynamics, positive feedback can be a dangerous thing, leading to food shortages and the spread of disease, whereas negative feedback helps prevent such density-dependent issues from getting worse and threatening a population.

In a positive feedback loop in populations, more offspring grow up and produce additional offspring which, in turn, produce even more offspring.

In a negative feedback loop, something such as a paucity of food or the introduction of a disease will regulate the population and slow down or reverse population growth.

Rabbits have a reputation for being able to quickly increase their population. Thanks to positive feedback, the population gets bigger but will eventually reach the carrying capacity for its environment.

C4.1.7 – Population growth

C4.1.7 – Population growth curves

Students should study at least one case study in an ecosystem. Students should understand reasons for exponential growth in the initial phases. A lag phase is not expected as a part of sigmoid population growth.

NOS: The curve represents an idealized graphical model. Students should recognize that models are often simplifications of complex systems.

Application of skills: Students should test the growth of a population against the model of exponential growth using a graph with a logarithmic scale for size of population on the vertical axis and a non-logarithmic scale for time on the horizontal axis.

In 1980, the ecosystems around Mount Saint Helens were destroyed when the volcano erupted in a sudden, violent blast. The devastation is shown in the picture on the left. Contrast this with the picture on the right, where a few of the dead tree trunks can still be seen, as well as smaller trees that have grown since the eruption.

339

In 1980, there was a major volcanic eruption of Mount Saint Helens on the west coast of the United States. After the massive event, little was left of the forest and rivers that had existed on and around the mountain. Forest fires and hot gases burned everything in sight. Volcanic ash rained down, smothering the destroyed forest and covering the carcasses of the animals that had died there. Many species that could escape fled the area.

Yet, within months of the eradication of an ecosystem, life was back. Seeds, dropped by birds or blown in by the wind, germinated in the fertile volcanic ash. Little by little, insects, then birds, then small mammals, moved in. Within a couple of decades, a grassland and shrub ecosystem had reappeared. Today, thousands of species flourish in what had been a desolate landscape.

If we look at the tree species now present on Mount Saint Helens, such as conifers and the red alder (*Alnus rubra*), their populations have increased over the decades, but some experienced a decrease in growth when the North American elk (*Cervus elaphus*) population started to thrive there too. Attracted by new food sources, elk browsed (ate) the young tree saplings of deciduous trees and conifers, slowing the growth of the tree populations. However, the elk droppings also helped improve the soil quality, so other vegetation could grow better. These complex interactions are part of the reason why an ideal population growth curve like the one shown in Figure 1 almost never exists in nature.

C4.1 Figure 1 An idealized population curve showing growth over time following a sigmoid shape. Phase 1 represents the exponential growth phase, phase 2 the transition phase and phase 3 the plateau phase or stationary phase, in which the population reaches its carrying capacity (K).

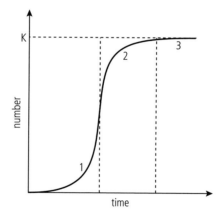

The sigmoid (S-shaped) curve of the graph in Figure 1 shows the three stages of population growth.

1. The **exponential phase**, also called the logarithmic phase, during which the number of individuals increases at a faster and faster rate.
2. The **transitional phase**, during which the growth rate slows down considerably; the population is still increasing but at a slower and slower rate.
3. The **plateau phase** or stationary phase, during which the number of individuals stabilizes, and there is no more growth.

So what causes the three different phases of a population growth curve?

The exponential phase

In ideal conditions, a population can double in size on a regular basis. Not counting mortality, for example, a population of bacteria can theoretically double its population every few hours: 1, 2, 4, 8, 16, 32, 64, 128, 256, 512, 1024, and so on. Without predators, introduced species, such as cane toads (*Bufo marinus*) in Australia, can take over habitats with uncontrolled population growth. The reasons for this first phase of exponential growth are:

- plentiful resources, such as food, space and (for photosynthetic organisms) light
- little or no competition from other inhabitants
- favourable abiotic factors, such as temperature and (for aquatic organisms) dissolved oxygen levels
- little or no predation or disease.

The transitional phase

Eventually, after the exponential increase in numbers of individuals, some of the factors listed above no longer hold true. This leads to the transitional phase. The causes of the transitional phase are:

- with so many individuals in the population, there is increasing competition for resources
- predators, attracted by a growing food supply, start to move into the area
- because of the large numbers of individuals living together in a limited space, opportunities for diseases to spread within the population increase.

The plateau phase

Consider the land around a volcano such as Mount Saint Helens, slowly being taken over by vegetation within months of a deadly eruption that wiped out all nearby life. Once all the fertile volcanic ground is covered with plants, the space available will be occupied to its maximum. Thus, there is gradually less and less available space for any seeds produced by the plants to germinate, and the number of plants stabilizes.

With increasing numbers of herbivores, there is a limited supply of food. In response to limited food supplies, animals tend to have smaller numbers of offspring. Some may leave the area and emigrate to a place where there is more abundant food.

Predators and disease increase mortality, and the growth curve tends to level off. In this phase, the number of births plus the number of immigrations is balanced by the number of deaths plus the number of emigrations.

Nature of Science

The population growth curve represents an idealized graphical model. Models are simplifications of complex systems. They help us conceptualize phenomena that otherwise are difficult to understand. They are useful in helping our brains grasp difficult concepts. But because they are simplifications, they have their limits.

What are the benefits of models in studying biology?

You can test the growth of a population against the model of exponential growth using a graph with a logarithmic scale for the size of the population on the vertical axis and a non-logarithmic scale for time on the horizontal axis. Online simulators exist that show how populations grow over time. Do an online search for "Howard Hughes Medical Institute population dynamics logistic growth model". It allows you to determine parameters such as the carrying capacity then launch the simulator. Other simulations show how predators and herbivores can modify a population. Do an online search for "Annenberg learner online ecology lab" to find a simulator of a food web that allows you to introduce plants, herbivores and predators into a habitat and watch what happens to a graph of their populations.

Many biologists, environmental groups, economists and governments wonder what the carrying capacity of Earth is for the human population. Will the number of people continue to increase or will disease, climate change or competition for resources lead to a transitional phase or a plateau? How reliable are mathematical models and what role could they play in shaping the way we make decisions about the future? Do different experts use mathematical models in different ways?

The world human population over the last twelve millennia. Compare this observed data with the expected sigmoid population curve.

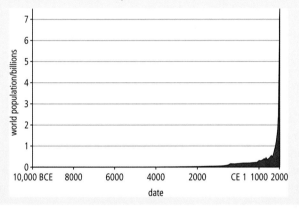

C4.1.8 – Modelling population growth

C4.1.8 – Modelling of the sigmoid population growth curve
Application of skills: Students should collect data regarding population growth. Yeast and duckweed are recommended but other organisms that proliferate under experimental conditions could be used.

Follow the downloadable activity in the eBook to model the sigmoid population growth curve using baker's yeast or by growing duckweed in pond water. Collect data to determine the population growth and discover whether the population will reach the carrying capacity of the environment that you design.

C4.1.9 – Communities

C4.1.9 – A community as all of the interacting organisms in an ecosystem
Communities comprise all the populations in an area including plants, animals, fungi and bacteria.

A **community** is a group of populations living and interacting with each other in an area. Examples include the soil community in a forest and the fish community in a river.

In ecology, the term "interacting" can mean one population feeding on another, or being eaten by another. It can mean that one species provides vital substances for another, as in the case of certain bacteria that can help plants obtain nitrogen from the air. It can also mean that one species is protected by another, as in the case of aphids protected by ants from predator attacks. Interacting can also mean that one species relies on another for its habitat, as is the case for parasites living on or inside the bodies of other animals.

Challenge yourself

1. Pick three organisms in the figure and determine how many other organisms each one depends on. Which organisms depend on them? What about environmental factors? Which ones does each organism contribute to, and which ones does each depend on?

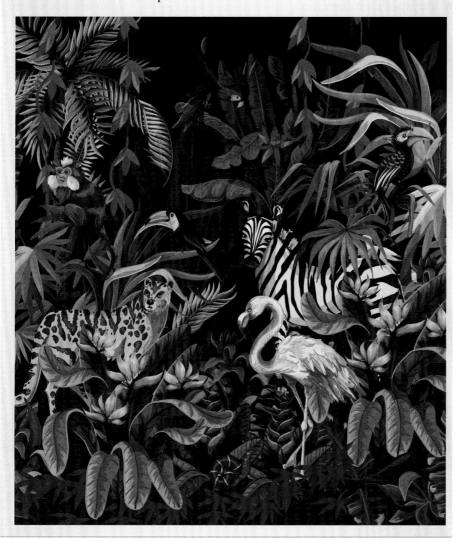

C4.1.10 – Intraspecific relationships

C4.1.10 – Competition versus cooperation in intraspecific relationships
Include reasons for intraspecific competition within a population. Also include a range of real examples of competition and cooperation.

Intraspecific relationships refer to those that occur between individuals of the same species. Intra means "within". Intraspecific relationships can involve cooperation (helping each other) or competition (competing with other members of the same species for the same resources). Cooperation can be thought of as a "win/win" situation, and competition as a "win/lose" or a "zero sum game", whereby if one wins, the other must lose.

In **intraspecific cooperation**, an individual of a species will help another individual from the same species so that survival is assured not only for the individual but for the group as well. An example is hunting as a pack, as wolves do, rather than hunting solo. Hunting cooperatively has a higher chance of success than hunting alone. Cooperation might mean caring for a neighbour's young while the mother is out looking for food, as vampire bats do. Taking turns helps the survival of both families. Cooperation can also mean that multiple individuals take turns discouraging intruders from approaching their territory, as lions do. If only a single female lion had responsibility for this, she would become exhausted more easily. Another way of thinking about intraspecific cooperation is teamwork; an individual will invest some of their time to help others, which will benefit the whole group.

In **intraspecific competition**, members of the same species compete for the same resources. The resources that are in demand could be space and sunlight for two oak trees growing near each other in a forest, or zones of grazing pastures for herds of bison. Other examples of competition within a species include male gorillas competing to become the alpha male and be able to mate with the females, or lizards competing with other members of their species for the best spot on a sunny rock to bask on. Another way of thinking about intraspecific competition is as a battle; one member of the species will succeed and get the resource, while the other will lose and not get it. But this analogy is not perfect because often there is no aggression or combat, as in the case of an individual finding food and eating it before others of its species can.

C4.1.11 – Interspecific relationships

C4.1.11 – Herbivory, predation, interspecific competition, mutualism, parasitism and pathogenicity as categories of interspecific relationship within communities
Include each type of ecological interaction using at least one example.

Interactions between different species in a community are called **interspecific relationships**. Inter means "between". Different types of interspecific interactions have different effects on the species involved, and can be positive, negative or neutral. Table 1 lists some of the possible ways in which populations can interact.

▲ Social animals such as these Barbary macaques (*Macaca sylvanus*) display intraspecific cooperation in many ways. Nature is not only about competition: often individuals look out for each other.

C4.1 Table 1 Types of interspecific interactions

Interaction	Examples
Herbivory: eating plant material	Snails eating lettuce leaves, sheep grazing on grass, giraffes eating acacia leaves. The photosynthetic organisms and their fruits or tubers are damaged or potentially killed in the process.
Predation: killing and eating prey or eating something that has recently died (scavenging)	Lions hunting and eating zebras, seahawks eating fish, vultures eating the abandoned carcass of a gazelle. In this relationship, only the predator survives: the prey is killed and its body parts benefit others.
Interspecific competition: two species struggle to get the same food resource	An oak tree and a balsam fir tree attempting to get the same soil minerals and sunlight. Spotted hyenas and lions hunting the same population of zebra for food. Humans overfishing a zone where sharks feed. Although one species may get a larger share of the resources, this kind of competition does not necessarily involve death or injury in the species that is less successful.
Mutualism: two species providing food or other resources where both benefit	Lichens are made up of an alga providing food using photosynthesis and a fungus providing minerals. Each one helps the other, and neither is hurt by the relationship; it is an example of mutual benefit. Mutualism is covered in more detail in Section C4.1.12. Lichens growing on a tree branch ▶
Parasitism: one species living on or in a host and depending on the host for food for at least part of its life cycle; the host can be harmed by the parasite	Parasites belonging to the genus *Plasmodium* cause malaria in humans. They reproduce in the human liver and red blood cells. Part of the life cycle of *Plasmodium* takes place in the body of the *Anopheles* mosquito. The mosquito transmits the malaria parasite from one human to another. ▲ An illustration of the parasite *Plasmodium*, a protist that can live in red blood cells and that causes malaria in humans.
Pathogenicity: the ability of microbes such as bacteria and viruses to cause disease in other species	Pneumonia is a transmissible disease caused by a pathogen, either a bacterium or a virus. The host of the pathogen will suffer and can potentially be killed by the microbe infecting it.

Mosquitoes are arguably the deadliest animal to humans, killing between 750,000 and one million people per year with the diseases they spread. When asked to name dangerous animals, we often come up with examples such as sharks. But there are only a handful of documented cases of people being killed by sharks worldwide each year. Mosquitoes are thousands of times more deadly to us than sharks are.

C4.1.12 – Mutualism

C4.1.12 – Mutualism as an interspecific relationship that benefits both species

Include these examples: root nodules in Fabaceae (legume family), mycorrhizae in Orchidaceae (orchid family) and zooxanthellae in hard corals. In each case include the benefits to both organisms.

Note: When students are referring to organisms in an examination, either the common name or the scientific name is acceptable.

Sometimes two species help each other survive and thrive. Neither organism is injured or destroyed. **Mutualism** is a type of interspecific cooperation that benefits both species. We will explore three examples of mutualism.

Plant root nodules and bacteria

Living organisms need the element nitrogen for the formation of amino acids and nucleic acids. Earth's atmosphere is nearly 80% nitrogen gas (N_2). Unfortunately, plants and animals cannot metabolize N_2 directly. The ability to turn gaseous nitrogen into usable nitrogen-rich molecules is called **nitrogen fixation**. One genus of bacteria that can fix nitrogen is *Rhizobium*. *Rhizobium* bacteria live in the **root nodules** of plants in the legume family, Fabaceae, which includes beans, lentils, peanuts and clover. The bacteria convert nitrogen gas from the air into ammonia (NH_3), an organic molecule that acts as a fertilizer for plants, helping them grow better. The bacteria have a mutualistic relationship with the plants: they are **symbiotic**. The *Rhizobium* bacteria receive carbohydrates and a favourable environment in the nodules of their host plant, and the plants receive usable nitrogenous compounds. Farmers and gardeners who want to fertilize their soil without using artificial chemical fertilizers can plant species that have root nodules to enrich the soil for part or all of a planting season.

The spheres protruding from this pea plant root are nodules hosting beneficial bacteria that provide usable nitrogen for the plant. Two species are cooperating for mutual benefit.

Mycorrhizae in Orchidaceae

Another example of mutualism is **mycorrhiza**, which occurs when a plant and a fungus help each other. Species of orchids (family Orchidaceae), for example, rely on fungi for one or more stages of their life cycle. Orchid seeds, which do not have enough energy and nutrients to germinate on their own, obtain the nourishment they need to produce a new plant from a fungus. The fungus in the mycorrhizal relationship takes nutrients from its environment and passes them on to the seed.

We usually think of all plants as being autotrophs (organisms that can produce their own food) because the vast majority of them make their own food via photosynthesis. There are a small number of orchid plant species that are non-photosynthetic and therefore qualify as heterotrophs (organisms that cannot produce their own food). They rely on fungi to decompose dead material and pass the nutrients to them via a

The network of fungal hyphae in the forest floor has been called the "wood wide web", in reference to how the internet's world wide web allows the transfer of resources such as information from locations that have it to those that need it. In a mycorrhizal network, it is food and minerals that are exchanged, although information in the form of chemical signals can be sent from one part of the forest to another, for example as a warning of insect attacks.

root-like system of **hyphae** throughout their adult lives. Hyphae are thin filaments produced by fungi to create a network in the soil that allows the transfer of nutrients from one place to another. Sometimes this network is used by trees and plants to transfer the sugars made by photosynthesis from one tree or plant to another.

At first glance, this might not seem to be of mutual benefit. Only the orchid is getting something from the relationship: taking food but not giving it back, whether only at the seed phase or in the adult phase too, depending on the species. Eventually, however, the orchid will die, and the fungi will benefit from the nutrients released as they decompose the orchids.

Fungal mycelia, branches of hyphae filaments, can spread out in soil and are used to transport nutrients not only to the fungi that produce them but between other species, plants and trees, as well.

Zooxanthellae in hard corals

Coral reefs are built by small animals called coral polyps, which are cnidarians, in the same phylum as sea jellies and sea anemones; they all possess stinging cells called nematocysts. Embedded in the tissue of their tentacles are single-celled photosynthetic algae of the genus *Symbiodinium*, a type of dinoflagellate that can photosynthesize and is referred to as zooxanthellae. These dinoflagellates coexist in a

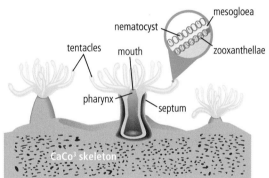

Zooxanthellae living inside the body tissue of the polyp that forms the hard coral. The two species are cooperating for mutual benefit.

symbiotic relationship with the coral polyps, giving them food in the form of the carbon-based energy molecules they make from sunlight. In exchange, the coral polyps give the zooxanthellae carbon dioxide and minerals, which they need to photosynthesize and grow. The coral polyps also provide the zooxanthellae a home on the coral reef. Both species live together for mutual benefit. Neither has to die or be injured for the other to flourish.

Dinoflagellates are colourfully pigmented creatures and are the source of the reds and greens we see in a coral reef. **Coral bleaching** is a phenomenon that happens when the zooxanthellae leave or die. The reef looks white because the calcium carbonate skeleton of the coral does not have any pigments. Bleaching is a sign of a very unhealthy or dead coral, and with changes in the temperature and pH of oceans it is being seen more frequently.

C4.1.13 – Endemic and invasive species

C4.1.13 – Resource competition between endemic and invasive species

Choose one local example to illustrate competitive advantage over endemic species in resource acquisition as the basis for an introduced species becoming invasive.

A species is considered to be **endemic** in an area if it is only found there and nowhere else in the world. **Invasive species** are those that have been introduced into a new area from a distant origin and their populations grow so well that they start to cause problems for the species that are already living there. Without any natural predators to keep their numbers in check, an introduced population can grow exponentially.

Galápagos tortoises are endemic to the Galápagos islands, and in nature are found nowhere else on Earth. The only other populations have been introduced by humans, such as those found in zoos.

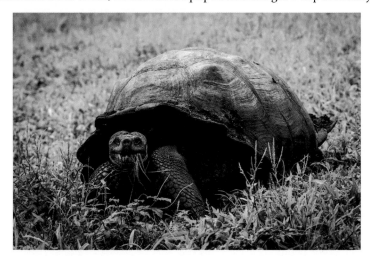

Charles Darwin came across the Galápagos tortoises in 1835. These giant tortoises are endemic to the Galápagos islands, and are not found in any other natural ecosystem in the world. Over many decades, the populations of giant tortoises have been decimated. There are many reasons for the decline of these species on the islands (including humans eating them), but one is competition with introduced invasive species such as goats. Goats compete with tortoises for grazing space and tend to destroy their habitats, such as forests, which provide shade and moisture for the tortoises. Goats breed more rapidly than tortoises, giving them an advantage. Local authorities noticed that the tortoises were losing the competition for space and food and were being quickly outnumbered.

Where did all the goats come from? For centuries, humans brought them to the islands to breed as a supply of meat. But some escaped and formed feral populations, which eventually grew to number tens of thousands. In an effort to correct the mistake of introducing goats to the islands and to save the endemic species, culling (reducing population numbers by killing off a certain percentage of the population) and eradication programmes have been implemented over several decades.

Carry out some research in the part of the world where you live or where you are from. What invasive plant or animal species are threatening local populations?

C4.1.14 – Interspecific competition

C4.1.14 – Tests for interspecific competition
Interspecific competition is indicated but not proven if one species is more successful in the absence of another. Students should appreciate the range of possible approaches to research: laboratory experiments, field observations by random sampling and field manipulation by removal of one species.
NOS: Students should recognize that hypotheses can be tested by both experiments and observations and should understand the difference between them.

One technique used to find out whether two species are competing with one another is to observe them in the field and see if they are present in the same zone at the same time. The presence of a species in a zone is recorded as a 1, and the absence of a species is recorded as a 0. A table of results called a **presence–absence matrix** is produced at the end of the study and can be mathematically analysed to see if any patterns exist. If the two species being studied are rarely or never found in the same zone, it might be because one species has out-competed the other.

Ecologists carried out a study of interspecific competition between birds in the Bismarck Archipelago, northeast of Papua New Guinea. Choosing a chain of islands for a study like this is a good way to ensure that the data being collected is from distinct zones that do not overlap. Choosing geologically young islands that were formed by relatively recent volcanic activity ensured that there were no species originally on the island for any arriving birds to compete with. The ecologists analysed which birds were found on which islands and recorded presence–absence matrices. Then they used mathematical models to predict the chances of finding birds in a particular zone by chance. By comparing these predicted values to the ones observed in nature, they could see whether the bird species were distributed by chance or not.

The traditional explanation for why one species is found in an area where similar species are not, is that it has out-competed its rivals. Although the results of this study concluded that interspecific competition could explain some of the results, there were many other factors that had an influence, and chance was one of them. For example, the history of each island was different: by chance, some bird species arrived on certain islands first, giving them an advantage.

Just because the population numbers of some introduced species increase, while the endemic species' population decreases, does not mean that the changes are due to interspecific competition for resources. In science, to find an answer related to ecological phenomena such as competition, we use multiple techniques, including field observations, field manipulation (e.g. removing a species from an area) and laboratory experiments.

A flower called the Antioch Dunes evening primrose (*Oenothera deltoides* subspecies *howellii*) is endemic to central California. Like many endemic species, its numbers are falling, and invasive species of plants have been introduced into its habitat. Originally it was thought that the decline in numbers of this subspecies was the result of competition for resources with the invasive species, but this is a good example of a hypothesis or an assumption rather than an idea established by scientific testing.

▲ The Antioch Dunes evening primrose (*Oenothera deltoides*) is endemic to central California.

In 2008, researchers led by Marc T. J. Johnson published their results after using various scientific approaches to test hypotheses about interspecific competition between the endemic primrose subspecies and an introduced species, smooth brome grass (*Bromus inermis*). Both plants are adapted to have seeds that germinate in disturbed ground, such as sand dunes that are displaced by wind or fields that have been ploughed or trampled by animals. They used two methods to see what would happen when they grew primrose alone, brome grass alone, and primrose and brome grass together.

Method 1 involved laboratory experiments. Taking seeds out of their natural habitat and planting them in a laboratory environment, in this case a greenhouse, allows researchers to control variables that would otherwise be uncontrolled in the wild. These variables include light (the researchers used natural sunlight complemented with electric lighting), rainfall (they measured how much water each plant received) and nutrients (the same type of soil was used for all the experiments and any fertilizer used was measured).

Method 2 involved field experiments. The researchers wanted to see what would happen if the primrose seeds were planted in disturbed soil. They wanted to see how well they would colonize freshly dug soil compared to competitors, some of which were invasive or introduced species. Once the plants started to grow, they measured whether the growth was impeded by the presence of other species nearby.

What they found was that primrose's presence reduced the number of plant species near it, and reduced the growth of plant species surrounding it in the wild.

By comparing the plants germinated and grown separately with those grown in the same container, and comparing germination and growth rates in disturbed soil, the researchers came to the conclusion that, although interspecific competition could explain some of the changes in the population, other factors, such as primrose's genetic diversity and the degree to which the soil was disturbed, played a large role in the plant's success. Interestingly, the statistics indicated that primrose could out-compete the brome grass because the grass grew less well in the presence of the primrose, whereas for most measurements the brome grass did not negatively affect primrose's growth.

There is a third possible method: field manipulation. By removing *Bromus* grass from around primrose in the wild, researchers could see whether this will increase or decrease the growth and population size of the primrose. If it increases, it suggests that the brome grass is competing with it for resources. If the population or growth rates continue to decline, something else, such as changes in temperature or presence of pollutants, might be the cause, rather than competition for resources.

TOK

Correlation and causation: just because two things are correlated, does not mean that one causes the other. Correlated means two phenomena are connected; when one happens, the other happens. For example, night follows day. But that does not mean that daytime causes night time, or that night causes day. There is a third factor to consider: the spinning of planet Earth on its axis, which causes both. When two things show causation, it is because it is possible to find a mechanism that bridges the correlated phenomena. "Every time I drop my pen, it falls to the floor" is a correlation that actually does show causation, because the laws of gravity explain why this correlation exists. If you drop a pen in outer space, it will float. There is no longer a correlation because there is no gravity. In the natural sciences, we test assumptions whenever possible rather than simply accepting them. But even if we come up with a valid explanation that makes sense, can we be confident that it is the best explanation? To what extent is certainty attainable?

Nature of Science

Notice from this example that hypotheses can be tested by both experiments and observations. Experiments in laboratory conditions can help isolate one factor and control all other variables, but the results might not translate into an explanation about the real world. Organisms do not always behave in the same way in laboratories or greenhouses as they do in natural environments.

C4.1.15 – The chi-squared test

C4.1.15 – Use of the chi-squared test for association between two species

Application of skills: Students should be able to apply chi-squared tests on the presence/absence of two species in several sampling sites, exploring the differences or similarities in distribution. This may provide evidence for interspecific competition.

In Section C4.1.3, we explored the use of quadrats for random sampling of sessile species (those that stay in one place). We can take this a step further. If we pick two plant species to count in each randomly sampled area of the quadrat, we can use a statistical test called the **chi-squared test of association** (also called a test of independence) to see whether the two species tend to occur together more often than they would by chance or if, on the contrary, they are never found coexisting or at least less frequently than would be expected by chance. This can help indicate whether the species are in competition with each other or not. The following worked example illustrates how the chi-squared test works.

Worked example

A group of students has been working in a prairie in late summer and they have noticed that there are two species of plant that seem to occur together: the New England aster (*Symphyotrichum novae-angliae*) and Canada goldenrod (*Solidago canadensis*). After learning to correctly identify each species, they used random sampling with 1 m² quadrats to gather data from 20 quadrat samples. If they found any goldenrod growing in their quadrat, they recorded a 1 in the goldenrod column, if not, they put a 0. They did the same for the aster column, 1 for present and 0 for absent at that sample site (see the table).

New England aster (*Symphyotrichum novae-angliae*) (top) and a bee collecting pollen from Canada goldenrod (*Solidago canadensis*) (bottom)

Quadrat	Goldenrod	Aster
1	0	1
2	1	1
3	1	0
4	1	0
5	1	0
6	0	1
7	1	0
8	1	1
9	0	0
10	1	0

Quadrat	Goldenrod	Aster
11	1	0
12	1	0
13	0	0
14	1	1
15	0	0
16	1	1
17	1	1
18	1	0
19	0	0
20	1	0

Quadrat data for two plant species. 1=presence of the species in a quadrat; 0=absence of the species in a quadrat.

Apply the chi-squared test to these data to decide whether the two species are associated with each other or occur independently. If you have never done a chi-squared test before or need a refresher, refer to the Skills for biology chapter for an explanation of what the test is, how it works, and what the values for the degrees of freedom should be.

1. State the null hypothesis for this calculation.

2. Determine the number of degrees of freedom for this statistical test.

3. Determine the critical value in order to obtain a 95% certainty that there is a statistically significant difference between these two sets of numbers.

4. Calculate the chi-squared value for these data.

5. Interpret this value. Does it mean we can reject or not the null hypothesis?

6. Is there a statistically significant difference between these two sets of data?

7. Are there enough data points to be confident of the results?

Solution

1. The null hypothesis is that "the two categories (species 1, the goldenrod, and species 2, the aster) are independent of each other". In other words, the distribution of these two species is random, there is no association between them. When one is found, we do not find the other any more than would be expected if it was randomly distributed.

2. Because there are two possible outcomes (plant present or plant not present), the number of degrees of freedom is $2 - 1 = 1$.

3. According to the chi-squared table (see the chi-squared eBook activity on page 616), the critical value in order to obtain a 95% certainty is 3.84. This value is found under the column for 0.05, which corresponds to a 95% certainty, and in the row that has a degree of freedom of 1.

4. The chi-squared value is calculated to be 4.91. This is obtained using the following values in a contingency table of observed and expected values.

	Observed goldenrod	Observed aster	Grand total
Absent	6	13	19
Present	14	7	21
Grand total	20	20	40

The expected value of 9.5 comes from the calculation $(20 \times 19) \div 40$, and the expected value of 10.5 comes from the calculation $(20 \times 21) \div 40$.

	Expected goldenrod	Expected aster	Grand total
Absent	9.5	9.5	19
Present	10.5	10.5	21
Grand total	20	20	40

See the Skills in the study of biology chapter for help with this calculation using observed and expected values.

5. Because 4.91 is greater than the critical value of 3.84, this means we can reject the null hypothesis.

6. Yes, the two categories are associated with each other. We can be 95% sure that there is a relationship between the presence of the goldenrod and the aster. In other words, it would be very unlikely (i.e. a 5% chance) that they are independent of each other.

7. Twenty quadrats sounds a bit small. In a random sample, there is always the chance that the sampling is not representative of the zone studied. If the zone in the prairie being studied was the size of a sports field, for example, it would have a surface area of approximately 5000 m². Twenty 1 m² quadrats represents 20 m² of that surface, meaning that only 0.4% of the field was actually sampled. That is the equivalent of finding two pieces of a 500-piece puzzle you have never seen before and declaring that you know what the image will be once the puzzle is complete.

> You can be asked about the chi-squared test in IB biology exams. Be sure you know when the chi-squared test can be used, the steps for doing it, and how to interpret the results.

C4.1.16 – Predator–prey relationships

C4.1.16 – Predator–prey relationships as an example of density-dependent control of animal populations
Include a real case study.

Because organisms rely on each other, it would be a disadvantage for one species in a community to completely take over. If an alga floating on the surface of a lake, for example, increased its population until it completely covered the surface, it would cut off light for the photosynthetic plants and phytoplankton below. If a rabbit population on a prairie increased to such a high population density that they were hopping over each other, food supplies would become more and more scarce, their large numbers would attract predators, and their population density would increase the chances of a disease spreading. Density-dependent factors like these affect large populations and small populations differently. A larger, denser population is more likely to experience food shortages, predation, disease or even emigration, because individuals in the overpopulated area may move to other areas in search of more food and space.

Predators provide a form of population control; their activity will increase as a population of prey increases. The Canada lynx (*Lynx canadensis*) and the snowshoe hare (*Lepus americanus*) are often cited as a classic example of a predator–prey interaction. The lynx preys on the hare. Changes in the numbers of the lynx population are followed by changes in the numbers of the hare population.

The graph in Figure 2 shows a simplified model of the relationship between predator (in this case the lynx) and prey (the hare). Before reading further, try to see if you can describe what each peak on the graph represents and when it occurs compared to the other peaks. Then try to explain the pattern, which repeats one cycle about every 10 years. Now read the paragraph below for an explanation.

The Canada lynx (*Lynx canadensis*), an elusive predator

▲
The snowshoe hare (*Lepus americanus*) gets its name from its large hind legs.

C4.1 Figure 2 The cyclical relationship between predator and prey ▶

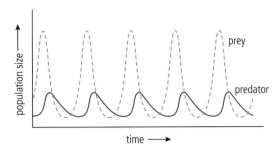

At the start of the graph in Figure 2, notice that the two lines go up and then down, but the maximum for the predator's line (the red line) is out of sync with the line for the prey (the dashed blue line), because the predator population always reaches its maximum after the prey population. Now focus on the first cycle from the lowest point of the predator line to the next lowest point of that line. At the beginning, the prey line increases to a maximum population for the hare because there are few lynx around to hunt them. This is shown by the predator line, indicating the lynx population, which is low at this point. Over time, as the population of hares increases, the lynxes have more and more food available, which allows them to have more offspring. Recall that the availability of resources such as food acts as a limiting factor for populations, so when more food is available, the population is not as limited.

As the lynx offspring grow and produce young of their own, the population increases to its maximum. However, we notice that by the time the two lines cross, the hare population has declined rapidly. Why? This is because there are so many lynxes hunting that many of the hares are being eaten. Shortly after the lynx population reaches its maximum, the hare population reaches its minimum. During this time we see the predator line going down as the lynx population decreases because of a lack of food. And then we are back at the start of the cycle. This is a density-dependent relationship.

More recent analysis has suggested that this model is oversimplified, however. For one thing, both species live in a food web that involves other species. Snowshoe hares are preyed upon by many other species, such as wolves, foxes, owls and hawks.

SKILLS

The online PhET Interactive Simulation called "Natural Selection" was originally designed to show how mutations could help a rabbit population survive and adapt in the presence of changes in the environment, such as the presence of food or of wolves. It is also a good simulation of predator–prey interactions when wolves are introduced. The simulation tracks the size of the population as you modify different parameters. Do a search for it online and give it a try. See what happens to the population with and without wolves.

C4.1.17 – Control of populations

C4.1.17 – Top-down and bottom-up control of populations in communities
Students should understand that both of these types of control are possible, but one or the other is likely to be dominant in a community.

Communities rely on limiting factors to make sure that no one species takes over completely. In Section C4.1.5 you learned about some limiting factors that define the carrying capacity of a habitat, such as availability of food, predation and disease.

Limiting factors can be top-down or bottom-up. **Top-down controls** are seen when a species' population can be reduced by other species feeding on it. Predation, as seen in the example with the lynx, is a top-down control. Herbivory is a top-down control for plant populations because the more they are eaten, the more their population goes down. **Bottom-up controls** are seen when a species' population can be reduced by a lack of resources such as food, sunlight (for photosynthetic organisms) or minerals. Recall that when there were not enough hares to sustain the lynxes, the lynx population went down.

Challenge yourself

A study of a tropical coral reef revealed the effects of top-down and bottom-up limiting factors. The bottom-up limiting factor was the nutrients that increased algal blooms, which negatively affected the coral. The top-down limiting factor was the fish that ate the algae, so keeping the coral reef healthy. Two study sites on the coral reef, 1 and 2, were isolated for 24 months and their conditions were manipulated. Controlled experiments were performed by pairing high and low herbivory (the amount of algae eaten by fish) with high and low nutrient levels. See the table.

	Study site 1 (low herbivory)		Study site 2 (high herbivory)		Significant differences ($p < 0.05$)
	Reduced nutrients A	Elevated nutrients B	Reduced nutrients C	Elevated nutrients D	
Crustose corallines	41.2 ± 4.6	1.8 ± 1.8	<0.1	71.7 ± 3.0	D > A > B, C
Frondose macroalgae	20.8 ± 4.3	63.7 ± 8.2	0.6 ± 0.3	16.9 ± 4.1	B > A, D > C
Algal turfs	37.1 ± 3.9	14.5 ± 4.7	<0.1	22.1 ± 2.9	A > D > B > C
Predicted dominants	Turfs	Macroalgae	Corals	Corallines	

Mean percentage cover (with standard error) of benthic functional groups colonizing clay diffusers following 24 months of reduced and elevated nutrients in low- and high-herbivory study sites ($n = 4$)

Three types of algae were included in the study, as shown in the table:

- crustose corallines, which are beneficial algae that help the coral build the reef
- frondose macroalgae, which are fleshy and filamentous, and can overgrow the coral and prevent healthy reef building because of their algal blooms
- algal turfs, which are microalgae and their blooms are also detrimental to reef building.

The herbivorous fish were parrotfish and surgeonfish.

The question posed by the study was how the effect of top-down herbivores and bottom-up nutrients affected the competition of harmful and beneficial algae. The percentage of reef cover by each type of algae was a measure of its success.

2. For study site 1, compare the mean percentage cover of all three alga types with reduced and elevated nutrients. What were the effects on the coral?

3. For site 1, the prediction was that macroalgae would be dominant in the competition for percentage cover with elevated nutrients. Was that prediction confirmed? Give evidence to support your answer.

4. Describe a benefit to the coral reef that occurred over the 24 months in part D of the experiment.

5. Explain the conditions under which elevated nutrient-induced microalgae blooms decreased the growth of the reef-building corals.

Parrotfish feed on algae

Although both top-down and bottom-up factors influence ecosystems, one is usually dominant. In marine ecosystems, the limiting factors are usually bottom-up. If phytoplankton are plentiful and able to photosynthesize efficiently, the rest of the ecosystem can usually flourish. Occasionally, because of overfishing, there can be periods of top-down limits to population growth.

C4.1.18 – Allelopathy and antibiotic secretion

C4.1.18 – Allelopathy and secretion of antibiotics

These two processes are similar in that a chemical substance is released into the environment to deter potential competitors. Include one specific example of each—where possible, choose a local example.

Sometimes competition between species generates survival adaptations such as camouflage to hide from predators, aggressive behaviour to defend a territory, or features that allow a plant or animal to obtain a resource before its competitors can. Instead of enhancing their own survival techniques, another way organisms can gain a competitive edge is to release molecules into the surrounding environment to make life difficult or impossible for their competitors. Whereas **primary metabolites** are molecules that are needed for the basic functions of life, such as energy and growth, molecules produced to impede or kill competitors are called **secondary metabolites**. Such molecules are used in a process called **allelopathy**, which is the production of secondary metabolites that influence the growth and success of other organisms.

Examples of allelopathy include:

- inhibiting seed germination in nearby competitors
- interfering with nutrient uptake in roots so that plants cannot grow nearby
- killing bacteria or inhibiting the growth of nearby bacteria.

Allelopathy in plants

Garlic mustard (*Alliaria petiolata*) is an introduced and invasive species, and allelopathy may have a role in its success in colonizing new territory. Garlic mustard produces a secondary metabolite called sinigrin. This is the same molecule responsible for the zing in the taste of mustard, horseradish and wasabi. Sinigrin can inhibit the germination of seeds from other plants, and can reduce the growth of roots of plants already growing in the area. Both properties of sinigrin give the garlic mustard a competitive advantage over other plants. Its seeds can germinate unhindered while its competitors' seeds are prevented from germinating, and it can grow its roots down into the soil to obtain water and nutrients, while its competitors' root growth is slowed down by the presence of sinigrin.

Does garlic mustard or any other such plant grow near you? Carry out some research to determine whether your area contains plants or trees that use allelopathy to fight off competition.

Allelopathy in microbes

Competition exists not only in large, visible organisms such as mustard plants and wildcats, but also in the invisible world of microbes. The single-celled fungus *Penicillium rubens* produces a molecule whose identity you have probably already guessed from its name: penicillin. Penicillins are produced by certain species of

Garlic mustard (*Alliaria petiolata*) has a chemical weapon against its competitors: sinigrin.

moulds and have the remarkable ability to stop the growth of bacteria. Molecules that can inhibit the growth of bacteria or that can kill bacteria are called **antibiotics**. In nature, this means that *P. rubens* can compete for space and food sources by releasing the allelopathic molecule penicillin into its surroundings. Antibiotics are used by many species of microscopic organisms to establish a bacteria-free zone around them, allowing their colonies to spread while inhibiting the spread of competing colonies.

Different antibiotics work in different ways. Penicillin blocks bacterial enzymes that are trying to link certain molecules together when a bacterium is building its cell wall. Because the bacterium's enzymes cannot do their job, the cell wall cannot form correctly. This allows water to leak into the cell, causing it to burst and killing the cell. Other antibiotics interfere with protein synthesis in bacteria.

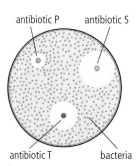

antibiotic P   antibiotic S

antibiotic T   bacteria

One way to test antibiotics is by putting discs of paper soaked in different antibiotic solutions in a Petri dish, to see whether a bacterial colony is inhibited by any. The bacteria-free zones around the three different types of antibiotics shown in this diagram indicate that the allelopathic molecules being tested can successfully suppress bacterial growth.

Guiding Question revisited

How do interactions between organisms regulate sizes of component populations in a community?

Within this chapter you have learned:

- to estimate a population of sessile organisms, random sampling using quadrats can be used
- for motile organisms, mark and recapture techniques can generate data to use with the Lincoln index
- populations have a maximum size, the carrying capacity, and this is regulated by positive and negative feedback including density-independent factors such as temperature and density-dependent factors such as the spread of disease
- population growth can be modeled using a sigmoid population curve.

Guiding Question revisited

What interactions within a community make its populations interdependent?

Within this chapter you have learned:

- when multiple populations interact, a community is formed (interactions include herbivory, predation, interspecific competition, mutualism, parasitism and pathogenicity)
- some populations compete for resources and others cooperate to help each other survive
- to find out if a species is competing with another, various techniques are used, including the removal of the competing species to see if the endemic species' population increases without its competitor
- statistical tests such as the chi-squared test of association can help determine whether two species tend to occur together or whether one tends to dominate and out-compete the other
- population numbers can change over time depending on the relationship between species, as seen in predator–prey relationships or top-down and bottom-up control of population numbers
- some organisms, such as the fungus that makes penicillin, keep away competitors by the use of allelopathy, secreting secondary metabolites into their environment.

Ever since the Nobel Prize winning work of Alexander Fleming, Howard Florey and Ernst Chain in the first half of the 20th century, penicillin has been used as an antibiotic to treat bacterial infections in humans. This is one of the most significant breakthroughs in the history of medicine. The original molecules used in antibiotic medications were extracted from moulds, but today they can be synthesized in the laboratory. In addition to penicillin, doctors today can prescribe dozens of molecules as antibiotics, and new antibiotics are being tested as candidates for future use. Unfortunately, new strains of bacteria that are resistant to the antibiotics we already have are emerging faster than we can find or design new ones.

Exercises

Q1. A group of individuals belonging to the same species and living in the same area at the same time is called:

    A   A community.

    B   Intraspecific competition.

    C   An ecosystem.

    D   A population.

Q2. Which population phase is most affected by positive feedback?

    A   The exponential phase.

    B   The transition phase.

    C   The plateau phase.

    D   The carrying capacity.

Q3. A species found in one geographic area and nowhere else in the world is said to be:

    A   Native.

    B   Endemic.

    C   Invasive.

    D   Mutualistic.

Q4. Which factors that control populations are density-independent?

    I.   Forest fires

    II.   Predation

    III.   Volcanic eruptions.

    A   I and II only.

    B   I and III only.

    C   II and III only.

    D   I, II and III.

Q5. Which are examples of mutualism?

    I.   Zooxanthellae in hard corals

    II.   Root nodules in Fabaceae

    III.   *Plasmodium* in *Anopheles* mosquitoes.

    A   I and II only.

    B   I and III only.

    C   II and III only.

    D   I, II and III.

Q6. Distinguish between the top-down and bottom-up control of populations in communities, giving an example of each.

C4.2 Transfers of energy and matter

Guiding Questions

What is the reason matter can be recycled in ecosystems but energy cannot?

How is the energy that is lost by each group of organisms in an ecosystem replaced?

There is a limited number of carbon and oxygen atoms available on Earth for organisms to use in their bodies. Although cosmic dust may rain down in small quantities from space, it is not enough to meet the needs of life on Earth. As a result, organisms need to reuse and recycle the matter that is available. In contrast, the Sun sends energy to Earth every day and therefore energy does not need to be recycled. In addition, organisms do not have mechanisms for storing heat energy efficiently, so it is lost to their environment. This lost energy is replaced when plants convert the energy from sunshine into food, and that food fuels cells that have lost heat energy through cellular respiration.

C4.2.1 – Ecosystems are open systems

C4.2.1 – Ecosystems as open systems in which both energy and matter can enter and exit
Students should know that in closed systems only energy is able to pass in and out.

In a tropical rainforest, energy enters the ecosystem in the form of sunlight and leaves in the form of heat, essentially from cellular respiration. When the seeds of a tree land in a river that flows through the forest, the river can transport the seeds far from the ecosystem and sometimes all the way to the ocean. Matter that was once in the soil (the minerals the tree used to grow the seeds) or in the air (the carbon dioxide the tree used in photosynthesis) can be removed from the forest ecosystem. Any birds or mammals that migrate into the forest bring with them matter that can be integrated into the ecosystem when they produce waste or die, leaving their carcass. A system that allows matter in and out is considered an **open system**. When humans cut down and remove trees to make paper or wood furniture, we are demonstrating the openness of the forest ecosystem.

In a **closed system**, matter does not enter or leave and so it must be recycled. Energy, however, can still enter and leave the system. An example of this is Earth when viewed on a planetary scale: energy enters in the form of sunlight and leaves in the form of heat, but matter must be recycled on Earth. Only a small quantity of matter enters this closed system, in the form of meteorites or cosmic dust from space, which lands on the surface. Although humans sometimes send equipment into outer space, made from metals, for example, mined from the rocks on Earth, this only represents a tiny fraction of the minerals and other matter recycled and reused on Earth by living organisms. Generally speaking, Earth is a closed system where only energy enters and leaves, while matter is recycled.

Systems theory provides an explanation of how systems interact with each other and with their environment. Scientists use systems theory to explain open and closed systems and to predict what would happen if something in a system or in the environment around a system changed. An ecosystem follows certain laws, such as the law of conservation of mass, which states that matter cannot be destroyed or created. There is also the law of thermodynamics, which describes the flow of energy through a system and how one form of energy can be transformed into another, such as light into chemical energy or chemical energy into heat.

Planet Earth can be considered a closed system because only relatively minute quantities of minerals enter or exit it from space. Energy, however, is lost as heat radiating into space, but is added to the system every day in the form of sunlight.

Nature of Science

In science, a **law** is a generalized principle that describes a natural phenomenon. We observe something and try to come up with a general rule that summarizes what is happening. There is no attempt to explain or give a reason why, only an effort to model what we observe. A **theory** does provide an explanation. A theory is used to explain a phenomenon by describing the underlying mechanisms that are responsible for producing it. Laws and theories help us understand the natural world and predict what will happen next. For example, systems theory is being used to predict climate change on Earth.

C4.2.2 – Sunlight sustains most ecosystems

C4.2.2 – Sunlight as the principal source of energy that sustains most ecosystems

Include exceptions such as ecosystems in caves and below the levels of light penetration in oceans.

NOS: Laws in science are generalized principles, or rules of thumb, formulated to describe patterns observed in living organisms. Unlike theories, they do not offer explanations, but describe phenomena. Like theories, they can be used to make predictions. Students should be able to outline the features of useful generalizations.

The best studied ecosystems are those found on Earth's surface, whether they are on land or in surface water. Such systems rely on sunlight, and they will be the main focus of this section. Be aware, however, that there are other, less well-studied, ecosystems that exist in total darkness, such as those in deep ocean water and those found in dark caves or deep underground; these ecosystems are not well understood because they are so difficult to access and study. One example is the ecosystem created by hydrothermal vents, which exist on the ocean floor below levels where sunlight can penetrate. Minerals dissolved in seawater that has infiltrated cracks in the ocean floor and been warmed by magma, rise to the surface of the ocean floor and provide a rich source of minerals for chemical reactions in microorganisms that use chemosynthesis instead of photosynthesis to make food. You can find out more about this in Section C4.2.7.

All life that you see around you on Earth's surface relies either directly or indirectly on sunlight. Photosynthetic organisms such as plants and phytoplankton take inorganic carbon dioxide (CO_2) and convert it into energy-rich sugar ($C_6H_{12}O_6$). The addition of minerals allows them to synthesize complex molecules such as cellulose, proteins and lipids, which allow them to build stems, leaves, fruit and seeds. Notice what is happening in this process: light energy from the Sun is being converted into chemical energy (food). Chemical energy refers to the fact that carbon compounds, such as carbohydrates, proteins and lipids, are rich in energy, thanks to the chemical bonds that exist between the carbon atoms and other atoms. This is what makes fruits, grains and vegetables good food sources. Consumers cannot "eat" sunlight and air, but they can eat carbohydrates, proteins and lipids. The chemical energy in these carbon compounds can be measured in calories, kilocalories, joules or kilocalories, which we are familiar with on food labels.

Sunlight is the initial source of energy for all ecosystems on the surface of Earth's land and oceans.

C4.2.3 – The flow of energy

C4.2.3 – Flow of chemical energy through food chains

Students should appreciate that chemical energy passes to a consumer as it feeds on an organism that is the previous stage in a food chain.

Photosynthetic organisms provide a remarkable link between the abiotic world (e.g. rocks, water and air) and the biotic world (e.g. plants, bacteria, fungi and animals). They transform air and water into food. This is why they are referred to as producers. By feeding on producers, consumers can utilize chemical energy to grow and stay healthy. For example, a cow (the consumer) grazing in a field of grass (the producer) is taking chemical energy from the grass and digesting the carbon compounds to help build meat or milk inside its own body. Humans can consume the meat or milk from the cow, to benefit from the chemical energy the cow has obtained from the grass. Such a pattern of feeding is called a **food chain**. The process of passing energy from one organism to another through feeding is referred to as the flow of energy through a food chain.

Can you identify the consumers and producers in this Peruvian scene?

C4.2.4 – Food chains and food webs

C4.2.4 – Construction of food chains and food webs to represent feeding relationships in a community

Represent relationships in a local community if possible. Arrows indicate the direction of transfer of energy and biomass.

When studying feeding habits, it is convenient to write down which organisms eat which by using an arrow. Thus "herring → seal" indicates that the herring is eaten by the seal. When more of the life cycle of the seal and herring are investigated, new organisms can be added to the chain: copepods (a common form of zooplankton) are eaten by the herring, and great white sharks eat seals. Lining up organisms with arrows between them is how food chains are represented. Here are three examples of food chains from three different ecosystems.

Grassland ecosystem:

> clover → grasshoppers → toads → snakes → hawk

River ecosystem:

> green algae → mayfly larvae → juvenile trout → kingfisher

Marine ecosystem:

> diatoms → copepods → herring → seals → great white shark

The definition of a food chain is a sequence showing the **feeding relationships** and energy flow between species. In other words, it answers the question "What eats what?" The direction of the arrow shows the direction of the flow of energy.

Look at the food chains in Figure 1 showing part of a river ecosystem. Notice how they link together into a **food web**. A food web shows multiple food chains and how they are connected. Often organisms such as juvenile trout eat not only caddis fly larvae but also the larvae of other species.

SKILLS

Can you observe any food chains in your local area? Think about organisms you see everyday and try to construct a food chain with three or four organisms, starting with a producer. If you live in a city, this might be challenging but think about sparrows eating caterpillars, which have fed on the leaves of trees lining the streets. Maybe a neighbour's cat succeeds in hunting and catching a sparrow for a snack.

A food chain is a single linear set of connections from producer to consumers with only one species at each trophic level, whereas a food web shows a more complete picture because often one producer is eaten by more than one consumer, and consumers eat more than one type of organism.

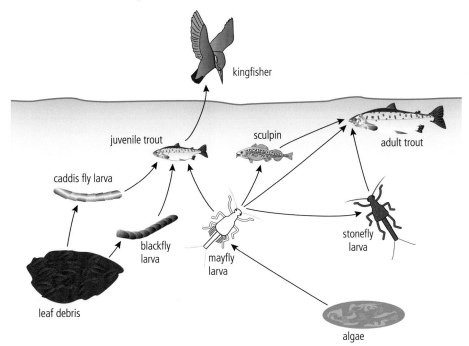

C4.2 Figure 1 A food web from a river ecosystem

C4.2.5 – Decomposers

C4.2.5 – **Supply of energy to decomposers as carbon compounds in dead organic matter**
Include faeces, dead parts of organisms and dead whole organisms.

An effective way of unlocking the precious nutrients stored in the cells of plants and animals is through decay. **Decomposers** (saprotrophs and detritivores) break down non-living food sources such as the faeces of organisms, entire dead bodies, or fallen leaves or the skin shed from a snake. Saprotrophs such as fungi secrete enzymes onto dead matter such as a fallen tree and absorb the nutrients. Detritivores such as the minotaur beetle have mouthparts to ingest dead matter and digest it inside their bodies. The digestive enzymes of decomposers convert the organic matter into a more usable form for themselves, and therefore for other organisms. For example, proteins from a dead organism are broken down into ammonia (NH_3) and, in turn, the nitrogen in ammonia can be converted into useful nitrates (NO_3^-) by bacteria.

In this way, decomposers recycle nutrients so that they are available to other organisms and are not locked inside the bodies or waste products of organisms in the ecosystem. Decomposers play a major role in the formation of soil, without which plant growth would be greatly impaired, if not impossible. The rich black layer of soil called **humus** is made up of organic debris and nutrients released by decomposers.

▲
The minotaur beetle (*Typhaeus typhoeus*) is a type of decomposer called a detritivore.

C4.2.6 – Autotrophs

C4.2.6 – **Autotrophs as organisms that use external energy sources to synthesize carbon compounds from simple inorganic substances**
Students should understand that energy is required for carbon fixation and for the anabolic reactions that build macromolecules.

Some organisms are capable of making their own organic molecules as a source of food. These organisms are called **autotrophs**, and they synthesize their organic molecules from simple inorganic substances. This process involves either photosynthesis or chemosynthesis.

Why do organisms need energy? Metabolic processes such as carbon fixation and other anabolic reactions that build complex molecules from building blocks require energy.

Photoautotrophs can take light energy from the Sun, combine it with inorganic substances (water and carbon dioxide), and obtain a source of chemical energy in the form of a carbon compound (glucose). This ability to convert inorganic carbon dioxide, which is unusable to consumers, into organic molecules that are useful for energy and growth is called **carbon fixation**. Once this is done, some of the molecules can be combined using anabolic reactions to make larger molecules. Plants use these reactions to make macromolecules such as cellulose.

Chemoautotrophs are able to take carbon dioxide and, using inorganic compounds such as hydrogen sulphide (H_2S) as an energy source, build more complex molecules that can be useful to them as food.

Because autotrophs make food that is useful for themselves but also consumed by other organisms, they are also called **producers**.

Examples of photoautotrophs include:

- cyanobacteria
- clover
- algae such as giant kelp
- pine trees.

Examples of chemoautotrophs include:

- sulfur-oxidizing bacteria
- nitrogen-oxidizing bacteria
- iron-oxidizing bacteria.

Electronic monitoring equipment at a hydrothermal vent on the Juan de Fuca Ridge in the Pacific Ocean. Red and white tube worms (*Riftia pachyptila*) host sulfur-oxidizing chemosynthetic bacteria that produce food for them. What looks like black smoke coming out of the vent is, in fact, hot water charged with dark minerals.

TOK

The first hydrothermal vent, and the community of surprising organisms such as giant tube worms surrounding it, was discovered in 1977 just west of the Galapagos Islands. Why had no-one noticed them before? Because they were over 2,000 m below the ocean surface. Think about the kinds of tools and technology that were necessary to make this discovery. How does this show the importance of material tools in the production and acquisition of knowledge for marine biologists?

C4.2.7 – Energy sources

C4.2.7 – Use of light as the external energy source in photoautotrophs and oxidation reactions as the energy source in chemoautotrophs

Students should understand that oxidation reactions release energy, so they are useful in living organisms. Include iron-oxidizing bacteria as an example of a chemoautotroph.

Electrons are needed to produce adenosine triphosphate (ATP) for a cell. During an oxidation reaction, electrons are removed (lost) from atoms. The donated electrons are free to participate in reactions in the cell, such as helping in the production of ATP. Light is used by photosynthetic organisms to oxidize water molecules by photolysis, in order to donate electrons and hydrogen (H^+) ions, and, in the process, the reaction releases oxygen gas. The ATP produced using the donated electrons and the hydrogen ions can then be used to generate the organic molecule glucose with carbon dioxide as the source of carbon.

Photoautotrophs get the energy they need for carbon fixation from sunlight, whereas chemoautotrophs get their energy from oxidation reactions that do not require a source of light energy.

How does the transformation of energy from one form to another make biological processes possible?

Iron-oxidizing bacteria such as *Mariprofundus ferrooxydans*, a bacterium that thrives near hydrothermal vents like those near Hawaii, hundreds of metres below the ocean surface and far from where sunlight can penetrate, are capable of obtaining electrons from iron. They do this by taking one form of iron, iron(II), otherwise written as Fe^{2+}, and removing an electron to transform it into iron(III), or Fe^{3+}. For every negative electron lost, iron will gain a positive charge. The lost (or donated) electron from this oxidation reaction can be used to generate ATP for the cell. These microbes are the producers for (and therefore the starting point of) food chains near the hydrothermal vents where they live.

C4.2.8 – Heterotrophs

C4.2.8 – Heterotrophs as organisms that use carbon compounds obtained from other organisms to synthesize the carbon compounds that they require

Students should appreciate that complex carbon compounds such as proteins and nucleic acids are digested either externally or internally and are then assimilated by constructing the carbon compounds that are required.

Heterotrophs cannot make their own food from inorganic matter, and must obtain organic molecules from other organisms. They get their chemical energy from autotrophs or other heterotrophs. Because heterotrophs rely on other organisms for food, they are called **consumers**. Heterotrophs ingest organic matter that is living or has been recently killed.

Examples of heterotrophs include:

- zooplankton
- sheep
- fish
- insects.

Organisms that are not capable of synthesizing their own food from inorganic components of their environment need to get their nourishment by ingesting (eating) parts of other organisms. Consumers take in the energy-rich carbon compounds, such as sugars, proteins and lipids, synthesized by other organisms in order to survive.

When heterotrophs consume food, they digest the proteins into amino acids, the lipids into fatty acids, and the DNA and RNA into nucleic acids. They can then synthesize their own proteins using the amino acids, make their own lipids using the fatty acids, and synthesize their own DNA and RNA using the nucleic acids. This process of integrating nutrients into useable substances in the tissues of the body is called assimilation. When a fox eats a chicken egg, it will digest the chicken protein and then assimilate the amino acids to build its own fox proteins.

The only component in our diet that we can synthesize using sunlight is vitamin D. There are precursors in our skin that absorb ultraviolet (UV) light waves and produce vitamin D. But in order to get all the other types of molecules needed to keep us healthy, we need to consumemolecules made by other living things.

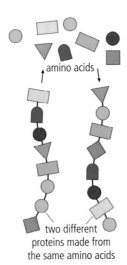

amino acids

two different
proteins made from
the same amino acids

◄ Proteins obtained from food
sources (e.g. chicken protein)
can be broken down into
amino acids and reconnected
in the consumer's cells to
make new proteins
(e.g. fox proteins).

C4.2.9 – The release of energy by cell respiration

C4.2.9 – Release of energy in both autotrophs and heterotrophs by oxidation of carbon compounds in cell respiration

Students are not required to be familiar with photoheterotrophs.

To release energy, carbon compounds such as glucose are oxidized. Cells need a constant supply of energy, which is the role of cellular respiration. Although we often only think about the autotrophs' ability to produce food for consumers, they still need to use the organic molecules they have synthesized as an energy source. So photosynthetic autotrophs and chemoautotrophs both carry out cellular respiration. For example, plants oxidize the glucose they have made in order to release energy for chemical reactions and growth.

In addition to making food for themselves, producers are a food source for heterotrophs, which also use cellular respiration to release energy from carbon compounds such as sugar, proteins and lipids, which the producers have synthesized in their cells. When a bird eats a seed, the carbohydrates and lipids in the seed will be oxidized to release the energy.

C4.2.10 – Trophic levels

C4.2.10 – Classification of organisms into trophic levels

Use the terms "producer", "primary consumer", "secondary consumer" and "tertiary consumer". Students should appreciate that many organisms have a varied diet and occupy different trophic levels in different food chains.

Biologists use the term **trophic level** to indicate how many organisms the energy in the system has flowed through. The first trophic level is occupied by the producers. The next trophic level is occupied by the **primary consumers** (organisms that eat the producers), and the trophic level after that is occupied by **secondary consumers** (organisms that eat primary consumers). If the secondary consumers are eaten by another organism, the next trophic level has been reached: **tertiary consumers**.

Although a food web provides a representative but complicated picture of what is being eaten in an ecosystem (see Figure 2), problems sometimes arise when determining trophic levels. Can you see the following difficulties when you look at the food web in Figure 2?

- An eagle is a tertiary consumer when eating rattlesnakes, but a secondary consumer when eating rabbits.
- A coyote is a primary consumer when it eats the fruit of a cactus, but a tertiary consumer when it eats a rattlesnake.
- A lizard is a tertiary consumer when it eats rattlesnake eggs, but a secondary consumer when it eats insects.

C4.2 Figure 2 A desert food web ▶

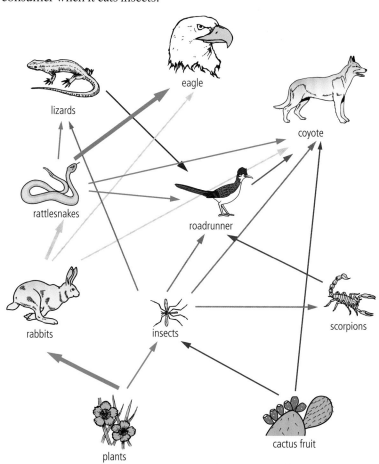

C4.2.11 – Energy pyramids

C4.2.11 – Construction of energy pyramids
Application of skills: Students should use research data from specific ecosystems to represent energy transfer and energy losses between trophic levels in food chains.

Be careful not to confuse pyramids of energy with pyramids of numbers: pyramids of numbers show the population sizes of each trophic level, not the energy.

A **pyramid of energy** is used to show how much and how fast energy flows from one trophic level to the next in a community (see Figure 3). The units used are energy per unit area per unit time: kilojoules per square metre per year (kJ m^{-2} yr^{-1}). Because time is part of the unit, energy pyramids take into account the rate of energy production, not just the quantity. Because energy is lost, each level is always smaller than the one before. It would be impossible to have a higher trophic level wider than a lower trophic level because organisms cannot create energy, they can only transfer it inefficiently.

T4 top carnivores — 4 kJ m^{-2} yr^{-1}
T3 carnivores — 40 kJ m^{-2} yr^{-1}
T2 herbivores — 400 kJ m^{-2} yr^{-1}
T1 producers — 4,000 kJ m^{-2} yr^{-1}

C4.2 Figure 3 A pyramid of energy. The green base is trophic level 1, the producers. The pink level is trophic level 2, the herbivorous primary consumers. The purple layer is trophic level 3, the carnivorous secondary consumers, and the top layer is trophic level 4, the top carnivores as tertiary consumers.

SKILLS

Use the data below from Silver Springs, Florida (USA), to draw a pyramid of energy. Typically, data for decomposers are not included in pyramids of energy, but you can draw them to one side of the pyramid. Draw the pyramid stepped rather than as a triangle and try to draw the levels to scale. As an extension to this activity, you could research similar data in your area of the world.

A **pyramid of energy** is used to show how much and how fast energy flows from one trophic level to the next in a community.

Energy flow measured in kJ m^{-2} yr^{-1}

1,700,000 kilojoules

solar energy not used
1,679,190 (98.8%)

20,810 (1.2%)

producers

energy still in organic wastes and remains

transferred to the next trophic level

energy lost as metabolic heat

4,245   3,368   13,197

herbivores

720   383   2,265

carnivores

90   21   272

top carnivores

5   16

decomposers

5,060

energy output   20,810 + 1,679,190
total annual energy flow   1,700,000 (100%)

C4.2.12 – Energy loss between trophic levels

> **C4.2.12 – Reductions in energy availability at each successive stage in food chains due to large energy losses between trophic levels**
>
> Decomposers and detritus feeders are not usually considered to be part of food chains. However, students should understand the role of these organisms in energy transformations in food chains. Consider the causes of energy loss.

No organism can use 100% of the energy present in the organic molecules of the food it eats. Typically, only 10–20% of the energy available is used from the previous step in a food chain. This means that as much as 90% is lost at each level.

The main reasons why not all of the energy present in an organism can be used by another organism in the next trophic level include the following.

- Not all of an organism is swallowed as a food source, some parts are rejected and will decay.
- Not all of the food swallowed can be absorbed and used in the body, for example undigested seeds can be found in the faeces of fruit-eating animals.
- Some organisms die without having been eaten by an organism from the next trophic level.
- There is considerable heat loss as a result of cellular respiration at all trophic levels (shown by the wavy arrows in Figure 4), although the loss of heat varies between different types of organism. Most animals have to move, which requires much more energy than a stationary plant needs. Endotherms (warm-blooded animals) need to use a considerable amount of energy to maintain their body temperature.

When an owl swallows a mouse, many of the body parts are retained and processed in the digestive system, but the bird cannot digest bones and hair and will spit them out in an oblong mass called a pellet. When the pellets of undigested material land on the ground, they are decomposed by detritivores and other decomposers.

Decomposers play a key role in an ecosystem. If the nutrients in a dead leaf, for example, stayed locked in the leaf instead of being released back to the soil, the living organisms that need those nutrients would not have access to them.

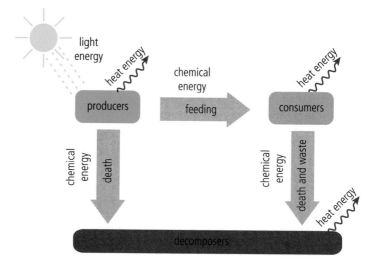

▲ **C4.2 Figure 4** How energy moves through an ecosystem in multiple forms, such as light energy, chemical energy and heat energy.

C4.2.13 – Heat loss from cell respiration

C4.2.13 – Heat loss to the environment in both autotrophs and heterotrophs due to conversion of chemical energy to heat in cell respiration
Include the idea that energy transfers are not 100% efficient so heat is produced both when ATP is produced in cell respiration and when it is used in cells.

When studying energy in organisms in a meadow or grassland, inside an animal such as a grasshopper, for example, chemical energy is used for cellular respiration. Glucose originally produced by the grass is converted by the grass into usable energy or, if it is eaten by the grasshopper, it will be converted into carbon dioxide and water. The transfer of energy from one form to another is not 100% efficient. Chemical reactions necessary for the production of ATP generate heat but heat is also generated whenever ATP is used. Any heat generated by cellular respiration is radiated away from the organism and lost to the air, soil or water in which the organism is living. Although this might be more obvious in mammals, which can give off considerable amounts of heat, even grass and grasshoppers will lose heat to the environment. If the grasshopper is eaten, some of the chemical energy in its body (in the form of protein, for example) is passed on to the next organism (a toad, for example). If the grasshopper dies and is not eaten, detritivores and decomposers will use its available energy.

The cells of decomposers also carry out cellular respiration and, as a result, the heat they produce will also be lost to the environment.

A thermogram is a photo taken with a special camera that can sense heat. Mammals such as this elephant produce heat that is radiated from the body. The red colour indicates hotter parts of the body, while blue indicates cooler parts of the body. Elephants pump blood to their ears to radiate heat and keep cool.

What does it mean when heat is "lost"? As you may already know from other science courses, there is a law stating that energy cannot be created or destroyed, only converted from one form to another. We have seen that light energy can be converted into chemical energy by the process of photosynthesis. We have also seen that, during the process of cell respiration, not all the energy is converted into useful energy (ATP) by the cell: some of it is converted to heat energy. Although this keeps mammals warm, once that heat leaves an organism's body it cannot be used again as a biological energy resource. So, for the organism and its ecosystem, this energy is "lost". It has not disappeared, however; it has simply been converted into a form that the organism can no longer use as a source of energy.

Is this a problem? Usually no, because the Sun is constantly providing new energy to producers. The energy is converted to chemical energy and passed on from one trophic level to the next. However, if, for some reason, sunlight could no longer reach

Earth's surface, because it is blocked by clouds or particles in the sky (as happens after large volcanic eruptions or large asteroid impacts), then the food chain is affected. This can be catastrophic for some organisms, as it was for dinosaurs, for example.

▶ Catastrophic events, such as the asteroid that smashed into Earth marking the end of the Cretaceous period 65 million years ago, can produce enough debris in the atmosphere to greatly reduce the intensity of sunlight energy arriving at the Earth's surface. As a consequence, 65 million years ago the lack of autotroph production led to mass extinctions, including the disappearance of all non-avian dinosaurs. Such major events illustrate just how vital sunlight is to most ecosystems.

C4.2.14 – The number of trophic levels

C4.2.14 – Restrictions on the number of trophic levels in ecosystems due to energy losses

At each successive stage in food chains there are fewer organisms or smaller organisms. There is therefore less biomass, but the energy content per unit mass is not reduced.

Although some food chains can have up to six trophic levels, most have four. The number of levels is limited by how much energy enters the ecosystem. Because so much is lost at each level, low energy at the start will be quickly transferred, whereas abundant energy at the start can sustain several trophic levels. The number of organisms in the chain, as well as the quantity of light available at the beginning of the chain, will determine how long a food chain is.

The **biomass** of a trophic level is an estimate of the mass of all the organisms within that level. Biomass is defined as the dry weight of an organism, because the actual mass of an organism includes a large amount of water. Water needs to be removed for the dry weight to be measured. It is expressed in units of mass, but also takes into account area or volume, for example gram per metre squared per year, $g\,m^{-2}\,yr^{-1}$. Although other factors are involved, the amount of sunlight reaching the photosynthetic producers strongly influences the biomass, so sunnier parts of the world can produce more biomass. Phytoplankton nearer the equator generate more biomass than phytoplankton further from the equator, for example.

Examine the following freshwater food chain:

green algae → caddisfly larvae → stickleback fish → pike fish

As we move backwards along a food chain or down an energy pyramid from the top, the number of organisms that occupies the lower trophic level increases. For every predator such as a pike, many dozens of sticklebacks will be eaten by it per year. The sticklebacks will eat hundreds of caddisfly larvae. This is another reason why food chains have a limited length, there is a limited amount of biomass production to start with at the first level, and for each subsequent level there is a continuing appetite for energy. If we examine the biomass, we see that, because there are fewer organisms each time we go up one trophic level, and because not all biomass is consumed or digested, there will be less biomass, and therefore less energy, at that next level. A stickleback weighs less than all the caddisfly larvae and other foods it eats.

It is important to note that, although biomass goes down and energy goes down from a lower trophic level to a higher trophic level, the energy values per unit mass (e.g. $J\,g^{-1}$) of the organisms do not go down. This can be seen in Table 1.

	Trophic level	**Energy per unit mass / J per g of wet mass**
Algae	1	3,439
Caddisfly larvae	2	3,760
Fish	3	5,341

C4.2 Table 1 Unlike the overall energy in the next level of a food chain or energy pyramid, energy per unit mass does not reduce as the trophic level increases.

C4.2.15 – Primary production

C4.2.15 – Primary production as accumulation of carbon compounds in biomass by autotrophs

The units should be mass (of carbon) per unit area per unit time and are usually $g\,m^{-2}\,yr^{-1}$. Students should understand that biomes vary in their capacity to accumulate biomass. Biomass accumulates when autotrophs and heterotrophs grow or reproduce.

In contrast to warmer places in the world, cooler **biomes**, or biomes with fewer hours of sunlight per year, have a lower biomass and therefore cannot support as many organisms. A biome is a large community of plants and animals, such as a desert, a tropical forest or a grassland. Biomes tend to occupy zones that cover wide expanses, often on a continental scale. The grasses in the temperate steppe biome of Mongolia, for example, could be the start of a food chain that includes lemmings, snakes and eagles.

But temperate steppes will accumulate less biomass in a year than a tropical rainforest, where trees, because of their massive size compared to the plants that occupy grasslands, will photosynthesize more and therefore fix more carbon. Bigger trees also produce more food and more habitat space for consumers and the rest of the food web. These heterotrophs will grow and reproduce to contribute to the overall biomass. Each biome varies in its capacity to produce biomass.

Primary production refers to the biomass generated by the activity of producers such as photosynthetic organisms when they fix carbon and make carbon compounds that can be used as a food source. This production is measured as the mass of carbon per unit area per unit of time, or grams per metre squared per year: $g\,m^{-2}\,yr^{-1}$.

 Do not confuse units of energy with units of mass in an ecosystem. In an energy pyramid, we use $kJ\,m^{-2}\,yr^{-1}$, whereas when referring to biomass production we use $g\,m^{-2}\,yr^{-1}$.

C4.2.16 – Secondary production

C4.2.16 – Secondary production as accumulation of carbon compounds in biomass by heterotrophs

Students should understand that, due to loss of biomass when carbon compounds are converted to carbon dioxide and water in cell respiration, secondary production is lower than primary production in an ecosystem.

Some molecules along the food chain cannot contribute to the accumulating biomass because they are lost in various forms, for example carbon dioxide is lost from organisms during cellular respiration, and waste products including urea are excreted. So, just as not all energy is passed on from one trophic level to the next, not all biomass is passed on either. When carbon compounds are assimilated by the next trophic level, it is often in the form of proteins and lipids. This conversion of one form of carbon molecule (e.g. glucose from producers) to another (e.g. lipids) inside consumers is called **secondary production**. Primary production is the generation of biomass in the first trophic level; secondary production is the addition of biomass in subsequent heterotrophic levels. As much of the energy is lost from one level to the next, biomass production is always lower in secondary than in primary production.

C4.2.17 – The carbon cycle

C4.2.17 – Constructing carbon cycle diagrams

Students should illustrate with a diagram how carbon is recycled in ecosystems by photosynthesis, feeding and respiration.

The element carbon is the cornerstone of life as we know it. Carbon is such a crucial element for living organisms that it is part of the definition of the term "organic". Hence, life on Earth is referred to as carbon-based life.

Not only is carbon found in the **biosphere** in organic molecules such as carbohydrates, proteins, lipids and vitamins, it is also found in the **atmosphere** as carbon dioxide and in the lithosphere as carbonates and fossil fuels in rocks. The biosphere refers to all the places where life is found; the atmosphere is where the gases that make up air are found; while the **lithosphere** refers to all the places where rocks are found. Petroleum, from which products such as gasoline, kerosene and plastics are made, is rich in carbon because it originates from partially decomposed organisms that died millions of years ago.

As shown in Figure 5, carbon is constantly cycled between living organisms and the inorganic processes that make the carbon available. The carbon atoms that make up the cells of the flesh and blood of a giraffe, for example, came from the vegetation the giraffe ate. Eating organic material provides newly dividing cells in the giraffe's body with a fresh supply of carbon-based energy-rich molecules with which the cells can carry out work. When cellular respiration is complete, carbon dioxide is released into the atmosphere and, when the giraffe dies, its body will be eaten by scavengers and the remains broken down by decomposers. Some of the carbon from the giraffe's body will go back into the atmosphere as carbon dioxide as the decomposers carry out cellular respiration. This section will look at some of the many different forms carbon can take as it is cycled by nature.

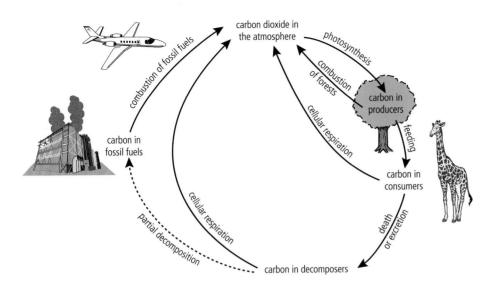

◀ **C4.2 Figure 5** Earth's carbon cycle

We will start with food. Photosynthetic autotrophs take carbon dioxide from the atmosphere and convert it into carbohydrates. The unbalanced chemical equation for photosynthesis is:

$$CO_2 + H_2O \rightarrow C_6H_{12}O_6 + O_2$$

The sugar on the right-hand side of the equation (in green) is a source of food, not only to the autotroph synthesizing it but also to the organisms that feed on the autotrophs. In its inorganic form on the left, as atmospheric carbon dioxide (in black), the carbon is not usable as a food source by the autotrophs or by any consumers. Few people realize how dependent the biosphere is on energy from the Sun for food production. And the biosphere includes us.

Carbon dioxide

Carbon dioxide is absorbed by photoautotrophs such as photosynthetic bacteria, phytoplankton, plants and trees. As you will recall, these producers are eaten by consumers, which then use the carbon in their bodies. Cellular respiration from all trophic levels, including decomposers, produces carbon dioxide, which is released back into the environment. This carbon dioxide diffuses into the atmosphere or into the water, depending on whether the organism is terrestrial or aquatic.

Methane

Other carbon compounds are produced by microbes such as archaea. Some archaea are anaerobic methanogens, meaning they do not require oxygen gas. When these methanogenic archaea metabolize food, they produce methane (CH_4) as a waste gas. You should be familiar with methane because it is the gas used in laboratories (for example the flame of Bunsen burners) and it makes the blue flame used in homes for cooking and heating.

These microbes are common in wetlands, where they produce marsh gas, which can sometimes glow at night, but they are also responsible for producing methane gas in the digestive tracts of mammals, including humans. With large herds of cattle being raised worldwide, there is a concern that the quantities of methane they produce are contributing to the runaway greenhouse effect, which will be discussed later.

SKILLS

Using the ideas listed below, draw a diagram showing the carbon cycle.

- Photosynthesis providing carbon to producers.
- Feeding providing carbon to herbivores.
- Cellular respiration by producers, consumers and decomposers releasing carbon into the atmosphere.
- Excretion and death providing carbon to decomposers.
- Partial decomposition by decomposers providing carbon to fossil fuels.
- Combustion of organic material releasing carbon into the atmosphere.

Compare your diagram with the one in Figure 5. How did you do?

C4.2.18 – Carbon sinks and sources

C4.2.18 – Ecosystems as carbon sinks and carbon sources

If photosynthesis exceeds respiration there is a net uptake of carbon dioxide and if respiration exceeds photosynthesis there is a net release of carbon dioxide.

A **carbon source** in an ecosystem is an organism that is a net producer of carbon dioxide. A **carbon sink** in an ecosystem is an organism that absorbs and holds more carbon than it releases. Plants that photosynthesize more than they respire and that hold their carbon in the form of roots, buds, stems, seeds and fruits are carbon sinks. Consumers in the form of herbivores or carnivores produce carbon dioxide through cellular respiration and release it into the atmosphere; they are carbon sources.

It is possible for some organisms to be either a source or a sink depending on the situation. For example, if a tree burns after a lightning strike or if a human cuts down a tree to burn it as firewood, the tree goes from being a carbon sink to being a carbon source.

TOK

Often people who choose to study ecology do so because of a profound love of nature, a respect for our planet and a desire to pass on a healthy ecosystem to future generations. They want to understand the things that they admire and love. But would this push them towards only wanting to study parts of the world that are considered remote, pristine or exotic, rather than studying their own local areas? And are they attracted to aesthetically pleasing organisms such as butterflies, wild cats or rare orchids, rather than less admired species such as spiders, parasitic organisms, cockroaches or slime moulds? In what ways do our values affect our pursuit and acquisition of knowledge?

C4.2.19 – The release of carbon dioxide during combustion

C4.2.19 – Release of carbon dioxide into the atmosphere during combustion of biomass, peat, coal, oil and natural gas

Students should appreciate that these carbon sinks vary in date of formation and that combustion following lightning strikes sometimes happens naturally but that human activities have greatly increased combustion rates.

One way to produce carbon dioxide from organic material and release it into the atmosphere is to burn organic matter. This can happen naturally or can be caused by human activity. Forest fires occur naturally when lightning strikes a forest, and it is

Plant material can be considered a carbon sink or a carbon source depending on what is happening. Combustion releases carbon dioxide into the atmosphere, so a forest on fire behaves as a carbon source (as seen on the left), but when the trees and other plants are growing, they are a carbon sink, absorbing more carbon dioxide than they are releasing (as seen on the right).

part of the natural cycle for dead fallen branches to be burned away and forest growth to start afresh. In fact, some seeds cannot germinate until they have been exposed to fire, such as those of the lodgepole pine (*Pinus contorta*). Fires can also be ignited naturally by volcanic eruptions and lava flows cutting through woodlands. However, forest fires can release a considerable quantity of carbon dioxide into the atmosphere.

Humans have changed the natural fire cycle; we ignite forest fires sometimes by accident but often on purpose to clear land for agriculture. Humans and our hominid ancestors have known how to use and control fire for cooking and tool making for at least a million years, but it is only in the most recent decades that there have been enough humans and enough forests burned to have a considerable impact on the quantity of carbon dioxide in the atmosphere. Traditionally, wood has also been used to heat our homes, whether they be 100,000-year-old caves or modern houses with a fireplace.

Since the start of the industrial era, we have used more and more sources of energy other than wood. Organic sources of carbon in the form of **biomass**, coal, peat, oil and natural gas are burned for many purposes, including generating energy for factories, electricity for homes, and fuels for vehicles such as automobiles, trains and aircraft.

Biomass in this context is organic waste and can come from agricultural practices such as growing crops, raising livestock and harvesting trees in forestry. The non-useable parts of plants or the excrement of farm animals can be burned directly for energy, or they can be fermented in large tanks called **biodigesters** and the methane gas produced used as fuel for combustion.

Wood and **peat** form over many decades and, if well managed, forests and peat bogs can be considered renewable resources if the extraction rate does not exceed the production rate. Peat is a form of waterlogged soil found in certain types of wetlands, such as mires and bogs. Although peat is a heterogeneous mixture of many things, at least 30% of its dry mass must be composed of dead organic material for it to be called peat. Because of the high acidity of the soil, the environment is difficult for decomposers, which is why so much energy-rich organic matter remains. In order for it to be usable as a fuel, cut peat is dried out to reduce its high levels of humidity. It is cut into slabs, granules or blocks, and moved to where it is needed for combustion.

Slabs of peat left to dry in Scotland, UK. Think about where peat fits into the carbon cycle.

Some of the sources of energy, such as coal, crude oil and natural gas, used for human activities are considered to be **fossil fuels** because they are mined from the ground and, once removed, they are not renewed. It takes millions of years for these petroleum products to form: partially decomposed organic matter is pushed underground by geological forces. Coal is a hard black rock that has to be mined by breaking apart the layers of sediments it is found in. Crude oil is a viscous black liquid that needs to be pumped out of the ground, where it is trapped between layers of rock. This naturally occurring substance is then refined and transformed into many manufactured products such as fuels and plastics. Natural gas, which has a very low density, bubbles towards the surface and, when it is trapped by a dome of impenetrable rock, can be collected by digging a well.

Type of resource	Time it takes to form	Examples of energy sources used by humans
Renewable	Replenished daily (or almost every day)	Sunshine, wind, river currents, wave motion, geothermal, biomass
Long-term renewable	Decades or hundreds of years	Well-managed hardwood forests and peat bogs
Non-renewable	Millions of years	Coal, crude oil, natural gas, uranium

C4.2.20 – The Keeling Curve

C4.2.20 – Analysis of the Keeling Curve in terms of photosynthesis, respiration and combustion

Include analysis of both the annual fluctuations and the long-term trend.

The longest continuous monitoring of the carbon dioxide concentration in the Earth's atmosphere has been carried out by the National Oceanic and Atmospheric Administration (NOAA) in the US on the Hawaiian island Mauna Loa. This site was chosen because it is in the middle of the ocean far from highly industrialized zones, and because of its high altitude. Mauna Loa is a volcano, so any carbon dioxide released from the volcano is subtracted from the data. The measurements were started under the direction of Charles Keeling, and the graph of the results is called the Keeling Curve.

Figure 6 shows the data from 1958, when Keeling started the measurements, to 2022. You can find the most recent values online. The black line shows the average trend. The wavy red line shows seasonal changes, reflecting the fact that photosynthetic organisms absorb more carbon dioxide in the summer and autumn (the downward pointing spikes under the solid black line) than in the winter and spring (the spikes pointing above the solid black line).

Atmospheric CO$_2$ at the Mauna Loa observatory

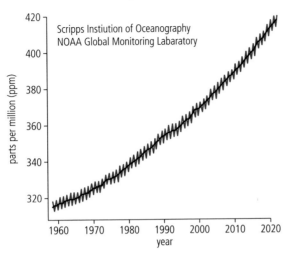

Scripps Instiution of Oceanography
NOAA Global Monitoring Labaratory

◀ **C4.2 Figure 6** The National
Oceanic and Atmospheric
Administration (NOAA)
data on carbon dioxide
levels in the atmosphere
1958–2022. The oscillating
pattern is caused by seasonal
fluctuations in activities such
as photosynthesis, but the
overall trend largely reflects
the results of human activities
such as combustion of fossil
fuels.

Challenge yourself

1. Using the solid black trend line in Figure 6, determine the atmospheric carbon dioxide concentration for 1960 and for 2020.

2. Calculate the percentage change in atmospheric carbon dioxide concentration between 1960 to 2020.

3. Explain why the measurements have a high point and a low point for each year.

So why has there been such a sudden increase in atmospheric carbon dioxide in less than the duration of one human lifetime? According to the International Panel on Climate Change (IPCC), it is mainly because of human activity such as the combustion of fossil fuels, notably for transporting goods and people within our global economy. This is discussed further in Chapter D4.3, which explores climate change.

Worked example

The figure below, taken from the NOAA website, shows the atmospheric carbon dioxide levels of recent years.

Recent monthly mean CO$_2$ at the Mauna Loa observatory

The pattern shown in red (the larger fluctuations) is caused by seasonal fluctuations in carbon dioxide levels. The smoother black line shows the trend corrected for those seasonal fluctuations.

1. (a) Determine how the years are divided up on the horizontal x-axis of the graph.

    (b) Estimate the level of atmospheric carbon dioxide in April 2017 and in April 2022 using the corrected values on the black trend line.

    (c) Look at the five lowest values on the red line. There is one per year. Determine the month of the year during which this low point most often occurs. Do the same for the high points.

2. (a) In terms of cellular respiration and photosynthesis rates in the northern hemisphere, explain the annual downward fluctuations from May to October.

    (b) Do the same for the upward fluctuations from October to May of the following year.

3. Describe the overall trend shown by the graph for the years shown, giving quantitative data in your description.

Solution

1. (a) The years are divided into quarters: January/February/March, April/May/June, July/August/September and October/November/December.

    (b) 406 p.p.m. and 417 p.p.m., respectively. It is important to include the units.

    (c) Lows are in October and highs in May.

2. (a) Because plants, phytoplankton and photosynthetic bacteria are generally more active in the summer and autumn months, more carbon dioxide is extracted from the atmosphere and levels drop. During this time, cellular respiration is contributing large quantities of carbon dioxide to the atmosphere, but not as fast as photosynthesis is taking it out.

    (b) Conversely, when photosynthesis is less intense during the winter and spring months, carbon dioxide levels rise and, although organisms are generally less active at colder times of the year, their cellular respiration rates put more carbon dioxide into the air than the photosynthetic organisms can remove.

3. The trend shows an increase from 406 p.p.m. at the beginning of 2017 to 417 p.p.m. in April 2022. This 11 p.p.m. increase represents a percentage change of +2.7% for the 5-year period shown.

What are the direct and indirect consequences of rising carbon dioxide levels in the atmosphere?

C4.2.21 – The dependence on atmospheric oxygen and carbon dioxide

C4.2.21 – Dependence of aerobic respiration on atmospheric oxygen produced by photosynthesis, and of photosynthesis on atmospheric carbon dioxide produced by respiration
The fluxes involved per year are huge, so this is a major interaction between autotrophs and heterotrophs.

If you look at the two summary equations for photosynthesis and aerobic respiration, you will notice that one is the reverse of the other:

Photosynthesis by autotrophs:

carbon dioxide $+$ water $\rightarrow$ glucose $+$ oxygen

Aerobic cellular respiration by heterotrophs:

glucose $+$ oxygen $\rightarrow$ carbon dioxide $+$ water

In order for aerobically respiring organisms to have oxygen, photosynthetic organisms need to produce oxygen gas. Life evolved on Earth for about a billion years without oxygen and without photosynthesis. Early bacteria and archaea relied on anaerobic respiration. Photosynthesis came after many hundreds of millions of years of respiration without oxygen. Aerobic cellular respiration was only able to evolve once oxygen gas was produced by photosynthesis.

How much carbon dioxide and oxygen are necessary to keep this cycle between producers and consumers going? The quantities are enormous. The quantity of carbon dioxide released into the atmosphere by all living organisms on Earth every year, for example, is estimated at around 200 GtC yr^{-1} (gigatonnes of carbon per year). A gigatonne is 1,000 million tonnes.

C4.2.22 – The recycling of chemical elements

C4.2.22 – Recycling of all chemical elements required by living organisms in ecosystems
Students should appreciate that all elements used by living organisms, not just carbon, are recycled and that decomposers play a key role. Students are not required to know details of the nitrogen cycle and other nutrient cycles.

We have focused on the recycling of carbon in the form of organic matter in organisms and carbon dioxide in the atmosphere, but all other elements that organisms need are cycled, too. One prominent example is nitrogen. Although it is present in huge quantities in Earth's atmosphere (roughly 79% of air is N_2 gas), like carbon dioxide it needs to be fixed in order to become usable by living organisms. Certain bacteria can fix nitrogen, and the usable molecules that they produce are passed on from one organism in a food chain to the next.

Other elements from the periodic table that are necessary for life and that are cycled and passed on through the food chain include hydrogen, oxygen, calcium, potassium, sodium, iron, phosphorus and many others. Because they cannot be synthesized by living organisms, and because they do not enter the system every day in the way solar energy does, these atoms need to be recycled. Your body is made up of atoms from the foods you have eaten throughout your lifetime. We are what we eat in the sense that we construct our living tissue using atoms borrowed from other sources. And when living organisms die, they pass on their atoms to the next trophic level, or to detritivores and decomposers.

Decomposers are a key part of any nutrient cycle because they make nutrients that are no longer needed by organisms available to others that do need them.

Chemical elements in minerals are absorbed into living trees and then returned to the soil when the trees die and decay.

1 minerals in soil

2 minerals incorporated into living tissue

3 minerals returned to soil when organism decays

Guiding Question revisited

What is the reason matter can be recycled in ecosystems but energy cannot?

In this chapter we have discussed how:

- without new supplies of minerals arriving on Earth in sufficient quantities, organisms need to reuse and recycle the matter that is already available
- decomposers play a key role in returning nutrients to an ecosystem
- carbon cycle diagrams show how carbon is absorbed by producers and released by consumers, or by human activity such as the burning of fossil fuels or deforestation
- food chains and food webs show how energy flows from one trophic level to the next
- energy pyramids show us that energy is lost at each trophic level, and part of this loss is in the form of heat as once the heat leaves an organism, it cannot be reused or recycled.

Guiding Question revisited

How is the energy that is lost by each group of organisms in an ecosystem replaced?

In this chapter you have learned:

- sunlight reaches Earth every day, replenishing the energy requirements of autotrophs
- therefore energy can leave Earth in the form of heat and does not need to be recycled.

Exercises

Q1. Which organism in a food chain is found on the third trophic level?

A	Secondary consumer.	**B**	Tertiary consumer.
C	Primary consumer.	**D**	Producer.

Q2. What is shown in the Keeling Curve?

  I   Annual fluctuations in atmospheric carbon dioxide concentration

  II   Primary production of photosynthesis in $kJ\,m^{-2}\,yr^{-1}$.

  III   An increase in atmospheric carbon dioxide in recent decades.

A	I and II only.	**B**	I and III only.
C	II and III only.	**D**	I, II and III.

Q3. Biosphere 2 is a large research facility, owned by the University of Arizona, USA, used to see whether humans can live for many months inside a sealed set of buildings where all the air, water and matter is recycled. Participants grow all their own food inside greenhouses and recycle all their waste. Explain whether this is an open system or a closed system.

Q4. State the names of the hydrocarbon-rich substances that are described below and used for fuel by humans.

(a) A kind of waterlogged soil found in wetlands and made of partially decomposed plant material.

(b) A hard black rock that can be burned to generate electricity or direct heat.

(c) A black viscous liquid trapped between layers of rock.

(d) Of all the commonly used petroleum products, this one has the lowest density.

Q5. The table shows data from two ecosystems, Cedar Bog and Lake Mendota in Wisconsin, USA.

	Cedar Bog		Lake Mendota	
Tropic level	Productivity/ cal cm^{-2} yr^{-1}	Efficiency /%	Productivity/ cal cm^{-2} yr^{-1}	Efficiency/ %
Solar radiation	119.000		119.000	
Plants	111	0.1	480	0.4
Herbivores	14.8	13.3	41.6	8.7
Carnivores	3.1	22.3	2.3	5.5
Higher carnivores			0.3	13.0

(a) Construct pyramids of energy for Cedar Bog and Lake Mendota.

(b) Compare and contrast the two energy pyramids.

(c) Describe the level of efficiency from one trophic level to the next.

(d) Suggest, giving a reason, which of the two could support the introduction of a new top predator to the ecosystem.

Q6. (a) From the following information, construct a food web.

- Grass is eaten by rabbits, grasshoppers and mice.
- Rabbits are eaten by hawks.
- Grasshoppers are eaten by toads, mice and garter snakes.
- Mice are eaten by hawks.
- Toads are eaten by hognose snakes.
- Hognose snakes are eaten by hawks.
- Garter snakes are eaten by hawks.

(b) From the food web you have drawn, what is the trophic level of the toad?

C4 Practice questions

1. Explain how a population of grasshoppers could be estimated using the capture–mark–release–recapture technique and outline the assumptions that must be considered.

(Total 6 marks)

2. Distinguish between parasitism and mutualism, giving an example of each.

(Total 2 marks)

3. This energy flow diagram for a temperate ecosystem has been divided into two parts. One part shows autotrophic use of energy and the other shows the heterotrophic use of energy. All values are kJ m^{-2} yr^{-1}.

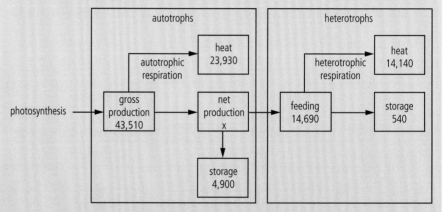

   (a) Calculate the net production of the autotrophs (i.e. what is left of the gross production once all the heat loss is subtracted). (1)

   (b) (i) Compare the percentage of heat lost through respiration by the autotrophs with the heterotrophs. (1)

       (ii) Most of the heterotrophs are animals. Suggest one reason for the difference in heat losses between the autotrophs and animal heterotrophs. (1)

   The heterotrophic community can be divided into food webs based upon decomposers and food webs based upon herbivores. It has been shown that of the energy consumed by the heterotrophs, 99% is consumed by the decomposer food webs.

   (c) State the importance of decomposers in an ecosystem. (1)

   (d) Deduce the long-term effects of sustained pollution that kills decomposers on autotrophic productivity. (2)

(Total 6 marks)

4. Seawater temperature has an effect on the spawning (release of eggs) of echinoderms living in Antarctic waters. Echinoderm larvae feed on phytoplankton. In this investigation, the spawning of echinoderms and its effect on phytoplankton was studied.

In the figure below, the top line indicates the number of larvae caught (per 5,000 l of seawater). The shaded bars below show when spawning occurred in echinoderms.

☐ = 0% to 25%
▨ = 25% to 75%
■ = 75% to 100%

The concentration of chlorophyll gives an indication of the concentration of phytoplankton. Note: the seasons in the Antarctic are the reverse of those in the northern hemisphere.

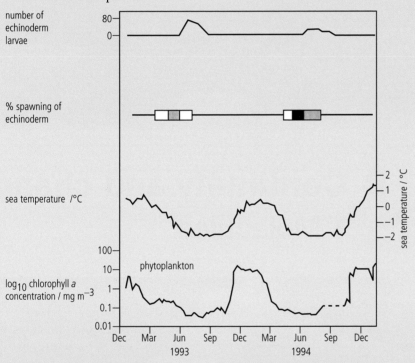

(a) State the trophic level of echinoderm larvae. (1)

(b) Identify the period during which the spawning of echinoderm lies between 25% and 75%. (1)

(c) Explain the relationship between the seasons and the concentration of phytoplankton. (2)

(d) (i) Outline the effect of sea water temperature on echinoderm larvae numbers. (2)

(ii) Using the data in the figure, predict the effect of global warming on echinoderm larvae numbers. (2)

(Total 8 marks)

THEME D Continuity and change

1 Molecules

We are at a critical junction today regarding life on our planet. A recently developed technology called CRISPR provides a revolutionary, inexpensive and effective technique for manipulating the code of life. Immediately, we think of the possibility of not treating but curing genetic diseases such a Huntington's disease, cystic fibrosis and Duchene muscular dystrophy. We see hope for the control of cancers we have had little success treating in the past. There are so many medical possibilities in our future based on this technology.

However, there is a deep concern that this technology will be used inappropriately. Should we allow embryos to be altered to produce the ultimate athlete, an individual with an IQ off the chart, someone with an immune system that can drastically decrease or even eliminate disease, or even someone with "perfect" physical attributes? Who benefits from this technology, the very wealthy? What will happen to the natural evolution of Homo sapiens? By changing one attribute, will we also affect others later in the life of an individual?

So many possibilities, and so many questions. If society is to decide the ethics and limits of the use of this technology, we all must have a basic understanding of the genetic code and how it controls life.

D1.1 DNA replication

Guiding Questions

How is new DNA produced?

How has knowledge of DNA replication enabled applications in biotechnology?

When the molecular structure of DNA was first presented by Francis Crick and James Watson in 1953 in the scientific journal *Nature*, it immediately stimulated research into the process of DNA replication. The production of DNA has to be accurate so that new cells and/or organisms could continue the functions of life. Just as important, this production of genetic material has to allow change, mistakes if you prefer, to account for the variation obvious in organisms over time.

Knowledge of this replication process has resulted in tremendous advances in biotechnology. We are seeing applications of DNA replication in many areas of medical science as well as in police investigations involving forensics. DNA analysis has led to revelations concerning family ancestry as well as predictions concerning the inheritance of medical conditions. By understanding the DNA production process, scientists have been successful in replicating extremely small amounts and, in some cases, very old samples of recovered DNA. We have now sequenced the DNA of a million-year-old woolly mammoth, and extracted cartilage from a 125-million-year-old *Caudipteryx* dinosaur. If DNA can be recovered from this cartilage, it might actually be possible to sequence a dinosaur's DNA.

D1.1.1 – The role of DNA replication

D1.1.1 – DNA replication as production of exact copies of DNA with identical base sequences

Students should appreciate that DNA replication is required for reproduction and for growth and tissue replacement in multicellular organisms.

DNA structure is discussed in Chapter A1.2. It is a **polymer** composed of **monomers** called **nucleotides** bonded together. Each nucleotide includes a 5-carbon sugar called deoxyribose, a phosphate group, and a nitrogenous base. There are four nitrogenous bases in DNA: adenine, thymine, cytosine and guanine. The basic structure of DNA is shown in Figure 1.

D1.1 Figure 1 The basic structure of a DNA molecule. Note that the bonding of adenine and thymine involves two hydrogen bonds, while the bonding between cytosine and guanine involves three hydrogen bonds. The outside strands are of alternating phosphate groups, shown as circles, and deoxyribose molecules, represented by shaded pentagons.

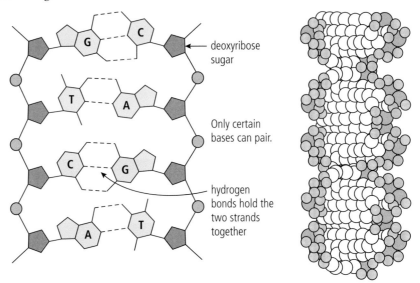

DNA provides the code that creates the uniqueness of all forms of life on our planet. This code is carried in the sequence of the four possible nitrogenous bases.

Because DNA is the code of life, it must be duplicated so that the reproduction of organisms is possible. DNA duplication is also essential for growth and tissue replacement in multicellular organisms. This duplication is known as **DNA replication**. Replication doubles the quantity of DNA and ensures that exact copies of each DNA molecule are made.

In 2014, a group of Scripps Research scientists introduced two new, unnatural nitrogenous bases into the genetic code of living bacteria. These new bases were incorporated into the DNA code for the bacteria. The research is ongoing: many feel that this approach could provide a means by which new medicines and vaccines can be developed. With six bases instead of four, many more products could be produced.

D1.1.2 and D1.1.3 – Semi-conservative replication and complementary base pairing

D1.1.2 – Semi-conservative nature of DNA replication and role of complementary base pairing
Students should understand how these processes allow a high degree of accuracy in copying base sequences.

D1.1.3 – Role of helicase and DNA polymerase in DNA replication
Limit to the role of helicase in unwinding and breaking hydrogen bonds between DNA strands and the general role of DNA polymerase.

The replication process of DNA is said to be **semi-conservative**. Each strand in the DNA double helix acts as a template for the synthesis of a new, complementary strand. Evidence for a semi-conservative replication process was provided by research using bacteria (see Figure 2).

Figure 2 shows how each daughter DNA double helix contains an old strand from the parental DNA double helix and a new strand. Thus the name semi-conservative replication.

In the nucleus of cells are two types of molecules that are particularly important for the process of DNA replication. They are:

- enzymes needed for replication, which include **helicase** and a group of enzymes collectively called **DNA polymerase**
- free nucleotides, which are nucleotides that are not yet bonded and are found floating freely in the nucleoplasm, some contain adenine, some thymine, some cytosine, and some guanine.

One of the first stages in DNA replication is the separation of the double helix into two single strands. You should remember that the double helix is held together by the hydrogen bonds between the complementary base pairs adenine (A) and thymine (T), and cytosine (C) and guanine (G). The enzyme that initiates this separation into two single strands is called helicase (Figure 3).

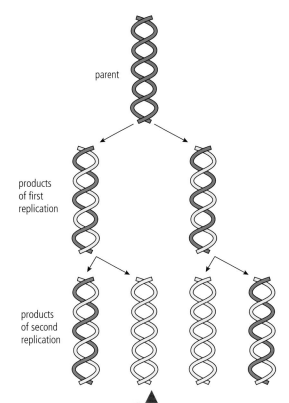

D1.1 Figure 2 A diagram demonstrating the general process of semi-conservative replication of DNA.

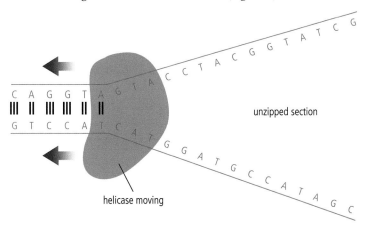

D1.1 Figure 3 The first step of DNA replication is helicase unzipping the double-stranded DNA molecule, forming a section with two single strands.

The unpaired nucleotides on each of these single strands can now be used as a template to create two double-stranded DNA molecules identical to the original. Some people use the analogy of a zipper for this process. When you pull on a zipper, the slide mechanism is like helicase. The separation of the two sides of the DNA molecule is like the two opened sides of a zipper.

Formation of two complementary strands

Once DNA has become unzipped, the nitrogenous bases on each of the single strands are unpaired. In the environment of the nucleoplasm, there are free-floating nucleotides available to form complementary pairs with the single-stranded nucleotides of the unzipped molecule. This does not happen in a random fashion. A free nucleotide locates one end of an opened strand, and then a second nucleotide

In the 1950s, Matthew Meselson and Frank Stahl carried out research using radioactive nitrogen that provided evidence for the semi-conservative replication of DNA.

The process of semi-conservative replication as described allows a high degree of accuracy in copying base sequences and provides stability when passing on the genetic code.

D1.1 Figure 4 DNA replication, showing the enzymes helicase and DNA polymerase at work. Helicase is opening the double helix while DNA polymerase is catalysing the formation of covalent bonds between adjacent nucleotides as complementary base pairing occurs.

What biological mechanisms rely on directionality?

How is genetic continuity ensured between generations?

Helicase has been found to catalyse the unzipping of DNA at a rate measured in hundreds of base pairs per second.

DNA polymerase has two very important roles in DNA replication. It catalyses the formation of covalent bonds between adjoining nucleotides in the growing DNA strand. It also serves in a proofreading role to ensure each new DNA strand is a near perfect copy of the original.

can arrive to join the first. To join, these first two nucleotides must become covalently bonded together, creating the beginning of a new strand. The formation of a covalent bond between two adjoining nucleotides is catalysed by one of the DNA polymerase enzymes, which is an important part of the process.

A third nucleotide then joins the first two, and the process continues in a repetitive way for many nucleotides. The other unzipped strand also acts as a template for the formation of a new strand. This strand forms in a similar fashion, but in the opposite direction to the first strand. In Figure 4, notice that one strand is replicating in the same direction as helicase is moving, while the other strand is replicating in the opposite direction.

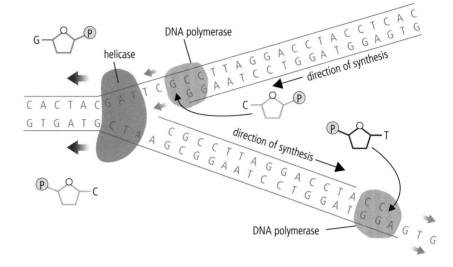

The significance of DNA replication is that it ensures two identical copies of DNA are produced from an original strand. Notice that in the area where replication has already taken place, the two strands are identical to each other. This is because the original double-stranded molecule had complementary pairs of nucleotides and it was complementary nucleotides that used the unzipped single-stranded areas as templates.

The role of helicase and DNA polymerase in DNA replication

The enzyme helicase is essential during DNA replication. It separates double-stranded DNA into single strands allowing each strand to be copied. Helicase begins at a point in or at the end of a DNA molecule and moves one complementary base pair at a time, breaking the hydrogen bonds so that the double-stranded DNA molecule becomes two separate strands.

DNA polymerase is another key enzyme in DNA replication. DNA polymerase adds nucleotides one by one to the growing DNA chain.

D1.1.4 – Amplifying and separating DNA

D1.1.4 – Polymerase chain reaction and gel electrophoresis as tools for amplifying and separating DNA

Students should understand the use of primers, temperature changes and *Taq* polymerase in the polymerase chain reaction (PCR) and the basis of separation of DNA fragments in gel electrophoresis.

The exploration and manipulation of DNA has been enhanced by the development of many different techniques during recent decades. In this section we will examine two of these techniques: **polymerase chain reaction (PCR)** and **gel electrophoresis**.

Polymerase chain reaction

PCR is a laboratory technique using a machine called a thermocycler, which takes a small quantity of DNA and copies all the nucleotides to make millions of copies of the DNA (see Figure 5). PCR is specific in that it only amplifies a targeted section of DNA.

small quantity of DNA goes in

large quantity of DNA comes out

thermocycler copies segments of DNA

D1.1 Figure 5 A polymerase chain reaction (PCR) is a method of generating enough DNA for analysis.

Various components are needed to run a PCR, including the following.

- **Primers**: single-stranded, short polymers of 15–20 nucleotides that are complementary to the nucleotides at one end of the target DNA to be copied. These primers provide a starting point for DNA synthesis because DNA polymerases can only attach new DNA nucleotides to an existing strand of nucleotides.
- *Taq* **polymerase**: *Taq* is a polymerase from a bacterium that lives in hot springs. This enzyme can withstand high temperatures.
- Free nucleotides: to synthesize the new strands of DNA.

Once all these components have been mixed in a tube, the tube is placed in the thermocycler, which controls the temperature of the reaction.

There are three steps to the PCR process.

1. **Denaturation**: the mixture is heated to a temperature between 92°C and 98°C, which breaks the hydrogen bonds holding the two strands of DNA together.
2. **Annealing**: the mixture is then cooled to between 50°C and 65°C, to allow the primers to bind with nucleotides on both strands at the ends of the target sequence.
3. **Elongation**: The *Taq* polymerase catalyses the building of new DNA strands by extending the primers. A temperature of 70–80°C is needed for this step.

This three-step cycle is repeated over and over until enough target DNA has been produced to accomplish the desired task.

Gel electrophoresis

This laboratory technique is used to separate fragments of DNA in order to identify its origin. Enzymes are used to chop up the long filaments of DNA into varying sizes of fragments. If two different sources of DNA are to be compared, the same enzymes are used to produce fragments. The DNA fragments are then placed into small wells (holes) in a gel placed in an electrophoresis chamber. Figure 6 on the next page illustrates an electrophoresis chamber.

D1.1 Figure 6 An electrophoresis chamber with power supply and electrodes. Notice the buffer solution in the holding tanks at the ends of the chamber. An agarose gel with sample wells is placed on the chamber bed, the middle region of chamber between the end tanks.

D1.1 Figure 7 This autoradiogram or autoradiograph shows banded lines that were formed from nine different DNA samples during electrophoresis. The black traces are left by the radioactivity of the materials used in marking the DNA samples. Smaller chains of DNA travel farther on the gel than larger chains, allowing characteristic banding patterns for a particular sample of DNA.

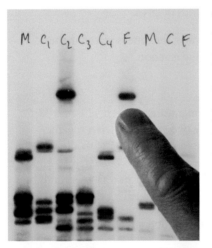

The gel, with samples of fragments in the wells, is exposed to an electric current, positive on one side and negative on the other. The biggest, heaviest and least charged particles do not move easily through the gel, so they get stuck very close to the wells they were in at the beginning of the experiment. The smallest, least massive and most charged particles pass through the gel to the other side with little difficulty. Intermediate particles are distributed in between. At the end of the experiment, the fragments leave a banded pattern of DNA, like the example shown in Figure 7.

D1.1.5 – Applications of amplifying and separating DNA

D1.1.5 – Applications of polymerase chain reaction and electrophoresis

Students should appreciate the broad range of applications, including DNA profiling for paternity and forensic investigations.

NOS: Reliability is enhanced by increasing the number of measurements in an experiment or test. In DNA profiling, increasing the number of markers used reduces the probability of a false match.

The process of matching an unknown sample of DNA with a known sample to see if they correspond is called **DNA profiling**. This is also sometimes referred to as **DNA fingerprinting** because there are some similarities with identifying fingerprints, although the techniques are very different.

If, after separation by gel electrophoresis, the pattern of bands formed by two samples of DNA fragments are identical, it means that both must have come from the same individual. If the patterns are similar, it means that the two individuals are probably related.

DNA profiling can be used in paternity suits when the identity of someone's biological father needs to be known for legal reasons. At a crime scene, forensic specialists can collect samples such as blood or semen, which contain DNA. Often such samples only contain very small amounts of DNA. In this case, PCR is carried out to amplify the DNA available for profiling.

Gel electrophoresis can be used to compare DNA collected from a crime scene with that of suspects. If they match, the suspect can be questioned further. If there is no match, the suspect is probably not the person the police are looking for. Criminal cases are sometimes reopened many years after a judgement was originally made, to consider new DNA profiling results. In the USA, this has led to the liberation of many individuals who had been sent to jail for crimes they did not commit.

The processes of PCR and DNA profiling have been used to sequence the bases of human DNA. This sequencing was the goal of the Human Genome Project.

suspect 1   suspect 2   suspect 3   blood stain   suspect 4   suspect 5   suspect 6

D1.1 Figure 8 These seven tracks were produced by gel electrophoresis to allow investigators to analyse and match DNA samples.

Challenge yourself

1. Using Figure 8, showing the DNA profiles from six suspects, can you identify which one matches the DNA profile of the blood stain found at the crime scene?

DNA profiling is used in other circumstances too, for example in studies of ecosystems, when scientists use DNA samples taken from birds, whales and other organisms to clarify relationships. This has helped establish a better understanding of social relationships, migration patterns and nesting habits, for example. In addition, the study of DNA in the biosphere has given new credibility to theories of evolution: DNA evidence can often reinforce evidence of common ancestry based on anatomical similarities between species.

Nature of Science

How do we decide when evidence is reliable or not? Often when DNA evidence is used in a courtroom trial, it has a certain credibility as scientific fact, yet we know from our own experience in laboratory work that there is a degree of error in any procedure. Whether it be in the laboratory or in a courtroom, it is difficult to imagine evidence that can be considered 100% certain. When a scientist comes up with new evidence, old theories can be challenged or even overturned. Reliability of data can be enhanced by increasing the number of measurements in an experiment or test. In DNA profiling, increasing the number of markers used reduces the probability of a false match. A marker refers to a particular sequence of DNA.

Guiding Question revisited

How is new DNA produced?

In this chapter we have discussed how:

- DNA replication is essential to produce new cells, both in unicellular and multicellular organisms
- semi-conservative replication results in every new copy of DNA possessing one strand from the parent DNA and one new strand
- complementary base pairing in the DNA replication process ensures a high degree of accuracy
- helicase is required to unwind and break hydrogen bonds so that the process of DNA replication can occur
- DNA polymerase controls the addition of free nucleotides to the developing DNA strand.

Guiding Question revisited

How has knowledge of DNA replication enabled applications in biotechnology?

In this chapter we have discussed how:

- polymerase chain reactions (PCR) are used to amplify DNA segments so that valid analyses can be carried out
- a thermocycler, *Taq* polymerase, primers and free nucleotides are necessary for the process of PCR
- gel electrophoresis uses an electrical field and agarose gel to separate segments of DNA
- PCR and gel electrophoresis are often used together in attempts to solve crimes where small amounts of DNA have been recovered
- gel electrophoresis is also used in the study of ecosystems
- paternity questions can often be solved using gel electrophoresis.

Exercises

Q1. Using complementary base pairing and the first letter of the nitrogenous base in nucleotides, write the letter of the nucleotide that would base pair with the exposed base in the following sequence.

ATG ACC GCT

Q2. What type of bonds are broken by helicase?

Q3. Rearrange the following steps in DNA replication in their proper order.

   I   Two new molecules of DNA are created.

   II   DNA polymerases attach the free nucleotides to the exposed nitrogenous bases.

   III Helicase begins to break the hydrogen bonds between nitrogen bases.

   IV Free floating nucleotides pair up with exposed nitrogen bases.

D1.2 Protein synthesis

Guiding Questions

How does a cell produce a sequence of amino acids from a sequence of DNA bases?

How is the reliability of protein synthesis ensured?

DNA controls the production of proteins in a cell. These proteins are specific to each type of cell. It is the sequence of DNA bases in a cell that determines the amino acids and their order in each protein produced. The production of proteins actually occurs in the cytoplasm at organelles called ribosomes. Even though the DNA does not leave the nucleus, it is able to send its message to the ribosomes, directing the synthesis of the exact proteins needed in that cell.

How does the diversity of proteins produced contribute to the functioning of a cell?

The DNA code produces the proteins by two processes: transcription and translation. Reliability is essential to the successful production of functional proteins, but so is the potential for a degree of variability, to allow evolution to proceed within a population. DNA and the process of protein synthesis fulfil these required characteristics of the genetic code.

D1.2.1 – The synthesis of RNA

D1.2.1 – Transcription as the synthesis of RNA using a DNA template

Students should understand the roles of RNA polymerase in this process.

DNA is sequestered (locked away) in the nucleus of a cell. The organelles essential to protein synthesis are the **ribosomes**, which are located in the cell cytoplasm. The DNA must communicate with the ribosomes to control the production of proteins. It does so by producing a code that is carried from the nucleus to the cytoplasm by **ribonucleic acid (RNA)**. **Transcription** is the synthesis of RNA using the base sequence in DNA as a template.

The set of ideas first proposed by Francis Crick in 1956, called the central dogma of molecular biology, states that information passes from genes (specific base sequences on the DNA) to the RNA copy. The RNA copy then directs the production of proteins at the ribosomes in the cytoplasm by controlling the sequence of amino acids. This mechanism is one-way and fundamental to all forms of life.

The sections of DNA that code for polypeptides (proteins) are called **genes**. Any one gene is a specific sequence of nitrogenous bases found at a particular location in a DNA molecule.

The process of transcription begins when an area of DNA for one gene becomes unzipped (see Figure 1). This is very similar to the unzipping process involved in DNA replication, but in this case only the area of the DNA where the gene is found is unzipped. The two complementary strands of DNA are now single stranded at the gene location. Recall that RNA is a single-stranded molecule. This means that only one of the two strands of DNA will be used as a template to create the **messenger RNA (mRNA)** molecule. This strand is called the **template strand**. An enzyme called **RNA polymerase** is essential to the process of transcription. Transcription can only begin when RNA polymerase binds to a **promoter sequence** near the beginning of a gene on the template strand. RNA polymerase is also involved in linking the RNA nucleotides to form an RNA strand.

D1.2 Figure 1 ▶ **D1.2 Figure 1** The process of transcription (the synthesis of an RNA molecule). RNA polymerase has a helicase-like role, as it plays a part in opening the DNA double helix. It also catalyses the addition and bonding of free RNA nucleotides to the growing messenger RNA (mRNA) strand.

D1.2.2 – Hydrogen bonding and complementary base pairing

D1.2.2 – Role of hydrogen bonding and complementary base pairing in transcription

Include the pairing of adenine (A) on the DNA template strand with uracil (U) on the RNA strand.

RNA polymerase acts as helicase does in DNA replication, to break the hydrogen bonds between the two strands of DNA, unzipping the double helix. As RNA polymerase moves along the DNA template strand, RNA nucleotides that exist in the nucleus float into place by complementary base pairing. The complementary base pairs are the same as in double-stranded DNA, with the exception that adenine (A) on the DNA is now paired with uracil (U) on the newly forming mRNA molecule. Hydrogen bonds form between the bases of the template strand of DNA and the RNA nucleotides undergoing complementary base pairing. The chemical structures of the bases and the number and location of the hydrogen bonds they form ensure that these bonds can only form between specific bases.

D1.2 Figure 2 Transcription involves RNA polymerase, complementary base pairing, and the formation of hydrogen bonds to produce a mRNA molecule. Notice that the RNA includes the base uracil, in place of the thymine in DNA. mRNA is one of several types of RNA produced by transcription.

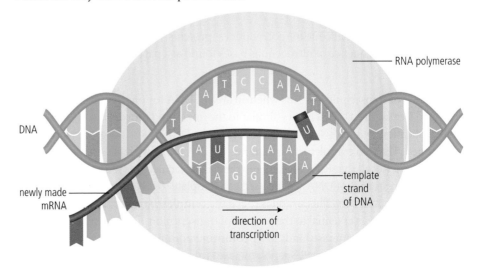

Study Figure 2 on the previous page and consider the following facts concerning transcription:

- the process of transcription produces all the RNA in a cell
- only one of the two strands of DNA is "copied", the other strand is not used
- RNA polymerase catalyses bonds between the RNA nucleotides as complementary base pairing occurs
- RNA is always single-stranded and shorter than the DNA that it is copied from, because it is a complimentary copy of only one gene
- the presence of thymine in a molecule identifies it as DNA (the presence of deoxyribose is another clue)
- the presence of uracil in a molecule identifies it as RNA (the presence of ribose is another clue).

What biological processes depend on hydrogen bonding?

D1.2.3 – DNA templates

D1.2.3 – **Stability of DNA templates**
Single DNA strands can be used as a template for transcribing a base sequence, without the DNA base sequence changing. In somatic cells that do not divide, such sequences must be conserved throughout the life of a cell.

Chapter A1.2 discusses the fact that DNA is a relatively stable molecule. It can be used as a template for the formation of RNA without undergoing any changes to its base sequence. This is especially important in **somatic** (body) cells such as nerve cells that do not participate in cell division but do produce RNA and need proteins throughout their life.

This stability of DNA may be compromised by, for example, the presence of free radicals, chemicals, cigarette smoke or exposure to ultraviolet (UV) radiation. Cells have specialized proteins that can detect and repair many instances of damage. Permanent changes to DNA are referred to as **mutations** and often negatively impact the cell's ability to produce essential proteins. However, not all mutations are harmful, and some may even contribute to the overall efficiency and/or survival of a species.

Mutations can lead to the development of cancer. An example of this occurs when genes coding for proteins involved in cellular growth mutate, resulting in cells growing and dividing out of control. Some of these mutations may even be heritable and passed on to future generations.

D1.2.4 – The expression of genes

D1.2.4 – **Transcription as a process required for the expression of genes**
Limit to understanding that not all genes in a cell are expressed at any given time and that transcription, being the first stage of gene expression, is a key stage at which expression of a gene can be switched on and off.

Transcription involves an area of DNA called a **gene**. A chromosome and, indeed, the genome of an organism, has thousands of genes. Not all genes go through transcription at the same time. Because of this we can say that not all genes are expressed simultaneously. Different genes are expressed at different times and at different developmental stages in an organism's life. Because transcription is the first step in the expression of a particular gene, it represents the mechanism by which gene expression in an organism can be turned on and off.

The DNA of the bacterium *Escherichia coli* has a total of about 4 million base pairs and nearly 3,000 genes, together known as its genome. Most bacteria have similar amounts of DNA and genes. Bacteria genomes are only 0.1% the size of ours, and they only have about 10% of the number of genes we do.

D1.2.5 – The synthesis of polypeptides

D1.2.5 – Translation as the synthesis of polypeptides from mRNA

The base sequence of mRNA is translated into the amino acid sequence of a polypeptide.

> A polypeptide is a polymer of amino acids linked by peptide bonds. A protein is usually more complex than a polypeptide, with folding of one or more polypeptide chains held together by non-covalent bonds.

The mRNA molecule produced by transcription represents a complimentary copy of one gene of DNA. The sequence of mRNA nucleotides is the transcribed version of the original DNA sequence. This sequence of nucleotides making up the length of the mRNA typically provides enough information to make one polypeptide. As you will recall, polypeptides are composed of amino acids covalently bonded together in a specific sequence.

Once the appropriate mRNA is produced from the DNA template by transcription, the process of producing the protein at the ribosomes can begin. **Translation** is the process by which the information encoded in mRNA directs the synthesis of specific amino acid sequences into polypeptides/proteins. We will be using the terms polypeptide and protein interchangeably as we describe translation.

The overall process that allows the production of a polypeptide/protein is represented by Figure 3.

D1.2 Figure 3 The template strand of DNA is transcribed to form mRNA, which is then translated to produce a polypeptide/protein. Transcription occurs in the nucleus, while translation occurs in the cytoplasm at the ribosome.

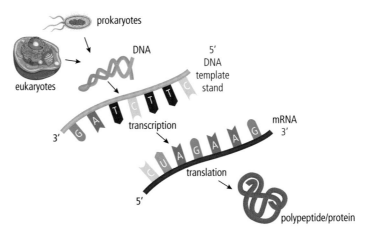

D1.2.6 – RNA and ribosomes

D1.2.6 – Roles of mRNA, ribosomes and tRNA in translation

Students should know that mRNA binds to the small subunit of the ribosome and that two tRNAs can bind simultaneously to the large subunit.

Three major types of RNA are involved in translation: **messenger RNA (mRNA)**, **transfer RNA (tRNA)** and **ribosomal RNA (rRNA)** (see Table 1).

D1.2 Table 1 The three major types of RNA

Type of RNA	Role in protein synthesis
Messenger (mRNA)	Carries the message from the DNA in the nucleus to the ribosomes in the cytoplasm
Transfer (tRNA)	Functions in the cytoplasm to carry amino acids to the ribosomes
Ribosomal (rRNA)	Combines with ribosomal proteins to construct the cytoplasmic ribosomes

All three types of RNA (Figure 4) are single stranded, and each is transcribed from a gene (a section of DNA). mRNA is a straight chain. rRNA along with proteins makes up the structure of the ribosome. tRNA has a unique shape.

messenger RNA (mRNA)   ribosomal RNA (rRNA)   transfer RNA (tRNA)

At the centre of the change from the language of DNA to the language of protein, i.e. translation, is the ribosome. Each ribosome consists of a large subunit and a small subunit. The subunits are composed of rRNA molecules and many distinct proteins. Chapter A2.2 discusses how prokaryotic ribosomes are smaller than eukaryotic ribosomes.

Other types of RNA have been found. Most are involved in protein synthesis, but there are a few that function in other ways. RNA could also have been the first genetic molecule on our planet.

Once an mRNA molecule has been transcribed, the mRNA detaches from the single-strand DNA template and floats free in the nucleoplasm. The mRNA will then pass through one of the many pores in the nuclear membrane and enter the cytoplasm. The translation process can then begin.

1. The mRNA binds with the small subunit of a ribosome.
2. A tRNA molecule with a specific amino acid attached now moves in, attaches to the large subunit of the ribosome and, through complementary base pairing, combines with mRNA.
3. A second tRNA with its amino acid follows the first tRNA, complementary base pairs with mRNA, and attaches to the large subunit of the ribosome. Two tRNAs can bind simultaneously to the large subunit.
4. An enzyme then catalyses a condensation reaction between the two amino acids, forming a peptide bond (Figure 5).

D1.2 Figure 5 A peptide bond, bracketed in the middle of this figure, forms when water is released. This process is called condensation. The amino acid on the right of the figure was the first amino acid brought to the ribosome.

5. The first tRNA then breaks free of its amino acid, detaches from the mRNA and floats away into the cytoplasm, where it can usually attach to another amino acid of the same type.
6. The ribosome then moves down the mRNA molecule.
7. The second tRNA molecule is now in the position that the first tRNA originally occupied.
8. A third tRNA floats in and pairs with the next sequence of bases on the mRNA.
9. Another peptide bond forms, and the process continues until the complete polypeptide is assembled.

Figure 6 summarizes the process of translation.

D1.2 Figure 6 Translation at the ribosome. Note the polypeptide chain being assembled, the mRNA, and the tRNAs involved in the process. The mRNA combines with the small subunit of the ribosome. Two tRNAs are capable of attaching simultaneously to the large ribosomal subunit.

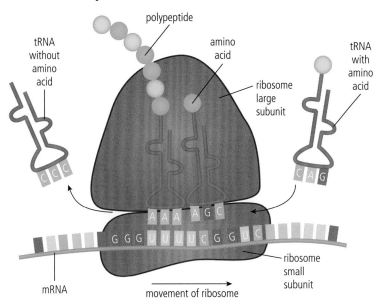

D1.2 Figure 7 The two-dimensional clover-leaf structure of tRNA, with three loops. The anticodon triplet is unique to each tRNA.

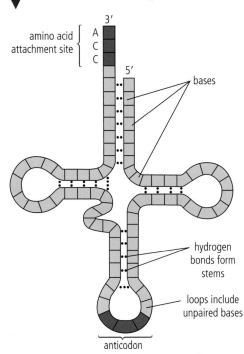

D1.2.7 – RNA complementary base pairing

D1.2.7 – **Complementary base pairing between tRNA and mRNA**
Include the terms "codon" and "anticodon".

The message written into the mRNA molecule determines the order of the amino acids when a protein is assembled.

A set of three bases provides the code for each of the 20 amino acids that make up proteins. Any set of three bases found in DNA that determines the identity of one amino acid is called a **triplet**. When a triplet is found in an mRNA molecule, it is called a **codon**.

DNA triplet → (transcription) → mRNA codon

Translation depends on complementary base pairing between codons on mRNA and anticodons on tRNA. Figure 7 shows a typical tRNA molecule. Notice that the three bases in the middle loop are called the **anticodon** bases and they determine which of the 20 amino acids is attached to the tRNA.

There are 20 different tRNA molecules, one for each of the different amino acids. Each tRNA can be differentiated by its anticodon. The anticodon of a tRNA is complementary to a codon of the mRNA. These match up during the translation process to produce the specific protein coded for by the mRNA.

D1.2.8 – The genetic code

D1.2.8 – Features of the genetic code
Students should understand the reasons for a triplet code. Students should use and understand the terms "degeneracy" and "universality".

Researchers have found experimentally that the genetic code is written in a language of three bases. In other words, a set of three bases contains enough information to code for each of the 20 amino acids. There are four possible bases in mRNA (A, U, C and G). In Chapter B1.2, which discusses proteins, it is explained that there are 20 possible amino acids in the human body. So, one base is inadequate to code for 20 different amino acids. If two bases coded for an amino acid, there would be only 16 different combinations possible. However, with three bases coding for an amino acid there are 64 possible combinations. This is an adequate number of combinations to provide a code for all the amino acids.

The genetic code is **degenerate**, which means that, for each amino acid, there may be more than one codon. Also, the genetic code is **universal**, which means that, with only a few minor exceptions, all organisms share the same code. It is this universal aspect of the code that allows us to insert genes from one species to another using genetic engineering techniques. Genetic engineering has enabled us to place the human insulin-coding gene into bacteria so that the bacteria can produce this protein for human use in the treatment of certain cases of diabetes. It is important to note that, even though the code is degenerate, it is not ambiguous or uncertain. A particular codon will always code for the same amino acid or a start or stop message.

A computer graphic of an insulin molecule. Insulin is a protein hormone produced by protein synthesis and is essential for the control of blood glucose levels. The complex structure of this molecule is dictated by DNA. The genetic code for the production of human insulin has been introduced into bacteria. As a result, the bacteria can produce human insulin because of the universality of the genetic code.

D1.2.9 – mRNA codons

D1.2.9 – Using the genetic code expressed as a table of mRNA codons
Students should be able to deduce the sequence of amino acids coded by an mRNA strand.

The genetic code can be expressed as a table of mRNA codons. Table 2 shows the meaning of the 64 different possible codons.

D1.2 Table 2 The genetic code. The first, second and third positions represent the base location in the codon. Twenty amino acids are coded for.

First position	Second position				Third position
	U	C	A	G	
U	Phenyl alenine	Serine	Tyrosine	Cysteine	U
U	Phenyl alenine	Serine	Tyrosine	Cysteine	C
U	Lencine	Serine	Stop	Stop	A
U	Lencine	Serine	Stop	Tryptophan	G
C	Lencine	Proline	Histidine	Arginine	U
C	Lencine	Proline	Histidine	Arginine	C
C	Lencine	Proline	Glutamine	Arginine	A
C	Lencine	Proline	Glutamine	Arginine	G
A	Isoleucine	Threonine	Asparagine	Serine	U
A	Isoleucine	Threonine	Asparagine	Serine	C
A	Isoleucine	Threonine	Lysine	Arginine	A
A	*Methionine	Threonine	Lysine	Arginine	G
G	Valine	Alanine	Aspartic acid		U
G	Valine	Alanine	Aspartic acid		C
G	Valine	Alanine	Glutamic acid		A
G	Valine	Alanine	Glutamic acid		G

*And start.

Several important points are illustrated in Table 2. There is a start codon (AUG) that signals the beginning of a polypeptide chain. This codon also encodes the amino acid methionine. Three codons are stop codons (UAA, UAG and UGA). These three codons have no complementary tRNA anticodon and signal the end of a polypeptide chain.

Challenge yourself

Deductions can be made using Table 2. Each amino acid has between 1 and 6 possible codons. Table 2 shows the position of the bases. The left-hand side of the table represents the first base in the codon, the second base is represented on the top of the table, and the third base is represented on the right-hand side of the table.

1. Using this information, deduce the sequence of amino acids coded for in the following example.

D1.2.10 – Producing a polypeptide chain

D1.2.10 – Stepwise movement of the ribosome along mRNA and linkage of amino acids by peptide bonding to the growing polypeptide chain

Focus on elongation of the polypeptide, rather than on initiation and termination.

The decoding of a strand of mRNA to produce a polypeptide occurs in the space between the two subunits of the ribosome. mRNA binds to the small subunit of the ribosome while tRNA with its attached amino acid binds to the large subunit of the ribosome (Figure 8). Essentially, the ribosome coordinates the functioning of the mRNA and the tRNA at the mRNA–ribosomal complex. The two ribosomal subunits hold the mRNA and tRNA close together so that amino acids can be connected by peptide bonds to produce the specific polypeptide required. This elongation process produces the polypeptide chain.

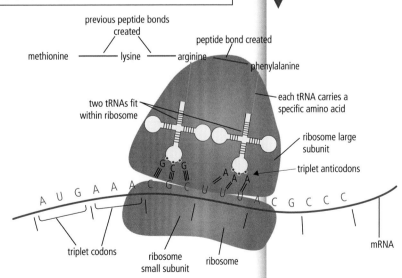

D1.2 Figure 8 The events of translation (synthesis of a polypeptide). The mRNA is held between the large and small subunits of the ribosome.

tRNAs carrying specific amino acids move sequentially through the binding sites of the ribosome as base pairing occurs between the tRNA anticodons and the mRNA codons, to create the exact sequence of amino acids needed. A continuous cycle of events occurs until the full polypeptide chain has been produced. To accomplish this, the mRNA is moved through the ribosome, one codon at a time.

D1.2.11 – Changing the protein structure

D1.2.11 – Mutations that change protein structure

Include an example of a point mutation affecting protein structure.

Normal red blood cells

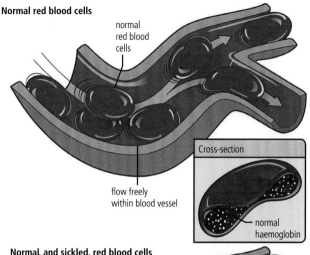

normal red blood cells

flow freely within blood vessel

Cross-section

normal haemoglobin

Normal, and sickled, red blood cells

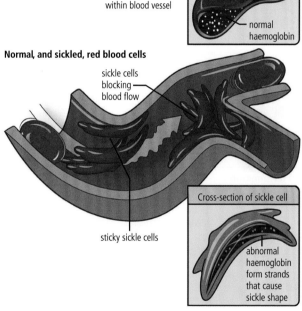

sickle cells blocking blood flow

sticky sickle cells

Cross-section of sickle cell

abnormal haemoglobin form strands that cause sickle shape

Mutations occur when permanent changes to DNA take place. There are many possible causes of mutations and they may be heritable. Even with high levels of accuracy, because of the sheer number of potential mistakes that could occur during DNA replication and protein synthesis, some errors or changes do happen. One type of mistake involving the genetic code is called a **point mutation**. A point mutation involves a change in only one base of a gene. This single nucleotide change alters the transcription and can change the specific amino acid produced and therefore affect protein structure.

A single point mutation is the cause of the genetic disorder known as sickle-cell disease. In this case a single base substitution causes a dramatic change in the shape of haemoglobin, the protein that carries oxygen in the blood. Normally, the shape of haemoglobin is such that red blood cells can move easily through our vascular system. However, with a point mutation of just one amino acid in the haemoglobin, the protein shape is changed. The change in shape of this protein leads to a change in the shape of the red blood cells. Figure 9 shows both normal red blood cells and sickle cell disease red blood cells.

◀ **D1.2 Figure 9** Sickle cell disease red blood cells and normal red blood cells. The abnormal shape of the red cells with the mutation means it is difficult for the cells to move through the body's blood vessels. This is a serious medical condition.

Guiding Question revisited

How does a cell produce a sequence of amino acids from a sequence of DNA bases?

In this chapter you have learned how:

- transcription and translation are involved in the process of producing proteins using a code carried by DNA
- three types of RNA are involved in the process of protein synthesis, mRNA, tRNA and rRNA
- there is a control mechanism for genes in a cell so that only those that need to be expressed are active at any given time
- the genetic code is both degenerate and universal

- transcription produces mRNA
- translation produces chains of amino acids known as polypeptides or protein
- transcription occurs in the nucleus while translation occurs in the cytoplasm at the ribosomes.

Guiding Question revisited

How is the reliability of protein synthesis ensured?

In this chapter we have discussed how:

- complementary base pairing is essential to the high reliability of the processes of transcription and translation, including the base pairing that occurs between codons and anticodons in the translation process
- a triplet code is necessary so that all possible amino acids can be utilized when producing a specific protein
- the sequence of amino acids in a protein can be deduced by examining the base code in an mRNA strand producing that protein
- amino acids are linked by peptide bonds in the elongating polypeptide chain being assembled at the ribosome during translation
- a mutation occurs when there is a change in the amino acid sequence of a protein being assembled at the ribosome
- a point mutation occurs when there is a change in only one base in a gene, sickle-cell disease is due to a point mutation that results in a dramatic change in the shape of haemoglobin present in red blood cells.

Exercises

Q1. Explain an example of a human disease caused by a point mutation.

Q2. Describe the functions of the three types of RNA involved in the translation process.

Q3. From the following DNA base sequence, determine the sequence of amino acids that would be assembled.

TACCGTCATAGAAAAATC

Q4. (a) On what structures are you most likely to find codons?

(b) On what structures are you most likely to find anticodons?

Q5. Describe two characteristics of the genetic code.

Q6. Compare the roles of hydrogen bonds and peptide bonds in protein synthesis.

D1.3 Mutation and gene editing

Guiding Questions

How do gene mutations occur?

What are the consequences of gene mutation?

Genes can be modified if one or more letters in a sequence is deleted, inserted or substituted for another letter. Sometimes mutations can be catastrophic for the organism and lead to severe health problems or death, but most have little effect. Some mutations make no difference at all to the organism while a few can be advantageous. Beneficial mutations create a new version of a gene that will result in a trait that gives the organism an advantage for survival.

D1.3.1 – Gene mutations

D1.3.1 – Gene mutations as structural changes to genes at the molecular level

Distinguish between substitutions, insertions and deletions.

A **mutation** is a random, rare change in genetic material. One type involves a change in the sequence of bases in DNA. If DNA replication works correctly, this should not happen (see Chapter D1.1). But nature sometimes makes mistakes. For example, the base thymine (T) might be put in the place of adenine (A) along the DNA sequence. When this happens, the corresponding bases along the messenger RNA (mRNA) are altered during transcription (see Chapter D1.2).

Three types of mutations are **substitutions**, **insertions** and **deletions**. A substitution happens when one letter is replaced by another. T instead of A, for example. An insertion happens when a letter is added, while a deletion happens when one letter is removed from a sequence. When only one nucleotide (base) is involved, it is called a **point mutation**. But sometimes two or more bases can be involved in a mutation, or even thousands of bases when a segment of a chromosome is involved. We will focus mostly on point mutations.

The **locus** (position) of the gene *GNAT2*, controlling a protein called transducin that enables colour vision in humans, is found on chromosome 1. A mutation of this gene stops an individual from being able to make the protein transducin properly. Transducin is needed to transmit information about colour from the eye to the brain; as a result of the mutation, a person cannot see in colour. This extremely rare genetic condition is called complete achromatopsia. When we say "the ability to see in colour is a genetic trait" we mean that one of two things can happen with a person's DNA: either that person has the DNA code for making colour vision possible, or the person does not have it. Figure 1 illustrates this concept.

A mutation is when one or more nucleotides of a DNA code are modified by mistake. In the case of a point mutation, a single letter can be switched during a substitution, removed during a deletion, or added during an insertion.

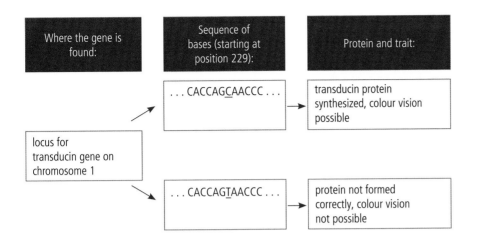

Worked example

Look carefully at the two sequences of DNA below. These sequences are from the coding strand of a section of genetic information that helps in the formation of haemoglobin, found in red blood cells. Identify the difference between the two sequences and complete the phrase below.

DNA sequence 1: GTG CAC CTG ACT CCT GAG GAG

DNA sequence 2: GTG CAC CTG ACT CCT GTG GAG

'Codon number __ along the first sequence has the letter __ in position number __, whereas the codon in the same position in sequence 2 has the letter __ instead.'

Solution

Codon number 6 along the first sequence has the letter A in position number 2, whereas the codon in the same position in sequence 2 has the letter T instead.

Worked example

Now look at the effect this has on the mRNA sequences produced from the template strand that is found opposite the coding strand when the DNA is unzipped for transcription:

mRNA sequence 1: GUG CAC CUG ACU CCU GAG GAG

mRNA sequence 2: GUG CAC CUG ACU CCU GUG GAG

Using the figure below, showing which codons are associated with which amino acids, and the mRNA sequences given above, fill in the names of the missing amino acids (a) to (h).

Codons and their associated amino acids. For example, the DNA code for lysine is AAA or AAG.

| Sequence 1: | valine | – | histidine | – | (a)_____ | – | (b)_____ | – | (c)_____ | – | (d)_____ | – | glutamic acid |

| Sequence 2: | valine | – | histidine | – | (e)_____ | – | (f)_____ | – | (g)_____ | – | (h)_____ | – | glutamic acid |

Using the figure and the mRNA sequences given above, can you find the missing amino acids?

Solution

| Sequence 1: | valine | – | histidine | – | leucine | – | threonine | – | proline | – | glutamic acid | – | glutamic acid |

| Sequence 2: | valine | – | histidine | – | leucine | – | threonine | – | proline | – | valine | – | glutamic acid |

Is this what you found? We will need these sequences later when we explore sickle cell disease.

Notice how the error of only one letter in the original DNA code changed the composition of amino acids in sequence 2. This would change the composition and the structure of the resulting protein, in the same way that changing the shape and composition of some of the bricks used to build a house would change the shape (and therefore the structural integrity) of the house. This kind of change in the DNA code is produced by a mutation.

D1.3.2 – Base substitutions

D1.3.2 – Consequences of base substitutions

Students should understand that single-nucleotide polymorphisms (SNPs) are the result of base substitution mutations and that because of the degeneracy of the genetic code they may or may not change a single amino acid in a polypeptide.

One base can make a big difference

The type of mutation that results in a single letter being changed is called a **base substitution mutation**. When a gene sequence is altered by one letter a **single-nucleotide polymorphism (SNP)** results. SNPs are of great interest to geneticists because they define different versions of genes and some can explain genetic diseases and cancers. The consequence of changing one base could mean that a different amino acid is placed in the growing polypeptide chain. This may have little or no effect on the organism, or it may have a major influence on the organism's physical characteristics (see Figure 2).

D1.3 Figure 2 Two possible consequences of a base substitution. In the first example, a different amino acid is coded for, which changes the shape and probably function of the protein. This is a **missence** mutation. In the second example, a stop codon is created that cuts off the gene translation, resulting in no functional protein being made. This is a **nonsense** mutation.

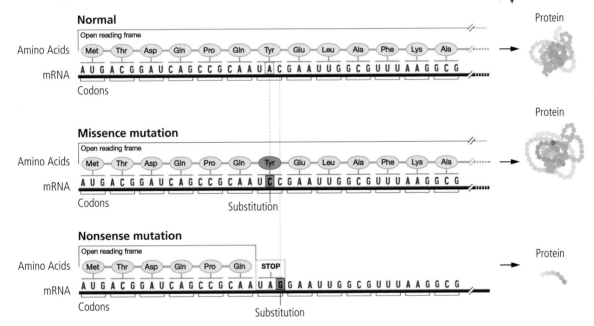

There is another possibility not shown in Figure 2. Sometimes changing a letter does not, in fact, change the resulting protein's composition, structure or resulting function. Because of the degenerate nature of the DNA code, the amino acid alanine can be coded for by any codon starting with CG_. CGA works, but if the codon is mutated to CGT, CGC or CGG, the amino acid alanine will still be translated in the final protein.

Sickle cell disease

In humans, a mutation is sometimes found in the gene that codes for haemoglobin in red blood cells. This mutation alters the shape of the haemoglobin molecule. The red blood cells do not look like the usual flattened disc with a hollow in the middle.

The mutated red blood cell has a characteristic curved shape, which made its discoverers think of a sickle (a curved knife used to cut tall plants). The condition that results from this mutation is therefore called **sickle cell disease**, also known as sickle cell anaemia.

The "sickle" shape of a red blood cell in someone with sickle cell anaemia.

Look at the sequences in Figure 3. The first is for the section of the haemoglobin gene's DNA that codes for standard-shaped red blood cells, and the second shows the mutation that leads to the sickle shape. In this case, one base is substituted for another so that the second codon in this sequence of haemoglobin, GAG, becomes GTG. As a result, during translation, instead of adding glutamic acid, which is the intended amino acid in the sixth position of the sequence, valine is added there instead. You can refer back to the Worked example to see this mutation.

D1.3 Figure 3 The cause of sickle cell diseasea

Because valine has a different shape and different properties compared to glutamic acid, the shape of the resulting polypeptide chain is modified. As a result of this, the haemoglobin molecule has different properties that cause the complications associated with sickle cell disease, such as weakness, fatigue and shortness of breath.

How does variation in subunit composition of polymers contribute to function?

Although sickle cell disease is a debilitating condition, those who have it are very resistant to malaria infection. Malaria is an infectious disease that occurs in tropical regions. A parasite of the genus *Plasmodium* is transmitted to human blood by an infected female mosquito of the genus *Anopheles* feeding on a human's blood. The parasite attacks the person's red blood cells and produces symptoms of high fever and chills, which can result in death.

D1.3.3 – Insertions and deletions

Consequences of insertions in the *HTT* gene

Huntington's disease is caused by a dominant allele (see Chapter D3.2 for how dominant alleles work). This genetic condition causes severely debilitating nerve damage, but symptoms do not show until a person is about 40 years old. As a result, someone who has the gene for Huntington's disease may not know that they have it when they are younger, before they have perhaps had their own children.

chromosome 4 — non-mutated HTT gene — non-mutated Htt protein — healthy neurons — healthy brain

10–26 CAG repeats

chromosome 4 — mutated HTT gene — mutated hHtt protein — neuron degeneration — atrophy of basal ganglia, enlargement of lateral ventricles

35 or more CAG repeats — expanded polyQ region

D1.3 Figure 4 Comparison of healthy neurons and a healthy brain to those of a person with Huntington's disease.

Huntington's disease is a life-limiting disease. Symptoms include difficulty walking, speaking and holding objects. Within a few years of the symptoms presenting, a person with Huntington's disease loses complete control of their muscles. This genetic disease is caused when a gene called huntingtin or *HTT*, found on chromosome 4, has an insertion mutation whereby multiple copies of three nucleotides, CAG, are added to the gene. This kind of mutation is known as a **trinucleotide repeat expansion** (see Figure 4).

The resulting mutated protein, mHtt, has an adverse effect on brain cells, causing the symptoms of the disease. Because the CAG trinucleotide repeats the code for the amino acid glutamine, there is much more glutamine in the mutated protein than the normal protein. The more trinucleotide repeats a person has, the more severe their symptoms can be. If a person has any more than 40 repeats, they will be affected. Huntington's disease is not the only polyglutamine disorder that can affect people's nervous system.

To help understand how mutations work, look at these variations of a sentence consisting of only three-letter words, similar to the three-nucleotide codons that code for amino acids:

1. **Sue did ask him why.** (The original, unmutated sentence.)
2. **Ued ida skh imw hy_.** (Deletion mutation: the first letter in the sequence has been removed.)
3. **ASu edi das khi mwh y_ _.** (Insertion mutation: the letter "A" has been added at the start.)
4. **Sue did not ask him why.** (Insertion mutation: the three letters "not" have been added in the middle.)
5. **She did ask him why.** (Substitution of one letter: an "h" has been added instead of "u".)

Grammatically sentences 1, 4 and 5 make sense. But removing or adding one letter and trying to keep the rule that all words must be three letters long results in nonsensical words in sentences 2 and 3 (see Frameshifts below). If there is an addition of letters in multiples of three, the code can still work, but it will be modified. Sentence 4 makes sense, but has a very different meaning compared to the original one. This is how some mutations can cause a genetic disease. A protein can be synthesized from the mutated code, but it might do something very different from the original code. Some substitution mutations have no effect on the result: sentence 5 means the same thing as sentence 1.

Frameshifts

A phenomenon that happens when an insertion or a deletion occurs in non-multiples of three is called a **frameshift**. Normally, the genetic code is read in triplets (codons), but if a letter is added to the sequence as a result of a mutation, the code is shifted. If the code is shifted by three new letters being added or three being deleted, all the other codons will remain unchanged. This is true for any multiple of three.

But if the insertion or deletion is not a multiple of three, the code changes drastically and can often end up not making sense anymore, like sentences 2 and 3 in the Hint for success. Or it could transform a normally coding codon into an unexpected stop codon. This kind of error can happen when the DNA proofreading system attempts to repair a mistake; the DNA polymerase can sometimes reattach in the wrong place along the sequence it is trying to repair.

Consequences of deletions

Leukocytes (white blood cells) need to move towards zones of infection in order to protect the body from disease-causing invaders such as viruses. There are chemical signals called **chemokines** that tell the leukocytes which way to go in order to find the invaders. Special proteins on the surface of the leukocytes act as receptors to pick up these chemicals and follow the message. A molecule that helps form such a receptor is a co-receptor molecule called **C-C chemokine receptor type 5**, or **CCR5**.

There is a virus that can use these receptors in a different way: as a point of entry to infect leukocytes called CD4 cells. This virus is HIV-1, the first of two types of human immunodeficiency virus. As a result, people who have a working set of *CCR5* genes on chromosome 3 and can make a fully functioning version of this protein for their CD4 cells are, unfortunately, at risk of the virus entering their cells if they are exposed to it. People who are HIV-positive slowly experience their leukocytes being destroyed by the virus, and without treatment they will eventually no longer be able to fight off other infections. At that stage, they have AIDS, acquired immunodeficiency syndrome.

There is a mutation of the *CCR5* receptor gene called the **delta 32 mutation**, or **CCR5-Δ32**. It is a deletion mutation whereby 32 nucleotides have been removed.

Because 32 is not a multiple of 3, this deletion causes a frameshift and a stop codon is accidentally formed where it should not be. Because the codon tells the ribosome to prematurely stop making the protein, people with this mutation cannot produce the functioning chemokine receptor protein that HIV needs to infect their leukocytes and, as a result, they are highly protected from HIV infection. Without a working CCR5-enabled doorway to infect the leukocytes, HIV cannot make a person sick.

 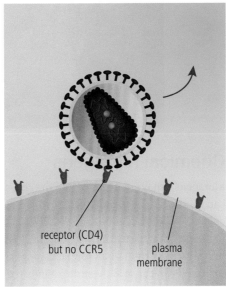

▲
How HIV can enter a cell when the CCR5 protein is present and how it is inhibited from entering the cell when it is absent.

Clearly, this mutation has a beneficial effect. HIV is a major health issue worldwide, and decades of research have not yet found a successful cure or vaccine. To have excellent protection against HIV-1 infections is good news for those who have the mutation. But the mutation can also make people more susceptible to other types of infection, such as the West Nile virus, which is transmitted by mosquitoes and can cause serious illness.

D1.3.4 – Mutagens and replication errors

D1.3.4 – Causes of gene mutation
Students should understand that gene mutation can be caused by mutagens and by errors in DNA replication or repair. Include examples of chemical mutagens and mutagenic forms of radiation.

In principle, DNA is not supposed to be modified during the lifetime of an individual. Normally, the code should be preserved. However, there are exceptions, and exposure to ionizing radiation (from ultraviolet light and radioactive substances) or to **mutagens** (chemicals that can cause a genetic mutation) can sometimes modify the code and cause serious health threats such as cancer. Over exposure to sunlight is linked to melanomas, for example.

An artist's conception of how DNA can be damaged by radiation.

Chemical mutagens

Mutagens can come from within the cell, such as certain enzymes that attack DNA and can transform the identity of a nitrogenous base. Or they can come from the outside environment, such as benzene (C_6H_6), a chemical used by industries to make other molecules that we use in our everyday lives, such as acetone (for nail polish remover), polystyrene (a rigid foam for packaging) and nylon fibres (for clothing). Benzene is toxic and can cause leukaemia, so industries must handle it carefully and minimize their employees' exposure to it, notably by providing masks that prevent its inhalation.

DNA and ionizing radiation

When radiation hits a DNA molecule, it can sometimes knock one or more base pairs out of place, modifying the genetic code. This causes a mutation that can sometimes be benign (not harmful) but at other times can be harmful to an organism. When the DNA mutation leads to cancer, as happened to Marie Curie, an organism's health is in jeopardy.

The world witnessed the terrifying effects of radiation poisoning when the city of Hiroshima in Japan was the target of the first atomic bomb used in warfare, in aAugust 1945. It is estimated that 100,000 people died at its impact or shortly after, but it is difficult to estimate how many in the city died later from the effects of radiation.

Marie Curie, who discovered the radioactive elements polonium and radium, did not benefit from the safety standards we have today, and died at the age of 66 from her exposure to radioactivity.

As long as nuclear power plants are safe and secure, there should not be any risk of radiation leaking into the environment. But the accidents at Chernobyl in 1986 and Fukushima in 2011 are examples of when people and the environment were exposed to radiation. In both cases, radioactive material leaked out and zones within a radius of tens of kilometres around the power plants were evacuated of all human populations. The disasters would have been much worse if it had not for the heroic efforts of scientists and workers to contain the situation as best they could, many of whom risked their own health and safety.

Scientists assessing the radiation levels in tree trunks near Chernobyl. Notice the dosimeter held by the person standing: it is used to measure radiation levels.

Ecologists are studying the area around Chernobyl to see how ecosystems have responded to the presence of radiation. In some instances, the scientists have been pleasantly surprised to find that nature seems to be doing fine despite the dangerously high radiation levels. In other instances, they have confirmed the presence of mutations in the plants and animals that have colonized the evacuated zone. Cancer studies in the peripheral zones where people are allowed to live, beyond 30 km from the shut-down Chernobyl reactor, suggest that there has been an increase in cancer frequencies. The nuclear power industry has made an effort to isolate the abandoned nuclear power plant at Chernobyl by encasing it in a dome of cement. The hope is that the cement will be thick enough to stop the radiation from continuing to escape into the environment.

Errors in DNA replication or repair

Mutations do not always have to be caused by toxic chemicals in the environment or by radioactive sources. Some mutations are purely random and can occur as easily as a spelling mistake when writing. During DNA replication, letters (nucleotides with nitrogenous bases) are added to the replicated strand and sometimes the wrong letter is added. Most of the time, the errors are corrected during a proofreading process carried out by DNA polymerase. The code is checked and incorrect bases are replaced with the expected ones in a fashion similar to the spellchecker on your phone or computer. Still, even after verification and correction, mistakes can happen. This is rare (in the order of one letter in a million) but does happen.

D1.3.5 – Location of mutations

D1.3.5 – Randomness in mutation

Students should understand that mutations can occur anywhere in the base sequences of a genome, although some bases have a higher probability of mutating than others. They should also understand that no natural mechanism is known for making a deliberate change to a particular base with the purpose of changing a trait.

Just as someone who buys two lottery tickets has twice as much chance of winning as someone who buys only one, some genes or sequences have more than one copy and therefore have a greater chance of being mutated. A gene that has two functioning copies can tolerate mutations in the extra copy without it having an effect on the cell, because the original copy still exists. Also, uncoiled DNA has a higher probability of suffering a mutation than DNA tightly coiled around histones, because the uncoiled DNA is more exposed.

When DNA (in darker blue) is packed closely with histones (in lighter yellow) it is better protected from mutations than when it is unravelled and more exposed to mutagenic forces.

Mutations can occur anywhere that sequences of nucleotides can be found. In humans that means they can occur on any of the 22 autosomal (non-sex) chromosomes or on the X or Y sex chromosomes. As only about 1–2% of our genome actually codes for proteins, it is far more likely for mutations to happen in the non-coding zones of the genome. However, such modifications can still sometimes have an effect on the production of proteins because these zones of the genome contain regulatory sequences that can turn coding sequences on and off. **Satellite DNA**, which can be found in the centromere, is an example of non-coding DNA and is used for structural purposes. Surprisingly high mutation rates occur in satellite DNA in humans (on average one mutation per 1,000 base pairs per generation) compared to coding DNA (about one mutation every 500,000,000 base pairs per generation).

Mutations can also be found in mitochondrial DNA and in RNA sequences. In addition to being found in humans, mutations are found in other animals, bacteria, fungi, archaea and plants, and in viruses, where they can be responsible for generating new strains of viruses that are more dangerous or can spread faster.

Another example of **mutation hotspots** (zones where mutations are more frequent) are places where the nucleotide cytosine is followed by guanine. These are called **CpG sites** and, when methylation happens, the C can mutate into a T. Places where these hotspots repeat are called **CpG islands** and they are more likely to generate mutations. Such mutations are associated with cancers such as colorectal cancer, and therefore researchers are interested in understanding them better.

C ⟶ T

cytosine
$C_4H_5N_3O$

thymine
$C_5H_6N_2O_2$

◀ Cytosine, when methylated, can sometimes be transformed into thymine when it is deaminated. This is a common mutation.

Can cells invent a mutation on purpose in order to improve?

It would be nice if cells could say, "If I change that C to a T, I could solve this problem I am having", in the same way that a cook changes a recipe or an architect changes a blueprint. No such intentional mechanism for genomic self-improvement has been detected in cells. The DNA repair system can proofread for errors and fix problems to bring the code back to its original sequence, but there is no system to introduce new variations on purpose. Instead, variations only arise as a result of mutations, and then those changes in the genetic sequence are selected for (or against) by natural selection.

TOK

One of the most common misconceptions about evolution is that it is somehow aiming for perfection, or that there is a force or desire to improve or to follow a plan. This is a tempting conclusion because we perceive ourselves as being the culmination of evolution. In his book *The Accidental Species*, Henry Gee points out that evolution has no plan: "natural selection cannot be seen as evolution's guiding hand. It has no personality, no memory, no foresight and no end in view". He uses the example of feathers, which did not originally evolve for flight but, because birds use them today for flying, we jump to a conclusion and say how logical and marvellous it was that feathers evolved in order to make flight possible. Gee reminds us of Stephen Jay Gould's thought experiment, imagining undoing millions of years of evolution and then pressing the "play" button to see what would happen if natural selection could start over: we would not see the same creatures evolve on Earth as we have today. How do our expectations and assumptions have an impact on how we perceive things?

D1.3.6 – Mutations in germ cells and somatic cells

D1.3.6 – Consequences of mutation in germ cells and somatic cells

Include inheritance of mutated genes in germ cells and cancer in somatic cells.

Mutations can have very different effects depending on which type of cell they occur in: **germ** cells or **somatic** cells. Somatic cells use mitosis for cell division to grow tissues and organs all over the body, whereas germ cells use meiosis to produce gametes (sperm cells or egg cells).

Germ cells use meiosis to produce egg cells in ovaries and sperm cells in testes. Mutations that happen in germ cells can be passed on to the next generation.

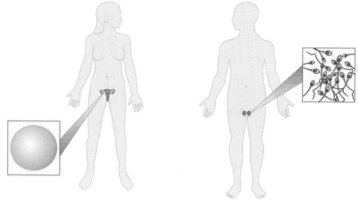

If a mutation happens in germ cells, it can be passed on to a future child. A mutation present in the cells making a male's sperm cells could potentially contribute that mutation to the first cell, the zygote, of a child. As all the child's other cells will be generated from that zygote, they will all contain the mutation. If a female child that has inherited a mutation grows up and has children, the germ cells in her ovaries will contain the mutation and will pass it on to the next generation. Cells involved in passing on genetic information to offspring make up the organism's **germ line**: eggs, sperm cells and zygotes. Each genetic disease that exists in the human population today, such as sickle cell disease, originated as a mutation in the past that was then passed down from generation to generation. Likewise, genetic mutations that were beneficial, such as those that gave us bigger brains or a better-functioning immune system, were also passed down. Remember, mutations are not automatically bad for our health.

If, on the other hand, a mutation happens in a somatic cell, it will not be passed on to the next generation. Mutations in somatic cells are associated with cancer and tumours. These mutations will affect the organ where the mutated cell is found and, as the tumour grows, the mutation will be found in all the cells of the tumour. If the cancer metastasises (spreads out), the cancerous cells can colonize other parts of the body, and each of them will contain the mutation. But the cells that already exist in the parts of the body being invaded by the cancer do not possess the mutation.

Gametes are sex cells, either sperm cells or egg cells. A zygote is a fertilized egg. Somatic cells are cells that are not gametes: they are the cells that make up the tissues and organs in the rest of the body.

If a mutation happens in a somatic cell, for example a skin cell, the mutation will be present in the daughter cells that are produced by mitosis. Such mutations, like the ones that can cause skin cancer, only affect the individual.

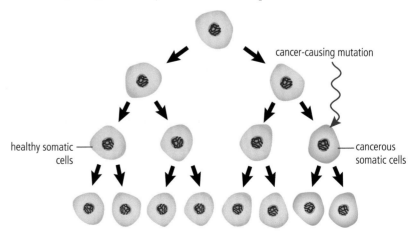

healthy somatic cells — cancer-causing mutation — cancerous somatic cells

- Somatic cell mutations only affect the individual; they are not passed on to future generations.
- Germ cell mutations affect the individual and their offspring; they can be passed on to future generations.

D1.3.7 – Genetic variation

D1.3.7 – Mutation as a source of genetic variation

Students should appreciate that gene mutation is the original source of all genetic variation. Although most mutations are either harmful or neutral for an individual organism, in a species they are in the long term essential for evolution by natural selection.

NOS: Commercial genetic tests can yield information about potential future health and disease risk. One possible impact is that, without expert interpretation, this information could be problematic.

Are mutations good or bad for us?

A mutation that provides an individual or a species with a better chance for survival is considered to be a beneficial mutation, and there is a good chance that it will be passed on to the next generation. In contrast, mutations that cause disease or death are detrimental mutations, and they are less likely to be passed on to future generations, because they decrease the chances of an individual's survival. In addition to beneficial and harmful mutations, there are neutral mutations that do not have an effect on a species' health or chances for survival.

When a mutation is successfully passed on from one generation to the next, it becomes a new **allele**: it is a new version of the original gene. See Chapter D3.2 for more about alleles. This is how new alleles are produced. You and everyone you know possess many mutations. Whether they are harmful, beneficial or neutral depends on what they are and what kind of environment you need to survive in.

It is important for new alleles to be produced in order for evolution to occur. If there were never any mutations, we would only be able to pass on existing genes and nothing new would appear. The problem is that nature and the environment are in constant flux: things are changing all the time and organisms need to adapt. Droughts, temperature changes, alterations in ocean salinity and pH, are all factors that can pose a threat to organisms on land or in the oceans, and having some individuals in a population that possess a slightly different genome can help species survive the changes.

A mutation to help digestion

For most of our existence, humans have been hunter-gatherers and our genes are generally well adapted for this lifestyle. Originally, as for all mammals, the only age at which we drank milk was when we were infants. By the time our ancestors reached adulthood, their bodies had stopped being able to digest milk; more precisely, they could not break down the disaccharide in milk called lactose. This continues to be the case for most people today: more than half of the human population has lactose intolerance, and such people can only digest lactose in their infancy.

In the past 10,000 years, however, many human populations have adopted a lifestyle based more on agriculture, raising animals for milk and consuming dairy products on a daily basis. In their genetic makeup, dairy-based agricultural societies show a higher frequency of the genetic code that allows humans to digest lactose throughout adulthood. From an evolutionary point of view, this advantage has increased humans' ability to survive harsh climatic conditions. As these European human populations spread out and established populations elsewhere, notably in North America, they took their lactose tolerance (and their livestock) with them.

▲ In parts of the world where dairy farming had been carried out for thousands of years, genes that enabled adults to digest lactose gave people there a survival advantage.

419

Nature of Science

How can you know if you possess a particular gene? In countries where this practice is allowed by law, commercial companies can offer genetic testing kits and provide genome information. Different companies provide different services, such as medical information about the presence of genes that could lead to diseases such as Huntington's or might increase a person's risk of cancer or cardiovascular disease. In association with a doctor who can help interpret what the information means, answer questions and provide advice and medical treatments, such information can be useful to have. But if the information is received via email or text message without a proper context and interpretation, it can be emotionally devastating; the person receiving the information might not understand how to interpret the results.

In addition to medical information, many of these companies provide genealogical data about a person's ancestry. Again, sometimes the news is unexpected and people might learn that their parents or other ancestors were not who they thought they were. This raises many ethical questions, which is one of the reasons why some countries, such as France, have passed laws against using such DNA testing services for paternity and ancestry.

▲ Getting a DNA test can reveal things about your health and origins. But people should consult their doctor to help interpret the medical information and get appropriate advice.

How can natural selection lead to both a reduction in variation and an increase in biological diversity?

Guiding Question revisited

How do gene mutations occur?

In this chapter you have learned that:

- gene mutations are random and can involve the addition of one or more nucleotides, or can involve deletions or substitutions
- mutations can happen during DNA replication but many are corrected in a proofreading process carried out by DNA polymerase
- mutations can also happen as a result of exposure to chemical mutagens or ionizing radiation.

Guiding Question revisited

What are the consequences of gene mutation?

In this chapter you have learned how:

- some mutations affect the germ line and can be passed down to future generations, whereas others only affect the somatic cells in one individual
- some mutations have a negative effect, some positive, and some have no effect at all on an organism
- natural selection will favour newly formed alleles that help an organism's survival, or reduce the frequency of those that are harmful.

Exercises

Q1. Which terms can be used to describe a mutation that leads to cancer?

    I.   Somatic.

    II. Germ line.

    III. Non-heritable.

  A I and II only.

  B I and III only.

  C II and III only.

  D I, II and III.

Q2. Which sentences can be used to describe the *CCR5 delta 32* mutation?

    I.   It is a single base substitution.

    II. It affects the immune system.

    III. It can protect someone from HIV.

  A I and II only.

  B I and III only.

  C II and III only.

  D I, II and III.

Q3. Mutations can happen in two types of cells, somatic cells and germ cells.

  (a) Distinguish between the two types of cells in terms of the type of cell division they are capable of and where they are found in the human body.

  (b) Distinguish between the consequences of a mutation in each type of cell.

Q4. Give an example of a chemical mutagen and one mutagenic form of radiation.

D1 Practice questions

1. Which of the following is an inherited disease as a result of a base substitution mutation in a gene?

   A   Trisomy 21.

   B   Sickle cell anaemia.

   C   AIDS.

   D   Type II diabetes.

   (Total 1 mark)

2. During the process of replication, which bond(s) in the diagram of DNA below is/are broken?

   A   3

   B   4, 5

   C   1, 2, 6, 7

   D   1, 7, 4, 5

   (Total 1 mark)

3. The diagram below shows a short section of DNA molecule before and after replication. If the nucleotides used to replicate the DNA were radioactive, which strands in the replicated molecules would be radioactive?

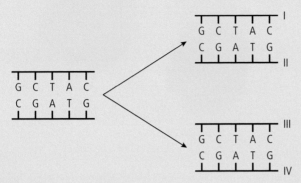

    A II and III only.

    B I and III only.

    C I and II only.

    D I, II, III and IV.

(Total 1 mark)

4. Outline the process of translation.

(Total 5 marks)

5. A certain gene codes for a polypeptide that is 120 amino acids long. Approximately how many nucleotides long is the mRNA that codes for this polypeptide likely to be?

    A 30

    B 40

    C 360

    D 480

(Total 1 mark)

6. Compare DNA transcription with translation.

(Total 4 marks)

7. Describe the genetic code.

(Total 6 marks)

8. Outline how single-nucleotide polymorphisms generate new alleles.

(Total 3 marks)

9. Explain the significance of insertions and deletions happening in multiples of 3.

(Total 4 marks)

THEME **D** **Continuity and change**
2 Cells

Cells need to be able to divide to ensure the growth and repair of tissue but also to prepare for the next generation. These cells are in various stages of cell division; their chromosomes have been dyed red.

One of the defining characteristics of life is the ability to grow. This requires making copies of cells. Cell division involves a series of steps to make sure that the DNA code is copied into each new cell. In the production of sperm and egg cells, only half of the parent's DNA is passed on, so a special version of cell division is necessary.

Cell division is just one of the many processes that cells carry out. In order for a cell to maintain all the essential processes of life, it needs to maintain the correct balance of water and solutes. The movement of water from an area of low solute concentration to an area of higher solute concentration through a semipermeable membrane is called osmosis. Experiments can be done in the laboratory that involve placing plant tissues in different solutions to see how water flows in or out of the cells. Understanding how water interacts with cells can be applied to medicine, such as knowing what concentration of solution to bathe transplant organs in.

D2.1 Cell and nuclear division

Guiding Questions

How can large numbers of genetically identical cells be produced?

How do eukaryotes produce genetically varied cells that can develop into gametes?

Cells can divide, and make copies of themselves, in two ways. Mitosis produces two genetically identical cells; this type of cell division is used for growth and repair. Most cells in the body replicate using mitosis. In a developing embryo, after one division there are two cells, but after six divisions there are 64 cells, and after 20 divisions there are over a million cells.

The other type of cell division is called meiosis; this type of cell division is only used for the production of one category of cells, the sex cells. When egg cells or sperm cells are generated, one cell divides twice to make four cells, but each of the four has two exceptional characteristics that make them different from other cells in the body and different from each other. Firstly, each sex cell only contains half the genetic information from the parent cell, and will not have a complete set of DNA until it fuses with another sex cell. Secondly, each of the four cells has a different combination of hereditary information from the parent cell. This increases the genetic variety in the offspring.

D2.1.1 – Generating new cells

D2.1.1 – Generation of new cells in living organisms by cell division
In all living organisms, a parent cell – often referred to as a mother cell – divides to produce two daughter cells.

In order to maintain the population of a single-celled organism, and in order to keep a multicellular organism growing and repairing itself, cells need to make copies of themselves. The process of producing two cells from one is called **cell division**. The role of cell division is to make sure genetic information is passed on to the next generation of cells along with copies of all the organelles necessary to make the cell function.

The cell that produces a copy is called the **parent cell** or **mother cell** and the two new cells that are generated are called the **daughter cells**. This is one of the principles of cell theory: all cells are made from pre-existing cells.

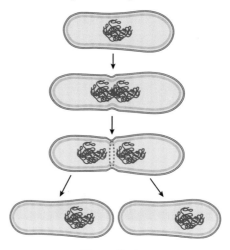

Binary Fission

▲

The process of binary fission in prokaryotes. The single DNA molecule is copied before the two cells split apart.

> Mitosis produces genetically identical daughter cells, which are necessary for organisms to grow and to repair themselves. Meiosis produces sex cells, which are needed for fertilization and have half the amount of DNA of any other cell in the body.

With only a single chromosome of DNA to replicate, prokaryotic cells divide by **binary fission**. During this process, the DNA is copied, the two daughter chromosomes become attached to different regions on the plasma membrane, and the cell divides into two genetically identical daughter cells. For organisms that have multiple chromosomes, the daughter cells are produced using **mitosis**. For sexually reproducing organisms, egg cells and sperm cells require two divisions that will produce four daughter cells in a process called **meiosis**.

D2.1.2 – Cytokinesis

D2.1.2 – Cytokinesis as splitting of cytoplasm in a parent cell between daughter cells

Students should appreciate that in an animal cell a ring of contractile actin and myosin proteins pinches a cell membrane together to split the cytoplasm, whereas in a plant cell vesicles assemble sections of membrane and cell wall to achieve splitting.

After a cell reaches a certain size, it needs to split in two, a process called **cytokinesis**. The process differs depending on the type of cell. In animal cells, cytokinesis involves an inwards pinching of the fluid plasma membrane to form **cleavage furrows**, a groove along the cell membrane. However, plant cells have a relatively rigid cell wall and during cytokinesis they form a **cell plate** instead. The cell plate is built up by vesicles that collect midway between the two poles of the cell and lay down cell membrane and cell wall cells, which then expand outwards towards the sides of the cell from a central region. Both types of cytokinesis result in two separate daughter cells that have genetically identical nuclei.

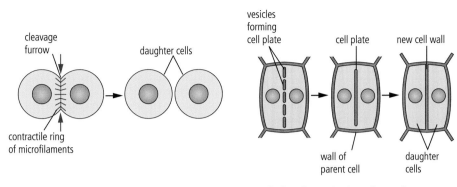

Cleavage of an animal cell   **Cell plate formation in a plant cell**

◀ Cytokinesis in animal and plant cells

D2.1.3 – Cytoplasm division

D2.1.3 – Equal and unequal cytokinesis

Include the idea that division of cytoplasm is usually, but not in all cases, equal and that both daughter cells must receive at least one mitochondrion and any other organelle that can only be made by dividing a pre-existing structure. Include oogenesis in humans and budding in yeast as examples of unequal cytokinesis.

In most instances of cell division, the two daughter cells are identical. This is necessary for the process of growth and repair. Each cell receives a full copy of the parent cell's DNA and some of the essential organelles. If all the mitochondria from the parent cell ended up in only one of the two daughter cells, the other would not be able to survive. So at least one mitochondrion from the eukaryotic parent cell needs to end up in each daughter cell to ensure its survival. The same is true of chloroplasts in photosynthetic eukaryotic cells.

But sometimes the daughter cells are not identical, and there is unequal sharing of the parent cell's resources. One example of this is found in the production of eggs, called **oogenesis**. Oogenesis produces four haploid cells. Three of the four cells donate their cytoplasm and organelles to the fourth cell and are not used as eggs because they are much too small to produce a viable **zygote** (fertilized egg). This **unequal cytokinesis** provides the zygote with the resources it needs to survive until it is implanted in the wall of the uterus.

Cytokinesis can also be unequal in yeast cells. Yeast cells divide using a process called **budding**, which involves generating a small cell from the parent cell. When the daughter cell becomes big enough to survive on its own, cytokinesis closes the cell membranes and each cell is an independent organism.

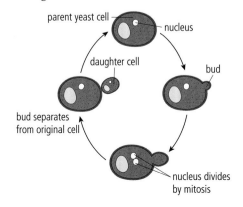

◀ The process of budding in yeast cells is an example of unequal cytokinesis. The daughter cell produced is smaller than the parent cell because of an uneven distribution of resources from the parent cell.

Before cells divide by either mitosis or meiosis, nuclear division must take place: the DNA must be replicated. After replication, each chromosome is made of two strands of DNA held together with a centromere.

D2.1 Figure 1 Chromosomes make a copy of themselves before cell division happens. In real life, before it is copied the single chromosome does not look as depicted here: it would not be coiled up, but would be unwound so that it can undergo replication.

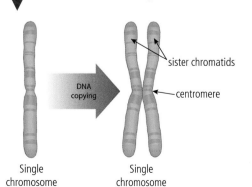

Single chromosome with one copy of DNA

DNA copying

Single chromosome with two copies of DNA

sister chromatids

centromere

D2.1.4 – Nuclear division

D2.1.4 – Roles of mitosis and meiosis in eukaryotes

Emphasize that nuclear division is needed before cell division to avoid production of anucleate cells. Mitosis maintains the chromosome number and genome of cells, whereas meiosis halves the chromosome number and generates genetic diversity.

Mitosis is a type of cell division that results in two daughter cells, each with identical nuclei. Imagine what would happen if a cell divides before it makes a copy of its nucleus. Only one of the daughter cells would have a nucleus. The other cell would be **anucleate**: it would not have the instructions for carrying out its functions or how to divide. The process of mitosis makes sure a full copy of the nucleus is made *before* cytokinesis happens. As a result, each daughter cell not only has the same number of chromosomes as the parent cell, but the same genome. All the genetic information is preserved.

In **gametes**, or sex cells, it is necessary for each sperm and egg cell to have only half the genetic information of the parent. Some organisms, notably certain plants, can thrive with extra copies of chromosomes, but usually animals cannot.

In order to receive 46 chromosomes in total, a human baby needs to receive 23 chromosomes from their mother and 23 from their father. This is why chromosomes come in pairs; one of each pair is from each parent. To ensure that egg and sperm cells only get half the genetic information, a special type of cell division is needed: **meiosis**. Meiosis is a type of cell division that results in four daughter cells that each has a nucleus containing only half the parent cell's DNA. And each of the four receives a different *combination* of genetic information. It is extremely rare, for example, to find two sperm cells from the same male that have the same genetic profile. By putting random combinations of chromosomes into each sperm or egg cell, meiosis helps generate the genetic variety we observe in offspring.

D2.1.5 – DNA replication

D2.1.5 – DNA replication as a prerequisite for both mitosis and meiosis

Students should understand that, after replication, each chromosome consists of two elongated DNA molecules (chromatids) held together by a centromere.

Whether it is in preparation for meiosis or mitosis, cell division cannot happen until a complete copy of the cell's DNA has been made. This process is called **DNA replication**, (see Chapter D1.1). It happens during a phase of the cell's life called the **S phase**, or synthesis phase. The replicated DNA is arranged in a chromosome as two sister **chromatids**, shown as the vertical elongated arms in Figure 1. The chromatid on the left side of the copied chromosome is *one* DNA molecule coiled up. The chromatid on the right side is another, identical, DNA molecule. This can be seen in the banding patterns, which match from top to bottom. The two sister chromatids are attached at the **centromere**. During division, the two sister chromatids are pulled apart. At that moment, they each become an individual chromosome, one for each daughter cell.

D2.1.6 – DNA condensation and chromosome movement

D2.1.6 – Condensation and movement of chromosomes as shared features of mitosis and meiosis

Include the role of histones in the condensation of DNA by supercoiling and the use of microtubules and microtubule motors to move chromosomes.

Preparing DNA for separation

During most of the life of a cell, the DNA is not bunched up as shown in Figure 1. Rather, it is spread out in long unwound chains, which might resemble a plate of spaghetti, inside the nucleus. In order not to misplace any DNA and to prevent the strands tangling up and breaking, condensation of the DNA is necessary before replication. Just as passengers need to gather on the correct platform to get on the right train, DNA molecules group together. The process of condensation (Figure 2) involves the DNA being wrapped around proteins called **histones**. Histones help organize the DNA, which is coiled and then **supercoiled**, so that the coils are stacked on top of each other and form a compact pair of chromatids. When DNA is associated with histone proteins, it is referred to as **chromatin**.

When the supercoiling is complete, the chromosome takes the shape we are familiar with in micrographs. This condensed structure ensures that different DNA molecules can be transported in one package rather than being spread out all through the nucleus.

D2.1 Figure 2 The familiar shape of chromosomes is the result of DNA (the strand) being coiled around histone proteins (the spheres) and then forming a supercoiled shape. Start on the right side of the figure with unwound DNA, and progress to the left to see how it is wound around the histones to become supercoiled.

▼

700 nm   300 nm   30 nm   10 nm   2 nm

Movement of chromosomes

The **centrosome** is a cell organelle described in Chapter A2.2. This organelle makes the microtubule spindle fibres that are needed to guide the chromosomes to the right place before the cells can divide. Microtubules can be constructed and disassembled as needed, and they have a directionality because one end has a negative charge and the other a positive charge.

Specialized molecules called **motor proteins** push or pull objects around a cell. They use the microtubules as tracks, or they can attach to two microtubules and get one to slide past the other (like muscle fibres do when they are contracting). Motor proteins use adenosine triphosphate (ATP) to produce a **conformational change** (a change in shape) that moves the microtubules.

D2.1 Figure 3 The role of microtubules and motor proteins in the movement of chromosomes during cell division

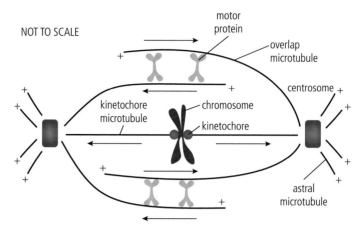

Figure 3 shows a chromosome in the middle of a cell, and three types of microtubule. The **astral microtubules** reach out from the centrosome. The **kinetochore microtubules** attach to the centromere of the chromosome, where the two sister chromatids are attached. The **overlap microtubules** are not attached to the chromosomes but rather pass beside them. Sandwiched between the overlap microtubules are **motor proteins**. These "walk" along the microtubules in such a way that the microtubules are pushed in opposite directions.

When a cell is ready to separate its chromosomes, the motor proteins between the overlapping microtubules become active. The action of two microtubules sliding past each other pushes the two poles of the centrosome away from each other. Because the two sister chromatids are attached to the poles via opposite-facing microtubules, they will be pulled away from each other. Each will be transported to one half of the cell, so that when the cell splits in two each newly formed cell has a copy of each chromosome. Once the chromatids have been separated, the microtubules are dismantled and the pieces of track recycled.

In order to ensure that the DNA is replicated successfully, DNA is condensed and coiled. Supercoiled chromosomes are moved around cells by a system of microtubules and microtubule motors.

D2.1.7 – Mitosis

D2.1.7 – Phases of mitosis

Students should know the names of the phases and how the process as a whole produces two genetically identical daughter cells.

When a cell is not going through cell division, it is in a phase called **interphase**. During interphase, the cell is performing its function in the organism as well as growing and preparing to divide. When the cell is ready, mitosis can begin. Mitosis is used to produce two identical cells, and involves four phases in the sequence:

- prophase
- metaphase
- anaphase
- telophase.

To help memorize the order of things, use mnemonics. You can use IPMAT for interphase and the four stages of mitosis. Make up a sentence with the first letter of each, such as "impatient people mostly arrived today".

Prophase

Figure 4 illustrates prophase. During prophase:

1. the chromatin fibres become more tightly coiled to form chromosomes
2. the nuclear envelope disintegrates and nucleoli disappear
3. a **mitotic spindle** forms as the centrosome builds new microtubules that will be used to pull the chromosomes into position
4. the **kinetochores**, a region in the centromere of each chromosome, attach to the spindle
5. the centrosomes move towards the opposite poles of the cell, as a result of lengthening microtubules.

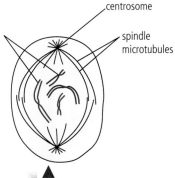

D2.1 Figure 4 This animal cell is in prophase. For clarity, only a small number of chromosomes is shown.

Metaphase

Figure 5 shows a cell in metaphase. During metaphase:

1. the chromosomes move to the middle or equator of the cell, which is called the **metaphase plate**
2. the centromeres of the chromosome align on the plate
3. the chromosomes move as a result of the action of the spindle, which is made of microtubules
4. the centrosomes are at opposite poles.

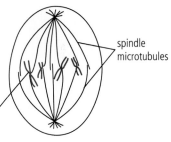

D2.1 Figure 5 The cell is now in metaphase. Again, only a small number of chromosomes is shown.

Anaphase

Figure 6 shows the cell in anaphase. This is usually the shortest phase of mitosis. It begins when the two sister chromatids of each chromosome move apart. During anaphase:

1. the chromatids, now each a chromosome, move towards the opposite poles of the cell
2. they move as a result of motor proteins pushing microtubules in opposing directions
3. because the centromeres are attached to the microtubules, they move towards the poles first
4. at the end of this phase, each pole of the cell has a complete, identical set of chromosomes.

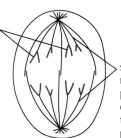

D2.1 Figure 6 The cell is now in anaphase. Again, for clarity, only a small number of chromosomes is shown.

Telophase

Figure 7 shows the cell in telophase. During telophase:

1. a set of chromosomes is located at each pole
2. a nuclear membrane (envelope) begins to re-form around each set of chromosomes
3. the chromosomes start to elongate
4. nucleoli reappear
5. the spindle apparatus disappears
6. the cell is elongated and ready for cytokinesis.

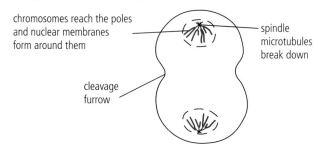

chromosomes reach the poles and nuclear membranes form around them

spindle microtubules break down

cleavage furrow

▲ **D2.1 Figure 7** Finally, the cell enters telophase.

The IB biology syllabus focuses on the main phases of cell division, but intermediate stages exist such as "prometaphase" between prophase and metaphase. This makes you wonder how we decide where to draw the line when separating the different steps of a repeating process. When is it better to adopt broader terminology and when is it more helpful to devise more specific phases? Are some types of knowledge less open to interpretation than others?

TOK

D2.1.8 – Identifying the phases of mitosis

What processes support the growth of organisms?

D2.1.8 – Identification of phases of mitosis
Application of skills: Students should do this using diagrams as well as with cells viewed with a microscope or in a micrograph.

The IB expects you to be able to identify the different stages of mitosis from both diagrams and photomicrographs. Try the following activities.

SKILLS

1. Name the stages labelled 2, 4, 5 and 6 in the left-hand figure below.
2. Identify which stage of mitosis the numbered cells in the micrograph in the right-hand figure below are going through.

   prophase__   metaphase __   anaphase __   telophase __

▲ A cell undergoing mitosis.

▲ A photo taken through a light microscope of cells going through various stages of mitosis.

See the eBook for a laboratory activity where you can observe cells undergoing mitosis in a root tip.

D2.1.9 – Meiosis

D2.1.9 – Meiosis as a reduction division

Students should understand the terms "diploid" and "haploid" and how the two divisions of meiosis produce four haploid nuclei from one diploid nucleus. They should also understand the need for meiosis in a sexual life cycle. Students should be able to outline the two rounds of segregation in meiosis.

D2.1 Figure 8 How the chromosome number is halved during meiosis. In this example, the organism has two pairs of chromosomes.

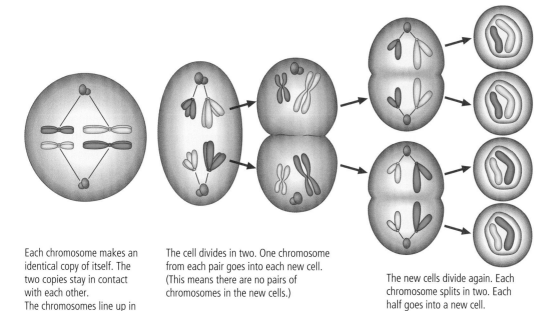

Each chromosome makes an identical copy of itself. The two copies stay in contact with each other.
The chromosomes line up in pairs. Each pair codes for the same characteristics.

The cell divides in two. One chromosome from each pair goes into each new cell. (This means there are no pairs of chromosomes in the new cells.)

The new cells divide again. Each chromosome splits in two. Each half goes into a new cell. This makes four haploid cells. These develop into gametes.

Gametes (sperm cells and egg cells) cannot contain a full set of pairs of chromosomes for the simple reason that, if they did, each new generation would double the chromosome number, creating an impossibly large amount of DNA to deal with. To avoid the problem of accumulating too many chromosomes, humans and other animals produce egg cells and sperm cells in such a way that the number of chromosomes in their nuclei is halved. Hence human sperm and eggs only contain 23 chromosomes, one from each pair, rather than 46 chromosomes arranged in 23 pairs. In order to make such special cells with half the chromosomes, a special type of cell division is needed. This type of cell division is called **meiosis**, and it is a **reduction division**.

Whereas mitosis produces **diploid** ($2n$) nuclei containing 46 chromosomes in humans (organized into 23 pairs), meiosis produces **haploid** (n) nuclei that contain 23 chromosomes, each representing half of one pair. Notice in Figure 8, from a single cell on the left, four cells have been produced on the right. Notice also that the number of chromosomes in this non-human example is four in the parent cell (so $2n = 4$) and two (so $n = 2$) in the daughter cells.

In the human testes and ovaries, respectively, meiosis produces haploid sperm and eggs, so that, when fertilization occurs, the zygote will receive 23 + 23 = 46 chromosomes; half from the mother, and half the father. This is how the problem of changing chromosome numbers is avoided. As a result, the human chromosome number of 46 is preserved by the sexual life cycle (Figure 9).

D2.1 Figure 9 How chromosome number is maintained in the sexual life cycle of humans.

23   23

fertilization

46

zygote – first cell
of a new baby

First round of segregation: halving the chromosome number

Meiosis involves two rounds of segregation (shown in Figures 10 and 11), which take place in a series of steps. During the first round of segregation (meiosis I), the chromosome number is halved.

Prophase I

1. Chromosomes become visible as the DNA is arranged around the histone proteins and becomes more compact.
2. **Homologous chromosomes** are attracted to each other and pair up. Homologous chromosomes are pairs of the same chromosome. One of the pair will originally have come from the individual's father, the other from the mother. Homologous chromosomes carry the same genes in the same order but the type of each gene (the **alleles**) may be different. Together, a pair of chromosomes is called a **bivalent**.
3. **Crossing over** occurs. During crossing over parts of homologous pairs can be swapped between chromosomes. This allows the mixing of alleles (see Section D2.1.11 for more on crossing over).
4. Spindle fibres form from microtubules.

Metaphase I

1. The homologous chromosomes line up across the cell's metaphase plate. They are randomly orientated, which means that either of the chromosomes from each pair is equally likely to end up at either pole.
2. The nuclear membrane disintegrates.

Anaphase I

Spindle fibres attach to the chromosomes and pull them to the opposite poles of the cell.

Telophase I

1. The spindles and spindle fibres disintegrate.
2. The chromosomes uncoil and new nuclear membranes form.

At the end of meiosis I, cytokinesis happens: the cell splits into two separate cells. The cells at this point are haploid because they contain only one chromosome of each pair. However, each chromatid still has its sister chromatid attached to it, so no S phase is necessary.

Many plants do not have a telophase I stage.

pair of homologous chromosomes

prophase I

cell's metaphase plate

metaphase I

anaphase I

telophase I

▲ **D2.1 Figure 10** The stages of meiosis I

Second round of segregation: separation of sister chromatids

During the second round of segregation (meiosis II), the sister chromatids are separated.

Prophase II

1. The DNA condenses into visible chromosomes again.
2. New meiotic spindle fibres are produced.

Metaphase II

1. Nuclear membranes disintegrate.
2. Individual chromosomes line up along the metaphase plate of each cell in no special order, i.e. random orientation.
3. Spindle fibres from opposite poles attach to each of the sister chromatids at the centromeres.

Anaphase II

1. The centromeres of each chromosome split, releasing each sister chromatid as an individual chromosome.
2. The spindle fibres pull individual chromosomes to opposite ends of the cell.
3. As a result of random orientation, the chromosomes can be pulled towards either of the newly forming daughter cells.

Telophase II

1. In animal cells, cell membranes pinch off in the middle. In plant cells, new cell plates form to make the cell membranes and cell walls of the four new cells.
2. The chromosomes unwind their strands of DNA.
3. Nuclear envelopes form around each of the four new nuclei. Then cytokinesis can take place.

prophase II

metaphase II

anaphase II

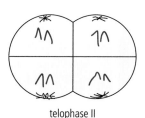

telophase II

D2.1 Figure 11 The stages of meiosis II

Mitosis has four phases: prophase, metaphase, anaphase and telophase. Meiosis goes through each of these phases twice and is a reduction division, which means that the amount of DNA in the daughter cells is half that in the original cell.

D2.1.10 – Non-disjunction

D2.1.10 – Down syndrome and non-disjunction
Use Down syndrome as an example of an error in meiosis.

Sometimes errors can occur during meiosis, and offspring can receive an atypical number of chromosomes, including extra or missing chromosomes. In humans, **Down syndrome** is caused by an extra copy of chromosome 21; the child has 47 instead of 46 chromosomes. The extra chromosome arises from a phenomenon called **non-disjunction**, which can happen at different times and on different chromosomes but for Down syndrome most often occurs when the 21st pair of homologous chromosomes fails to separate during anaphase I. If this happens in the future mother's ovary, the egg then carries two 21st chromosomes instead of one. When a sperm cell subsequently fertilizes the egg, the total number of 21st chromosomes is three.

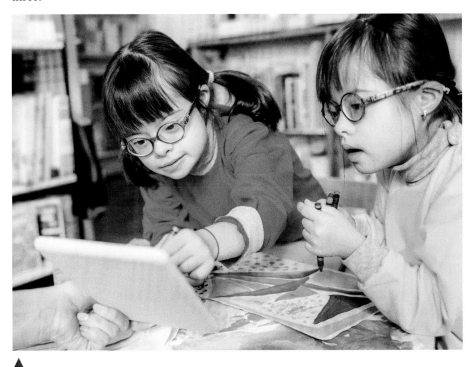

▲
The two boys in this photo are fraternal twins. The child on the right received an extra 21st chromosome and has Down syndrome.

Nature of Science

Researchers wanted to find out what influences the frequency of Down syndrome. Studies collected statistics on the many different characteristics of the parents and families of children born with Down syndrome. Such studies are called **epidemiological studies**, and they look at trends in populations, often examining thousands of cases. The incidence of Down syndrome increases with the age of the mother, particularly over the age of 35 (see the figure). Such data can help doctors and future parents assess the risks.

An extra chromosome for the 21st pair can occur during meiosis I or meiosis II, which is why the graph shows both, but the majority of cases happen during meiosis I.

D2.1.11 – Genetic diversity

D2.1.11 – Meiosis as a source of variation
Students should understand how meiosis generates genetic diversity by random orientation of bivalents and by crossing over.

Pairing of homologous chromosomes and crossing over

Meiosis is a step-by-step process during which a diploid parent cell produces four haploid daughter cells. Before the process begins, DNA replication allows the cell to make a complete copy of its genetic information during **interphase**. This results in each chromatid having an identical copy, or sister chromatid, attached to it at the centromere.

In order to produce a total of four cells, the parent cell must divide twice: the **first meiotic division** (meiosis I) makes two cells, and then each of these two cells divides again during the **second meiotic division** (meiosis II) to make a total of four cells.

One of the characteristics that distinguishes meiosis from mitosis is that, during the first step (prophase I) there is an exchange of genetic material between non-sister chromatids in a process called **crossing over** (see Figure 12). This exchange of chromosomal material happens when sections of two homologous but non-sister chromatids break at the same point, twist around each other, and then each connects to the other's initial position.

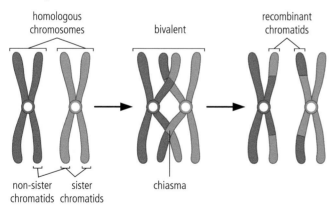

▲
D2.1 Figure 12 Crossing over occurring in a bivalent, a pair of homologous chromosomes.

Crossing over allows DNA from the maternal chromosomes to mix with DNA from the paternal chromosomes. In this way, the **recombinant chromatids** that end up in the sperm or the egg cells are a mix of the parent cells' original chromatids. This helps increase variety among offspring from the same parents, and so increases the chances of survival of at least some offspring if one combination of alleles proves to be more favourable for survival than others.

Random orientation

Figure 13 shows that, during metaphase I, the homologous pairs of chromosomes line up along the centre of the cell. Which of the pairs ends up at which pole is down to chance, which is why it is called **random orientation**. As with crossing over, this is another adaptation that increases variety in the offspring. The result of random orientation is that a male will only very rarely produce two sperm cells that are identical. Likewise, for a female, it is highly unlikely that she will ever produce two eggs with identical chromosomes in her lifetime.

Crossing over and random orientation explain why there is variation between siblings.

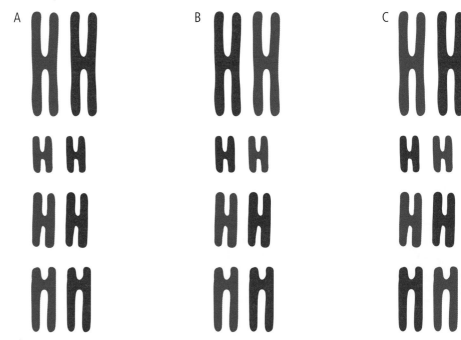

▲
D2.1 Figure 13 Rows A, B, and C show three of the sixteen possible orientations for four pairs of homologous chromosomes. In humans, with 23 pairs of chromosomes, there are more than 8 million possible orientations.

 The only way for a couple to have two children with the same combinations of chromosomes is to have identical twins. This is because identical twins form from a single zygote. Non-identical twins or any other siblings will be the product of an egg cell with a unique combination of half the mother's chromosomes and a unique combination of half the father's chromosomes.

How does the variation produced by sexual reproduction contribute to evolution?

How much variation can be generated by random orientation? In each haploid cell (n) the calculation is 2^n because there are two possible chromosomes in each pair (maternal and paternal) and there are n chromosomes in total. For humans, the number is 2^{23} because there are 23 chromosomes in each gamete. So the probability that a woman could produce the same egg twice is 1 in 2^{23} or 1 in 8,388,608. Even this calculation is an oversimplification, however, because it does not take into consideration the additional variety that results from crossing over. In addition, the calculation 2^n only considers one gamete. To produce offspring, two gametes are needed.

Guiding Question revisited

How can large numbers of genetically identical cells be produced?

In this chapter you have learned that:

- in prokaryotic cells, binary fission is used to make genetically identical cells
- cytokinesis is the splitting of cells and occurs after nuclear division, and the cytoplasm and cell organelles are usually split evenly between the daughter cells
- in multicellular organisms, mitosis is used to make two genetically identical daughter cells, which are diploid, $2n$.

Guiding Question revisited

How do eukaryotes produce genetically varied cells that can develop into gametes?

In this chapter you have learned:

- in multicellular organisms, meiosis is another type of cell division that is exclusively used for the production of gametes (sex cells)
- sperm cells and egg cells are haploid, n, containing half of an organism's genetic information
- at the end of meiosis, four cells are produced, but each sperm and each egg has a different combination of chromosomes as a result of random orientation and crossing over, which ensures variety in the offspring
- sometimes chromosomes do not separate correctly, and this non-disjunction of chromosomes can lead to genetic conditions such as Down syndrome.

Exercises

Q1. State two reasons why mitosis is necessary.

Q2. A chemical called colchicine disrupts the formation of microtubules. What effect would this drug have on a cell going through mitosis?

Q3. If a parent cell has 24 chromosomes, how many chromatids would be present during metaphase of mitosis?

Q4. Explain when cytokinesis occurs within the cell cycle.

Q5. Compare cytokinesis in plant and animal cells.

Q6. Describe how meiosis generates genetic diversity.

D2.3 Water potential

D2.2 is not included as it is for HL students only.

Guiding Questions

What factors affect the movement of water into or out of cells?

How do plant and animal cells differ in their regulation of water movement?

Many factors affect the movement of water into and out of cells. Plants must accomplish the seemingly impossible task of getting water from the roots to the top of the plant without a circulating organ. Unlike plants, animals have circulating organs that transport water to all the cells in the organism. It is important to remember, however, that in both plants and animals there are mechanisms that allow water to move through the cell membrane. The environment of organisms is subject to rapid changes, often resulting in varying water needs. If an organism does not adapt to these changes, its life will be at risk.

Plants and animals have different means of regulating water movement. Water delivery to plant cells is influenced by cell walls, which animal cells do not possess. However, the same underlying principles of water movement apply to all life on our planet. All life forms are dependent on water and the substances dissolved in it. Plant cells also require water for support and to maintain shape. The transportation of water and its dissolved content throughout an organism is essential for the maintenance of life.

D2.3.1 – Water as a solvent

D2.3.1 – Solvation with water as the solvent

Include hydrogen bond formation between solute and water molecules, and attractions between both positively and negatively charged ions and polar water molecules.

Water is the solvent of life and is known as the **universal solvent**. Living cells typically exist in an environment where there is water both within the cell (as cytoplasm) and outside the cell (as intercellular fluid, freshwater or saltwater, etc.). We refer to solutions as **aqueous** solutions if water is the solvent, no matter what mixture of substances make up the solutes. Thus cytoplasm and the ocean are both aqueous solutions.

The structure of water molecules

The basic chemical structure of water is presented in Chapter A1.1. In summary:

- water molecules are polar covalent molecules
- the oxygen region of a water molecule has a slight negative charge, while the two hydrogen regions each have a slight positive charge
- water molecules form large numbers of weak hydrogen bonds with other water molecules
- the hydrogen bonds are individually weak, but collectively are strong enough to account for most of the unique properties of water that make it so important to life
- when water is in a liquid state, hydrogen bonds are continually breaking, reforming and moving around

* polar substances, such as salts, alcohols and acids, dissolve in water quite readily
* non-polar substances, such as fats and oils, do not dissolve in water.

Solvation with water as the solvent

Solvation is the interaction of a solvent with a dissolved solute. Many refer to solvation as hydration. There are three steps to solvation:

1. the particles of a solute separate from each other
2. the water particles separate from each other
3. the separated solute and water particles combine to make a solution.

Water molecules have hydrogen bonds between them. In order to dissolve a solute the hydrogen bonds between the water molecules must break, as must the bonds between the solute molecules. The water molecules then surround the solute molecules and new hydrogen bonds form. Figure 1 shows the electrical attraction between solute and polar water molecules.

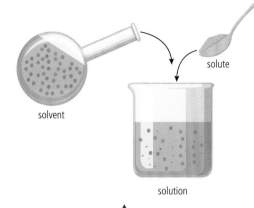

solute

solvent

solution

The process of solvation

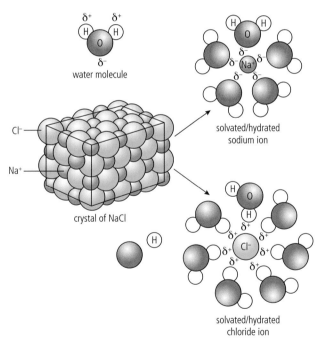

water molecule

Cl⁻

Na⁺

crystal of NaCl

solvated/hydrated sodium ion

solvated/hydrated chloride ion

D2.3 Figure 1 Attractions between positively and negatively charged ions and polar water molecules

What are the implications of solubility differences between chemical substances for living organisms?

Sodium chloride is solvated by polar solvents such as water. The small positive charges on the hydrogen regions of the water molecules surround the negatively charged chloride ions. The small negative charges on the oxygen region of the water molecule are attracted to the positively charged sodium ions. Hydrogen bonds form between the water molecules and the ions. Sodium chloride is relatively soluble in water but other polar molecules may be less soluble. Non-polar molecules such as lipids will not dissolve in water. Each substance has unique chemical properties that affect its solubility in water.

D2.3.2 – Water movement in relation to solute concentration

D2.3.2 – Water movement from less concentrated to more concentrated solutions

Students should express the direction of movement in terms of solute concentration, not water concentration. Students should use the terms "hypertonic", "hypotonic" and "isotonic" to compare concentration of solutions.

What variables influence the direction of movement of materials in tissues?

As biologists, we generally look at water movement in relation to either cells or the total organism. The cell membrane is very important when considering the movement of substances into and out of cells. It is imperative that this movement of materials is constantly monitored and controlled so that the best possible conditions for life are maintained for the cell and/or organism.

The terms described in Table 1 are introduced in Chapter B2.1, and now is a good time to review them.

D2.3 Table 1 Solution concentration terms

Term	Solute concentration in environment	Water concentration in environment
Hypertonic	Higher	Lower
Hypotonic	Lower	Higher
Isotonic	Equal	Equal

In order to work out which way water is moving between cells and their surroundings, we need to be able to talk about the concentrations of the solutions inside and outside cell membranes. We always talk about the concentration of the solution (not the concentration of the water).

If we say a cell is in a **hypertonic** environment, we mean it is surrounded by a solution that is higher in solutes and lower in water relative to the cytoplasm inside the cell membrane. A cell in a **hypotonic** environment is surrounded by a solution that is lower in solute particles. A cell in an **isotonic** environment has equal concentrations of solute inside and outside the cell.

D2.3 Figure 2 Describing solution concentrations. Notice the direction that the water is flowing in for the three different solution concentrations. The cell membrane is selectively permeable, allowing water to flow through it but not the solute particles.

Solution is hypotonic — Lower concentration of particles (or more water molecules) outside the cell.

Solution is isotonic — Equal concentration of particles (and water molecules) inside and outside the cell.

Solution is hypertonic — Higher concentration of particles (or less water molecules) outside the cell.

Figure 2 shows water movement in relation to solute concentration. Water moves from an area with a lower solute concentration to an area with higher solute concentration. In the case of the isotonic solutions, water movement is equal in both directions because there are equal concentrations of solute particles on both sides of the cell membrane. In this isotonic condition, water molecules continue to move in both directions as a result of the kinetic energy they possess. There is no net movement of water molecules.

D2.3.3 and D2.3.4 – Hypotonic and hypertonic solutions and osmosis

D2.3.3 – Water movement by osmosis into or out of cells
Students should be able to predict the direction of net movement of water if the environment of a cell is hypotonic or hypertonic. They should understand that in an isotonic environment there is dynamic equilibrium rather than no movement of water.

D2.3.4 – Changes due to water movement in plant tissue bathed in hypotonic and those bathed in hypertonic solutions
Application of skills: Students should be able to measure changes in tissue length and mass, and analyse data to deduce isotonic solute concentration. Students should also be able to use standard deviation and standard error to help in the analysis of data. Students are not required to memorize formulae for calculating these statistics. Standard deviation and standard error could be determined for the results of this experiment if there are repeats for each concentration. This would allow the reliability of length and mass measurements to be compared. Standard error could be shown graphically as error bars.

The passive transport processes of diffusion and osmosis are discussed in Chapter B2.1. The key points about these processes are:

- both diffusion and osmosis are examples of passive transport because they do not require energy in the form of cellular adenosine triphosphate (ATP) to occur
- in both processes, particles of a substance move along a concentration gradient
- osmosis requires a cell membrane, whereas diffusion can occur with or without a cell membrane
- osmosis involves the movement of water across a selectively permeable membrane
- a concentration gradient of water during osmosis is the result of the solute particle concentrations on either side of the cell membrane.

It is important to remember that water molecules are polar, making it difficult for them to pass through the hydrophobic region of a cell membrane. Chapter B2.1 discusses **aquaporins**. Aquaporins are essentially hydrophilic tunnels through the cell membrane that polar water molecules can pass through. Without aquaporins, there is limited water movement across a cell membrane. In osmosis, water moves along a concentration gradient through the aquaporins of the selectively permeable cell membrane. Osmosis will proceed if there is a hypotonic or hypertonic exterior cell environment. When the cell is in an isotonic environment, the water molecules still move but there is no net movement of the water molecules. The cell is said to be in **dynamic equilibrium**.

Challenge yourself

We are going to look at some examples of water movement in and out of cells by osmosis. Consider the three scenarios, and answer the following questions.

1. Raisins in pure water.
2. Human red blood cells in a solution with a high solute concentration.
3. Gargling with saltwater to relieve a sore throat.

For each of these examples, answer the following questions.

(a) What term best describes the external cell environment?
(b) Which way does water move and why?
(c) What is the result of osmosis?

The following activity will allow you to study water movement in plant tissue when it is placed in solutions of varying tonicity. After reading the overall procedure in the downloadable laboratory file accessed from this page of the eBook, write a hypothesis including an explanation.

Follow the directions to determine the osmolarity of potato tissues. The samples will be bathed in hypotonic, isotonic and hypertonic solutions. Note that tissues from plants other than potatoes may be used.

D2.3.5 – Water movement without cell walls

D2.3.5 – Effects of water movement on cells that lack a cell wall

Include swelling and bursting in a hypotonic medium, and shrinkage and crenation in a hypertonic medium. Also include the need for removal of water by contractile vacuoles in freshwater unicellular organisms and the need to maintain isotonic tissue fluid in multicellular organisms to prevent harmful changes.

Not all cells have a cell wall. Animal cells lack cell walls, which means that osmosis has some particular effects on these cells. To look at this in detail we can use the example of human red blood cells placed in a solution with a high solute concentration (a hypertonic medium). Because of the relatively high number of solute particles outside the cell, water molecules move out of the cell. With no cell wall present, the results are almost immediate, and the cell visibly shrinks, ultimately resulting in **crenation** (extreme shrinkage where the cell becomes crinkled). Figure 3 shows crenation of a red blood cell.

Conversely, if red blood cells are placed in a hypotonic cell environment (for example pure water), water will move into the cell and the solute particle concentrations will become closer to equal. This results in swelling of the red blood cells in the solution, possibly to the point of bursting the cell membrane.

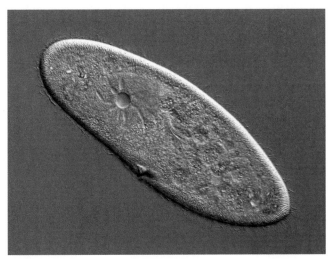

A contractile vacuole in *Paramecium caudatum*, visible as a circle with radiating canals, in the middle left region of the organism. The contractile vacuole functions in controlling water concentrations within this single-celled organism.

Adaptations have evolved in both single-celled and multicellular organisms to prevent harmful swelling and shrinkage of cells. Aquatic single-celled animals often possess a specialized organelle known as a **contractile vacuole**. This is a regulatory organelle that collects excess water from the interior of the cell and periodically empties it into the surrounding environment. This prevents cells swelling when there is a hypotonic environment.

Although extremely rare, very rapid water consumption can cause a condition called **hyponatremia**. This condition occurs when water dilutes the blood sodium levels below a certain level. The result is an increased uptake of water into cells, potentially harming those cells.

Humans experience the effects of cell shrinkage if they become dehydrated. If dehydration is severe, the cells of the body will begin to shrink and malfunction. In some cases, light-headedness and fainting can occur. Dehydration can cause a build-up of waste products within the body, including in the tubules of the kidneys. Blood flow throughout the body may be diminished, possibly causing muscle cramping and organ malfunctions. To overcome these potential results, it is essential that we drink adequate amounts of water.

Challenge yourself

4. The human body is about 60% water, but that water is not quite as salty as seawater. Knowing that our body's water salinity levels are less than seawater, what would be the effect of drinking seawater?

D2.3.6 – Water movement with cell walls

D2.3.6 – Effects of water movement on cells with a cell wall

Include the development of turgor pressure in a hypotonic medium and plasmolysis in a hypertonic medium.

The presence of a cell wall leads to some interesting effects when water moves into and out of plant cells. Most plant cells are hypertonic relative to their environment, and this includes their large central vacuole. Because of this, water tends to move into plant cells. The result is a high **hydrostatic pressure**. Hydrostatic pressure is the pressure that a fluid exerts in a confined space, against the boundary of the space. In plant cells, the pressure is exerted against the cell walls and is called **turgor pressure.** Incoming water swells the cell and presses the cell membrane against the rigid cell wall. Most plants depend on this turgor pressure to maintain their shape and remain upright. Figure 4 represents the different conditions a plant cell may experience as a result of water movement.

D2.3 Figure 4 The cell wall of a plant makes it difficult to see the many changes that occur inside as a result of water movement. The rigid cell wall resists changes in shape. However, the cell membrane and cell contents are affected by water moving into and out of the cell.

Osmosis in a plant cell

When water is lost from a plant cell, the cell membrane and the cytoplasm shrink away from the cell wall (look at the right-hand cell in Figure 4). This process of cytoplasmic shrinking in plant cells is called **plasmolysis**. When turgor pressure is lost in a plant, plasmolysis occurs and the plant visibly wilts. Large scale and long-lasting plasmolysis can result in cell death and, of course, plant death.

Halophytes are plants that thrive in conditions involving high salinity. They can concentrate large amounts of solutes in compartments so that they may take up water from their salty surroundings. Recent findings also indicate an ability to sequester toxic ions and salts within their cells and especially in leaves which eventually fall off the plant. Some even have salt glands which can excrete excess salt from plant tissues. Present research is looking into the gating action of aquaporins (see B2.1.5) in halophytes. Many parts of the world would benefit from the genetic development of plants that thrive in hypertonic conditions or that flourish while being irrigated with sea water. These plants could be used to provide our world's ever-increasing need for food.

Many people use natural weed killers in their gardens rather than commercial weed killers. These natural weed killers are a mixture of common table salt and vinegar, and a small amount of detergent, in water. They work by causing water to move out of a plant's cells, resulting in massive plasmolysis and death of the weed.

▲
These mangroves on the Florida coast of the Gulf of Mexico represent promise for increased food production in the future due to their ability to thrive in a high saline environment.

D2.3.7 – Isotonic solutions

D2.3.7 – **Medical applications of isotonic solutions**
Include intravenous fluids given as part of medical treatment and bathing of organs ready for transplantation as examples.

Food, blood products and medication can be administered through a line (catheter) placed in a peripheral vein. These intravenous (IV) fluids are used in many medical treatments for conditions such as haemorrhaging, surgery, cancer and dehydration. IV fluids must maintain a balance of the solutes that exist in the solutions in our cells, called intracellular fluids (ICF), and outside our cells, called extracellular fluids (ECF).

If ECF have a greater solute concentration than ICF, water will leave the body cells, resulting in shrinking and possible crenation. If ECF have a lower concentration of solutes than ICF, water will enter the cells, possibly resulting in swelling and even bursting. When ICF and ECF are isotonic, a dynamic equilibrium is maintained and, while water does move in and out of the cells, there is no net movement. IV solutions of different solute concentrations are used to treat particular medical conditions. Most IV solutions are isotonic so that they do not to cause excess water movement in or out of the body cells.

For organ transplants, there are many steps involved in the preparation and storage of a suitable organ. The organ is immersed in isotonic solutions during the period before it is transplanted, to avoid any cell damage.

Guiding Question revisited

What factors affect the movement of water into or out of cells?

In this chapter you have learned that:

- water is a polar molecule that must pass through the cell membrane via specialized channels called aquaporins

- the hydrogen bonds that water forms with ions and other charged particles contribute to the transport of essential materials into a cell and waste products out of a cell
- solvation is the interaction of a solvent with a dissolved solute
- water will move out of cell when it is placed in a hypertonic environment
- water will move into a cell when it is placed in a hypotonic environment
- the cell membrane contributes to the control of materials moving into and out of the cell
- osmosis requires a selectively permeable membrane to occur.

Guiding Question revisited

How do plant and animal cells differ in their regulation of water movement?

In this chapter you have learned that:

- turgor pressure (pressure potential) plays a large role in controlling water movement in and out of plant cells
- because animal cells do not have a cell wall, turgor pressure is not a large factor in water movement in and out of animal cells
- human red blood cells and all cells without a cell wall, when placed in a hypotonic solution, will swell and possibly burst
- human red blood cells and all cells without a cell wall, when placed in a hypertonic solution, will shrink and undergo crenation
- plant cells, when placed in a hypotonic solution, will develop high turgor pressure, as the cell contents push against the cell wall, which maintains plant shape
- cells possessing a cell wall will undergo plasmolysis when placed in a hypertonic environment
- when cells are placed in an isotonic environment, there is dynamic equilibrium rather than no movement of water.

Exercises

Q1 Describe the water environment surrounding roots in which plants are most likely to wilt. Explain why.

Q2. Explain the role of aquaporins in the transport of water in and out of cells.

Q3. Explain the consequences of fertilizing a plant too often.

Q4. Compare what would happen to plant and animal cells when placed in a hypotonic environment.

Q5. Which of the following is most likely to increase water uptake by a plant cell?

    A   Placing a plant cell in hypertonic environment.

    B   Placing a plant cell in an isotonic environment.

    C   Placing a plant cell in a hypotonic environment.

    D   Increasing the turgor pressure within the plant cell.

Q6. Explain the role of the contractile vacuole in freshwater unicellular organisms.

D2 Practice questions

1. Explain the effect of placing the following types of cells in an environment of pure distilled water.

   (a) Red blood cell (2)

   (b) Plant cell (2)

   (c) Unicellular freshwater dwelling organism (2)

   (Total 6 marks)

2. Fluid replacements are essential for individuals who have undergone a period of strenuous exercise and heavy perspiring.

   (a) Describe the solution of the preferable replacement. (1)

   (b) Explain the reason for your answer. (2)

   (Total 3 marks)

3. Suggest what the effects of a defective contractile vacuole would be on a freshwater unicellular animal.

   (Total 3 marks)

THEME

D Continuity and change

3 Organisms

◀ An illustration of a human foetus within the uterus. In this chapter we will explore how organisms reproduce, with a brief overview of asexual reproduction but a focus on the mechanisms of sexual reproduction. An integral part of sexual reproduction is combining the genetic material of two parents into an offspring. Thus, a study of genetics will be a portion of this unit, leading to an understanding of how pairs of chromosomes (one from each parent) interact with each other to give molecular instructions for an entire, complex organism. Finally, in this unit you will study how organisms maintain a homeostatic balance of processes in their bodies by feedback systems involving the nervous system and hormones produced by the endocrine system.

D3.1 Reproduction

Guiding Questions

How does asexual or sexual reproduction exemplify themes of change or continuity?

What changes within organisms are required for reproduction?

Reproduction of living organisms enables both continuity and change. New generations of species allow the continuation of favourable characteristics that have existed in some form for potentially eons, while encouraging new traits through the process of evolution. Asexual reproduction is possible by several mechanisms but only ever requires one parent. New individuals are formed by mitotic cell division. Very little genetic variation is produced, but if an environment is relatively unchanging this strategy is very efficient. Sexual reproduction requires two parents and the union of gametes that each have only half the chromosomes characteristic of the species. A great deal of variation is possible both in the chromosome composition of the gametes and the events associated with choice of mates and biological selection of which gametes are joined together. Offspring show a great deal of genetic variation, and thus sexual reproduction is favoured in environments that change over time.

All organisms must replicate their genetic material (DNA) before reproduction. In sexually reproducing organisms, a cell division called meiosis is used to reduce the chromosome number to half that of the adult. The resulting cells are called gametes and must be joined together to restore the full chromosome number. Often, organisms must reach a certain stage of maturity before they can produce gametes. In humans, this stage is called puberty.

D3.1.1 – Sexual and asexual reproduction

D3.1.1 – Differences between sexual and asexual reproduction

Include these relative advantages: asexual reproduction to produce genetically identical offspring by individuals that are adapted to an existing environment, sexual reproduction to produce offspring with new gene combinations and thus variation needed for adaptation to a changed environment.

Sexual reproduction requires both a male and a female parent. Each parent contributes some, but not all, of their genes to an offspring. This creates a unique genetic makeup that did not exist before. **Asexual reproduction** requires only one parent and results in multiple organisms from that single parent that all have the same genetic makeup.

Some of the major differences between asexual and sexual reproduction ▶

Sexual reproduction	Asexual reproduction
Gametes (usually a sperm and an egg) fertilized	Organism makes a copy of itself
Two parents required	Only one parent required
Offspring are genetically unique compared to both parents	Offspring are genetically identical to parent
Provides new gene combinations and thus promotes genetic variation	Provides no new gene combinations and relatively little genetic variation
Allows adaptations for a changing environment	Promotes little change in adaptations but may be beneficial in an existing non-changing environment

Asexual reproduction can be accomplished by a variety of mechanisms, for example those listed below.

- **Binary fission**: bacteria and unicellular eukaryote cells grow, replicate their DNA and divide into two cells. In an environment well suited for growth this can occur more than once an hour in some species of bacteria.
- **Budding**: new genetically identical organisms grow directly from an existing organism.
- **Fragmentation**: the body of an existing organism breaks up into several fragments, each growing into a complete organism.
- **Vegetative reproduction**: common in many plants, with new plants emerging from roots, bulbs, tubers or shoots.
- **Parthenogenesis**: in an animal species, growth and development of an egg cell without the involvement of a male gamete.

Hydra is a genus of small freshwater animals called cnidarians. This *Hydra* is reproducing asexually by growing a bud. The bud will eventually separate from the original organism and become an entirely separate, but genetically identical, organism.

It is incorrect to consider asexual reproduction as a less important or inferior form of reproduction. It is true that sexual reproduction provides much greater genetic variation, but asexual reproduction has its own place in nature. Asexual reproduction is almost always faster, requires less expenditure of energy, and a single organism can colonize a new area relatively easily. The process of evolution has favoured both types of reproduction in different circumstances. If either sexual or asexual reproduction was always advantageous, evolution would have eliminated the less fit mechanism.

D3.1.2 – The role of meiosis and gametes

D3.1.2 – Role of meiosis and fusion of gametes in the sexual life cycle

Students should appreciate that meiosis breaks up parental combinations of alleles, and fusion of gametes produces new combinations. Fusion of gametes is also known as fertilization.

Meiosis produces haploid cells by a process often described as **reduction division,** because the number of chromosomes in each gamete is reduced to one-half of the original number. In a diploid cell, chromosomes exist in homologous pairs. Each chromosome of a homologous pair can be traced back to one of the two parents. In other words, each homologous pair has a **maternal** chromosome and a **paternal** chromosome. Each new individual produced by sexual reproduction will represent a brand new combination of chromosomes.

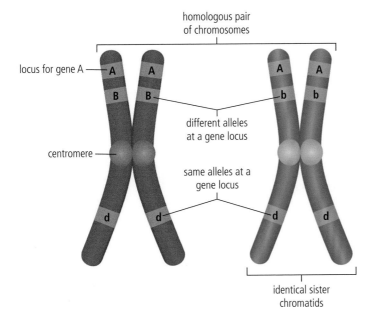

An illustration showing one homologous pair of chromosomes after DNA replication. One of the chromosomes in the homologous pair is from the mother of this individual and one is from the father. Both chromosomes of the homologous pair have the same gene locations (loci) but may or may not have different alleles for any one gene. This is exemplified by genes A, B and D. DNA replication has produced the identical sister chromatids. In a human gonad cell preparing to create a gamete, there would be 22 other homologous pairs all with their own gene loci. As gametes only contain one chromosome from each homologous pair, there is enough genetic material in a gonad cell to make four gamete cells.

Before meiosis begins DNA replication takes place, producing a pair of chromatids from each chromosome. The cells produced at the end of meiosis do not contain a random split of the chromosomes. Each cell has only one chromosome from each **homologous pair**, so is referred to as a **haploid** (n) number of chromosomes. Each chromosome from a homologous pair will have the same genes in the same order, but may contain different alleles. The resulting cells are gametes: **eggs** in females and **sperm** in males. In humans the haploid number of chromosomes is 23. Each sperm and egg has 23 chromosomes, one of each type of chromosome.

Gamete genetic variation is provided by:

- crossing over, leading to recombination of alleles during metaphase I
- random orientation and division of chromosomes during metaphase I (often called independent assortment of chromosomes)
- random fertilization, as there are millions of sperm and egg possibilities even for just one set of parents.

Fertilization occurs when gametes fuse into a single cell. The purpose of fertilization is to restore the **diploid** (2n) number of chromosomes. This will also restore the chromosomes as homologous pairs. This is necessary as some genes require the interaction of both chromosomes of a homologous pair for gene expression. The diploid number of chromosomes in humans is 46, existing as 23 homologous pairs.

Genetic variation in sexually reproducing organisms is the result of the many chromosome combinations possible during the formation of gametes, and the many possible gamete combinations during fertilization.

D3.1.3 – Male and female gametes

D3.1.3 – Differences between male and female sexes in sexual reproduction

Include the prime difference that the male gamete travels to the female gamete, so it is smaller, with less food reserves than the egg. From this follow differences in the numbers of gametes and the reproductive strategies of males and females.

Even though the process of meiosis is identical in males and females, the resulting morphology of the gamete cells is quite different. Sperm are often motile and have a **flagellum** to help them move towards the egg. They are also very small, to increase swimming efficiency. When a sperm fertilizes an egg, it contributes nothing towards the food reserves for the early embryo. Sperm are adapted to provide an efficient delivery system for a haploid nucleus.

In comparison to sperm cells, an egg is huge. Human eggs are thousands of times larger in volume than human sperm An egg contains all the nutrients needed for early embryonic growth. In some species the egg provides all the nutritional requirements until the young animal hatches from an egg encased in a shell. In humans and other placental mammals, the egg provides the initial source of nutrition for the developing embryo, followed by nutrition from the uterus and then the placenta. In addition, the eggs contain the initial organelles, such as mitochondria, that are needed as the cells grow and divide during development.

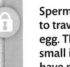

Sperm are designed to travel to the egg. They are very small in size, and have multiple mitochondria and a flagellum, making them efficient "swimmers".

A coloured scanning electron micrograph (SEM) of a human egg with multiple sperma attempting fertilization. Eventually one sperm cell will penetrate the plasma membrane of the egg. A chemical reaction will then occur to prevent any other sperm from fertilizing the egg.

Some animals produce large numbers of eggs, especially if the fertilization process is external. Human females typically release only one egg during the menstrual cycle, although two or more are possible. Human males produce millions of sperm each day, with many of the cells dying. The cellular components of these cells are biologically recycled if they are not used. A single **ejaculation** contains millions of sperm cells in a fluid called **semen**. Males produce such large numbers of sperm cells because the vast majority will never find the egg. Of the millions of sperm that begin swimming towards the egg after an ejaculation, only 100–200 will actually reach the egg and only one will fertilize it.

In many species females have a greater impact on mate selection than the male. This is especially noticeable in bird species where males are more brightly coloured and carry out elaborate mating rituals to attract females. Often it is the female that chooses a mate. Females produce fewer gametes and are often more heavily invested in caring for the young.

The female reproductive strategy is based on producing one or very few, very large stationary eggs, whereas the male strategy is to produce huge numbers of very small and motile sperm.

D3.1.4 – Male and female reproductive systems

D3.1.4 – Anatomy of the human male and female reproductive systems

Students should be able to draw diagrams of the male-typical and female-typical systems and annotate them with names of structures and functions.

The IB requires you to draw diagrams of the male and female reproductive systems and annotate those diagrams with the names of the structures and their functions. Study Figures 1 and 2 and Tables 1 and 2 and practise making your own diagrams.

The structures of the male and female reproductive systems are adapted for the production and release of gametes. In addition, the female reproductive system ensures a suitable location for fertilization and provides an environment for the growth of the embryo/foetus until birth.

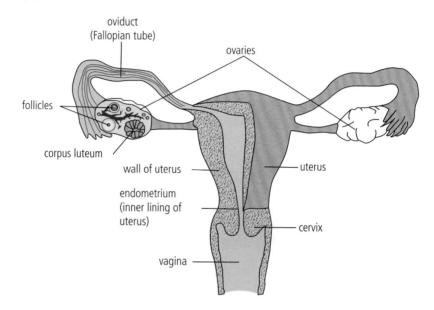

D3.1 Figure 1 A frontal view of the human female reproductive system.

D3.1 Table 1 Anatomy and function of the female reproductive system

Anatomical structure	Function(s)
Ovaries	The female gonads. These organs produce and secrete oestrogen. They also produce and release eggs (in the form of **secondary oocytes**). The area where ovulation occurs grows into the **corpus luteum**, which temporarily produces the hormone progesterone
Fallopian tubes (oviducts)	Ducts that carry the egg (or early embryo) to the uterus
Uterus	A muscular structure where the early embryo implants and develops if a pregnancy occurs
Endometrium	The highly vascular inner lining of the uterus
Cervix	The lower portion of the uterus, which has an opening to the vagina that allows the sperm to enter for fertilization and provides a pathway for childbirth
Vagina	A muscular tube that leads from the external genitals to the cervix; semen is ejaculated here during sexual intercourse

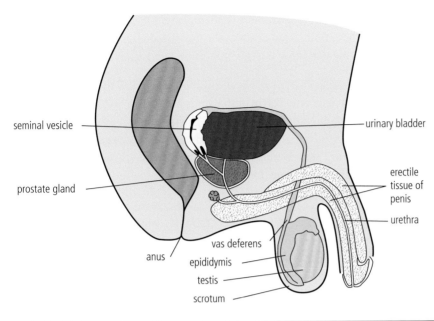

D3.1 Figure 2 A lateral (side) view of the human male reproductive system. The urinary bladder is shown because of its close anatomical proximity to the reproductive structures.

Anatomical structure	Function(s)
Testes	The male gonads. Sperm are produced here in small tubes called **seminiferous tubules**
Epididymis	The area where sperm are received, become mature, and become capable of a swimming motion via movement of their flagella
Scrotum	A sac that holds the testes outside the body cavity so that sperm production and maturation can occur at a temperature cooler than body temperature
Vas deferens	A muscular tube that carries mature sperm from the epididymis to the urethra during an ejaculation
Seminal vesicles	Small glands that produce and add seminal fluid to the semen
Prostate gland	A gland that produces much of the seminal fluid, including carbohydrates for the sperm
Penis	An organ that becomes erect as a result of blood engorgement in order to facilitate ejaculation
Urethra	After all the glands have added fluids, this is the tube via which the semen leaves the penis

D3.1 Table 2 Anatomy and function of the male reproductive system

D3.1.5 – Hormonal control of the menstrual cycle

D3.1.5 – Changes during the ovarian and uterine cycles and their hormonal regulation

Include the roles of oestradiol, progesterone, luteinizing hormone (LH), follicle-stimulating hormone (FSH) and both positive and negative feedback. The ovarian and uterine cycles together constitute the menstrual cycle.

At puberty, human females begin a hormonal cycle known as the **menstrual cycle.** Each cycle lasts, on average, 28 days. The purpose of the menstrual cycle is to time the release of an egg (**ovulation**) for possible fertilization and later **implantation** into the inner lining of the uterus. This implantation must occur when the uterine inner lining (the **endometrium**) is rich with blood vessels (i.e. highly vascular). The highly vascular endometrium is not maintained if there is no implantation. The breakdown of the blood vessels of the endometrium leads to the menstrual bleeding (**menstruation**) of a typical cycle. This menstruation is a sign that no pregnancy has occurred.

The part of a female's brainstem known as the hypothalamus is the regulatory centre for the menstrual cycle. The hypothalamus produces a hormone known as **gonadotropin-releasing hormone (GnRH)**. The target tissue of GnRH is the nearby pituitary gland, and results in the pituitary producing and secreting two hormones into the bloodstream. These two hormones are the **follicle-stimulating hormone (FSH)** and **luteinizing hormone (LH)**. The target tissue for both these hormones is the ovaries.

The hormones FSH and LH have several effects on the ovaries. One is to increase the production and secretion of another reproductive hormone by the ovaries' follicle cells. This hormone is **oestradiol**. Like all hormones, oestradiol enters the bloodstream. Its target tissue is the endometrium of the uterus. One effect of oestradiol is an increase in the density of blood vessels in the endometrium; in other words the endometrium becomes highly vascular. Because of a positive feedback loop, oestradiol stimulates the pituitary gland to release more FSH and LH.

The increase in FSH and LH results in the production of structures called **Graafian follicles**. Within the ovaries are cells known as **follicle cells**, and the reproductive cells (possible future eggs) are at a stage of development called **oocytes**. Under the chemical stimulation of FSH and LH, the somewhat randomly arranged follicle cells and oocytes now take on a particular cellular arrangement known as a Graafian follicle.

> The increase of oestradiol secretion by the follicle cells of the ovary stimulates a spiked release of FSH and LH by the pituitary gland. This is an example of positive feedback control.

A light micrograph of a section of human ovary. Two Graafian follicles are shown with an oocyte at the centre of each. The oocytes are the large, nearly perfect circles, surrounded by ring of very small follicle cells.

oocyte
inner ring of follicle cells
fluid-filled cavity
outer ring of follicle cells

A spike in the level of FSH and LH leads to **ovulation** (as a result of the positive feedback loop between FSH, LH, the pituitary gland and oestradiol). The oocyte is accompanied by the Graafian follicle's inner ring of follicle cells. This structure is referred to as a single follicle, and typically enters a Fallopian tube soon after ovulation. The Graafian follicle's outer ring of follicle cells remains within the ovary. These follicle cells begin to produce and secrete another hormone, **progesterone**. The cells of the outer ring begin to divide and fill the "wound" area left by ovulation. This forms a glandular structure known as the **corpus luteum**. The corpus luteum is only hormonally active (producing progesterone) for 10–12 days after ovulation if fertilization does not occur. Progesterone is a hormone that maintains the thickened, highly vascular endometrium. As long as progesterone continues to be produced, the endometrium will not break down and an embryo will be able to implant. In addition, the high levels of both oestradiol and progesterone at the same time provide a negative feedback signal to the hypothalamus and prevent production of GnRH.

Negative feedback control is exemplified by high levels of oestradiol and progesterone inhibiting the production of GnRH. Without GnRH, FSH and LH are inhibited. As soon as the corpus luteum degenerates, another cycle can begin because the progesterone levels are not being maintained.

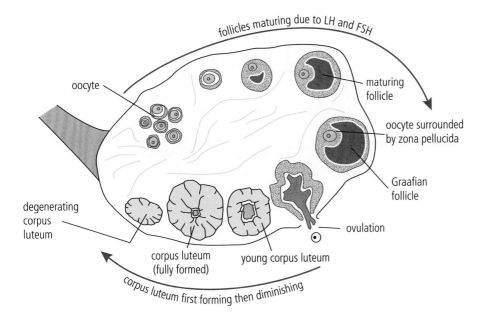

follicles maturing due to LH and FSH

oocyte

maturing follicle

oocyte surrounded by zona pellucida

Graafian follicle

degenerating corpus luteum

ovulation

corpus luteum (fully formed)

young corpus luteum

corpus luteum first forming then diminishing

The sequence of events that occur within an ovary during a single menstrual cycle. Twenty-eight days of ovarian events are shown within this single ovary as a time lapse sequence.

Because the hypothalamus does not produce GnRH when oestradiol and progesterone levels are high, FSH and LH remain at low levels, which is not conducive to the production of another Graafian follicle. Assuming there is no pregnancy, the corpus luteum begins to break down after 10–12 days, and this leads to a decline in both progesterone and oestradiol levels. As both of these hormone levels fall, the highly vascular endometrium can no longer be maintained. The capillaries and small blood vessels begin to rupture and **menstruation** begins. The drop in progesterone and oestradiol also signals the hypothalamus to begin secreting GnRH, and thus another menstrual cycle begins. Because the menstrual cycle is a cycle, there is no true beginning or ending point. The first day of menstruation is designated as the first day of the menstrual cycle simply because this is an event that can be recorded fairly easily (see Figure 3).

D3.1 Figure 3 The events that occur during a 28-day menstrual cycle. Note that these events are all aligned on the same time scale. Ovulation and possible fertilization occur near the middle of the cycle. The ovarian and uterine cycles together constitute the menstrual cycle.

When studying Figure 3, use the arrow indicating the time of ovulation as an important marker. Ask yourself: What events led up to and resulted in this ovulation? What will now happen after ovulation?

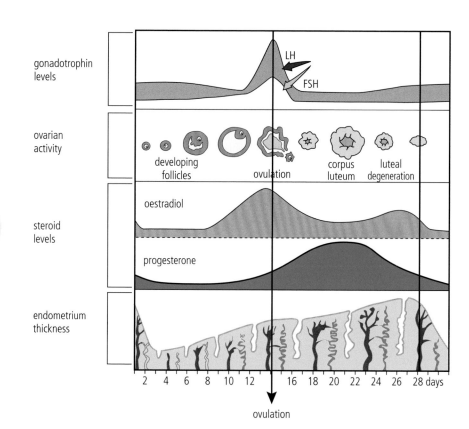

D3.1 Figure 3 The events that occur during a 28-day menstrual cycle. Note that these events are all aligned on the same time scale. Ovulation and possible fertilization occur near the middle of the cycle. The ovarian and uterine cycles together constitute the menstrual cycle.

The overall menstrual cycle can be considered as two components. The ovarian cycle controls the production and release of eggs and the cyclic release of oestradiol and progesterone. The uterine cycle controls the preparation and maintenance of the lining of the uterus to receive a fertilized egg. The two components must be timed synchronously because the endometrium must be highly vascular and ready to receive an early embryo when the embryo enters the uterus from a Fallopian tube.

D3.1.6 – The process of fertilization

D3.1.6 – Fertilization in humans

Include the fusion of a sperm's cell membrane with an egg cell membrane, entry to the egg of the sperm nucleus but destruction of the tail and mitochondria. Also include dissolution of nuclear membranes of sperm and egg nuclei and participation of all the condensed chromosomes in a joint mitosis to produce two diploid nuclei.

As a result of sexual intercourse, millions of sperm are **ejaculated** into a female's vagina. The motile sperma absorb some of the sugar in semen in order to have enough "fuel" for their potentially long journey. Some of the sperma find their way through the cervical opening (the **cervix** separates the vagina and the uterus) and gain access to the uterus. They begin swimming up the endometrial lining, and some enter the openings of the two Fallopian tubes. If the female is near the middle of her menstrual cycle, there may be an egg within one of the two Fallopian tubes. The reason for millions of sperma in each ejaculate becomes clear when you consider that only a very small percentage of the motile sperma will ever reach the location of an egg.

The typical location for fertilization is within one of the Fallopian tubes. No single spermatozoon can accomplish the entire act of fertilization because it takes many sperm to penetrate the follicle cell layer and a coating called the **zona pellucida** (a gel composed of glycoproteins) surrounding the egg. Several sperma release hydrolytic enzymes contained in their acrosomes to help penetrate the egg's plasma membrane. One spermatozoon will reach the plasma membrane of the egg first and gain entry to the egg. This is the sperm cell that fertilizes the egg.

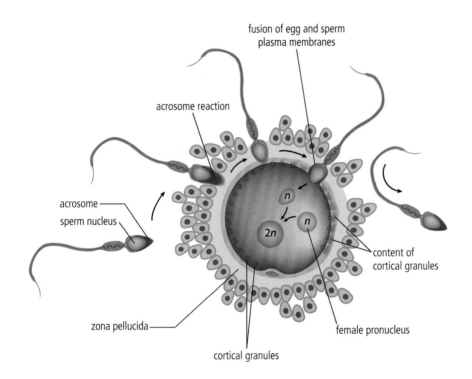

The events of fertilization, including the acrosome reaction, fusion of the plasma membranes of the egg and sperm, and fusion of the two haploid nuclei.

Vesicles are released from the egg that destroy the sperm flagellum and mitochondria. The haploid set of chromosomes from the sperm and the egg are now both within the cytoplasm of the egg. The paternal and maternal sets of chromosomes remain separate for a period of time, and membranes form around each. These haploid structures are each called a **pronucleus**. While in this pronuclei stage, the DNA in each undergoes replication in preparation for mitosis. The two pronuclei then come together and the temporary nuclear membranes dissolve. A spindle apparatus typical of mitosis forms as the chromosomes prepare for the first mitotic division of the newly formed diploid cell. It takes about 30 hours after fertilization for this first mitotic cell division to be completed. Subsequent mitotic cell divisions increase in frequency, and by the end of three days 16 cells will have been formed.

D3.1.7 – In vitro fertilization

D3.1.7 – Use of hormones in in vitro fertilization (IVF) treatment

The normal secretion of hormones is suspended, and artificial doses of hormones induce superovulation.

There is a wide variety of possible reasons for infertility, including:

- low sperm counts in males
- impotence (failure to achieve or maintain an erection) in males
- inability to ovulate normally in females
- blocked Fallopian tubes in females.

Reproductive technologies have been developed to help people overcome the problem of infertility. One of the most common of these technologies is **in vitro fertilization** (IVF).

As part of the IVF procedure, eggs are "harvested" from the female's ovaries. In order to ensure the proper timing for this, and to maximize the number of available eggs, the female undergoes hormone therapy. During the initial stages of the therapy she is treated with a drug that suspends the hormones associated with her natural menstrual cycle. Subsequently she takes hormone injections that include FSH. This ensures that she will produce many Graafian follicles in each ovary and therefore many potential eggs for harvesting. The production of many more eggs than is typical of a normal menstrual cycle is called **superovulation**.

When the time is right, many eggs are harvested surgically. To obtain the sperm cells that are needed for fertilization, the male ejaculates into a container. Harvested eggs are mixed with the sperm cells in separate culture dishes. Microscopic observation reveals which eggs are fertilized, and whether the early development appears normal and healthy. Between one and three healthy **embryos** are later introduced into the female's uterus for implantation. Any healthy embryos from the culturing phase that are not used for the first implantation can be frozen and used later if another implantation procedure is needed.

TOK

During IVF, pre-implantation genetic screening is often carried out on embryo cells taken at the **blastocyst** stage. Chromosome disorders and some genetic diseases can be diagnosed and the prognosis used to decide which embryos to implant. What ethical constraints should there be on the pursuit and application of knowledge?

D3.1.8 – Sexual reproduction in plants

D3.1.8 – Sexual reproduction in flowering plants

Include production of gametes inside ovules and pollen grains, pollination, pollen development and fertilization to produce an embryo. Students should understand that reproduction in flowering plants is sexual, even if a plant species is hermaphroditic.

The gametes of flowering plants are produced within structures called **ovules** (female) and **pollen grains** (male). Meiosis gives rise to the ovules and pollen grains, while the actual gametes are produced by mitosis. Because the reproductive structures are already haploid, reduction division is not required to produce haploid gametes.

Many species of flowering plants produce flowers that are **hermaphroditic**, with both male and female structures. Some hermaphroditic species, including orchids and sunflowers, **self-pollinate**. Self-pollination and fertilization are a form of sexual reproduction because the gametes are produced by meiosis and there is a fusion of gamete nuclei to form an embryo. The disadvantage to this method of reproduction is the loss of genetic variation that is a natural part of combining the chromosomes from two separate individuals.

Cross-pollination is the transfer of pollen produced on one plant with another plant. There are many flower adaptations that facilitate the transfer of pollen from one plant to another. Flowers that have petals often use shapes, markings and colours to attract specific pollinating animals. The magnificent shapes and colours of many species' flowers are not intended to please humans, but to attract animals that can transfer pollen from one plant to the next.

Pollen develops within structures called **anthers**. Anthers are often positioned in a flower so that pollinators can come into contact with them without even realizing it. When that pollinator moves on to a different flower, some of the pollen will be transferred to another structure, called the **stigma**, that is often held upright in the female part of the flower. The stigma is sticky and pollen grains can easily adhere to it.

Pollen that adheres to a stigma will then begin to grow into a structure called a **pollen tube**. In some respects this is equivalent to animal sperm that swim. Rather than swimming, pollen grains grow into tubes that penetrate other parts of the flower, to take male reproductive nuclei to the ovule where female nuclei await fertilization.

A pollen tube growing from a pollen grain of a lily plant, one of many species within the genus *Lilium*. The growth of the pollen tube is the plant's mechanism for getting the male gametes (nuclei) to the female gametes (nuclei) during a double fertilization process.

One pollen tube carries two male nuclei, and each results in a fertilization, hence plants use a double fertilization process. Within the **ovule** there are three haploid nuclei. One pollen nucleus fertilizes one ovule nucleus to create a zygote. The other pollen nucleus fertilizes the other two nuclei within the ovule to create a tissue called **endosperm**. Because three nuclei were used for this fertilization, the endosperm has the unusual chromosome number of $3n$: it is a **triploid** tissue. Growth of this triploid tissue produces the nutritive endosperm within the seed that will nourish the early plant embryo.

The nutritive value of seeds, such as peas and corn, that we exploit primarily comes from the seed's endosperm tissue.

D3.1.9 – Insect pollination

D3.1.9 – Features of an insect-pollinated flower

Students should draw diagrams annotated with names of structures and their functions.

There are many pollinators of different flower species, including insects, birds, bats and even some mammals. Here we will focus on flowers that are specialized for insect pollination. Common insect pollinators include bees, wasps, flies, butterflies and moths. Insect-pollinated flowers tend to be large and brightly coloured. They give off a strong scent to attract insects, and they offer a reward in the form of nectar at the base of the flower. The stamens of these flowers are often deep inside the flower, so that insects drinking nectar will brush up against the pollen grains. The pollen is often sticky and has numerous spikes in order to adhere easily to the legs or body of the insect. The stigma is also sticky so that the pollen can be transferred from the insect to the stigma as the insect visits other flowers.

How can interspecific relationships assist in the reproductive strategies of living organisms?

A bee-pollinated flower of a peach tree (*Prunus persica*). Peach flowers have many anthers but only one stigma, style and ovary. Peach fruits have only one ovule in their ovary, and thus the fruit has only one seed.

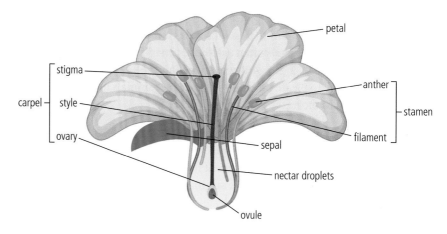

The function of different parts of a flower

Flower part	Function
Sepal	Protecting the developing flower while it is inside the bud
Petal	Often colourful, to attract pollinators
Anther	The part of the stamen (the male portion of a flower) that produces the male sex cells (pollen)
Pollen	Contain the male nuclei used for fertilization
Filament	The stalk of the stamen that holds up the anther
Stigma	The sticky top of a carpel (the female portion of a flower), on which pollen lands
Style	The part of the carpel that supports the stigma
Ovary	The base of the carpel, containing one or more ovules
Ovule	The chamber within an ovary where the female nuclei develop

 If you are asked to draw and annotate a diagram, be sure to list the functions of the parts you draw. This can be done directly within your diagram labels rather than providing a separate table, unless you are directed otherwise.

D3.1.10 – Cross-pollination in plants

D3.1.10 – Methods of promoting cross-pollination

Include different maturation times for pollen and stigma, separate male and female flowers or male and female plants. Also include the role of animals or wind in transferring pollen between plants.

The common primrose (*Primula vulgaris*) produces two different kinds of flowers on different plants. The design of the two flower types ensures that the pollen picked up from one type of flower can only be deposited on the other type. This promotes cross-pollination in this species.

Plants benefit from genetic variation within populations just as much as animals do. Even in plant species where flowers are hermaphroditic, a variety of mechanisms have evolved to promote cross-pollination. Such mechanisms include the following.

- Some plant species have different maturation times for the pollen and ovules of the same flower. Maturation at differing times ensures that self-pollination cannot occur.
- In some species, the pollen and stigma of flowers of the same plant use chemical self-incompatibility mechanisms. If pollen lands on the stigma of a flower on the same plant, the pollen tube does not grow because of the chemical incompatibility.

- Some flowering species produce flowers that only have male parts or only have female parts.
- In some species, an entire plant is either male or female and can only produce flowers of their own sex.
- The pollen of some species is transferred by wind, which often takes pollen away from the parent plant.

Pollination and fertilization in plants are different processes and rely on different mechanisms.

D3.1.11 – Self-incompatibility mechanisms

D3.1.11 – Self-incompatibility mechanisms to increase genetic variation within a species

Students should understand that self-pollination leads to inbreeding, which decreases genetic diversity and vigour. They should also understand that genetic mechanisms in many plant species ensure male and female gametes fusing during fertilization are from different plants.

In Section D3.1.8, you learned that when pollen lands on the stigma a pollen tube begins to grow. The pollen tube takes the male nuclei to the ovule, where fertilization can occur. Many plants use the growth of pollen tubes as a mechanism to control self-pollination. Each plant has a set of genes that controls the growth of the pollen tube. When the pollen of a plant lands on the stigma of a flower of the same plant, protein interactions occur that reduce or stop growth of the pollen tube. This is called a **self-incompatibility mechanism.** The specifics of the mechanism differ among plant species, but include the following.

- The pollen grain on the stigma fails to germinate into a pollen tube.
- The pollen grain germinates but does not enter through the stigma into the style.
- The pollen tube enters the ovule but the pollen nuclei degenerate before fertilization can occur.
- Fertilization occurs but the plant embryo degenerates during early growth.

The most successful pollination occurs when the pollen is from one plant and the stigma is in the flower of a completely different plant of the same species. This promotes genetic variation in and healthy growth of the new plant (called **vigour**). Self-pollination leads to inbreeding and a decrease in genetic diversity and vigour.

D3.1.12 – The role of seeds

D3.1.12 – Dispersal and germination of seeds

Distinguish seed dispersal from pollination. Include the growth and development of the embryo and the mobilization of food reserves.

Seed dispersal

Once a successful double fertilization has occurred, a seed will begin to develop. Each fertilization within an ovary will lead to a different seed, and the flowers of some species promote the development of many seeds from the same flower. If a flower has a single ovule within its ovary, then only one seed will develop. If there are many ovules within an ovary, then many seeds will develop. The ovary itself grows (ripens) and becomes a fruit. The number of seeds inside a fruit is an indication of how many ovules the ovary contained.

In biology, the term vegetable has no meaning. Many items commercially marketed as vegetables are really fruits. These include cucumbers, peppers and tomatoes. All were once ovaries of flowers and contain seeds. Other vegetables that do not have seeds include the roots, stems or leaves of a plant.

Plants invest a great deal of chemical energy in the production of fruits. The reward for the plant is that the fruit provides a means of dispersing the seeds away from the parent plant. Many fruits have evolved to be attractive to animals as a source of food. After ingestion and digestion by an animal, the seeds are often still protected within their seed coats, and are deposited in the animal's faeces potentially far away from the parent plant. Successful seed dispersal also depends on available light as well as water and nutrients within the soil. A location will not allow germination or early plant growth unless environmental conditions support that growth. Some seeds, such as a coconut, use water to float to a new location before geminating. Other seeds have structures that allow them to be easily carried by the wind for dispersal. A few species develop pods that dry out as the seeds become ripe. Once the pod is dry enough, it pops open explosively, releasing the seed away from the parent plant.

The seeds of a dandelion plant (*Taraxacum officinale*) develop filamentous structures that allow them to be easily dispersed by the wind.

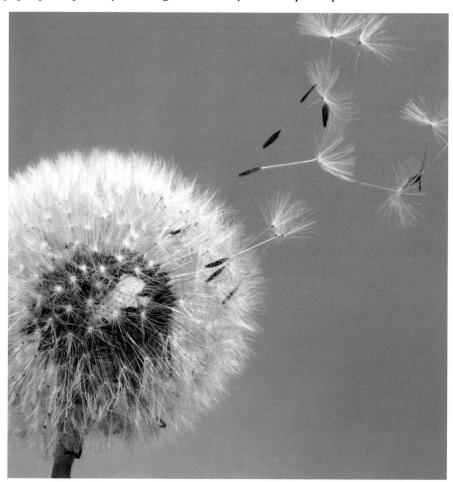

Seeds can stay dormant for very long periods of time under the right conditions. Seed banks, where seeds are stored by promoting dormancy, have been established in many countries in order to preserve the genetic diversity of plant species.

Seed germination

Once seeds have formed within a flower's ovary, they usually begin a period of dormancy. This requires the seed to lose most of its water content. During dormancy, the seeds display a very low metabolism, with no growth or development. The dormancy period is variable for different types of seeds, but many can remain dormant for years. This represents an adaptation feature to overcome harsh, but potentially temporary, environmental conditions, because it allows seeds to wait and germinate when conditions may be more suitable for growth.

D3.1

When conditions become more favourable, seeds may **germinate**. Germination is the early growth of a seed as it develops into a plant. Several general conditions must be met for a seed to germinate:

- water is needed, to rehydrate the dried seed tissues
- oxygen is needed, to allow aerobic respiration to produce adenosine triphosphate (ATP)
- an appropriate temperature for the seed is necessary.

A seed contains a small plant embryo and the food reserves it needs for early growth. The food reserves are called endosperm tissue, and are transferred to the plant embryo through structures called **cotyledons**. As the plant embryo grows larger, the reserves of endosperm tissue become depleted.

Seeds begin germination by absorbing water in a process called **imbibition**. This activates the biochemistry of the seed. The rate of cell respiration and protein synthesis greatly increases following imbibition, as the embryonic plant prepares to emerge from the seed coat. In most plant species, the portion of the embryo that emerges first is the initial root structure, called the **radicle**. The radicle is positively influenced by gravity and grows down into the soil.

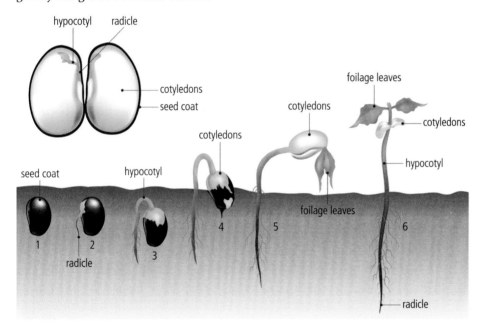

The process of germination, showing the early growth of the radicle and slightly later growth of the shoot.

In many plant species the first structure to appear above ground is called the **hypocotyl**. This is a curved portion of the plant shoot that is found below the cotyledons. The hypocotyl grows in the opposite direction to the force of gravity, and therefore grows up. Once above ground, the early shoot will straighten. This orientates the growing young plant perpendicular to the ground. The first real leaves then develop and begin photosynthesis as the endosperm tissue in the cotyledons is depleted. The root structures during this time continue to develop, forming **secondary roots** as well as **root hairs**. All plant growth from this point on will occur at areas called **meristem tissue** located at the tips of shoots and roots.

Guiding Question revisited

How does asexual or sexual reproduction exemplify themes of change or continuity?

In this chapter we have described the following:

- asexual reproduction provides nearly genetically identical offspring well adapted to a stable environment
- sexual reproduction provides genetic diversity in populations necessary for adaptations in a changing environment
- genetic diversity is provided by independent assortment of chromosomes during gamete production
- genetic diversity is also provided by the many possibilities inherent in the fertilization of gametes
- genetic variation in flowering plants is enhanced by cross-pollination.

Guiding Question revisited

What changes within organisms are required for reproduction?

In this chapter we have described the following:

- specific reproductive cells in organisms undergo meiosis in order to produce gametes
- gametes have half the number of chromosomes (n) compared to other body cells ($2n$)
- human females, after puberty, undergo menstrual cycles of approximately 28 days to coordinate ovulation with preparation of the uterus for embryo implantation
- flowering plants undergo regular cycles of flower production, often including asynchronous timing of pollen production and stigma maturation to avoid self-pollination
- seed germination is often delayed until environmental conditions can support a young plant.

Exercises

Q1. Arrange the following structures in the sequence that sperma would follow during the process of fertilization:

cervix, urethra, Fallopian tube, testis, vas deferens, uterus, epididymis, vagina.

Q2. Which of these is not an advantage associated with asexual reproduction?

A Typically requires less expenditure of energy

B Promotes genetic variation

C Fewer organisms can quickly populate a new environment

D May be beneficial in a stable environment

Q3. What changes does the uterus undergo during a typical menstrual cycle?

Q4. Where in a female's body does fertilization take place?

Q5. Which of these events is not associated with in vitro fertilization (IVF)?

A Hormone therapy that suspends the action of the female's own hormone cycle.

B Use of FSH to induce superovulation.

C Use of hormones to induce males to produce large numbers of sperm.

D Testing of embryos following Petri dish fertilization.

D3.2 Inheritance

Guiding Questions

What patterns of inheritance exist in plants and animals?

What is the molecular basis of inheritance patterns?

Sometimes a heritable trait is controlled by a single gene, but more often it is controlled by more than one. Certain traits show up more often in males than females depending on which chromosome the gene is located. Some versions of genes mask others, so sometimes a characteristic can skip a generation. With the ABO blood type in humans, there are multiple versions of genes rather than just two, and they can mix in such a way that someone can have type AB blood. In some flowering plant species, plants with red flowers crossed (bred) with plants with white flowers produce plants with either only red flowers or only white flowers, while some species can produce pink flowers.

Genetic variety in humans can be explained by whether or not the gene is found on the first 22 chromosomes or on the X or Y chromosome, what nucleotides are present in the different alleles, and which ones code for which proteins.

D3.2.1 – Haploid gametes and diploid zygotes

D3.2.1 – Production of haploid gametes in parents and their fusion to form a diploid zygote as the means of inheritance

Students should understand that this pattern of inheritance is common to all eukaryotes with a sexual life cycle. They should also understand that a diploid cell has two copies of each autosomal gene.

The term **diploid** is used to describe a nucleus that has chromosomes organized into pairs of homologous chromosomes. Most cells in the human body are diploid cells, and in such cells the nucleus contains a set of 23 chromosomes from the mother and 23 from the father (see Figure 1). There is a category of cell that only contains 23 chromosomes in total: the sex cells, also called **gametes**. Because the chromosomes in sperm and egg cells do not come in pairs, but rather only have a single chromosome from each pair, they are said to be **haploid**. The adult form of animal cells is rarely haploid, but there are exceptions, for example male bee, wasp and ant cells are haploid.

The variable n represents the haploid number, and it refers to the number of sets of chromosomes that a nucleus can have. For a human egg cell, $n = 23$. When an egg cell is fertilized by a sperm cell (a sperm is also haploid and therefore contains 23 chromosomes), a **zygote** is formed and the two haploid nuclei fuse together, matching up their chromosomes into pairs. Hence humans generally have a total of 23 + 23 = 46 chromosomes. This means that, in humans, $2n = 46$, so diploid cells in humans have 23 pairs of chromosomes making a total of 46 chromosomes.

D3.2 Figure 1 Two haploid gametes meet during fertilization to produce a diploid zygote.

23   23

fertilization

46

zygote – first cell of a new baby

What biological processes involve doubling and halving?

D3.2.2 – Genetic crosses in flowering plants

D3.2.2 – Methods for conducting genetic crosses in flowering plants

Use the terms "P generation", "F1 generation", "F2 generation" and "Punnett grid". Students should understand that pollen contains male gametes and that female gametes are located in the ovary, so pollination is needed to carry out a cross. They should also understand that plants such as peas produce both male and female gametes on the same plant, allowing self-pollination and therefore self-fertilization. Mention that genetic crosses are widely used to breed new varieties of crop or ornamental plants.

Mendel's experiments with pea plants

In 1865, an Austrian monk named Gregor Mendel published the results of his experiments on how garden pea plants passed on their characteristics. At the time, the term "gene" did not exist (he used the term "factors" instead) and the role that DNA played would not be discovered for nearly another century. For thousands of years people have used artificial pollination techniques, deliberately placing pollen from the male parts of one flower on the female part of another flower in order to produce seeds for the next generation. Mendel used artificial pollination to get the sperm cells in the pollen of pea plants into the ovum cells inside the ovaries of other pea plants. He covered the flowers so that bees or other pollinating insects did not interfere with his work.

Table 1 shows some of the characteristics Mendel tried to cross. The × in the first column indicates a cross between one variety of pea plant and another, i.e. two varieties were bred together. As we will see later in this chapter, the expected ratio after two generations of crosses is 3:1 (for every 3 of the first type of plant, we expect 1 of the other type): look how close Mendel got.

Gregor Mendel studied the garden pea (*Pisum sativum*).

D3.2 Table 1 Mendel's results

Characteristics in parents	First generation produced	Second generation produced	Ratio of results seen in second generation
Round × wrinkled seeds	100% round	5,474 round 1850 wrinkled	2.91:1
Yellow × green seeds	100% yellow	6,022 yellow 2,001 green	3.01:1
Green × yellow pods	100% green	428 green 152 yellow	2.82:1
Tall × short plants	100% tall	1,787 tall 277 short	2.84:1

We will consider in more detail one of the crosses that Mendel carried out with his garden pea plants. He took **purebred** tall plants and crossed them with purebred short plants. Purebred means that the tall plants' parents were known to be all tall, and the short plants' parents were known to be all short. In other words, he knew that none of the plants had a mix of short and tall traits. He wanted to find out whether he would get all tall plants, some tall and some short plants, or all short plants when he crossed tall and short plants.

The answer took months for Mendel to confirm, but theoretical techniques can now be used to get the answer in seconds: the result was 100% tall plants. Why? Because in garden pea plants the **allele** (version of the gene) for tall is dominant over the allele for short plants, thus masking the short trait in **heterozygotes**. A heterozygote is an organism that possesses one dominant allele and one recessive allele for a particular trait, e.g. **Tt**. To show this a **Punnett grid** can be used.

Constructing a Punnett grid

Figures 2 and 3 on the next page show a Punnett grid. A Punnett grid can be used to show how the alleles of parents are split between their gametes and how new combinations of alleles can show up in their offspring. The purpose of a Punnett grid is to show all the possible combinations of genetic information for a particular trait in a **monohybrid cross**. A monohybrid cross is one in which the parents have different alleles, and it shows the results for only one trait. We will see how this works with Mendel's tall and short pea plants. In order to set up a Punnett grid, you need to follow a series of steps.

1. Choose a letter to represent the alleles.

   Use the capital (uppercase) and lowercase of the letter to represent the different alleles. Usually, a capital letter represents the **dominant allele** while the lowercase letter represents the **recessive allele**. Recessive alleles can be masked by dominant ones. For example:

   - **T** = dominant allele, for example tall pea plants is a trait that is dominant over short pea plants
   - **t** = recessive allele, the trait that results in short pea plants.

2. Determine the parents' genotypes.

   To be sure that no possibilities are overlooked, write out all three possibilities (based on the possible combinations of two traits) and decide by a process of elimination which genotype or genotypes fit each parent.

   The three possibilities here are:

   - **TT** – the genotype is **homozygous dominant**, and in this example the phenotype presents as tall plants
   - **Tt** – the genotype is heterozygous (with one of each allele), and in this example the phenotype presents as tall plants (heterozygotes carry the recessive allele but display the dominant trait, and they can potentially pass the recessive allele on to their offspring)
   - **tt** – the genotype is **homozygous recessive**, and in this example the phenotype presents as short plants.

   The easiest genotype to determine by simply looking at a pea plant is **tt**. The other two are more of a challenge. To determine whether a plant is **TT** or **Tt**, we have to look for evidence that the recessive gene was received from a short parent or was passed on to the plant's offspring. The only way to produce a short plant is for each parent to donate one **t**.

   If a cross is to be made between **TT** and **tt** purebred parents, they represent the **parental generation**, or **P generation**. The name given to the first generation produced by such a cross is the **first filial generation**, usually referred to as the

Get used to saying "big T" and "little t" when reading alleles and genotypes. Also, do not mix letters: for example, you cannot use **T** for tall and **s** for short. Once you have chosen a letter, write down what it means so that it is clear which allele is which.

F1 generation. Crossing two members of the F1 generation produces the **second filial (F2) generation**.

3. Determine the gametes that the parents could produce.

   An individual with a genotype **TT** can only make gametes with the allele **T** in them.

   Heterozygous carriers can make **T**-containing gametes or **t**-containing gametes. Obviously, individuals whose genotype is **tt** can only make gametes that contain the **t** allele. So you can record and label with **T** or **t** all the possible gametes.

4. Draw a Punnett grid.

   Once all the previous steps have been completed, drawing the actual grid is simple. The parents' gametes are placed on the top and side of the grid. As an example, consider a cross involving a female plant **Tt** crossed with a male short plant **tt**.

   Figure 2 shows a Punnett grid for the parents' gametes.

   Now you can fill in the empty squares with each parent's possible alleles by copying the letters from the top down and from left to right. When letters of different sizes end up in the same box, the big one goes first, as shown in Figure 3.

5. Work out the chances of each genotype and phenotype occurring.

   In a grid with four squares, each square represents one of two possible statistics:

   - the chance that these parents will have offspring with that genotype, each square representing a 25% chance
   - the probable proportion of offspring that will have the resulting genotypes, although this only works for large numbers of offspring.

   The grid in Figure 3 for Mendel's tall and short plants can be interpreted in two ways:

   - there is a 50% chance of producing tall offspring and a 50% chance of producing short offspring
   - 50% of the offspring will be tall and 50% of the offspring will be short.

   All the tall plants are heterozygous.

   In a real experiment, it is unlikely that exactly 50% of the offspring will be short plants. The reason is essentially due to chance. For example, if 89 F2 peas were produced and all of them were planted and grew into new plants, there is no mathematical way that exactly 50% of them would be short. At the very most, mathematically 45 out of the 89 plants would be short, which is 50.6%; that is as close to 50% as is possible in this case.

   Even if a convenient number of plants was produced, such as 100 plants, farmers and breeders would not be surprised if they got, for example, 46, 57 or even 63 short plants instead of the theoretical 50. If the results of hundreds of similar crosses were considered, however, the number would probably be very close to 50%.

▲ **D3.2 Figure 2** A Punnett grid showing the parent plants' gametes.

▲ **D3.2 Figure 3** A Punnett grid with all the possible genotypes filled in.

 Be careful when choosing letters when writing them by hand. Nearly half the letters of the alphabet should in fact be avoided because the capital and lowercase versions are too similar, which can lead to confusion. Try not to use Cc, Ff, Kk, Oo, Pp, Ss, Uu, Vv, Ww, Xx, Yy or Zz. If in an exam one of these letters is used in a question, be sure to clearly differentiate between lowercase and capitals when writing your answer.

Summarizing the five steps of the Punnett grid method:

1. **Choose a letter.**
   T = allele for a tall plant
   t = allele for a short plant

2. **Define the parents' genotypes.**
   TT for the purebred tall parent. **tt** for the purebred short parent.

3. **Determine the gametes.**
   The purebred tall parent can only provide **T**. The purebred short parent can only provide **t**.

4. **Draw a Punnett grid.**

	t	t
T	Tt	Tt
T	Tt	Tt

5. **Interpret the grid.** 100% will be Tt and therefore will be tall, so 0% will be short.

Self-pollination

Some plants, such as peas, have flowers that can produce both male pollen and female ova. If they prepare gametes at the same time, it is possible for them to **self-pollinate**. Self-pollination is when a plant's pollen lands on flowers it has produced itself. As a result, it is possible for **self-fertilization** to happen. This will result in less genetic diversity than cross-pollination, which allows genes from different individuals of a species to breed.

When farmers want plants with the same characteristics as previous generations, they can use self-pollination techniques, but when they want to create new varieties with combinations of traits not seen before, they can use cross-pollination techniques. Modern wheat, for example, was produced by crossing breeds that have stalks that are a convenient height for higher yields, that do not drop their seeds before harvest time, and are resistant to disease. It took many years of scientific research and thousands of cross-breeding trials over many years to achieve the desired combination of traits.

D3.2.3 – Combinations of alleles

D3.2.3 – Genotype as the combination of alleles inherited by an organism

Students should use and understand the terms "homozygous" and "heterozygous", and appreciate the distinction between genes and alleles.

Alleles: versions of genes

An **allele** is one specific form of a gene, differing from other alleles by one or a few bases. For example, some people are born without the ability to see any colours because of an inability to produce the protein transducin in their retinas. This is caused by a single base pair difference between the most common allele (with a C at position 235) and the rare mutated allele (with a T at position 235). These different forms allow a single trait, such as the trait for the ability to see in colour, to have variants, in this example either colour or grey-scale vision.

Chapter D1.2 discusses transcription and translation of DNA, and shows how important it is for each letter in the genetic code to be in a specific place. If, for whatever reason, one or more of the bases (A, C, G or T) is misplaced or substituted for a different base, the results can be dramatic. The difference between one version of a gene and another (e. g. the mutated and non-mutated alleles of the transducin gene) can mean the difference between fully functional organs and impaired organs.

Genotype

The **genotype** is the symbolic representation of the pair of alleles possessed by an organism, typically represented by two letters. All eukaryotes that reproduce sexually will inherit one allele from the female parent and one from the male parent.

Examples: **Bb**, **GG**, **tt**.

Homozygous refers to having two identical alleles of a gene (see Figure 4).

Examples: **AA** is the genotype of someone who is homozygous dominant for that trait, whereas **aa** is the genotype of someone who is homozygous recessive.

Heterozygous refers to having two different alleles of a gene (see Figure 5). This results when the paternal and maternal alleles are different.

Example: **Aa** is a heterozygous genotype.

A **carrier** is an individual who has a recessive allele of a gene that does not have an effect on the phenotype.

Example: **Aa** carries the allele for albinism but has pigmentation, which means an ancestor must have been albino and some offspring might be albino. If both parents are unaffected by a recessive condition yet both are carriers, some of their offspring can be albinos (because they would be **aa,** like the penguin in Figure 6).

A gene is a DNA sequence that codes for a protein that will give an organism a specific trait, whereas alleles are versions of a gene. In peas, for example, all plants have a gene that determines their height, but one allele of the gene is for tall and the other allele is for short.

▲
D3.2 Figure 4 The **locus** of a gene on a pair of chromosomes. The locus is the particular position of a gene on homologous chromosomes. Sometimes, as in this example, the allele we inherit from each parent is the same.

▲
D3.2 Figure 5 In this instance, the individual has inherited **A** on the maternal chromosome and **a** on the paternal chromosome.

◄ **D3.2 Figure 6** An albino does not have the genes necessary to produce pigmentation. That is why this Cape penguin (*Spheniscus demersus*) has no black coloration.

D3.2.4 – Phenotype

D3.2.4 – Phenotype as the observable traits of an organism resulting from genotype and environmental factors

Students should be able to suggest examples of traits in humans due to genotype only and due to environment only, and also traits due to interaction between genotype and environment.

The **phenotype** is the observable characteristics or traits of an organism. Examples include being colour blind and having type O blood. But phenotype is not always 100% the result of genes. A person is born with a certain skin colour but that colour can change with exposure to sunlight. A suntan or burn is an acquired trait and is caused by the ultraviolet (UV) rays present in the environment. Can you think of examples of characteristics produced only by genetics? Only by the environment? What about traits that are a mix of genetics and environment? Below are some suggestions.

Phenotypes produced exclusively by genetics include:

- ABO blood type
- genetic conditions such as Huntington's disease, cystic fibrosis and colour blindness.

Phenotypes produced exclusively by the environment include:

- learned behaviour (e.g. birds learning a new song, humans learning mathematics)
- acquired physical traits (such as a scar or large muscles from weightlifting).

Phenotypes produced by an interaction between genes and the environment include:

- height in humans (while the maximum height is genetic, a poorly nourished person might not reach their maximum potential)
- cancer (which can have a genetic component but is often triggered by cancer-causing substances in the environment).

D3.2.5 – Dominant and recessive alleles

D3.2.5 – Effects of dominant and recessive alleles on phenotype

Students should understand the reasons that both a homozygous-dominant genotype and a heterozygous genotype for a particular trait will produce the same phenotype.

A **dominant allele** is an allele that has the same effect on the phenotype whether it is paired with the same allele or a different one. Dominant alleles are always expressed in the phenotype.

Example: the genotype **Aa** gives rise to the dominant **A** trait because the **a** allele is masked; the **a** allele is not transcribed or translated during protein synthesis.

A **recessive allele** is an allele that has an effect on the phenotype only when no dominant allele is present to mask it.

Example: **aa** gives rise to the recessive trait because there is no **A**.

An individual with **Aa** will have the same phenotype as an individual with **AA**.

Co-dominant alleles are pairs of alleles that both affect the phenotype when present in a heterozygote.

Example: a parent with curly hair and a parent with straight hair can have children with different degrees of hair curliness, because both alleles influence hair condition when both are present in the genotype.

D3.2.6 – Phenotypic plasticity

D3.2.6 – Phenotypic plasticity as the capacity to develop traits suited to the environment experienced by an organism, by varying patterns of gene expression

Phenotypic plasticity is not due to changes in genotype, and the changes in traits may be reversible during the lifetime of an individual.

Phenotypic plasticity is an organism's ability to express its phenotype differently depending on the environment. Not surprisingly, this is an effective way of adapting, and natural selection is all about either adapting or dying out. Birds can activate genes that produce more of the digestive enzyme maltase when fewer insects are available to eat and there are more grains in their diet. A plant can activate genes that produce growth hormones to make thicker leaves when it senses that there is more light available. Animals can modify their foraging behaviour depending on the types of foods available in their environment from one season to the next. Cyclical events such as seasonal food availability are referred to as **phenological**. Phenotypic plasticity can generate changes in physiology (as in the enzyme example), in morphology (the leaf example, or the snail shells we shall look at next), in behaviour (e.g. foraging) or in phenology (e.g. adapting to seasons). Notice that some are permanent changes, such as how an organism grows, but others can change within the organism's lifetime, such as physiological and behavioural changes.

The freshwater snail *Physa virgata* normally has an elongated shell. However, this snail can sense cues from its predator (the bluegill fish, *Lepomis macrochirus*) and express its genes slightly differently in order to grow a less conical shape and a more rounded, shorter shell that is more difficult for its predator to crush. The genotype itself does not change, so this is not a mutation. The snail still has the same alleles. But by expressing and silencing alleles in different combinations, the shell shape can grow differently and produce a modified phenotype.

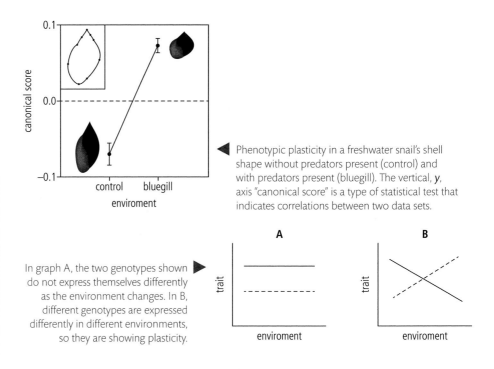

Phenotypic plasticity in a freshwater snail's shell shape without predators present (control) and with predators present (bluegill). The vertical, *y*, axis "canonical score" is a type of statistical test that indicates correlations between two data sets.

In graph A, the two genotypes shown do not express themselves differently as the environment changes. In B, different genotypes are expressed differently in different environments, so they are showing plasticity.

3.2.7 – Recessive genetic conditions

D3.2.7 – Phenylketonuria as an example of a human disease due to a recessive allele

Phenylketonuria (PKU) is a recessive genetic condition caused by mutation in an autosomal gene that codes for the enzyme needed to convert phenylalanine to tyrosine.

How is it possible for two healthy parents to have a child who is affected by a genetic disease? The disease is caused by a recessive allele and both healthy parents must be carriers of the version of the gene that causes the disease. For example, the genetic disease called **phenylketonuria**, abbreviated to **PKU**, is caused by mutations in the autosomal *PAH* gene, which results in low levels of the enzyme phenylalanine hydroxylase. This enzyme converts the amino acid phenylalanine into tyrosine, which is not toxic. If there is a large quantity of protein in a child's diet, phenylalanine levels can become toxic and can impair brain development. Children diagnosed with PKU are recommended a diet that omits foods rich in phenylalanine, such as eggs, chicken and nuts. The problem is such foods are rich in protein, which is needed for healthy growth and development. As a result, dietary supplements are sometimes needed.

Let us call the allele that produces a functioning enzyme **F**, and the allele that can cause PKU **f**. In Figure 7, showing a family that has the PKU allele, the parents James and Iyana are carriers (**Ff**). The only way to have the disease is to have the genotype **ff**, so James and Iyana do not suffer from PKU but they can pass it on to their children. If you set up a Punnett grid for these parents, you will see that there is a 1 in 4 chance (25%) that they will have a child with PKU, and there are three possibilities for the genotypes in their children: Malik is **Ff**, Jada is **ff**, and Kasim is **FF**.

parents' alleles

James   Iyana

children's alleles

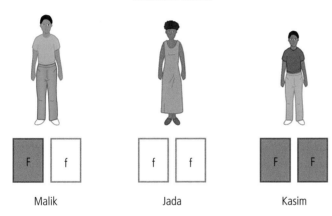

Malik   Jada   Kasim

◀ **D3.2 Figure 7** How an autosomal gene can be passed on to offspring: Malik is a carrier like his parents, Jada has the autosomal disease caused by the recessive allele, and Kasim not only does not have PKU but cannot pass the recessive allele on to his future children.

Such diseases are called **autosomal recessive diseases** because they are caused by recessive alleles, and the locus of their gene is found on one of the first 22 pairs of chromosomes but not on the sex chromosomes X or Y. Parents can find out if their child has PKU shortly after birth thanks to blood tests that can screen for it. The following are other examples of autosomal recessive diseases, which are all rare in the human population:

- albinism
- cystic fibrosis
- sickle cell disease
- Tay Sachs disease
- thalassemia.

Students sometimes get the impression that genetics is only about diseases. This is not true. It is just that more is known about disease-causing genes than about genes for characteristics such as eye colour, because researchers spend time and funds studying topics that can help society. Studying diseases and discovering their genetic causes is more useful to medicine than studying eye colour. Governments and university laboratories investing money in research want their work and their discoveries to lead to healthier lives for people. Getting a return on their investment also motivates them. Fundamental research ("I would like to study this just to find out how it works") does not attract as much funding as applied research ("I would like to find out how this disease is caused so that we can develop better medical treatments for it").

D3.2.8 – Single-nucleotide polymorphisms and multiple alleles

D3.2.8 – Single-nucleotide polymorphisms and multiple alleles in gene pools
Students should understand that any number of alleles of a gene can exist in the gene pool but an individual only inherits two.

So far, only two possibilities have been considered for a gene: dominant, for example **A**, and recessive, **a**. With two alleles, three different genotypes are possible, which can produce two different phenotypes. However, genetics is not always this simple; sometimes there are three or more alleles for the same gene. This is the case for the alleles that determine the ABO blood type in humans, which we will look at in detail in Section D3.2.9. In the immune system, there are genes called HLA genes, some of which have dozens of alleles and some thousands. This makes sense when you think of all the different pathogens our bodies have to fight off.

Single-nucleotide polymorphisms, also known as **SNPs**, occur when a nucleotide of the genetic code, such as T, is not found where it is expected, and another, such as C, is found at that position instead. If such a difference is found in a coding part of the genome, it can potentially cause another amino acid to be coded for, and this could modify the structure and properties of the protein that is translated. Such a variation in a nucleotide indicates a different allele.

The ability to taste bitter substances is important to a mammal's survival because bitterness in nature is often a warning that something is toxic. The molecule phenylthiocarbamide (PTC) can be perceived as tasting bitter if a person possesses a particular taste receptor. People with one version of the gene *TAS2R38* can taste PTC while people with a different version cannot taste it. The protein that is encoded by this gene, taste receptor 2 member 38, allows the transduction of a chemical stimulus (bitterness) into a nerve signal for the brain to interpret. If the protein cannot be produced correctly, no signal or a much weaker signal will be passed. Several SNPs have been found on this gene and some account for an allele that is different enough to not allow the person with two copies of the allele to be able to taste PTC. This trait is an example of one that has multiple alleles.

For a person to be able to taste PTC, only one functioning allele needs to be present. It is thought to be a dominant allele, although there is debate about this in the scientific community because in some cases it does not show complete dominance. For our purposes, we will assume it is dominant. Most people can taste PTC and other thiourea compounds similar to it. It is customary to use **T** for "taster" and **t** for "non-taster", to show that three genotypes are possible, **TT**, **Tt** or **tt**. Only the last one would be unable to taste thiourea compounds. In reality, the recessive alleles have more than one version depending on which SNPs they have. But no matter how many alleles are available in the gene pool, each person receives two: one from their mother and one from their father.

D3.2.9 – ABO blood groups

D3.2.9 – ABO blood groups as an example of multiple alleles

Use I^A, I^B and i to denote the alleles.

The ABO blood type system in humans has four possible phenotypes: A, B, AB and O. To create these four blood types there are three alleles of the gene. These three alleles can produce six different genotypes.

The gene for the ABO blood type is represented by the letter **I**. To represent more than just two alleles (**I** and **i**) superscripts are introduced. As a result, the three alleles for blood type are written as follows: I^A, I^B and **i**. The two capital letters with superscripts represent alleles that are co-dominant:

- I^A = the allele for producing proteins called type A antigens, giving type A blood
- I^B = the allele for producing proteins called type B antigens, giving type B blood
- **i** = the recessive allele that produces neither A nor B antigens, giving type O blood.

Crossing these together in all possible combinations creates six genotypes that give rise to the four phenotypes already listed:

- $I^A I^A$ or I^A**i** gives a phenotype of type A blood
- $I^B I^B$ or I^B**i** gives a phenotype of type B blood
- $I^A I^B$ gives a phenotype of type AB blood (because of codominance, both types of antigens are produced)
- **ii** gives a phenotype of type O blood.

Notice how the genotype $I^A I^B$ clearly shows codominance. Neither allele is masked: both are expressed in the phenotype of type AB blood. The person makes proteins that include type A antigens and type B antigens.

Worked example

Is it possible for a couple to have four children, each child with a different blood type?

Solution

There is only one way this can happen: one parent must have type A blood but be a carrier of the allele for type O blood, and the other parent must have type B blood and also be a carrier of the allele for type O blood, as shown in the figure below (if necessary, remind yourself of the blood group alleles listed above).

The cross would be $I^A i \times I^B i$ and the corresponding Punnett grid is shown below. See if you can determine the phenotype of each child before reading on.

Would it be possible for the same couple to have four children and all of them have type AB blood? In theory, yes, but it would be unlikely. This question is similar to asking "Could a couple have four children, all of them girls?" It is possible but statistically less likely than having boys and girls.

How ABO blood groups can be inherited.

	I^A	i
I^B	$I^A I^B$	$I^B i$
i	$I^A i$	ii

A Punnett grid for blood type alleles.

D3.2.10 – Intermediate and dual phenotypes

D3.2.10 – Incomplete dominance and codominance

Students should understand the differences between these patterns of inheritance at the phenotypic level. In codominance, heterozygotes have a dual phenotype. Include the AB blood type (I^AI^B) as an example. In incomplete dominance, heterozygotes have an intermediate phenotype. Include four o'clock flower or marvel of Peru (*Mirabilis jalapa*) as an example.

Note: When students are referring to organisms in an examination, either the common name or the scientific name is acceptable.

The four o'clock flower (*Mirabilis jalapa*), also known as marvel of Peru, gets its name because it tends to open its flowers in the later part of the afternoon. Unlike the ABO blood type in humans, which can show codominance with I^AI^B expressing both phenotypes, this plant shows **incomplete dominance** in its genetics regarding flower pigmentation, resulting in neither one phenotype nor the other but something in between. Here is how it works.

The system of letters for showing colour in four o'clock flowers uses a prefix **C**, which refers to the gene that codes for flower colour, plus a superscript, which refers to the specific colour, **R** (red) or **W** (white).

So the alleles for flower colour are:

- C^R for red flowers
- C^W for white flowers.

The genotypes and their phenotypes are:

- C^RC^R makes red flowers
- C^WC^W makes white flowers
- C^RC^W makes pink flowers.

In a cross of purebred four o'clock flowers, white × red = pink (shown in Figure 8). Notice how pink is not a colour present in either of the parents. It is a mix of the two parents' alleles and each one contributes to the phenotype but neither is masked. This is why we call it incomplete dominance.

D3.2 Figure 8 With incomplete dominance, white and red parent flowers can generate pink offspring.

D3.2.11 – Sex determination

D3.2.11 – Sex determination in humans and inheritance of genes on sex chromosomes

Students should understand that the sex chromosome in sperm determines whether a zygote develops certain male-typical or female-typical physical characteristics and that far more genes are carried by the X chromosome than the Y chromosome.

Sex determination

The chromosomes in the 23rd pair in humans are called the sex chromosomes because they determine whether a person is a male or a female. The X chromosome is longer than the Y chromosome, and contains many more genes. Unlike the other 22 pairs of chromosomes, this is the only pair in which it is possible to find two chromosomes that are very different in size and shape (Figure 9).

D3.2 Figure 9 Human chromosomes: the majority (grey box) = autosomes, while one pair (shown in a separate purple box) = sex chromosomes.

In human females there are two X chromosomes. When females produce gametes, each egg will contain one X chromosome. Human males have one X chromosome and one Y chromosome. When males produce sperm cells, half of them contain one X chromosome and half contain one Y chromosome. As a result, when an egg cell meets a sperm cell during fertilization, there is always a 50% chance that the child will be a male and a 50% chance that the child will be a female (see Figure 10):

- XX = female
- XY = male.

D3.2 Figure 10 How sex is determined: will the baby be male or female?

parents

gametes

offspring

The chances remain the same no matter how many males and females a family already has. The genetics of being male or female depends on whether you inherit an X or a Y chromosome from your father. If a sperm cell containing an X chromosome fertilizes an egg, a female is produced. Conversely, if a sperm cell containing a Y chromosome fertilizes an egg, a male is produced. Embryos of both sexes are virtually identical until about the eighth week following fertilization. Alleles that interact on both of the X chromosomes of female embryos then result in relatively high oestrogen and progesterone production, resulting in the prenatal development of female reproductive structures. Genes located on the single Y chromosome are responsible for early testes development and relatively high testosterone production, resulting in male reproductive structures during subsequent foetal development.

Inheritance of genes on sex chromosomes

In humans, because the Y chromosome is significantly smaller than the X chromosome, it has fewer loci and therefore carries fewer genes (about 70 in total) than the X chromosome (which carries about 800). This means that most of the alleles present on the X chromosome have nothing to pair up with. For example, a gene whose locus is at an extremity of the X chromosome would have no counterpart on the Y chromosome because the Y chromosome does not extend that far from its centromere.

D3.2.12 – Haemophilia

D3.2.12 – Haemophilia as an example of a sex-linked genetic disorder
Show alleles carried on X chromosomes as superscript letters on an uppercase X.

Sex linkage

Any genetic trait that has a gene locus on the X or the Y chromosome is said to be a **sex-linked trait**. Often genetic traits that show sex linkage affect one sex more than the other. One example of such a genetic trait is **haemophilia**.

Haemophilia is a disorder in which blood does not clot properly. For most people, a small cut or scrape on their skin stops bleeding after a few minutes and eventually a scab forms. This process is called **clotting**. People with haemophilia have trouble with blood clotting and are at risk of bleeding to death from what most people would consider to be a minor injury such as a bruise, which is a rupture of many tiny blood vessels. Bleeding can also occur in internal organs. Medical treatments give people affected by haemophilia a better quality of life.

Alleles and genotypes of sex-linked traits

The alleles for haemophilia, represented by **H** and **h**, are found only on the X chromosome:

- X^h = allele for haemophilia
- X^H = allele for the ability to clot blood
- **Y** = no allele present on the Y chromosome.

The letters X and Y refer to chromosomes and not to alleles, so terms such as dominant and recessive do not apply. X and Y should be considered as entire chromosomes rather than alleles of a gene. For sex-linked alleles, the letter that indicates the allele is the superscript after the X or Y. An absence of a superscript means that no allele for that trait exists on that chromosome.

As there is no allele on the Y chromosome, Y is written alone without any superscript. Here are all the possible genotypes for haemophilia:

- $X^H X^H$ gives the phenotype of a non-affected female
- $X^H X^h$ gives the phenotype of a non-affected female who is a carrier
- $X^h X^h$ gives the phenotype of an affected female
- $X^H Y$ gives the phenotype of a non-affected male
- $X^h Y$ gives the phenotype of an affected male.

Notice how only one sex can be a carrier.

Carriers of sex-linked traits

Sex-linked recessive alleles such as X^h are rare in most populations of humans worldwide. For this reason, it is unlikely for offspring to inherit one and much less likely to inherit two such alleles. This is why so few females have haemophilia: their second copy of the gene is likely to be the dominant allele and will mask the recessive allele.

As you have seen, there are three possible genotypes for females but only two possible genotypes for males. Only females can be heterozygous, $X^H X^h$, and, as a result, they are the only ones who can be carriers.

Because males do not have a second X chromosome, there are only two possible genotypes, $X^H Y$ or $X^h Y$, in relation to haemophilia. With just the one recessive allele, **h**, a male will have haemophilia. This is contrary to what you have seen up to now concerning recessive alleles: usually people need two to have the trait, while with one they are carriers. In this case, the single recessive allele in males determines the phenotype. Males cannot be carriers for X-linked alleles.

Other examples of sex-linked traits include:

- colour blindness in humans
- Duchenne muscular dystrophy in humans
- white eye colour in fruit flies
- calico–tortoiseshell fur colour in cats.

D3.2.13 – Pedigree charts

D3.2.13 – Pedigree charts to deduce patterns of inheritance of genetic disorders

Students should understand the genetic basis for the prohibition of marriage between close relatives in many societies.

NOS: Scientists draw general conclusions by inductive reasoning when they base a theory on observations of some but not all cases. A pattern of inheritance may be deduced from parts of a pedigree chart and this theory may then allow genotypes of specific individuals in the pedigree to be deduced. Students should be able to distinguish between inductive and deductive reasoning.

The term "pedigree" refers to the record of an organism's ancestry. Pedigree charts are diagrams that are constructed to show biological relationships. In genetics, they are used to show how a trait can pass from one generation to the next. Used in this way for humans, a pedigree chart is similar to a family tree, complete with parents, grandparents, aunts, uncles and cousins.

To build such a chart, symbols are used to represent people. Preparing a pedigree chart helps with the use of Punnett grids for predicting the probable outcome for the next generation. As an example, we will look at the inheritance of Huntington's disease.

Huntington's disease (Huntington's chorea) is caused by a dominant allele that we will refer to by the letter **H**. This genetic condition causes severely debilitating nerve damage but the symptoms do not show until a person is about 40 years old. As a result, someone who has the gene for Huntington's disease may not know they have it until after they have had their own children.

Huntington's disease is life-limiting. Symptoms include difficulty walking, speaking and holding objects. Within a few years of starting to display symptoms, the person loses complete control of their muscles. Because it is dominant, all it takes is one **H** allele in a person's genetic makeup to cause the condition.

Worked example

A pedigree chart showing members of a family affected by Huntington's disease.

1. Describe the six individuals in the figure, stating who is affected and who is not. See the Key fact box for what the use of shape and colour means.

2. State the genotype for each individual.

Solution

1. The symbols indicate that the unaffected members of the family are the mother, the first child (a girl) and the fourth child (a boy). Those who are affected are the father, the second child (a boy) and the third child (a girl).

2. To work out whether the father is **HH** or **Hh**, consider the fact that some of his children do not have the trait. This indicates that he must have given one **h** to each of them. Hence, he can only be **Hh** and not **HH**. The mother is not affected so she must be **hh**. This is also true for the first daughter and the youngest son. Since the mother always gives an **h**, the two middle children must have at least one **h**, but, because they are affected, they are **Hh**.

From the pedigree shown in the Worked example above, it should be clear why it is taboo, discouraged or illegal in most societies for siblings or other close relatives to have children together. Rare diseases are rare in the general population but among closely related relatives they are much more frequent. In the pedigree example shown, half the individuals (50%) suffer from Huntington's disease, whereas in the general population only one in about 10,000 (or about 0.01%) suffer from this disease. Long before scientists understood genetics, groups of humans forbade intermarriage between close relatives, undoubtedly because they saw that unfavourable traits or diseases were produced in disturbingly higher frequencies when this was allowed.

The symbols used in pedigree charts are:

○ **empty circle = female**

□ **empty square = male**

● **filled-in circle = a female who possesses the trait being studied**

■ **filled-in square = a male who possesses the trait being studied**

| **vertical line = the relationship between parents and offspring**

– **horizontal line between a male and a female = the parents of the offspring.**

Worked example

The pedigree chart below shows how red, white and pink pigmentation is passed on in four o'clock flowers. For codominant traits, grey is used in pedigree charts rather than black or white. For allele symbols and genotypes, refer to Section D3.2.10.

1. Using the pedigree chart below, state the genotypes for all the plants A to K.

2. What evidence is there that genetic characteristics can sometimes skip a generation?

This pedigree chart shows how pink flowers can arise in purebred four o'clock (*Mirabilis jalapa*) flowers. Black shapes represent plants with red flowers, white shapes represent white-flowered plants and grey shapes represent plants with pink flowers.

Solution

1. A and H produce red flowers and must be homozygous for red, $C^R C^R$, because any other combination would give pink or white. B and K produce white flowers and must be homozygous for white, $C^W C^W$, because any other combination would give pink or red. C to G as well as I and J are pink and must be heterozygous, $C^R C^W$, because they have one of each allele from each parent plant.

2. It would be impossible for either the colour red or the colour white to be in the middle (F1) generation in this figure. These colours skip a generation and show up again in the last row, the F2 generation.

Nature of Science

When we use a pedigree chart to determine how a gene is passed down, we need to use both inductive and deductive reasoning. Inductive reasoning is used when a conclusion or theory is worked out by looking at samples of evidence of a phenomenon. Deductive reasoning is when we apply well-established knowledge about a phenomenon to reach a conclusion or theory to explain what is happening. In the pedigree chart above, we can observe which plants have which colour in the phenotype of their flowers, and use inductive reasoning when we notice that all the offspring show a phenotype that is different from the parents (pink flowers instead of red or white). We use deductive reasoning when we apply our knowledge of the phenomenon of inheritance of codominant genes in the genotype, $C^R C^W$. In a similar fashion we use both inductive and deductive reasoning to explain how the red and white traits can skip a generation.

D3.2.14 – Continuous variation

D3.2.14 – Continuous variation due to polygenic inheritance and/or environmental factors

Use skin colour in humans as an example.

Application of skills: Students should understand the distinction between continuous variables such as skin colour and discrete variables such as ABO blood group. They should also be able to apply measures of central tendency such as mean, median and mode.

Polygenic inheritance

Polygenic inheritance involves two or more genes influencing the expression of one trait. With two or more allelic pairs found at different loci, the number of possible genotypes is greatly increased. It is believed that most human traits are too complex and show too many combinations to be determined by one gene.

This could partly explain the difficulty in finding out which genes are responsible for traits whose genetic components are poorly understood, for example mathematical aptitude, musical talent or susceptibility to certain illnesses.

Continuous and discrete variation

With dominant and recessive alleles of a single gene, the number of possible phenotypes is limited. For example, a person either has PKU or not. When multiple alleles are introduced, the number of possibilities for a single trait increases accordingly. For example, the ABO blood type has three alleles and four possible phenotypes.

When a second gene is introduced, the number of possible genotypes increases dramatically. If three, four or five genes determine the phenotype, the number of possibilities is so big that it is impossible to see the difference between certain genotypes in the phenotype. When an array of possible phenotypes can be produced, it is called **continuous variation**. The colour of skin in humans is an example of continuous variation and it is thought that the intensity of pigment in our skin is the result of the interaction of multiple genes.

In humans, continuous variation can also be seen in the genetic components of traits such as height, body shape and intellectual aptitude. Each of these is also influenced by environmental components. A person's height, for example, is determined by whether they inherit genes for tallness, but it also depends on their nutrition as they grow.

Polygenic characteristics

When there are many intermediate possibilities in a phenotype, then the trait shows continuous variation. If the results are plotted as a graph, it will produce a bell-shaped distribution curve. There is a smooth transition between the groups of frequencies (see the figure in the Skills box).

Discontinuous variation:

- in distinct categories which have no transition between them
- the order does not matter
- best plotted as a bar chart
- can determine a mode but not a mean as there is no central tendency.

Continuous variation:

- not in distinct categories
- the order matters
- smooth transition from one value to the next with no abrupt jumps
- best plotted as a curve or histogram
- can determine a mean to show central tendency.

To help you decide whether or not a trait shows continuous variation, imagine a questionnaire for recording phenotypes. In general, if it is possible to tick "yes" or "no" for a trait, that trait does not show continuous variation, for example grey-scale or colour vision. The same is true for a trait where the possibilities can be represented by just a few choices, such as blood type (A, B, AB or O).

When variation is not continuous, it is referred to as **discrete variation** or **discontinuous variation**. The data for discontinuous variation can be displayed as bar charts (see the figure below). An unbroken transitional pattern from one group to another is not present.

▲
Blood type is an example of discontinuous variation.

▲
Height in humans is an example of a trait that shows continuous variation.

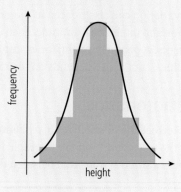

▲
Height in humans is an example of continuous variation, with an even distribution around a mean.

Is it fair to compare heights of humans from different parts of the world? On the one hand, we can argue that we are all the same species and therefore we are comparable. On the other hand, there has been a certain amount of isolation of gene pools over thousands of years, and we also know that different people grow up in very different environments. We are all the same. We are all different.

D3.2.15 – Box-and-whisker plots

> **D3.2.15 – Box-and-whisker plots to represent data for a continuous variable such as student height**
>
> **Application of skills**: Students should use a box-and-whisker plot to display six aspects of data: outliers, minimum, first quartile, median, third quartile and maximum. A data point is categorized as an outlier if it is more than 1.5 × IQR (interquartile range) above the third quartile or below the first quartile.

One way to show data for a trait that has continuous variation is by using a **box-and-whisker plot**. To generate the graph, we need to calculate some values that help describe the data set (see Figure 11).

Quartiles can be determined by entering the data points in a graphing calculator or a spreadsheet program. They each have a number and a relationship to the **median** value in the data set or to the maximum or minimum value. **Quartile 1** or the first quartile is the middle value between the median and the lowest value in the data set, otherwise expressed as the **25th percentile**. A percentile is a way of expressing the number of data points found below this level, so data in the 25th percentile means a quarter of the data points can be found below this level.

Quartile 2 is the median or the data point at the 50th percentile. **Quartile 3** or the third quartile is the middle number between the median and the highest value in the data set, the **75th percentile**. The **interquartile range**, **IQR**, measures the spread of the data and is defined as the difference between the 75th and 25th percentiles of the data. The IQR will be the box part of the plotted data. Multiplying the IQR by 1.5 gives us the length of the whiskers to add to the box. Any data points that are beyond the whiskers are considered **outliers**. Although outliers can sometimes be the result of faulty data collection, they can also sometimes reveal unexpected results that lead to a whole new investigation.

D3.2 Figure 11 This diagram shows which numbers are used to make the box and which are used to make the whiskers in a box-and-whisker plot. This plot is presented horizontally, but box-and-whisker plots can be displayed vertically.

Worked example

Using the data from the Centers for Disease Control and Prevention (CDC) shown below, construct a box-and-whisker plot showing the height of 17-year-old females in the US. Would a height of 183 cm be considered an outlier?

A summary of the data shown in the figure, rounded to the nearest cm, is:

25th percentile = quartile 1 = 159 cm

50th percentile = quartile 2 (the median) = 163 cm

75th percentile = quartile 3 = 167 cm

Solution

Calculate the interquartile range (IQR) as well as the IQR × 1.5:

IQR = 167 − 159 = 8

IQR × 1.5 = 12

Calculate where the whiskers should be by subtracting the IQR × 1.5 value from quartile 1 to get the bottom whisker, and adding the IQR × 1.5 value to quartile 3 to get the top whisker:

bottom whisker = 159 − 12 = 147 cm

top whisker = 167 + 12 = 179 cm

Using the numbers you have generated, construct the box-and-whisker plot. The figure below presents the plot vertically. A height of 183 cm would be considered an outlier because it is beyond the end of the top whisker.

A box-and-whisker plot of the height of females.

The box-and-whisker plot, also known as a **boxplot**, shows the IQR as well as the highest and lowest values in the data set. It shows the spread of the data and whether or not the data is evenly distributed around the median or **skewed** (biased) towards one side or other of the median. Data points that are outside the IQR zone can be considered outliers.

Guiding Question revisited

What patterns of inheritance exist in plants and animals?

In this chapter you have learned:

- genetic crosses show how haploid gametes can join to form diploid zygotes with various combinations
- dominant alleles mask recessive ones, so usually two copies of a recessive allele are needed in order to show the trait
- with the ABO blood type, there are multiple alleles rather than just two and they show codominance, whereby both alleles are expressed in the phenotype of someone with AB type blood
- in flowering plant species, red flowers crossed with white ones often give either only red flowers or only white, but some species can produce pink flowers, showing incomplete dominance whereby neither allele is completely dominant
- sometimes a heritable trait is controlled by a single gene, such as the colour of the peas Mendel experimented on, whereas others are caused by multiple genes
- haemophilia in humans is caused by sex-linked genes, whereas other conditions, such as Huntington's disease and PKU, are autosomal
- pedigree charts can be used to track and predict patterns of inheritance in a family.

Guiding Question revisited

What is the molecular basis of inheritance patterns?

In this chapter you have learned:

- the X chromosome is much larger than the Y chromosome so it carries more genes
- the X and Y chromosomes from the sperm and egg will determine the sex of the future baby
- polygenic inheritance leads to continuous variation in a trait
- single-nucleotide polymorphisms explain why different alleles can have different outcomes, for example with a C in the place of an A, and the amino acid coded for can therefore be different and give rise to a protein that has different properties.

Exercises

Q1. Explain why more human males are affected by colour blindness than females.

Q2. Using the C^R and C^W alleles for codominance in four o'clock flower colour, show how two plants could have some white-flowered offspring, some pink-flowered offspring and some red-flowered offspring within one generation.

Q3. Draw a pedigree chart for the two generations described in question 2.

Q4. Look at the grid below showing the chances that a couple's children might have haemophilia.

(a) State the genotypes of the mother and father.

(b) State the possible genotypes of the children.

(c) State the phenotypes of the children.

(d) Who are carriers in this family?

(e) What are the chances that the parents' next child will have haemophilia?

	X^H	Y
X^H	$X^H X^H$	$X^H Y$
X^h	$X^H X^h$	$X^h Y$

Q5. Use the pedigree chart on the next page to answer the following:

(a) Using the numbers, identify the carriers who are male in the pedigree chart.

(b) Using **F** for the functioning version of the gene and **f** for the allele with the PKU mutation, deduce the genotypes of 4, 8 and 11.

(c) Determine the percentage chance that the offspring of 7 and 8 could have a child with PKU.

(d) What are the chances that 11's offspring will receive at least one **f** allele?

(e) Explain why in most societies it is taboo or illegal for close relatives to marry.

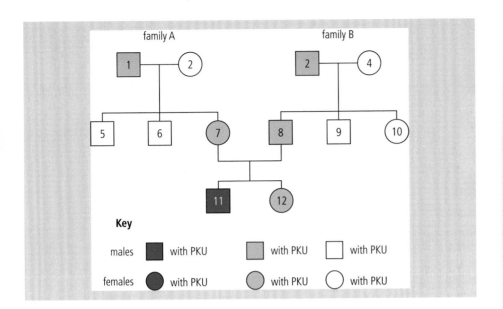

Key

| males | ![dark square] with PKU | ![grey square] with PKU | ![white square] with PKU |
| females | ![dark circle] with PKU | ![grey circle] with PKU | ![white circle] with PKU |

D3.3 Homeostasis

Guiding Questions

How are constant internal conditions maintained in humans?

What are the benefits to organisms of maintaining constant internal conditions?

How does your body stay warm, but not too warm? How do you maintain a healthy amount of water to support your body's needs? Your behaviour influences the answers to these questions, but your body does an amazing job even if your behaviour is not helping. The internal environment of an organism includes regulatory mechanisms to maintain variables within preset limits. We call this maintenance homeostasis. Homeostasis refers to an organism's ability to regulate various physiological processes to keep internal states at or near limits that are optimal. The regulation of these processes takes place mostly without our conscious awareness.

Temperature, blood glucose and water all have set points that are maintained. When a level begins to deviate from a set point, homeostasis works to correct it. Many body systems interact to keep physiological parameters within these set points, including the endocrine, nervous, excretory and circulatory systems. This chapter will look at how these homeostatic mechanisms work.

D3.3.1 – Maintaining the body's internal environment

D3.3.1 – Homeostasis as maintenance of the internal environment of an organism

Variables are kept within preset limits, despite fluctuations in external environment. Include body temperature, blood pH, blood glucose concentration and blood osmotic concentration as homeostatic variables in humans.

Humans and other species have evolved regulatory mechanisms to keep certain physiological factors within preset limits. This is called **homeostasis**. These preset limits are maintained even though our external environment can change considerably. The following are examples of parameters that are maintained by homeostasis:

- internal body temperature
- pH of the blood
- blood glucose concentration
- blood osmotic concentration.

Each of these physiological parameters is influenced by our environment and by what we ingest. For example, the air temperature that we experience can fluctuate from below freezing to very hot. Despite those possible extremes our body uses regulatory mechanisms to keep our internal temperature at or very close to 37°C.

D3.3.2 – Negative feedback mechanisms

D3.3.2 – Negative feedback loops in homeostasis
Students should understand the reason for use of negative rather than positive feedback control in homeostasis and also that negative feedback returns homeostatic variables to the set point from values above and below the set point.

The physiological processes that bring a value back towards a set point are called **negative feedback** mechanisms. Think of negative feedback control as working like a thermostat. The thermostat triggers one set of required actions when a value rises above its set point, and another set of actions when a value falls below its set point. In other words, negative feedback functions to keep a value within the narrow range that is considered normal. Positive feedback mechanisms are not appropriate for homeostasis because that type of feedback amplifies a response.

Negative feedback control in the body works like a thermostat. If the controlled variable (temperature) goes above a set point, it will trigger a response (in this case a cooling response). If the variable goes below a set point, the opposite response will be triggered (heating). All actions work towards the set point.

The nervous and endocrine systems often work cooperatively in order to ensure homeostasis. Many of the homeostatic mechanisms initiated by your nervous system are under the control of your autonomic nervous system. The **endocrine system** consists of numerous glands that produce a wide variety of hormones. Each hormone is transported by the bloodstream from the gland where it is produced to the specific cell types, called target cells, in the body that are influenced by that particular hormone.

Endocrine glands produce a hormone(s). The hormone always enters the blood for dispersal to all body cells. The body cells that have protein receptors for that hormone are called the target cells of that hormone. The response of the target cell will depend on the specifics of the hormone and the cell type.

D3.3.3 – The role of hormones

D3.3.3 – Regulation of blood glucose as an example of the role of hormones in homeostasis
Include control of secretion of insulin and glucagon by pancreatic endocrine cells, transport in blood and the effects on target cells.

The pancreas is the only gland in the body that is an endocrine gland and an exocrine gland. Its endocrine secretions (the hormones insulin and glucagon) are secreted into the blood. Its exocrine secretions (digestive fluids) go through ducts into the small intestine.

Insulin and **glucagon** are hormones that are both produced and secreted by the pancreas. In addition, they are both involved in the regulation of blood glucose levels. Cells rely on glucose for the process of cell respiration. Cells never stop cell respiration and thus are constantly lowering the concentration of glucose in the blood. Many people eat three or more times a day, including foods containing glucose, or carbohydrates that are chemically digested to glucose. This glucose is absorbed into the bloodstream in the capillary beds of the villi of the small intestine, and thus increases blood glucose levels. One factor that causes our blood glucose levels to fluctuate is simply that our blood does not receive constant levels of glucose because our ingestion of foods varies by time and content of carbohydrates. The increase and decrease in blood glucose levels caused by eating and digestion goes on 24 hours a day, every day of your life. However, even though blood glucose is expected to fluctuate slightly above and below the homeostatic normal level, it must be maintained reasonably close to the body's set point for blood glucose, and negative feedback mechanisms ensure this.

A transmission electron micrograph (TEM) of a cardiac muscle cell. Glycogen granules can be seen as small black dots. Glycogen is stored in muscle and liver cells.

In the pancreas there are cells known as β **(beta) cells** that produce the hormone insulin. Insulin is then secreted into the bloodstream and, because all body cells communicate chemically with blood, all cells are exposed to insulin. Insulin's effect on body cells is to open protein channels in their plasma membranes. These channels allow glucose to diffuse into the cell by the process known as **facilitated diffusion**.

There is another important effect attributed to insulin. Insulin stimulates muscle cells and liver cells to take in glucose (a monosaccharide) and convert it to glycogen (a polysaccharide). The glycogen is then stored as granules in the cytoplasm of these cells. The ultimate effect of insulin is to lower blood glucose levels.

Blood glucose levels begin to drop below the set point if someone does not eat for many hours or exercises vigorously. In either situation, the body needs to use the glycogen made and stored by the liver and muscle cells, as shown in Figure 1. Under these circumstances, α **(alpha) cells** of the pancreas begin to produce and secrete the hormone glucagon. The glucagon circulates in the bloodstream and stimulates hydrolysis of the granules of glycogen stored in hepatocytes (liver cells) and muscle cells. The hydrolysis of glycogen produces the monosaccharide glucose. This glucose then enters the bloodstream. The effect is to increase the glucose concentration in the blood and make that glucose available to body cells.

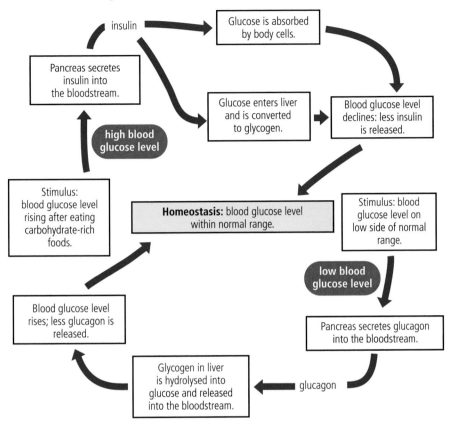

D3.3 Figure 1 A summary of the negative feedback control of blood glucose.

D3.3.4 – Type 1 and type 2 diabetes

D3.3.4 – Physiological changes that form the basis of type 1 and type 2 diabetes
Students should understand the physiological changes, together with risk factors and methods of prevention and treatment.

Diabetes is a disease characterized by **hyperglycaemia** (high blood glucose). The disease exists in two forms.

- Type 1 diabetes is an autoimmune disease where the immune system mistakenly destroys the α cells of the pancreas. These are the cells that produce insulin.
- Type 2 diabetes is a result of body cell receptors that do not respond properly to insulin and the cells do not take in sufficient glucose.

Type 1 diabetes

Symptoms of type 1 diabetes usually present in children or young adults, but can present in people of any age. Once type 1 diabetes is diagnosed, treatment is based on a controlled diet and injections of insulin as needed. People with type 1 diabetes regularly test their blood to monitor their glucose levels and plan their diet control and timing for insulin injections. Risk factors for type 1 diabetes include family history and age.

For what reasons do organisms need to distribute materials and energy?

Type 2 diabetes

The incidence of type 2 diabetes is almost ten times that of type 1 diabetes.

Type 2 diabetes is the result of body cells no longer responding to insulin as they once did. This is known as **insulin resistance**. Initially, the pancreas continues to produce a normal amount of insulin, but this level may decrease after a period of time. Type 2 diabetes is the most common form of diabetes. Risk factors for developing type 2 diabetes include family history, obesity and lack of exercise. You cannot change your family history but developing good eating habits and regular exercise can prevent or delay the onset of type 2 diabetes.

Low- and middle-income countries account for more than 80% of all deaths related to diabetes.

Uncontrolled diabetes of either type can lead to many serious effects, including:

- damage to the retina of the eye, leading to blindness
- kidney failure
- nerve damage
- increased risk of cardiovascular disease
- poor wound healing.

D3.3.5 and D3.3.6 – Body temperature control

D3.3.5 – Thermoregulation as an example of negative feedback control

Include the roles of peripheral thermoreceptors, the hypothalamus and pituitary gland, thyroxin and also examples of muscle and adipose tissue that act as effectors of temperature change.

D3.3.6 – Thermoregulation mechanisms in humans

Students should appreciate that birds and mammals regulate their body temperature by physiological and behavioural means. Students are only required to understand the details of thermoregulation for humans.

Include vasodilation, vasoconstriction, shivering, sweating, uncoupled respiration in brown adipose tissue and hair erection.

Many animals are **ectothermic**, meaning that their internal temperature equalizes with their environment. The air and water temperatures of their habitat greatly affect the geographical boundaries for many of these ectothermic animal populations, and greatly affect their behaviour. On cold days you may see ectothermic animals sunning themselves to gain body heat. Some fish and marine invertebrates have migratory patterns that help them remain in suitable water temperatures. The advantage of being ectothermic is that these animals do not have to metabolize foods to generate body heat, and as a result do not have to eat as much food.

Caimans (*Caiman yacare*) are ectothermic and often lie in the sun to increase internal body temperature.

Birds and mammals are **endothermic** and maintain a steady internal temperature that is almost always warmer than the environmental temperature. This requires extra nutrition specifically to generate internal body heat.

The location of the hypothalamus in the brainstem. The hypothalamus has control over many autonomic nervous system functions, including many thermoregulatory mechanisms.

There are two ways that humans and other endotherms can experience an increase of body temperature above their set point. One is by being in an environment that is warmer than their set point. The other is the result of muscular activity that generates internal body heat. Muscular activity during exercise or work activities often raises the internal body temperature.

The only way to decrease the body temperature below the set point is to be in an environment that is cooler than the body's set point temperature. Once the internal temperature of an endotherm begins to increase or decrease away from its set point, temperature-regulating negative feedback mechanisms are activated.

The two primary temperature sensing tissues in many animals are thermoreceptors located in the skin and a portion of the brain called the hypothalamus. The hypothalamus uses thermoreceptors to sense the temperature of the blood as it passes through that area of the brain. In addition, skin thermoreceptors send impulses to the hypothalamus. The hypothalamus responds by initiating cooling mechanisms or heating mechanisms, as shown in Figure 2.

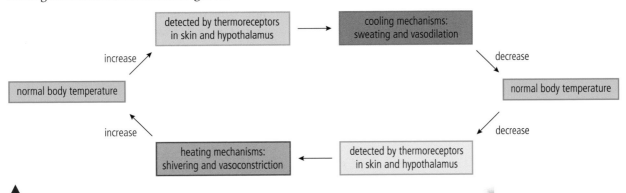

D3.3 Figure 2 A summary of the negative feedback mechanisms involved in thermoregulation. Increased internal body temperature leads to sweating and vasodilation of small blood vessels near the skin. A decrease in internal body temperature leads to shivering and vasoconstriction of small blood vessels near the skin.

The set point temperature for humans is 37°C. Any internal temperature that is higher or lower than 37°C will initiate negative feedback mechanisms for temperature regulation.

Skin structure adaptations to environmental temperatures that are above (shown on the left of the diagram) or below (shown on the right) the body's set point of 37°C.

The regulation of body temperature is called **thermoregulation**. We will consider two possible situations that require thermoregulation in humans.

If human body and blood temperatures rise above 37°C

When the internal temperature of the body begins to increase above 37°C the hypothalamus sends impulses to the arterioles near the skin that result in vasodilation. This results in more blood travelling through the capillaries in the skin and more heat being released into the surrounding air. The hypothalamus also initiates perspiration. The act of producing perspiration does not cool the body, but evaporation of the sweat from the skin provides evaporative cooling. The heat of the body is transferred to the water in sweat, initiating the phase change of evaporation.

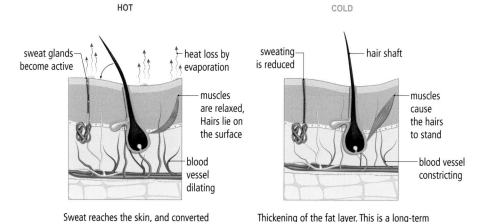

Sweat reaches the skin, and converted from a liquid to a vapour

Thickening of the fat layer. This is a long-term response to the cold enviroment

If human body and blood temperatures fall below 37°C

In colder environments, the hypothalamus sends impulses that result in the vasoconstriction of arterioles near the skin. More blood can then be shunted to the internal organs, with less to the skin. This is a protective measure for survival. The capillaries of the skin will receive less blood, and less heat will be released to the environment. In addition, vital internal organs will receive more warm blood. The release of epinephrine (adrenaline) in cold environments results in goosebumps or raised hairs on the skin. In furred mammals, this response creates an insulating layer of air between the fur and the skin to protect the body from the cold air. Humans no longer have enough skin hair for this to be very useful but we still have the **autonomic** response.

An elephant seal (*Mirounga angustirostris*) is an example of an animal that uses blubber as an insulator against the cold air and water typical of its environment in Arctic waters. ∨

When peripheral thermoreceptors in the skin sense cold, the pituitary gland is also stimulated to release hormones that activate the thyroid gland to release a hormone called **thyroxine**, which increases the metabolic rate of all body cells in order to generate heat. Shivering is another autonomic response to cold that is initiated by the hypothalamus. The act of shivering uses muscles and therefore generates body heat.

Some animals, especially marine mammals living in cold waters, have evolved to use blubber as an insulating layer between the water (or air) and their internal organs. The blubber is a form of adipose tissue and helps to retain the warmth generated by the internal metabolic activities of the animal.

Newborns have brown adipose tissue to generate heat when needed

When cold, adults use the rapid muscle contractions of shivering to create heat. Newborns are unable to shiver but instead have a higher proportion of fat that is called **brown adipose** tissue. Brown adipose is visibly darker than other adipose tissue because it contains many more mitochondria compared to other fat cells. When needed, the brown adipose cells use their mitochondria to begin cell respiration that is uncoupled from adenosine triphosphate (ATP) production. Glucose is oxidized for the sole purpose of generating body heat. Even though brown adipose tissue is mainly found in infants, adults retain a small amount of brown adipose tissue.

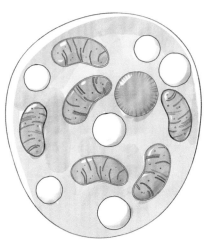

White adipose cells are designed for storing triglycerides (fat). Brown adipose cells have multiple mitochondria and are capable of cell respiration that generates heat uncoupled from ATP production. Newborns have a much higher proportion of brown adipose tissue compared to adults.

What biological systems are sensitive to temperature changes?

When you are cold, not all of your responses may be at the subconscious level: your cerebrum will also help by initiating conscious behaviours that will help you to stay warm. You may decide to move around more to generate muscle heat, dress warmer, or stay inside. Conscious behaviors also help you to stay cool. You may seek shade or air conditioning, wear light clothing and stop exercising. These are all behaviours that help prevent overheating.

Guiding Question revisited

How are constant internal conditions maintained in humans?

Within this chapter you have learned:

- homeostasis maintains many body conditions near preset limits despite changes in the external environment
- homeostatic variables include body temperature, blood pH, blood glucose concentration and blood osmotic concentration
- negative feedback control is used to maintain homeostasis
- blood glucose homeostasis is maintained by two pancreatic hormones, insulin and glucagon
- failure to control blood glucose levels is characteristic of both type 1 and type 2 diabetes
- birds and mammals are endothermic and control a set internal temperature by various mechanisms
- receptors inform the nervous system of current values of body conditions that are maintained by homeostasis
- the nervous system and endocrine systems often work cooperatively to ensure body conditions are maintained within set limits
- maintenance of homeostasis is primarily controlled by autonomic functions integrating the nervous and endocrine systems.

Guiding Question revisited

What are the benefits to organisms of maintaining constant internal conditions?

Within this chapter you have learned:

- homeostasis allows an organism to live in a wide range of environments and still maintain vital physiological parameters within set limits
- negative feedback control provides mechanisms for a near steady blood glucose level even when food is not readily available
- maintenance of blood glucose levels within a healthy range avoids the damage to body tissues characteristic of type 1 and type 2 diabetes
- endotherms can remain active in cold and hot temperature extremes because of internal mechanisms that return the internal temperature to a set point.

Exercises

Q1. Which of these statements best describes negative feedback control?

    A Mechanisms that lower a body variable until it is within physiologically normal levels.

    B Mechanisms that allow an organism to change its internal environment to match its outside surroundings.

    C Mechanisms that regulate a body variable to keep it within the limits of a set point.

    D Mechanisms that are used to change behaviours to adapt to differing environmental challenges.

Q2. Identify the pancreatic hormone that would be secreted under each of these conditions.

    (a) Shortly after eating a sugar-filled dessert.

    (b) One hour into a very active sports workout.

Q3. Type 1 diabetes is an autoimmune disease.

    (a) What cells in the body are destroyed by the immune system in someone with type 1 diabetes?

    (b) Identify two specific tissues that are damaged when diabetes is not controlled.

Q4. Identify whether each of these actions has a "warming" or "cooling" effect during thermoregulation.

    (a) Perspiration.

    (b) Shivering.

    (c) Secretion of epinephrine (adrenaline).

    (d) Vasodilation of skin arterioles.

Q5. Mitochondria are known for oxidizing glucose to create ATP molecules. What is the purpose of oxidizing glucose in mitochondria found in brown adipose tissue?

Q6. Briefly explain why birds and mammals must consume considerably more calories per mass of body tissue as compared to other animals.

Q7. It is well known that the human body will use the autonomic nervous system to vasoconstrict arterioles at and near the skin in very cold environments. This activates peripheral thermoreceptors to send signals to the brain with the information that the skin is cold. Why is this done or more precisely, why not send more blood to the skin to make you feel warmer?

Q8. Turtles, snakes, and alligators are among the animals that are often seen "sunning" themselves on cold mornings. Briefly explain why they exhibit this behaviour.

D3 Practice questions

1. The figure shows a pedigree chart for the blood groups of three generations.

   (a) Deduce the possible phenotypes of individual X. (1)

   (b) Describe ABO blood groups as an example of codominance. (1)

   (Total 2 marks)

2. In the pea plant (*Pisum sativum*), the allele for tall plants is A and the allele for short plants is a. The allele for green plants is B and the allele for yellow plants is b.

   Determine the phenotype of Aabb.

   (Total 1 mark)

3. Outline how the human body responds to high blood glucose levels.

   (Total 5 marks)

4. Hormones are distributed throughout the body by the blood. Outline the roles of **two** reproductive hormones during the menstrual cycle in women.

   (Total 2 marks)

5. Draw a labelled diagram of the human adult male reproductive system.

   (Total 5 marks)

6. Diabetes is often associated with the failure of the β (beta) cells in the pancreas, but it is unclear what actually causes this failure. FoxO1 is a protein which acts as a transcription factor to regulate the expression of genes involved in cell growth. FoxO1 also regulates increase in number and differentiation in cells such as pancreatic β cells.

   A study was conducted using mice lacking the gene for FoxO1 in β cells (IKO) as well as normal (control) mice. Blood glucose levels after fasting were compared for four groups of mice: young (3 months old) male mice, young (3 months old) female mice, older females (who have had several pregnancies) and aging males (16–20 months).

   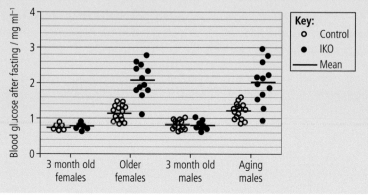

(a) Compare blood glucose levels after fasting in young control mice and young IKO mice without FoxO1. (2)

(b) Aging and having pregnancies are considered to be physiological stresses. Deduce the effect of stress on blood glucose levels. (2)

(c) Outline the relationship between blood glucose levels after fasting and lack of FoxO1 in the mice studied. (2)

The levels of pancreatic hormones and β cell mass in older female control mice and older female IKO mice lacking FoxO1 were then investigated.

(d) Calculate the percentage difference in β cell mass of the IKO mice compared to the control mice. (2)

(e) State the correlation between lack of FoxO1 and pancreatic hormones in mice. (2)

(Total 10 marks)

THEME **D** Continuity and change

4 Ecosystems

◀ With the choices we make every day, we can either contribute to climate change and the reduction of biodiversity or we can contribute to the sustainability of ecosystems.

Healthy ecosystems contain diversity; this gives them the ability to adapt to gradual change and can lead to changes in species. If the environment changes, populations will need to adapt. Through natural selection, organisms with higher fitness have a better chance of survival and therefore a higher chance of passing on their genes. Ecosystems are naturally self-sustaining, even when humans harvest some of the resources. However, the impact of humans on ecosystems has escalated in the last century to such an extent that natural processes can no longer maintain a balance. The consequences of human activity are climate change, extinction and pollution. By changing our daily behaviours such as reducing how much non-renewable energy we use as well as rewilding and setting up ecological reserves, we can help reduce climate change.

D4.1 Natural selection

Guiding Questions

What processes can cause changes in allele frequencies within a population?

What is the role of reproduction in the process of natural selection?

Some types of genes help organisms survive better, while others can be detrimental. Organisms that survive are more likely to reproduce. Their offspring are more likely to survive if they have inherited favourable traits from their parents. Natural selection can modify the frequency of alleles over time if the environment changes or if predators or diseases arrive in a population.

Natural selection depends in the production of viable offspring, so organisms must reproduce successfully. Certain physical characteristics can make a mate more attractive and this will help in the production of offspring. Sometimes physical or behavioural traits are used as an indicator of overall fitness. There is a balance to be struck, however, because traits that include bright showy colours, for example, may also attract the attention of predators.

D4.1.1 – Evolutionary change

D4.1.1 – Natural selection as the mechanism driving evolutionary change
Students should appreciate that natural selection operates continuously and over billions of years, resulting in the biodiversity of life on Earth.
NOS: In Darwin's time it was widely understood that species evolved, but the mechanism was not clear. Darwin's theory provided a convincing mechanism and replaced Lamarckism. This is an example of a paradigm shift. Students should understand the meaning of the term "paradigm shift".

Arguably, once evolution by natural selection is understood, many of the mysteries of nature are revealed. It was Charles Darwin and Alfred Russel Wallace who suggested natural selection as a mechanism for evolution. Here is a quick overview of the process of natural selection:

Evolution is the change in heritable characteristics of a population over time. Natural selection explains how the changes occur through a struggle for resources and differential survival, allowing some individuals to pass on their genes but not others.

The Museum of Comparative Anatomy in Paris, France. One indication that life evolves is the absence of live dinosaurs on Earth today.

- overproduction of offspring
- variation within the population, as a result of meiosis, sexual reproduction and mutations
- struggle for survival, because there are not enough resources for all members of the population to survive
- differential survival, those individuals which are the best fit for their environment tend to survive better
- reproduction, those who survive can pass on their genes to the next generation.

It is through these steps that populations evolve. Remember that, even though the changes can be observed in individuals from generation to generation, what is of importance is what happens at the level of populations rather than at the individual level.

TOK

Evolution is a law of nature because it describes what we observe. Natural selection is a theory because it explains the phenomenon. Laws cannot be modified over time because they simply describe what nature is doing, but theories can be modified over time as we understand mechanisms better.

Nature of Science

When a new idea allows us to see a phenomenon in a different way, it is considered a **paradigm shift**. Before Darwin, the accepted theory explaining evolution was that of Lamarck (see Section D4.1.6).

D4.1.2 – Sources of variation

D4.1.2 – **Roles of mutation and sexual reproduction in generating the variation on which natural selection acts**

Mutation generates new alleles and sexual reproduction generates new combinations of alleles.

There are three main sources of variation in a species:

- mutations in DNA
- meiosis
- sexual reproduction.

Mutation

Changes in DNA often have no effect on the phenotype of the organism. This might be because the mutation takes place in a piece of DNA that is not used to produce a protein. At other times mutations can produce genes that lead to genetic diseases, and can have devastating effects on the survival of some individuals in a species. However, sometimes a mutation can produce a characteristic that is advantageous. Mutation rates are generally low in populations, so sexual reproduction is a much more powerful source of variation in a population because thousands of genes are mixed and combined. But sexual reproduction is only possible thanks to meiosis.

Meiosis

Meiosis enables the production of haploid cells to make gametes (sperm cells and egg cells). At the end of meiosis, four cells are produced that are genetically different from each other and only contain 50% of the parent cell's genome.

The variety in gametes arises mainly from the process of random orientation during metaphase I (see Figure 1). The lining up of chromosomes in a random order is like shuffling a deck of cards, and it greatly promotes variety in the egg cells and sperm cells produced. In addition to this, the process of crossing over helps shuffle the genetic material and increase the genetic variety further (meiosis is discussed in Chapter D2.1).

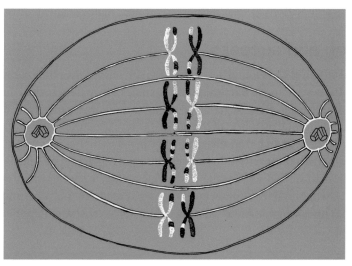

D4.1 Figure 1 Random orientation during metaphase I and crossing over (shown by dissimilar banding on sister chromatids) promote variety in the gametes. Each sister chromatid will separate into a different haploid cell at the end of meiosis.

One of the causes of the Great Famine in Ireland in the mid-1800s was that the potatoes had been produced asexually and were all clones, making them all susceptible to the same infection by a microorganism that causes potato blight.

Sexual reproduction

In an asexually reproducing population, all the members of the population are identical. There may be rare example of mutations or gene transfer, but overall such populations remain identical generation after generation. If there is no variety in a population, there is only a very limited number of outcomes in the event of a change in the environment: the whole population either survives or it dies. Variety in a population allows some individuals to be better adapted to whatever change in the environment is harmful to others. Meiosis means that offspring are varied because of the potential mix of alleles, but sexual reproduction itself also produces variety. Of the many sperm cells that may be present during sexual intercourse, only one will penetrate the egg. Each egg is also different. In flowering plants, which bees will land on which flowers, for example, is also a matter of chance.

Which of these yellow pollen grains on the bee's body will pollinate the next flower it visits?

Although mutation, meiosis and sexual reproduction generate variation, there is another aspect to natural selection that has little to do with chance and allows systematic accumulations of small changes to produce highly adapted forms of life.

Variation and success

There are three main sources for variation in a population:
- mutations in DNA
- meiosis
- sexual reproduction.

How frequently an allele is found in a population can change over time because of changes in the environment. But this is only possible if there is more than one form of a gene, i.e. different alleles. In bacteria, for example, there are essentially no differences within a population: all members of the population are genetically identical copies of each other. This means that if an adverse change happens in the environment, such as a change in pH, if one bacterium is susceptible to the change in pH and dies, they in fact all die because they all have the same vulnerability. In species where there is variation, a change in the environment will eliminate some but not all members of the population. This is why variation is a strength and not a weakness in a population.

D4.1.3 – Overproduction and competition

D4.1.3 – Overproduction of offspring and competition for resources as factors that promote natural selection
Include examples of food and other resources that may limit carrying capacity.

Darwin noticed that plants and animals produce far more offspring than could ever survive. Plants often produce hundreds or thousands more seeds than necessary to propagate the species. Mushrooms produce millions more spores than ever grow into new mushrooms. A female fish lays hundreds or thousands of eggs but only a handful survive to adulthood. This overproduction of offspring allows natural selection to occur, the selection of individuals that are fittest to survive.

Too many offspring and not enough resources cause a problem of supply and demand. If there is high demand for resources such as water, space, nutrients or sunlight, but a limited supply, then there will be competition between individuals to obtain those resources. If there are not enough resources for a growing population, the population will be limited as those who are outcompeted will not survive. The maximum number of individuals that an environment can provide for is called the **carrying capacity**. For example, less food in an area will reduce the carrying capacity, whereas an increase in food supply would raise the carrying capacity. The consequence of supply and demand is competition for resources in order to stay alive. This is called the **struggle for survival** and is a key component of natural selection.

What mechanisms minimize competition?

D4.1.4 – Selection pressure

D4.1.4 – Abiotic factors as selection pressures
Include examples of density-independent factors such as high or low temperatures that may affect survival of individuals in a population.

The main driving force of evolution is change in the environment. If nothing in a population's environment changes, there is no need to adapt. A **selection pressure** is a factor that can influence the success of parts of a population, and thereby influence changes in allele frequencies.

The mass extinction event that occurred 66 million years ago and wiped out so many dinosaurs and other species on Earth is an extreme example of a selection pressure. The environment changed rapidly when the collision of an asteroid with Earth released dust particles that then blocked out much of the Sun's radiation. Producers could no longer be as productive and, as food chains and food webs were disrupted, ecosystems collapsed all over the world. This sudden change in light levels, and subsequently temperature, was a selective pressure on populations. Many species did not make it. It is estimated that three-quarters of all life forms became extinct as a result.

An **abiotic factor** in an environment is something that is not living. Abiotic factors can be part of the physical environment, such as temperature, humidity or the availability of light. They can also be parts of the chemical environment that can affect organisms, such as the availability of certain minerals, the pH of water or soil, and the concentrations of gases such as oxygen and carbon dioxide in the atmosphere or in water. These factors are considered **density-independent factors** because they affect the population no matter how big or small the population is. An increase in the acidity of ocean water (a decrease in pH) will be as harmful to a large coral reef as it will be to a small one.

Magellanic penguins (*Spheniscus magellanicus*) live along the coasts of the southern tip of South America and are well adapted to surviving in the snow and cold. The penguin chicks are covered with down feathers that allow them to maintain a healthy body temperature by trapping warm air near their bodies. If snow lands on them, they can shake it off and keep dry and warm. They are not as well adapted for rain, however. Rain causes the downy feathers to stick to the chicks' bodies and the thermal insulation effect is lost because no more warm air surrounds their bodies. Chicks then die of hypothermia because of their inability to keep warm. With global climate change, higher temperatures in winter means that it now rains more frequently in parts of the range of Magellanic penguins, causing a decrease in some populations.

A Magellanic penguin (*Spheniscus magellanicus*) with young.

 Do not confuse density-independent factors with density-dependent factors. A density-dependent factor is one that affects a population more when the population numbers are higher. Disease, for example, will spread faster in a highly populated area where many organisms are close together, but more slowly in a smaller population that is spread over a wider area. With density-independent factors, it makes no difference if the population density is dense or sparse.

Snow crabs, of the genus *Chionoecetes*, thrive in cold water at northern latitudes of the Atlantic and Pacific Oceans. In the Labrador Sea, fishermen have observed an increase in populations of snow crabs in recent years that could be being caused by cooling water temperatures. Although it might seem counterintuitive that cold is better, colder water can dissolve more oxygen, and therefore can potentially provide resources for more individuals of species that can tolerate colder water. Increased dissolved oxygen availability and colder temperature are abiotic factors that can greatly increase the chances of survival of species that thrive under both conditions.

D4.1.5 – Intraspecific competition

D4.1.5 – Differences between individuals in adaptation, survival and reproduction as the basis for natural selection

Students are expected to study natural selection due to intraspecific competition, including the concept of fitness when discussing the survival value and reproductive potential of a genotype.

The noun "adaptation" and the verb "to adapt" are used freely when talking about evolution. However, the terms have very precise meanings within the framework of natural selection and should not be confused with other uses of the term, notably for human behaviour. For example, humans can consciously decide to adapt to a situation: think of a student learning the language of a country they have just moved to. This is a conscious adaptation made by an individual. In nature, the vast majority of adaptations referred to in evolution are unconscious, non-intentional adaptations that, although they take place on an individual level, are only meaningful if they affect the population.

Adaptation and survival

Exactly which individuals survive and which ones do not is not based on chance alone but determined by their surroundings and the compatibility of their characteristics with those surroundings. Competition between individuals of the same species is called **intraspecific competition**.

An organism that has characteristics that means that it is well adapted for its environment is said to be **fit** for its environment. Organisms with **high fitness** (those which possess characteristics that work well in their environment) have a higher chance of survival than those with **low fitness** (those which possess characteristics poorly suited for their environment). Natural selection tends to eliminate individuals from the population that show low fitness, whereas the fittest individuals in a population have a higher likelihood of surviving. Although there are rare exceptions, individuals are usually incapable of changing themselves to adapt to a particular factor. For example, a hummingbird with a short bill that cannot reach the nectar at the base of a flower cannot force itself to intentionally grow a longer bill. In offspring:

- useful variations allow some individuals to have a better chance of survival (e.g. hiding from predators, fleeing from danger or finding food)
- harmful variations make it difficult to survive (e.g. inappropriate colour for camouflage, heavy bones for birds, having such a big body size that there is not enough food to survive).

Because they survive to adulthood, successful organisms have a better chance of competing successfully with other members of the population, reproducing and passing on their successful genetic characteristics, their genotype, to the next generation. Over many generations, the accumulation of changes in the heritable characteristics of a population results in evolution.

How do intraspecific interactions differ from interspecific interactions?

An organism that is well adapted to its environment is not guaranteed success, it simply has a higher probability of survival than another that is less well adapted. Dinosaurs such as the sauropods were amongst the biggest, strongest animals ever to walk the planet. But they did not survive the environmental changes that drove them to extinction. In fact, the fossil record indicates that more than 99.99% of all life that has ever existed on Earth is now extinct.

D4.1 Figure 2 Plover eggs show adaptations that have been acquired by natural selection. Their colour and speckles help camouflage them from predators.

In Figure 2, the colours and speckles of the plover eggs act as effective camouflage, making these eggs difficult to spot by predators. Plover chicks are also speckled for camouflage. Such adaptations are good examples of traits that have **survival value**. If a mutation caused a shell to be bright white and/or the chicks to be bright yellow, the mutation would be unlikely to confer an advantage to this species and would have low survival value because these colours would attract the attention of a predator.

Only producing two eggs in a generation is risky. To increase their **reproductive potential**, organisms such as fish produce hundreds of eggs per generation. Reproductive potential is the maximum number of offspring an organism can produce in the absence of offspring mortality. Hypothetically, if 300 fish eggs are produced and they all develop into adults, the reproductive potential will have been reached. In reality, predation of eggs and young will reduce the survival rate and only a handful will reach adulthood.

It is crucial that you remember Darwin's steps of how natural selection leads to evolution:

- overproduction
- variation within the population
- struggle for survival
- survival of those best suited to the environment
- reproduction.

It is through these steps that populations evolve. Remember that, even though the changes can be observed in individuals from generation to generation, what is of importance is what happens at the level of populations rather than at the individual level.

D4.1.6 – Heritable traits

D4.1.6 – Requirement that traits are heritable for evolutionary change to occur

Students should understand that characteristics acquired during an individual's life due to environmental factors are not encoded in the base sequence of genes and so are not heritable.

In all the examples of traits given so far, you may have noticed that they are all heritable. **Heritable** means that the trait is encoded in the organism's DNA and can, therefore, be passed on to the next generation. Something that is acquired during the lifetime of an organism is considered to be an **acquired characteristic** and is not coded in the DNA. Acquired characteristics of organisms do not result in evolutionary changes. They only affect the individual and not their offspring.

Flamingos, also known as pink flamingos, are not pink because of pigments generated by their DNA. The pink colour comes from ingesting plankton that are rich in a molecule called beta carotene, the same one that contributes to the coloration of sweet potatoes and carrots. A flamingo that has a diet poor in beta carotenes will not turn pink. The offspring of a deeply coloured flamingo like the one in the image will not inherit the colour. It will have to find sufficient beta carotene in its diet if it wants to be as brightly pigmented as its parent.

Before Darwin, the accepted theory explaining evolution was that of Lamarck. Lamarckism states that organisms can inherit acquired characteristics from their parents and that there is a trajectory towards complexity and improvement. Lamarck

▲
The pink pigmentation in flamingos (*Phoenicopterus ruber*) is not due to genetics and DNA, so it is not heritable. The colour comes from eating plankton rich in beta carotene.

explained his idea with the use and disuse of body parts. For example, ruminants who use their horns in combat will develop bigger stronger horns and pass that trait on to their offspring, whereas the mole, living underground, has eyes weakened by disuse and weaker eyes are passed on to the next generation. These are compelling and plausible explanations but they have not been supported by experiments that can be replicated. In science we keep theories that stand up to testing and reject those that do not. Lamarck's theory has been refuted and rejected but Darwin's has been repeatedly confirmed for over a century and a half.

◀ How would the two theories, Lamarckism and Darwinism, explain the atrophied (weakened) eyes of the European mole (*Talpa europaea*)?

D4.1.7 – Sexual selection

D4.1.7 – Sexual selection as a selection pressure in animal species

Differences in physical and behavioural traits, which can be used as signs of overall fitness, can affect success in attracting a mate and so drive the evolution of an animal population. Illustrate this using suitable examples such as the evolution of the plumage of birds of paradise.

One of the best criteria for measuring long-term success in populations is reproductive success: more offspring means more success, less offspring means less success. Higher numbers in the present generation means more individuals that can reproduce and produce the next generation. Since natural selection is constantly removing members of the population, the ability to maintain healthy numbers is a sign of success.

There are different selection pressures that act upon organisms within a population. For example, lack of water may be a selection pressure in arid areas. **Sexual selection** occurs when the reproductive success of an individual results in more offspring compared to others in the population who do not have as much success in finding a mate. Sexual selection drives the evolution of a population in the same way than any other selection pressure does.

In birds, exceptionally brightly coloured or shiny feathers can be a way of showing potential mates that they are in such good health that they can afford to use some of their nutritional resources on pageantry. Birds of paradise are remarkable in this way. There are often striking differences between the male and female of the same species. Females have **cryptic** coloration, meaning they have dull or dark colours that blend in with the shadows of the forest and are not ornate or eye-catching. This helps them avoid being noticed by predators.

▲
The lesser bird of paradise (*Paradisaea minor*) lives in the forests of Papua New Guinea. This is a male with ornate colours. The female of this species is brown with a white underside and has no brightly coloured feathers.

When a bird has exaggerated, colourful and long tail feathers, it makes it difficult for the bird to fly, and proteins and minerals have to be invested in pretty colours and extravagance instead of being used for the bird's immune system or for energy. Any individuals who produce ornate plumage are clearly healthy because they can use valuable resources to produce good looks and they are agile enough to overcome any difficulties in flight that the bigger feathers create. This indicates that the bird has strong and healthy genes. The females of the species find this irresistible.

When there is a morphological difference between males and females like this, it is called **sexual dimorphism**. The differences are not visible in the young but appear when secondary sex characteristics develop. Competition between males for access to females is referred to as **intrasexual competition**. Male lesser birds of paradise with duller colours or shorter tails will not be as successful in attracting a mate.

In addition to colourful feathers, males can use behavioural characteristics to attract females. These could be building elaborate nests or performing courtship rituals that might include well-rehearsed choreographies. You may have seen humorous videos online of the male western parotia (*Parotia sefilata*) dancing with head bobs, fancy footwork and twirls to impress the females. Most of its feathers are jet black but some are colourful and reflective and can be orientated to suddenly flash into view as part of the choreography. Such a show sends a clear signal to the female that this individual is very healthy.

It is very tempting to choose human examples for natural selection but try not to when answering exam questions. For example, it is easy to imagine people with money, fame, dance moves and fancy cars attracting members of the opposite sex who are looking to pass on healthy and successful genes, but such clichés should not be used as examples in arguments for your coursework or for exams. Such ideas are useful analogies but should not replace examples from the natural world, such as birds with impressive feathers or courtship rituals.

D4.1.8 – Modelling selection pressures

D4.1.8 – Modelling of sexual and natural selection based on experimental control of selection pressures

Application of skills: Students should interpret data from John Endler's experiments with guppies.

In the 1970s, John Endler carried out a series of experiments to see if the presence of predators modified the bright colours seen in guppies (*Poecilia reticulata*). Guppies live in streams on the island of Trinidad off the coast of Venezuela in South America. The genes for the bright colours you can see in Figure 3 are only expressed in the males. Female guppies, like birds of paradise, show cryptic coloration that provides camouflage. The male coloration is highly diverse and made up of a mosaic of spots. Endler studied the fish for the following pigments: yellow, red, blue, iridescent and black. (Iridescent means that light glitters and is reflected in different ways depending on the angle at which it is viewed, and can look like different colours depending on the way light strikes the surface.) He carried out experiments in two ways: in the field (in the streams of Trinidad) and in 10 artificial ponds in a greenhouse at Princeton University, USA, where more variables could be controlled than in nature. In the field and in the greenhouse, some populations of guppies were kept in water that had predatory fish present that could eat them, while others were in water with fish that were not their predators and therefore harmless.

One of the reasons guppies thrive in the streams of Trinidad is that the sloped landscape is broken up by ledges of rock that form ponds and waterfalls. The waterfalls are like steps up the slope of the island; some fish species can swim upstream from the ocean, up the waterfalls, whilst other fish are blocked by them. Endler's hypothesis was that guppies in pools protected from predatory fish would show more ornate colours, whereas those in pools where predators were present would be less colourful because their ability to hide would lead to better chances of survival.

There are two separate and opposing selective pressures on male guppies.

- Predation: In this case high fitness includes more cryptic colours that blend in with the background. Brightly coloured males tend to be noticed more easily by predators and are selected against. The allele frequency for bright colours will decrease over time.
- Sexual selection: In this case the males need to stand out in order to be attractive to females, so high fitness includes bright conspicuous colours that make them stand out against the background. The allele frequency for bright colours will increase over time.

How did he test this hypothesis? Endler's independent variable was the presence of predators. In the greenhouse ponds, he introduced predatory fish species that eat guppies to some of the bodies of water and in others he introduced fish that were not predators. Then he waited to see what would happen over 15 generations. The predator was *Crenicichla alta*, a type of cichlid; the non-predatory fish was *Rivulus hartii*. In the field, he found ponds along streams where either *C. alta* or *R. hartii* was present, and introduced guppy populations that had the same level of genetic variety as used in the greenhouse experiments. To see the results of the greenhouse and field experiments, see the data interpretation exercise below.

SKILLS

Look at the graph of Endler's results below and answer the questions. Use the following legend to understand his graph:

K = ponds with no other fish

R = ponds with *R. hartii*, the harmless species

C = ponds with *C. alta*, the dangerous predator

Shaded bars = artificial ponds in the greenhouse at the university

Unshaded bars = natural ponds in the field (on Trinidad)

Error bars = two standard errors

▲
Results from John Endler's study of guppies indicating the number of spots per fish of the following colours: black, red, yellow, blue and iridescent.

1. Identify the group with the most spots per fish.
2. Explain why there is only one shaded bar for the results in group K.
3. Compare the number of spots on fish in ponds with *R. hartii* and ponds with *C. alta*.
4. Think about the controlled variables of the experiments carried out in the greenhouse and explain why it would be a poor idea to set up the ponds outdoors instead of inside the greenhouse.

Guiding Question revisited

What processes can cause changes in allele frequencies within a population?

In this chapter you have learned that:

- which versions of genes (alleles) are present in a population and the proportions in which they are found can change over time
- if the environment changes or there is another selective pressure on a population, the frequencies of alleles can be modified by natural selection
- abiotic changes, such as in temperature, humidity or pH, can contribute, as can biological factors, such as the presence of predators.
- organisms within a species compete with each other for resources, which is intraspecific competition
- the individual that is best suited to the environment is more likely to survive to reproduce and pass on its genes.

Guiding Question revisited

What is the role of reproduction in the process of natural selection?

In this chapter you have learned that:

- if something such as ornate colours in male birds or fish makes them more attractive to females, they have a higher chance of reproducing and passing on their genes
- which males mate with which females will determine which genes are passed on and therefore which alleles are present in a population
- organisms with higher fitness tend to be able to pass on their genes more frequently than those with lower fitness.

Exercises

Q1. Mutation is responsible for increasing variation in a population. List two other processes that also increase variation.

Q2. Explain how natural selection drives evolution.

Q3. Ground-nesting birds such as grouse lay their eggs in a nest made on the ground. The eggs of this species are generally speckled dark brown. If a mutation occurred causing the eggs to be brightly coloured, how would the change in colour affect their chances of survival?

Q4. Explain how a population of insects could develop resistance to the insecticides sprayed on them.

Q5. In his study about guppies, Endler said this about the males, "The color patterns in a particular place represent a balance between selection for crypsis by predators and selection for conspicuousness by sexual selection." Using the results in the skills box in Section D4.1.8, explain what he meant by this.

D4.2 Stability and change

> **Guiding Questions**
>
> What features of ecosystems allow stability over unlimited time periods?
>
> What changes caused by humans threaten the stability of ecosystems?

Ecosystems on Earth are remarkably stable when left free from the influence of humans. Evolutionary processes have taken place over eons of time and the result is sustainable and stable ecosystems. Stable ecosystems cycle nutrients efficiently, contain genetically diverse organisms, and have the means to collect sufficient energy through photosynthesis. A well-established ecosystem can also help keep abiotic factors within a tolerable range, for example transpiration from forests can influence rainfall in their area. Some ecosystems have existed for millions of years. The Daintree rainforest on the northeast coast of Queensland, Australia, is considered to be the oldest surviving ecosystem, at 180 million years old. The Amazon rainforest by comparison is approximately 55 million years old.

Unfortunately, human intervention in ecosystems can be devastating. Fuelled by the exponential growth of the human population, disruption of ecosystem processes is often getting worse rather than better. Each day many thousands of hectares of Amazon rainforest are burned in order to plant crops. The result is short-term profit but a devasting loss of biodiversity. Our reliance on fossil fuels continues to add carbon dioxide into the atmosphere. The resulting climate change as a result of global warming is measurable, and the consequences include severe weather, forest fires and crop failures. Plastic pollution is having a negative impact on both terrestrial and marine environments. Science is sounding the alarm but humans need to learn to listen and, most importantly, we need to change our wasteful habits and actions that are rapidly damaging our environment.

D4.2.1 – Stability of natural ecosystems

D4.2.1 – Stability as a property of natural ecosystems
Illustrate ecosystem stability with evidence of forest, desert or other ecosystems that have shown continuity over long periods. There is evidence for some ecosystems persisting for millions of years.

Evolution does not lead to unchanging ecosystems. There is no location on Earth where the living organisms have remained static for millions of years or even any time period that is close to that. However, there are many places on Earth where ecosystems have been healthy for very long periods of time. This does not mean that these ecosystems have not changed; it means that there have been slow adaptive changes and living organisms have thrived. A sustainable ecosystem is one that supports itself without any outside influences. Everything that the organisms within the ecosystem need is provided.

The Amazon rainforest

The Amazon rainforest in South America is the largest rainforest in the world. There is evidence that most of the area that is now rainforest was a marine lake as recently as

14 million years ago. Since then, the area has gained sediments and elevation and transitioned into tropical rainforest. What remains of the original lake is now the Amazon River.

At some point the rainforest largely became its own source of moisture, as transpiration from such a vast area of dense plant life became sufficient to provide water for the rainforest. The process of forming this amazing natural wonder has been happening for millions of years. Evolution has led to the amazing diversity of life forms that exists there today. Estimates of its biodiversity include 2.5 million insect species, 40,000 plant species and 1,300 bird species. These species are not static, they have evolved and will continue to evolve over time. This illustrates the importance of protecting sustainable ecosystems. Individual species are the genetic source for life in the future, and loss of biodiversity, and especially the loss of an ecosystem, reduces the possibilities of what that life can become.

D4.2.2 – Requirements for stability

D4.2.2 – Requirements for stability in ecosystems
Include supply of energy, recycling of nutrients, genetic diversity and climatic variables remaining within tolerance levels.

In order for an ecosystem to be stable in the long-term, specific requirements need to be met.

- A sufficient supply of energy
  The energy supply for ecosystems originates with light, which is harnessed by photosynthesis. Thus plants or algae must be abundant and productive. Energy can then pass through food chains and food webs in the form of organic molecules, until it reaches all trophic levels.

- Nutrient recycling
  Every ecosystem has a finite supply of nutrients. Since existing life forms are temporary, the nutrients they contain must be recycled. Nature has provided for this through specific nutrient cycles, such as the carbon, phosphorus and nitrogen cycles. If an organism is removed from an ecosystem, the nutrient stores within that ecosystem are reduced. For example, if a tree is logged in the Amazon and then removed from the forest, the carbon, phosphorus and nitrogen within the tree are no longer available to the forest ecosystem.

- Genetic diversity
  Healthy populations of a species need genetic diversity. Genetic diversity within a species provides protection against cataclysmic events decimating an entire population. A varied gene pool provides alleles that may allow a species to survive disease, harsh climatic events or a sudden increase in predators.

- Response to climatic change
  Ecosystems have responded to changes in climate for millions of years. A high genetic diversity within a species can help the survival of that species as long as the variables that contribute to the climate remain within the species' **tolerance levels**. Currently, human activities are causing global climate changes that are pushing climatic variables outside many tolerance levels, e.g. severe droughts and temperature increases.

A jaguar (*Panthera onca*), an apex predator found in the Amazon rainforest. This one was photographed in a rainforest area of Peru.
▼

Apex predators are also called top predators because they are at the top of a food chain. This trophic level does not have a high population as each level of a food chain has a declining biomass and energy availability.

Healthy sustainable ecosystems undergo changes over time; however, they are self-supporting.

Nutrient recycling is highly reliant on species of bacteria within soil.

523

D4.2.3 – Tipping points

| **D4.2.3 – Deforestation of Amazon rainforest as an example of a possible tipping point in ecosystem stability** |
| Include the need for a large area of rainforest for the generation of atmospheric water vapour by transpiration, with consequent cooling, air flows and rainfall. Include uncertainty over the minimum area of rainforest that is sufficient to maintain these processes. |
| **Application of skills:** Students should be able to calculate percentage change. In this case the extent of deforestation can be assessed by calculating the percentage change from the original area of forest. |

Severe deforestation of the Amazon rainforest has been occurring for the last 60 years. Some of that has been sanctioned by countries and some has been illegal. The motivation to cut and clear the rainforest is the short-term realization of money from logging, agriculture and ranching. At least 17% of the original Amazon rainforest has already been cleared.

Forest Loss in the Amazon Rainforest

● Forest lost from 2001 to 2020   ● Forest remaining from 2000

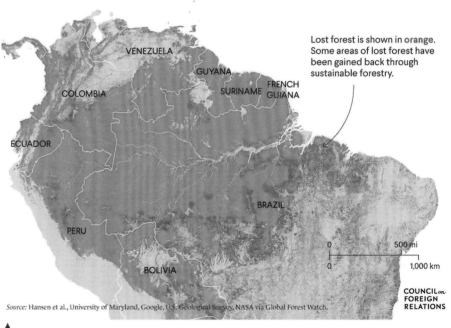

Lost forest is shown in orange. Some areas of lost forest have been gained back through sustainable forestry.

Source: Hansen et al., University of Maryland, Google, U.S. Geological Survey, NASA via Global Forest Watch.

COUNCIL on FOREIGN RELATIONS

A graphic showing the areas of the Amazon rainforest that have suffered deforestation since 2001. The spread of deforested areas occurs as new edges are exposed to create new areas to cut or burn.

No one yet knows how much of the Amazon rainforest can be destroyed before a tipping point is reached where the ecosystem will no longer be able to sustain itself. Past that point the ecosystem may also not be able to recover.

An incredibly large ecosystem like the Amazon rainforest influences its own weather and climate. For example, the rainforest vegetation produces huge amounts of water vapour by transpiration. This evaporation of water produces a cooling effect in the same way that evaporation of perspiration cools your body. This natural cooling effect influences the air flow and rainfall of the entire region. Researchers have already correlated increased temperatures and lower volumes of rainfall with deforestation. The effect is more pronounced near the edges of deforested areas.

What is not known is how much rainforest can be cleared before it reaches a tipping point, after which it will not be able to sustain itself as an intact ecosystem. No one knows how much of the rainforest must remain in order to maintain its self-propagating climate.

Reaching the tipping point may also remove the ability of the forest to re-establish itself. The Amazon rainforest appears to be capable of restoring itself at the edges of destroyed areas, but is not capable of regrowing from anywhere near the centre of deforestation. It must use the remaining edge vegetation to slowly grow back over areas that have been cut or burned. At some point that healing process may no longer be possible, especially if human activities and settlements are fighting for land in the opposing direction.

> ## Challenge yourself
>
> In 1970, the estimated remaining forest cover of the Brazilian Amazon rainforest was 4,100,000 km^2.
>
> 1. Using the data in the table below, determine the extent of deforestation for each of the given decades by calculating the percentage change compared to the 1970 area of rainforest cover.
>
> | Decade | Remaining forest cover (km^2) |
> |--------|-------------------------------|
> | 1980 | 3,845,000 |
> | 1990 | 3,692,000 |
> | 2000 | 3,524,000 |
> | 2010 | 3,359,000 |
> | 2020 | 3,290,000 |

D4.2.4 – Mesocosms

D4.2.4 – Use of a model to investigate the effect of variables on ecosystem stability

Mesocosms can be set up in open tanks but sealed glass vessels are preferable because entry and exit of matter can be prevented but energy transfer is still possible. Aquatic or microbial ecosystems are likely to be more successful than terrestrial ones.

NOS: Care and maintenance of the mesocosms should follow IB experimental guidelines.

A **mesocosm** is a self-contained system that provides a living environment for organisms. If taken to an extreme, air, water and food would all be self-generated by the organisms living in the mesocosm. There are several types of mesocosms; they can be set up in open tanks, but loosely sealed glass or plastic vessels are preferable. The aim is to produce something that allows energy transfer but prevents the entry and exit of matter. Aquatic or microbial ecosystems will probably be more successful than terrestrial ones. Directions for making a microbial mesocosm known as a **Winogradsky column** (see Figure 1) are given in the eBook. This type of mesocosm creates an environment that encourages the growth of different types of bacteria in various layers. Over time (approximately 2 months), different environments will form as a result of the combination of resource availability and the waste products released by different species of bacteria.

Fires are now quite common in the Amazon rainforest. Many fires are purposely set to clear the forest for farming and cattle ranching. Brazil is the largest beef exporter in the world, and much of the cleared land there is used to grow soy, a large percentage of which is used to feed the cattle.

SKILLS

A Winogradsky column can be used to demonstrate how bacterial ecosystems can become sustainable based on environmental resources. Think of the many variables that could be altered in order to make this laboratory exercise experimental. For example, you could vary the items you mix into the mud, vary the lighting conditions, vary environmental temperature; there are many other factors that could act as the independent variable in your investigation.

You will find a laboratory for creating your own mesocosm in the eBook.

SKILLS

For example, sulfur compounds produced by sulfate-reducing bacteria will accumulate in the lower portions of the column. Oxygen will be used up quickly in the lower portions of the container, as air will not be able to penetrate to replace the used oxygen.

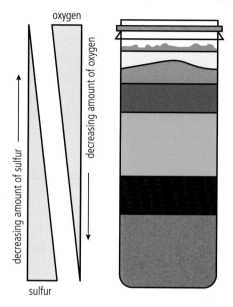

▲

D4.2 Figure 1 A schematic showing the environmental gradients that you can expect over time within a Winogradsky column. The bacterial species more suited to the bottom of the column will be those that use sulfur to produce hydrogen sulfide. These are **anaerobic bacteria**, as opposed to the **aerobic bacteria** that will grow closer to the top. Some algae will probably grow in the water at the top of the column. The colours shown in this artwork are not meant to be the exact colours that you will see, as every column is a little different. Be patient as you wait for the coloured layers to emerge, it may take 6–8 weeks.

Depending on the bacteria in your mud sample, and the temperature and lighting conditions, your results after a few weeks of growth may resemble those shown in Table 1. All sulfur-oxidizing layers will produce a smell like rotten eggs.

D4.2 Table 1 Characteristics of the layers in a Winogradsky column, from the top down. ▶

Bacteria type	Visual indication
Aerobic and photosynthetic	Green or red-brown layer
Non-photosynthetic sulfur oxidizers	White layer
Non-sulfur oxidizers	Red, purple, orange or brown layer
Purple sulfur oxidizers	Purple or purple-red layer
Green sulfur oxidizers	Green layer
Sulfate reducers	Black layer
Methanogens	Various dark colours, and small bubbles of methane

D4.2.5 – Keystone species

D4.2.5 – Role of keystone species in the stability of ecosystems
Students should appreciate the disproportionate impact on community structure of keystone species and the risk of ecosystem collapse if they are removed.

Keystone species are organisms of any type that play an important role in the biodiversity of their ecosystem. The effect they have is not caused by their numbers, but by their impact on the prevalence and population levels of other species within their community. One way to determine whether an organism is a keystone species is to perform a removal experiment (see Figure 2). Ecologist Robert Paine was the first to apply this method. He was studying an intertidal area of western North America. When Paine removed the sea star *Pisaster ochraceous* manually from the intertidal area, a mussel, *Mytilus californianus*, was able to take over the rocky area and exclude algae and other invertebrates from that zone. The mussel simply used all the space available when there was no sea star to keep it in check. It was evident that it was the sea star that limited the number of mussels that could reproduce and attach to the rocks. Paine collected data that showed that, when sea stars were present, 15–20 different species of invertebrates and algae were present. Without the sea star, the diversity rapidly declined to less than five species. This supported the hypothesis that the sea star was a keystone species. When it was present, it controlled the diversity of the community. When it was absent, diversity was lost.

A keystone species has a major impact on other species of an ecosystem regardless of their own population size.

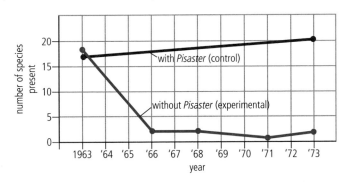

D4.2 Figure 2 Testing a keystone species hypothesis: the effect of removing the sea star *Pisaster ochraceous* from an intertidal area over a 10-year period. Notice that the **species richness** was greatly enhanced when the sea star was not excluded from the tidal area.

Loss of a keystone species can lead to collapse of an ecosystem. Wolves (*Canis lupus*) were largely eradicated from many areas of the western United States by bounty hunters. One area that suffered ecological effects as a result of this was Yellowstone National Park in north-western Wyoming. The park's last natural wolf was killed in 1926. The effect of losing the wolf as a keystone predator species resulted in what is called a **trophic cascade**. The wolves were predators of herbivorous species such as elk. When the wolves were eradicated, the herbivore populations increased. The increase in herbivores greatly decreased the availability of grasses and other

527

vegetation, including young willow and aspen trees. Without the young trees, songbird populations began to decline and beavers had no wood to build dams. Riverbanks started to erode without the roots of trees to stabilize them.

The grey wolf (*Canis lupus*) ▶

In mid-1990s, the US National Park Service decided to reintroduce 31 wolves to Yellowstone. Recently, the wolf count in Yellowstone was over 400. The herbivore population has been reduced, trees are regrowing and the entire ecosystem appears to be on its way to becoming sustainable once again.

D4.2.6 – Sustainable harvesting of natural resources

D4.2.6 – Assessing sustainability of resource harvesting from natural ecosystems

Sustainability depends on the rate of harvesting being lower than the rate of replacement. Include one terrestrial plant species and one species of marine fish as examples of renewable resources and how sustainability of harvesting can be assessed.

Natural resources include non-living and living materials that can be used by humans. This includes everything from trees to marine organisms. The problem is that we harvest many natural resources in such a way that the resource virtually disappears. Sustainability depends on the rate of harvesting being lower than the rate of replacement. Too many times humans have learned this the hard way. Two examples are given below of natural resources that have suffered high rates of harvesting but are now being closely monitored and harvested sustainably to provide a renewable resource.

Chilean sea bass

In the 1970s, a seafood merchant discovered a delicious fish that was being sold in a market in Chile. The fish was called the Patagonian toothfish (*Dissostichus eleginoides*). Knowing that this name was unappealing to consumers, he renamed it the Chilean sea bass. In only a few years the fish gained a huge market in many areas of the world.

In longline fishing, all of these hooks are baited and put into the water connected to a single fishing line. The hooks and bait are left in the water for a period of time, and then brought up one at a time to remove the caught fish. ▶

By the 1990s, the wild population of Chilean sea bass was collapsing. Consumer demand for the fish resulted in overfishing and much of it was not regulated. In addition, the fishing technique used was **longlining**. This technique sets hundreds of hooks and bait on a single line. Marine birds, like the albatross, often try to eat the bait and die after becoming ensnared by the hooks. The bad publicity surrounding the catch of these fish became so negative that many chefs refused to serve it and some seafood markets refused to sell it. However, illegal catches of the Chilean sea bass continued to be a serious problem until the year 2000.

Since then, the fishing industry that specializes in catching Chilean sea bass has agreed to make changes to its fishing techniques and to be monitored by an organization that certifies sustainable fishing practices that have minimal effects on the bycatch (species that are not wanted but are caught at the same time as the target species). As part of the certification process, a number of regulations detailing how many fish can be caught, the permitted age distribution of the fish in the catch, and when fishing can take place, have been applied to make the Chilean sea bass catch sustainable. Strict limits on damage to seabirds have also been applied, and changing the fishing season has drastically lowered the number of seabirds affected. The fish population is monitored to make sure sufficient repopulation is taking place. An observer from the certification organization is required to be on board the fishing vessels to document the fishing practices and make sure they are sustainable. The fleet of uncertified, illegal fishing vessels has been reduced to zero.

Overfishing is a global problem and cannot be tackled by just a few countries. There are countries that enforce fishing restrictions in their own territorial waters, but it is a difficult to apply them outside those boundaries. There are a variety of international non-profit organizations that attempt to regulate and improve fishing practices. One of the most effective ways appears to be educating the public about purchasing seafood from responsible sources that use sustainable fishing practices. The Marine Stewardship Council (MSC) is an example of an international non-profit organization that oversees fishing practices and provides certification to more than 25,000 seafood products that meet their standards. Responsible buyers of marine products look for their label or other certifications of sustainable fishing practices.

Seafood packaging with the Marine Stewardship Council label indicating that sustainable fishing practices were used to obtain the seafood. There are other non-profit and governmental organizations that also provide certifications.

Sustainable harvesting is promoted by informed consumer demand. Many people will avoid purchasing a product if they are convinced the product is not being harvested or grown using sustainable practices.

Black cherry trees

The black cherry (*Prunus serotina*) is a hardwood tree species that grows in many areas of North America including parts of New York, Pennsylvania, Ohio and West Virginia, collectively known as the Allegheny. Black cherry is usually found in hardwood forests mixed with other hardwood species. The light red of its wood makes it highly sought after for furniture and cabinet making.

Mixed hardwood forests are ecologically diverse communities. They are frequently so dense that availability of light for new growth is only provided when a mature tree falls or is removed as a result of age, disease, windstorm or selective logging.

Large areas of hardwood forest in the Allegheny region were once clear-cut to make farming fields. This type of logging completely removes almost all the trees in an area. The hardwood forests currently found in the Allegheny region are mostly second growth forests that are regrowing in some of the once clear-cut areas. Today, the logging of black cherry trees is usually done by selection of large specimens from the mixed forest. Current logging practices do not remove the surrounding trees. In most cases, an **arborist** will select the trees that can be removed sustainably. Removing a single tree leaves light gaps in the forest; black cherry and other seedlings can grow and establish in the light gaps left by logged specimens and naturally fallen trees. Black cherry trees produce a fruit that is consumed by a variety of bird species; germination and seed dispersal are enhanced by passage of the fruit through a bird's digestive tract and subsequent deposition in the bird faeces.

Hardwoods of all species are slow growing trees. It is imperative that selective logging occurs at a pace that does not exceed the growth of the remaining trees. The Forest Stewardship Council (FSC) was formed in 1993 by environmental, business and community leaders to regulate and certify sustainable logging practices. This organization provides regulations that promote sustainability and is relied on to certify wood supplies as being harvested by renewable logging practices.

The logging of black cherry trees is only sustainable if:

- selective logging is used, not clear-cut logging
- trees are selected rather than cut down randomly
- enough trees are left to produce fruit and seeds for the next generation
- data is collected to compare the quantity of wood being removed with current growth
- logging, processing companies and public buyers all appreciate and abide by the certification awarded by the Forest Stewardship Council or other organizations dedicated to sustainable harvesting.

D4.2.7 – Sustainability of agriculture

D4.2.7 – Factors affecting the sustainability of agriculture

Include the need to consider soil erosion, leaching of nutrients, supply of fertilizers and other inputs, pollution due to agrochemicals, and carbon footprint.

The goal of sustainable agriculture is to meet the food and textile needs of the world today without endangering the ability of future generations to do the same. In many areas of the world, agriculture has evolved from small-scale operations to huge corporate enterprises. However, whether a farm is family operated or a large business, agriculture must be sustainable.

To make agriculture sustainable, the most pressing issues that need to be addressed are soil erosion, leaching, fertilizer use, pollution and the carbon footprint.

Soil erosion

Loss of the upper soil layer of a field greatly reduces its productivity. The upper layer is called **topsoil** and should be rich in organic nutrients. Erosion of topsoil can be caused by excess rainwater and sometimes wind. Soil erosion is most often a problem when a field is left bare without a crop, because there will be no plant roots to hold on to the soil. Farmers sometimes plant rye or clover crops simply to cover the soil when the weather is not suitable for other crops. These cover crops reduce the penetrating force of heavy rainfall and their roots help hold the soil in place. Later, the cover crop can be ploughed back into the soil to increase the organic matter in the topsoil.

Leaching of nutrients

The nutrients needed by plants must be water soluble for the plants to use them. **Leaching** occurs when rain or irrigation water dissolves nutrients (usually nitrogen and phosphorus compounds) in the soil and then carries them away from the root zone of a crop. These dissolved chemicals often end up in the water supply of the area. Leaching cannot be prevented but it can be minimized by applying appropriate amounts of fertilizer and irrigation water at optimum times, taking the seasons and crop requirements into account.

Fertilizer supply

Many farms use chemical fertilizers to enrich their soils with nitrogen, potassium and phosphorus compounds. Chemical fertilizers are often a farm's largest single expense. The sources of fertilizers are limited and demand often exceeds supply. The manufacture of chemical fertilizers is also energy intensive. This drives up the cost of the fertilizers and that cost is then passed on to consumers. Some farmers minimize the use of chemical fertilizers by planting crops of beans or clover (legumes) that have nitrogen-fixing bacteria within their roots.

Pollution from agrochemicals

Water runoff and the leaching of both fertilizers and chemical pesticides results in pollution of water bodies. This is especially true when crops are overfertilized.

Carbon footprint

The carbon footprint is the total amount of greenhouse gases (including carbon dioxide and methane) that is generated by an activity. Currently the estimated percentage of greenhouse gases attributable to agriculture is about 12% of the total, although that number is thought by many to be too low. The carbon footprint of agriculture includes:

- the use of petroleum products to run farm machinery
- the addition of fertilizers, which often are made from petroleum products
- clearing natural forest land and other ecosystems to create farmland
- the transportation of crops grown in one area of the world to another.

At the start of the 1930s, a combination of drought conditions and poor farming practices led to the loss of tons of topsoil from land in the central United States. Huge windstorms blew away soil that had previously been held in place by the natural grasses of the prairie. The 1930s became known as the "dust bowl" years.

This graphic shows that the crop yield of grain is improved by adding nitrogen fertilizers, but only up to a certain point. The application of any excess nitrogen is not only expensive but is likely to be leached from the soil because it has not been taken up by the crop.

Many foods are shipped around the globe in refrigerated containers on huge cargo ships. As many as 740 million bananas can be shipped in 15,000 containers on some of the largest cargo ships. ▼

It is estimated that at any given moment there are over 50,000 cargo ships of various sizes carrying goods across the world's oceans.

D4.2.8 – Eutrophication

D4.2.8 – Eutrophication of aquatic and marine ecosystems due to leaching
Students should understand the effects of eutrophication resulting from leaching of nitrogen and phosphate fertilizers, including increased biochemical oxygen demand (BOD).

Eutrophication is the entire process that begins with water being over-enriched with nutrients, leading to the overgrowth of algae. It also includes the phases of algae die-off and bacterial decomposition, which deplete oxygen in the body of water leading to the death of aerobic organisms.

Eutrophication of a body of water is often measured indirectly by digital meters that measure dissolved oxygen in the water.

The thick algae growth on this pond is evidence of eutrophication.

Fertilizers containing phosphorus (phosphates) and nitrogen (nitrates) must be water soluble in order to be absorbed by plants. However, plants do not always absorb all the nutrients added to the soil. Irrigation and rainwater can leach these compounds through the soil and eventually into nearby streams, rivers and lakes. Sometimes excess minerals are washed into a saltwater (marine) environment. In both freshwater and saltwater, an influx of fertilizers can stimulate algae to grow excessively. This is the start of the process of **eutrophication.** The greatest growth occurs at or near the top of the water, because that is where there is the most sunlight. It does not take long for the algae to form a continuous thick layer across the water's surface. At that point very little light is able to penetrate through the water, and algae growing lower down in the water column begin to die. Aerobic bacteria begin decomposing the excess algae growth.

There is always some dead organic matter within all bodies of water. Decomposition by aerobic bacteria is normally a process that helps keeps an aquatic environment healthy. This decomposition uses oxygen dissolved in the water. The oxygen needed by bacteria in a body of water is called the **biochemical oxygen demand** or **BOD**. The need for extra decomposition caused by excess algal growth increases the BOD, and the amount of dissolved oxygen may become depleted. During the process of eutrophication, all the aerobic organisms in the water may die because the oxygen content of the water is no longer sufficient to support them. The only way to stop the eutrophication process is to stop nitrate and phosphate compounds from entering the system.

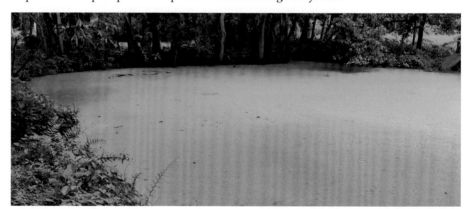

D4.2.9 – Biomagnification

D4.2.9 – Biomagnification of pollutants in natural ecosystems
Students should understand how increased levels of toxins accumulate in the tissues of consumers in higher trophic levels. Include DDT and mercury as examples.

Biomagnification is the phenomenon that occurs when harmful substances in the environment build up in the organisms towards the top of a food chain. Each trophic level typically consumes many organisms from the previous trophic level. If the organisms that are consumed contain a toxin that does not break down, then the substance becomes more concentrated in the living tissues of the organisms that eat them. Two examples of biomagnification are discussed below.

Mercury

Mercury is a naturally occurring element found throughout the world. Food chains displaying mercury accumulation are rare and almost always are located near an industrial source that emits mercury compounds. However, it now appears that mercury levels are increasing in organisms towards the top of aquatic food chains, and the source of the mercury seems to be the burning of coal and the production of cement. Mercury compounds released by these processes enter the atmosphere and are then washed into the oceans.

Microorganisms at the bottom of an aquatic food chain convert elemental mercury and inorganic mercury compounds into **methyl mercury**. As methyl mercury moves up through an ocean food chain, it becomes more and more concentrated in the tissues of animals.

Humans are exposed to high levels of mercury by eating fish that are long-lived and at or near the top of a food chain. Mercury accumulation in the body can have many severe health effects, including neurological damage. Figure 3 shows the relationship between size of fish and mercury content in their tissues.

DDT

DDT (dichloro-diphenyl-trichloroethane) is a synthetic insecticide developed in the 1940s. It was widely used for many years because it was effective, inexpensive to produce and long lasting (persistent). It was usually used against mosquitos and other insects that acted as vectors for diseases such as malaria and typhus. However, DDT was indiscriminate in which insects it killed.

After more than a decade of use, problems began to emerge from the widespread application of DDT. Beneficial insects were being killed and mosquitos were becoming resistant to it. Because DDT was often mass-sprayed from aircraft, some entered bodies of water in water runoff. In marine environments, DDT was absorbed by phytoplankton. As in the mercury example, DDT became more concentrated at each level of the food chain, and collected in the fatty tissues of consumers at higher trophic levels.

In 1956, an epidemic of a disease of the central nervous system was reported in an area of Japan called Minamata Bay. A local company there produced acetaldehyde using a mercury compound as a catalyst. The mercury and many other heavy metal compounds were being dumped as wastewater into Minamata Bay. Bacteria in the water transformed the inorganic mercury into an organic form, methyl mercury. Methyl mercury was incorporated into the local food chains, and became concentrated by biomagnification in the local shellfish and fish. Local people depended on shellfish and fish as a major component in their diet. It may never be known exactly how many people died or became severely disabled because of the mercury levels in this body of water, but it was in the many thousands.

D3.2 Figure 3 In the Madre de Dios river basin of Southeastern Peru small-scale gold mining uses mercury as part of its mining operations. A study was done to measure the mercury content of various freshwater fish in the immediate area. As seen in this graphic, larger fish correlated with higher mercury content within their tissues. Larger and older fish have more time to accumulate higher levels of mercury from their diet.

DDT being sprayed for mosquito control on a beach in Long Island, New York, USA, in 1945.

TOK What criteria should be used when weighing the risk of using DDT against the benefit of controlling a deadly disease such as malaria?

The greatest documented effect was on birds such as brown pelicans, osprey and bald eagles. Each of these species feeds on fish as a primary food source. The DDT did not immediately kill the birds, but it did alter their metabolism of calcium. The result was production of thin-shelled eggs that could not withstand the mass of a parent bird during incubation of the eggs. Researchers began to notice that the age structure of the birds was changing, with very few young birds in the populations.

Regulatory actions regarding DDT use began in the 1950s and by 1972 the Environmental Protection Agency in the United States banned the use of DDT for almost all applications, and the populations of predatory birds affected by DDT largely recovered. The United Nations Environment Programme negotiated a treaty to ban the use of DDT world-wide, with the exemption of areas where malaria is known to be a greater health concern than the effects of DDT.

D4.2.10 – Microplastic and macroplastic pollution

D4.2.10 – Effects of microplastic and macroplastic pollution of the oceans

Students should understand that plastics are persistent in the natural environment due to non-biodegradability. Include examples of the effects of plastic pollution on marine life.

NOS: Scientists can influence the actions of citizens if they provide clear information about their research findings. Popular media coverage of the effects of plastic pollution on marine life changed public perception globally, which has driven measures to address this problem.

Plastics are not **biodegradable**. A plastic object disposed of in an environment will remain there virtually forever. The object may be broken down mechanically, but that just creates smaller pieces of plastic. Plastic pollution is a global problem, especially the plastic objects of various sizes that end up in our oceans. Some plastic is washed up on shorelines, but some floats on the water surface and some sinks to the ocean floor. Plastic debris can be categorized into two main types based on size.

Macroplastics are any plastic debris that is larger than 5 mm. This includes water bottles, grocery bags, food containers and numerous other relatively large plastic objects. Many of these items are single-use items that are carelessly discarded.

Microplastics are plastic debris smaller than 5 mm. Some of this debris was originally a macroplastic object that has since been mechanically broken down into smaller pieces. Other microplastic pollution arises from the very small plastic granules used in face scrubs, cosmetics and hand cleansers. Many countries have now prohibited the use of microplastics in these products because they have been linked to cancer and shown to stimulate some harmful genetic conditions in humans, both the original products and as pollutants.

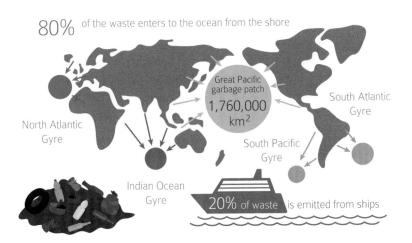

◀ **D4.2 Figure 4** The five major ocean garbage gyres formed by ocean currents. Each is composed of both macroplastics and microplastics.

80% of the waste enters to the ocean from the shore

Great Pacific garbage patch
1,760,000 km²

North Atlantic Gyre

South Atlantic Gyre

Indian Ocean Gyre

South Pacific Gyre

20% of waste is emitted from ships

Plastic items in the oceans tend to get caught up in large ocean vortices known as **gyres**. Each gyre is formed by prevailing winds and ocean currents, and creates a swirling mass of debris that is funnelled towards a centre. The largest plastic gyre is the Great Pacific Garbage Patch, which has two vortex centres. One is off the coast of California, USA, and the other off the coast of Japan. Five major garbage plastic gyres are shown in Figure 4.

Plastic pollution in the oceans is not just ugly, it is killing wildlife.

- Sea turtles sometimes eat plastic bags thinking they are jellyfish.
- Plastic rings from six-packs of canned drinks entrap sea birds and other wildlife.
- Albatrosses pick up plastics from the ocean surface and feed it to their chicks.
- Plastic fishing nets are often lost by fishing boats and are a death trap for a variety of fish, sea turtles and marine mammals.
- Microplastics are filling the stomachs and intestines of marine organisms after accidental ingestion.

Nature of Science

Scientists can influence the actions of others if they provide clear information about their research findings. Media coverage of the effects of plastic pollution on marine life has changed public perception globally, which has led to measures being taken to address the problem.

▲ The skeletal remains of a young Laysan albatross (*Phoebastria immutabilis*), interspersed with plastics. It is likely that the plastic items were fed to the young bird by its parents. It is also likely that the plastics led to the death of the animal, as many dead young birds have been found with plastics in their bodies.

D4.2.11 – Rewilding

D4.2.11 – Restoration of natural processes in ecosystems by rewilding

Methods should include reintroduction of apex predators and other keystone species, re-establishment of connectivity of habitats over large areas, and minimization of human influences including by ecological management. Include the example of Hinewai Reserve in New Zealand.

Rewilding activities are conservation efforts aimed at restoring and protecting natural processes and wilderness areas. Rewilding is a form of ecological restoration that leaves an area to nature, as opposed to active natural resource management. Successful long-term rewilding projects require little ongoing human attention, because successful reintroduction of keystone species can create a self-regulated and self-sustaining stable ecosystem.

Methods of rewilding include:

- the reintroduction of **apex predators** and other keystone species (see Section D4.2.5)
- establishing wildlife corridors, to connect habitats over larger areas
- stopping agriculture and resource harvesting such as logging and hunting
- minimizing human influences on an ecosystem, including using ecological management techniques.

Hinewai Reserve in New Zealand

Hinewai Reserve is a 30-year-old rewilded area on the South Island of New Zealand. The reserve started as 109 hectares of land that had previously been farmland. Subsequent land acquisitions have expanded the reserve to 1,250 hectares. The reserve is privately owned by a trust but is freely open to the public via numerous walking paths.

What is the distinction between artificial and natural processes?

The goal of the reserve is to foster regeneration of native vegetation and wildlife. Minimal intervention was used to remove a few invasive tree and animal species. The strategy is to allow nature to take its own course. The goal is for native vegetation to repopulate the area, along with many native species of animals. The Hinewai Reserve continues to be a work in progress but is an example of rewilding success.

One of many waterfalls within Hinewai Reserve, New Zealand.

Guiding Question revisited

What features of ecosystems allow stability over unlimited time periods?

Within this chapter you have learned that healthy, sustainable ecosystems must have:

- the capability to change gradually over time
- a constant supply of energy
- the ability to recycle limited nutrients
- genetic diversity
- a reasonably stable climate
- keystone species that give structure to the ecosystem
- freedom from human interference so that a tipping point is not reached, where the ecosystem becomes incapable of healing itself
- resource harvesting by humans that does not exceed the rate at which the resources are replaced.

Guiding Question revisited

What changes caused by humans threaten the sustainability of ecosystems?

Within this chapter you have learned that humans are threatening the sustainability of ecosystems through activities such as:

- the depletion of natural resources
- lowering the genetic diversity within ecosystems
- pollution by chemical wastes that can build up within food chains and result in the loss of apex predators
- pollution of aquatic ecosystems by macroplastics and microplastics
- the addition of extra nutrients to watercourses resulting in eutrophication and the subsequent depletion of dissolved oxygen (which, in turn, leads to loss of aquatic organisms)
- deforestation of areas for logging and agriculture
- the removal of keystone species
- rapid climate change caused by the release of greenhouse gases.

Exercises

Q1. A keystone species is one best described as:

   A   the most abundant species in an ecosystem

   B   a species that has the highest primary productivity

   C   a species that has a dramatic impact on its ecosystem no matter what its population size is

   D   a species that should be removed from an ecosystem.

Q2. What is BOD and how is it correlated with eutrophication of a body of water?

Q3. Why do some species of oceanic fish accumulate more mercury in their body compared to other species?

Q4. Explain why a farmer might plant a cover crop such as clover.

Q5. Which of these is *not* characteristic of ecological restoration by rewilding?

   A   Reintroduction of keystone species.

   B   Establishment of wildlife corridors between sections of preserves.

   C   Minimal active management.

   D   Minimal logging and hunting.

Q6. Ecologists say that the loss of Amazon rainforest could reach a "tipping point". What does the term tipping point mean in this context?

Q7. At one time DDT was often used as an inexpensive, persistent, and effective insecticide. State three reasons why DDT was banned in most areas of the world.

Q8. In order for an ecosystem to be sustainable it must have a constant input of energy. What is the source of that energy for most ecosystems?

D4.3 Climate change

Guiding Questions

What are the drivers of climate change?

What are the impacts of climate change on ecosystems?

It is crucial you understand that climate change is an enormous issue: our survival and existence as a species depends on what we decide now. Rarely in the human story has there been such a pivotal moment. Earth's climate is complex because it involves multiple systems interacting with each other. A change in one part of the system can have consequences across the globe. Energy transfers involving sunlight and heat determine the climate on Earth. Adding greenhouse gases to the atmosphere as a consequence of human activities such as burning fossil fuels modifies how much heat is retained. Melting snow and ice decreases the surface area of Earth that is reflecting sunlight. The ground and ocean surface that used to be covered with snow and ice is warming up more than usual. This is warming the atmosphere and modifying ocean currents. Changes in atmospheric temperature are leading to more drought in some areas, and increasing the number of forest fires. When climate changes, ecosystems are impacted. Changes to coral reef systems, breeding grounds, migration patterns and the availability of food can greatly disrupt food chains and ecosystems.

D4.3.1 – Human activity and climate change

D4.3.1 – Anthropogenic causes of climate change

Limit to anthropogenic increases in atmospheric concentrations of carbon dioxide and methane.

NOS: Students should be able to distinguish between positive and negative correlation and should also distinguish between correlation and causation. For example, data from Antarctic ice cores shows a positive correlation between global temperatures and atmospheric carbon dioxide concentrations over hundreds of thousands of years. This correlation does not prove that carbon dioxide in the atmosphere increases global temperatures, although other evidence confirms the causal link.

The role of carbon dioxide and methane

We live at the bottom of an ocean of air we call the **atmosphere**. The atmosphere plays a vital role in regulating the temperature of Earth's surface. The **greenhouse effect** refers to the way that the atmosphere retains heat and keeps the planet warm even when no sunlight is reaching the surface. When sunlight hits an object, some of its energy is absorbed, and the object warms up. The energy is re-radiated in the form of **infrared radiation**, which has longer wavelengths than energy in the form of visible light. The glass in a greenhouse traps infrared radiation.

Greenhouse gases (**GHGs**), such as carbon dioxide and methane in Earth's atmosphere, can be thought of as the glass of a greenhouse, although, like many models, this is not a very accurate representation of the natural phenomenon. GHGs

A greenhouse protects plants from the cold in a similar way to the atmosphere stabilizing Earth's temperature.

have the ability to absorb and radiate infrared radiation. When such gases are present, they keep the atmosphere near Earth's surface warm by converting some of the short-wave radiation from the Sun into long-wave radiation, and radiating it in all directions, including back down towards the surface.

Levels of some of the main greenhouse gases have increased as a result of human activities. Carbon dioxide is produced when fossil fuels are burnt to generate electricity or power vehicles. Natural carbon sinks such as forests are being cleared, which means that carbon dioxide is not being absorbed at the same rate. Methane is produced by livestock such as cattle and in some types of agriculture such as rice cultivation. The increased levels of greenhouse gases are causing the atmosphere to retain more and more heat. Something caused by human activity is said to be **anthropogenic**. According to the International Panel on Climate Change (IPCC) the climate change we are seeing in recent decades is not a natural phenomenon but instead is anthropogenic in origin.

Global climate change

Climate refers to the patterns of temperature and precipitation, such as rainfall, that occur over long periods of time. Whereas weather can change from hour to hour, climate changes generally occur over thousands or millions of years. Climatologists and palaeoclimatologists collect data about atmospheric conditions in recent decades and the distant past, respectively. As thermometers have only been around for a few hundred years, temperatures on Earth from thousands or millions of years ago have to be inferred from **proxies** (see the Nature of Science box).

▲ A summary of the greenhouse effect: short-wave radiation (the yellow arrow on the left) hits the surface and much of it bounces off the atmosphere and surface of the planet, but some is converted into long-wave radiation (heat, shown as the red arrow on the right) as continents and oceans warm. Some of this infrared heat escapes into space, but some is radiated to the surface by greenhouse gases that can trap and emit the heat.

Nature of Science

A proxy is a measurement that is used in place of another one. Because it is impossible to go back in time and measure the temperature of the atmosphere 15,000 years ago, climatologists use proxies such as tree rings, coral reef growth and the presence of certain fossils to estimate what the climate was like many thousands of years ago.

Layers found in thick sheets of ice that have been formed by annual snowfall can also provide information on past climates. By drilling into the ice and taking cylinder-shaped samples, called **ice cores**, scientists can study the substances trapped in the layers, such as air bubbles from the year

▲ An ice core being removed from its drilling apparatus.

when the layer was deposited. Researchers at Vostok Station in Antarctica have collected layers of ice from more than 3,000 m down, yielding climate information going back more than 400,000 years.

But we must always be careful about making the distinction between correlation and causation. Just because we see a positive correlation between carbon dioxide (CO_2) levels and global temperatures we cannot assume that a rise in one *causes* a rise in the other. Correlation is not the same as causation. In the case of climate change, there is further evidence to link carbon dioxide levels to an increase in global temperatures, and the greenhouse effect also provides a possible mechanism.

A positive correlation means as one factor goes up, the other also goes up. In contrast, a negative correlation would mean that as one variable increases, the other decreases.

Do not confuse the greenhouse effect with the depletion of ozone. Although both are the result of human activity, and both influence the atmosphere, they are not interchangeable phenomena. They have different causes and different effects on the environment.

Do you want to see the data for yourself? One of the principles of science, especially research funded by taxpayers, is to make data available to the public. This is to allow verification, critique and sharing of data, so that scientists with many different approaches can combine their findings and advance our understanding of the topics being studied.

One organization that does this is the National Oceanic and Atmospheric Administration (NOAA) in the US. If you go to the NOAA Earth System Research Laboratory Global Monitoring Division's website, you will find maps, graphs and databases of measurements of carbon dioxide and other atmospheric gases over many decades. Check out the section called Products.

As shown in Figure 1, there is a correlation between temperature increase and carbon dioxide increase. As discussed earlier, it is clear that an increase in carbon dioxide levels will lead to warming of the atmosphere, because there will be an increase in the greenhouse effect. Having said this, closer inspection of the data shows that the increase in temperature happens first and then the carbon dioxide concentration rises. This lag time is partly explained by the fact that, as oceans warm up, they release carbon dioxide, because gases dissolve less well in warm water than in cold water. This leads to further increases in temperatures over time: warmer temperatures → more carbon dioxide → even warmer temperatures → even more carbon dioxide, and so on.

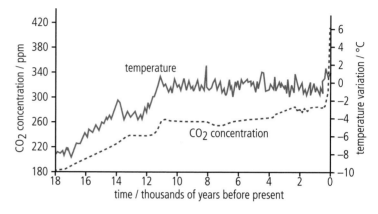

▲ **D4.3 Figure 1** 18,000 years of atmospheric changes in temperature and carbon dioxide (CO_2) concentrations measured mainly from air bubbles trapped in ice

Challenge yourself

Figure 1 shows the results of collected data representing thousands of years trapped in ice core samples.

The lower (dashed red) line shows carbon dioxide concentrations that were measured from air bubbles trapped in the ice.

The upper (continuous blue) line shows fluctuations in temperature, with zero representing pre-industrial climatic conditions.

1. Is there a strong or a weak correlation between carbon dioxide levels and atmospheric temperatures over the last 18,000 years?

2. Can scientists conclude that there is causality from the data in Figure 1, i.e. that rising carbon dioxide levels cause global temperatures to go up?

3. What further evidence would be necessary to confirm or refute causality?

Earth has shown many fluctuations in global temperatures over millions of years, long before humans started producing excessive greenhouse gases. The changes being observed now are alarming scientists because they are happening so quickly.

Climate change, especially as portrayed in the press or on social media, raises many issues about how science works, notably concerning the scope and limitations of science.

The fact that there are sceptics and critics of the IPCC reports on global climate change is a good thing. Science encourages constructive criticism and verification, and is open to modification if the criticisms are valid. But some critics will look at a few inaccurate predictions that have been made and say, "See? Your models are wrong. Therefore, no one should listen to you". This is an example of **cherry picking**. Cherry picking is a form of **confirmation bias** caused by only looking at the evidence supporting your side of the argument, and ignoring or downplaying the evidence that hurts your argument. To what extent is objectivity possible in the production or acquisition of knowledge?

D4.3.2 – Global warming

D4.3.2 – Positive feedback cycles in global warming

Include release of carbon dioxide from deep ocean, increases in absorption of solar radiation due to loss of reflective snow and ice, accelerating rates of decomposition of peat and previously undecomposed organic matter in permafrost, release of methane from melting permafrost and increases in droughts and forest fires.

Release of carbon dioxide

When a marine organism dies, it sinks, and either its corpse will be eaten on the way down or it will reach the bottom, called the **benthic zone**. Decomposers, archaea and bacteria digest and break down the organic material, and in so doing can produce carbon dioxide. If surface waters are warmer, phytoplankton can produce more biomass, which can then be passed on to the rest of the marine food chain. And when those organisms die, more organic matter will reach the ocean depths and the organisms living in the benthic zone will produce even more carbon dioxide.

A system that amplifies an effect like this is said to be a **positive feedback loop**. The more something happens, the more it allows the next step to happen, which, in turn, encourages the first step to keep going. Notice that the word "positive" does not necessarily mean it is a good thing, it just means the effect is amplified. Negative feedback loops also exist. A combination of positive and negative feedback loops helps a human maintain steady internal conditions such as the correct body temperature and blood sugar levels. Earth also maintains a balance using feedback loops. Unfortunately, an increase in carbon dioxide levels is increasing the influence of positive feedback loops, throwing the system off balance.

Increase in absorption of solar radiation

The ability of a surface to reflect light is called its **albedo**. Light-coloured objects, such as ice and white sand, have a higher albedo, so very little light is absorbed, meaning such objects do not heat up as much as dark objects. Think about walking barefoot on light-coloured cement on a hot and sunny day, compared to walking barefoot on black asphalt on the same day. Dark-coloured substances such as asphalt have a lower albedo, and can absorb lots of light and convert it into heat.

Snow and ice have a high albedo and tend to reflect so much light that they do not heat up very much. Soil, rocks and open ocean water have a lower albedo and tend to warm up more than snow and ice when exposed to sunlight. When Arctic ice that is floating on ocean water starts to melt because the water is too warm, the exposed ocean surface heats up more. This causes even more melting of the ice floating on the water. Can you see the feedback loop?

The darker zones between the ice sheets that are melting and breaking up have a lower albedo and so the water warms up much faster than it did before, when high-albedo ice was covering it.

Positive feedback loops intensify or exacerbate a situation. The message is "Yes, keep going. Amplify the effect!" Negative feedback loops prevent a process from going too far. The message is "Do not overdo it! Slow down. No need to keep amplifying the effect".

Decomposition of organic matter

Healthy soil hosts huge numbers of small invertebrates and microbes such as worms, fungi and bacteria that help decompose organic matter. If the temperature is cold, they are less active because their enzyme activity and other metabolic processes slow down at lower temperatures. Because of climate change, some parts of the world are becoming warmer. When peat bogs are in such zones, their decomposition rates increase and the faster decomposition releases more carbon dioxide than before. If undisturbed, peat bogs are considered to be **carbon sinks**. They trap organic matter and the conditions prevent it from being fully decomposed. But when climate change results in decomposition happening faster, peat bogs can become carbon sources, releasing carbon dioxide into the atmosphere.

A peat bog can switch from being a carbon sink to a carbon source if temperatures rise.

The same phenomenon is happening in the coldest zones of the world, where we find the **cryosphere**, places on Earth where water is in solid form, such as in **permafrost**. Permafrost is a type of soil that exists in very cold climates and is frozen solid. Sometimes, layers above the permafrost can thaw out for a few weeks or months in a year. Generally only low-lying plants with shallow roots can grow in such soil. Tonnes of organic matter are locked up in the permafrost because the microbes that carry out the decomposition cannot survive at such low temperatures.

Permafrost under the surface soil creates these bumps on the surface, called hillocks, when ice, which is less dense than water, rises towards the surface, pushing the soil up. The growing season is so short that plants do not have time to grow very tall.

In zones where temperatures are rising, the permafrost is melting. As it warms up, microbes can become active and decompose the trapped organic matter, and as a result generate more carbon dioxide. The carbon dioxide that is released into the atmosphere is now able to act as a greenhouse gas and contribute to even more warming. Again, you should notice the positive feedback loop here, because the further increase in temperature will cause the permafrost to melt to an even greater depth, allowing even more carbon dioxide to be released (see Figure 2).

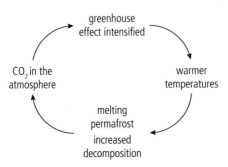

D4.3 Figure 2 Climate change creates a positive feedback loop when permafrost melts and releases greenhouse gases.

In addition to carbon dioxide, permafrost can release methane. This is because some of the microbes in the soil are **methanogenic archaea**, meaning that they generate methane. As seen previously, methane is also a greenhouse gas and so releasing more of it into the atmosphere will create a positive feedback loop similar to the one for carbon dioxide. Warmer temperatures will allow the methane-producing microbes to be even more active.

Methane is an invisible gas but one way to observe it is in frozen lakes, where it bubbles up from the sediments and gets trapped in the ice.

Global climate change can also lead to extreme weather events that cause droughts and more frequent forest fires. Forest fires release carbon dioxide into the atmosphere.

Drought conditions occur when there is not enough rainfall to sustain the water demands of an ecosystem or the needs of humans, notably for agriculture.

When fire destroys a forest, the carbon in the trees is released as carbon dioxide into the atmosphere. This is yet another positive feedback loop, exacerbating the problem of excess greenhouse gases being added to the atmosphere.

D4.3.3 – Tipping points

D4.3.3 – Change from net carbon accumulation to net loss in boreal forests as an example of a tipping point

Include warmer temperatures and decreased winter snowfall leading to increased incidence of drought and reductions in primary production in taiga, with forest browning and increases in the frequency and intensity of forest fires, which result in legacy carbon combustion.

We usually think of forests as getting water from rainfall, but in many parts of the world melted snow provides most of the water that the trees absorb. One consequence of global climate change is that, where temperatures are getting warmer, winters are shorter and there is less snow. Less snow in the winter means less snowmelt in the spring and summer, and therefore less water for **boreal forests**, otherwise known as **taiga**. Such forests are blanketed in snow for much of the year and are the coldest land biome, with average temperatures ranging from −5°C to 5°C.

A frozen river cutting through a boreal forest

The lack of water in the soil from reduced snowmelt water means that the trees cannot photosynthesize as much as they could before. Primary production in the forest is reduced, which means less carbon dioxide is being removed from the air. This is another example of a positive feedback loop. If coniferous trees cannot get enough water for photosynthesis, their needles lose their green pigment, turn brown and fall off. When this happens on a massive scale, it is called **forest browning**, and if it continues without any drought relief, the trees will die. This is a form of deforestation.

When coniferous trees in the taiga do not get enough water, they start to lose their needles and die in a process called forest browning.

Drought conditions and dying wood are perfect conditions for forest fires. When a wildfire takes hold in a boreal forest, the carbon that has accumulated and been locked up for centuries in the trees' organic matter (as wood and needles) is suddenly released as carbon dioxide into the atmosphere. There is carbon locked up in the soil, too. This carbon is from past ecosystems and can be hundreds or thousands of years old. It is called **legacy carbon**.

Normally, the organic matter holding legacy carbon cannot decompose because of the prevailing cold temperatures, but if a fire burns the organic material trapped in the soil, it can release the carbon into the air in a process called **legacy carbon combustion**.

Usually, forests are considered to be carbon sinks because they remove carbon from the atmosphere during photosynthesis, which converts the carbon dioxide into organic molecules that are locked up in the trees for centuries. But when thousands of decaying trees or forest fires release the carbon back into the atmosphere, the forest becomes a carbon source. The switch from sink to source because of an imbalance in the system is considered a **tipping point**. A tipping point is a certain threshold that if reached or surpassed causes massive changes, some of which could be irreversible.

D4.3.4 – Polar habitat change

D4.3.4 – Melting of landfast ice and sea ice as examples of polar habitat change

Include potential loss of breeding grounds of the emperor penguin (*Aptenodytes forsteri*) due to early breakout of landfast ice in the Antarctic and loss of sea ice habitat for walruses in the Arctic.

Note: When students are referring to organisms in an examination, either the common name or the scientific name is acceptable.

The emperor penguin (*Aptenodytes forsteri*) lives along the coasts of Antarctica and migrates dozens of kilometres over the ice to reach its breeding grounds each winter. Emperor penguins prefer to breed on **sea ice** (ice formed when ocean water freezes) that is connected to the mainland. These **landfast** sheets of ice start to break up and detach from the shore when the Antarctic summer arrives in December and warms up the waters. By then, the eggs, which have been incubated for two months by the males, have hatched and the chicks will have grown big enough to find their own food.

With global climate change, some zones of sea ice are being exposed to warmer temperatures and starting to break up and separate from the shore earlier in the breeding season, making it impossible for the penguins to raise their young. A few penguin colonies have moved to ice found on land, but the reduction of landfast ice means there are concerns about the future of the emperor penguin population.

At the other pole, Arctic ice is also melting and having an impact on ecosystems. Walruses (*Odobenus rosmarus*) also prefer sea ice shelves as breeding grounds and a place to rear their young. The mother walruses suckle their babies then dive into the water to get more food. The fact that the sea ice is on the water makes it convenient, because they never have to leave their young for very long. With melting ice, however, reduced breeding grounds and less space to rear young has meant that either populations have to move poleward (in this case nearer the north pole) or find places other than sea ice to rear their young. In the latter case, the mothers are not as close to the water and need to leave their babies for longer periods of time This, in turn, makes the young more vulnerable to attack by predators such as polar bears, who are also having trouble finding food as a result of climate change and melting ice.

▲ The habitat of emperor penguins (*Aptenodytes forsteri*) is being modified by climate change.

◄ Walruses (*Odobenus rosmarus*) in the Canadian Arctic

D4.3.5 – Ocean current change

D4.3.5 – Changes in ocean currents altering the timing and extent of nutrient upwelling

Warmer surface water can prevent nutrient upwelling to the surface, decreasing ocean primary production and energy flow through marine food chains.

The world's oceans are in perpetual motion. Currents move water that has been warmed near the equator to the poles where the water is cooled and sinks. The water is moved around by prevailing winds and also spins as it moves due to the **Coriolis effect**, which is caused by the rotation of Earth on its axis.

> Low pressure systems tend to form cyclones that spin air masses. These turn in a clockwise fashion in the southern hemisphere but anticlockwise in the northern hemisphere (when observed from a weather satellite). The direction is determined by Coriolis forces generated by the rotation of planet Earth.

> The world's oceans have currents that transport heat, oxygen and nutrients between ecosystems. The ribbon-like arrows in the figure can be thought of as conveyor belts.

Some currents exist year round and others can change or intensify with the seasons. Some do not necessarily happen every year, such as the **El Niño Southern Oscillation**, which occurs in the Pacific Ocean between Indonesia and South America. During years without an El Niño event, prevailing trade winds blow air towards the west of the Pacific Ocean. As coastal surface water is pushed seaward by the wind, the water below it is pulled upwards to take its place. This cold water is full of nutrients. Movement of nutrient-rich water from deeper parts of the ocean towards the surface is called **nutrient upwelling** and it plays an important role in bringing nutrients to the food webs that exist along the coasts. In years when there is no El Niño event, large schools of sardines can be found off the coast of South America. The sardines are feeding off organisms that are thriving because of the nutrients that have been pulled up from the deeper water.

When an El Niño event occurs, however (roughly every 2 to 7 years), the currents are modified. Winds arriving from Southeast Asia push against the normal prevailing winds and create an upwards air movement in the middle of the Pacific Ocean. This causes the warm water along the Central and South American coast to stagnate instead of being pushed seaward.

Normal conditions

El Niño conditions

Conditions in the Pacific Ocean in years without (top) and with an El Niño event (bottom). Upwelling is cancelled and vital nutrients do not reach the surface during an El Niño event.

When there is an El Niño event, instead of a nutrient upwelling along the Central and South American coast, water is pushed down (downwelling). As a result, the marine ecosystems that depend on cool water upwelling do not receive the nutrients they otherwise would. Primary production is the production of organic molecules by living organisms. In the oceans, algae are the producers. Without the nutrients that are provided by cold water, primary production falls, and less energy flows through the ecosystem. The whole food chain is affected, and fewer fish are seen during El Niño years.

An example of what can happen when nutrient upwelling occurs: plankton use the nutrients and proliferate in blooms so big they are visible from space, as here in the Bay of Biscay. This primary production will nourish the rest of the marine ecosystem in the area. Without the upwelling, not only will phytoplankton numbers be lower, the rest of the ecosystem that depends on it will also suffer.

549

Studies indicate that climate change is increasing the severity of El Niño events, as well as producing warmer surface temperatures in many other places in the oceans, which could reduce nutrient upwelling and throw marine ecosystems off balance.

D4.3.6 – Range shifts

D4.3.6 – Poleward and upslope range shifts of temperate species

As evidence-based examples, include upslope range shifts for tropical-zone montane bird species in New Guinea and range contraction and northward spread in North American tree species.

Poleward range migration of trees

Tree species in northern latitudes have been monitored over decades to see if their ranges are moving, for example towards the poles. This process is called **range migration**, and the fossil record shows that such a range migration happened just after the most recent glaciation.

Generally what stops a tree population in northern latitudes from expanding its northern upper range limit is extreme cold. Many seeds or saplings cannot tolerate the cold that intensifies nearer the North Pole, but as global temperatures increase and winters are becoming less harsh at many northern latitudes, it is hypothesized that some tree populations are spreading poleward, like they did when the Earth warmed after the most recent ice age.

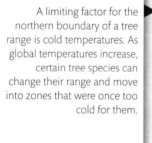

A limiting factor for the northern boundary of a tree range is cold temperatures. As global temperatures increase, certain tree species can change their range and move into zones that were once too cold for them.

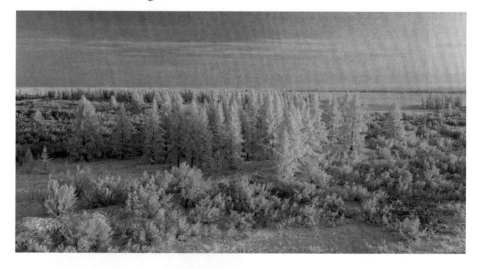

Figure 3 shows the results of a study published in 2014 by the Ecological Society of America, which compared the ranges of 11 species in North America between the 1970s and the early 2000s. To understand the graph, you need to know what a **percentile** is. When collecting data, if 50% of the data points can be found below a certain value, we say that value is the 50th percentile. So if half the trees are found below a certain latitude, that latitude is the 50th percentile latitude. The graph shows that black spruce trees (*Picea mariana*) in the 50th percentile latitude have shifted 11 km to the north (shown in green). The 90th percentile latitude for black spruce trees has a similar value, with a 10 km northward shift. However, the 90th percentile for the same species' saplings show an overall southward range migration of similar magnitude, which makes it difficult to draw decisive conclusions about what is happening to the black spruce's range in recent decades.

D4.3 Figure 3 Latitudinal range shifts (in kilometres) for four species. Saplings shown in blue are at the 50th percentile of latitudinal distribution; saplings shown in red are at the 90th percentile; mature trees shown in green are at the 50th percentile; and trees shown in purple are at the 90th. An asterisk indicates where the shifts are statistically significant.

A map showing the distribution of black spruce (*Picea mariana*) trees. The pairs of lines towards the top represent the 90th percentile of latitude and the lines towards the bottom show the 50th percentile latitude for this species, meaning 90% and 50% of the trees sampled were below these latitudes. Both pairs of lines show a northward migration of about 10 km over the decades of the study.

Several challenges and limitations make it difficult to answer the question of whether or not tree ranges are moving poleward as a result of human-induced climate change. One is that, for the same species, the pattern can change as we move from west to east: some zones might show a poleward range change and some might show no significant change or even a southward change. Patterns can be contradictory for a species when looking at the statistics for both full-grown trees and saplings. Another is that the range changes could be caused by factors other than temperature, such as ice storms, forest fires or logging activity.

TOK

Scientists use data and measurable evidence to reach conclusions. But are there some types of evidence that are less reliable than others? What features of evidence in the natural sciences make them convincing and valid, and what features make them unreliable? Often when evaluating the reliability of data and assessing the level of confidence, scientists look at the overall characteristics of the study, including sample size, range of results or the methodologies used, such as whether or not the data was measured directly or by proxy. What features of evidence in the natural sciences make them convincing and valid and what features make them unreliable?

The white-winged robin (*Penoethello sigillatus*) in Papua New Guinea has moved its habitat range upslope in recent decades.

Upslope range migration of birds

Papua New Guinea is a mountainous country in which some habitats are temperate and others tropical. The country is home to some visually spectacular and unique birds, including colourful birds of paradise, parrots and cockatoos. In the 1960s, detailed surveys of **montane** bird species (birds that live on mountains) were carried out. All birds have a certain altitude range they live in; some montane species prefer lower altitudes, some prefer living high up near the peaks, and others are found somewhere else along the slope of the mountain. In 2014, nearly 50 years after the original survey, a second survey looked at the zones where the birds were found. A few bird species had stayed in the same ranges, and some were observed at lower altitudes, but a surprising number had increased their range upslope (see Figure 4).

One such bird, the white-winged robin (*Penoethello sigillatus*), lives on Mt Karimui in the Eastern Highlands, and was found to have moved upslope more than 100 m since the previous survey. Tropical birds are very sensitive to temperature, so if their habitat warms up, they could go to higher altitudes to find the relatively cooler temperatures they are better adapted to. Since both climate change and human activity (such as hunting tropical birds for their ornate feathers) are possible explanations for birds changing their range, it is difficult to say which one is the main cause of the upslope modification.

D4.3 Figure 4 The results of the part of the study looking at upper elevational limits. Each montane bird species is represented by a dot. The *x*-axis shows the highest elevation at which the bird was observed back in 1965, and the *y*-axis shows the change since then. Although a few have stayed at the same altitude or moved down a few dozen metres, more species have moved upslope in recent decades, some by over 50 m. The average was more than 100 m upslope.

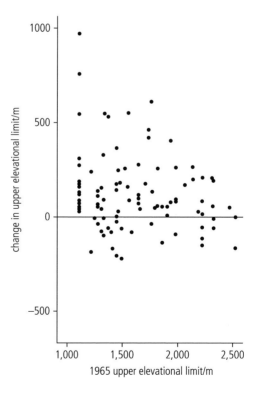

This phenomenon is much more extensively studied in birds living in temperate zones of the world and we know with more certainty that climate change is a major contributor to alterations in range both poleward and upslope.

D4.3.7 – Ecosystem collapse

D4.3.7 – Threats to coral reefs as an example of potential ecosystem collapse

Increased carbon dioxide concentrations are the cause of ocean acidification and suppression of calcification in corals. Increases in water temperature are a cause of coral bleaching. Loss of corals causes the collapse of reef ecosystems.

The organisms that build coral reefs are very sensitive to water temperature, water acidity and the depth of the water. Unfortunately, all three factors are changing in the world's oceans as a result of human activities. Increased carbon dioxide concentrations in the air lead to increased dissolved carbon dioxide in the oceans (see Figure 5). This lowers the pH of seawater because, as the carbon dioxide dissolves in the water, it forms **carbonic acid**:

$$CO_2 + H_2O \rightarrow H_2CO_3 \text{ (carbonic acid)}$$

$$H_2CO_3 \rightarrow H^+ + HCO_3^- \text{ (hydrogen carbonate ion)}$$

D4.3 Figure 5 Increased dissolved carbon dioxide will lower the pH of ocean water and increase the concentration of hydrogen carbonate ions.

When dissolved in water, the carbonic acid forms **hydrogen ions** (H$^+$ ions), which reduces the pH. **Calcium carbonate (CaCO$_3$)** is a key component in the skeletons of coral reefs and the hard shells of marine organisms. It is formed when **calcium ions** (Ca^{2+}) dissolved in ocean water are combined with **carbonate ions** (CO$_3^{2-}$):

$$Ca^{2+} + CO_3^{2-} \rightarrow CaCO_3$$

This process is called **calcification** and it is as important to the growth of coral reefs as the growth of bones is to us. Acid can dissolve calcium carbonate, so if calcification is disrupted by low pH levels, either the reef's growth rate or its ability to form strong, dense skeletons will be reduced. Beyond certain thresholds, ocean acidification and changes in temperature can lead to the death of coral polyps and algae, and when they die the reefs no longer thrive. As a result, the colour of the reef changes from being richly multi-coloured to as white as bone (see Figure 6).

D4.3 Figure 6 A healthy coral reef on the left and a dead reef on the right

The United Nations' Sustainable Development Goal 14, entitled "Life below Water", lists reducing ocean acidification as one of its targets because ongoing acidification of ocean water will continue to disrupt and destroy marine ecosystems. Coral reefs are also discussed in Chapter B4.1.

What are the impacts of climate change at each level of biological organization?

This coral reef death is called **coral bleaching** and it interrupts the reef's food chain, causing many of the organisms that live there to seek food and shelter elsewhere or die out. A bleached coral reef can no longer support the rich ecosystem that once lived there. This is an example of **ecosystem collapse**.

TOK

The scope and methodologies of scientists are different from those of politicians, and they do not always share the same perspectives. This is not a new phenomenon. The astrophysicist Carl Sagan, for example, testified before a US Senate hearing in 1985, warning the government that action needed to be taken to avoid irreversible changes to our planet's climate. The 2022 IPCC report on climate change was about 4,000 pages long. It is unlikely many politicians read the whole document. In addition to scientists, other experts have tried to warn governments of the dangers. Indigenous knowledge could hold possible solutions to bring nature back into balance, starting with rebuilding a relationship with the planet and showing more respect and humility. How can we decide between the judgements of experts if they disagree with each other?

D4.3.8 – Carbon sequestration

D4.3.8 – Afforestation, forest regeneration and restoration of peat-forming wetlands as approaches to carbon sequestration

NOS: There is active scientific debate over whether plantations of non-native tree species or rewilding with native species offer the best approach to carbon sequestration. Peat formation naturally occurs in waterlogged soils in temperate and boreal zones and also very rapidly in some tropical ecosystems.

One way to mitigate and potentially reverse global climate change is to find ways of removing the carbon we have added to the atmosphere. This process, called **carbon sequestration**, involves taking atmospheric carbon dioxide and locking it up somewhere. Although engineers have proposed many high-tech solutions using costly industrial technologies for carbon sequestration, some low-cost solutions are natural and use systems that have been proven to work for millions of years: forests and wetlands.

Forest regeneration involves planting saplings and ending tree felling, to let a new forest grow where a previous forest has been cut down or burned.

For many thousands of years humans have cut down trees to use the wood, or burnt forests to clear land for agriculture and other land uses. Humans have also become good at draining wetlands for agriculture or building. **Reforestation** is the process of replanting deforested areas, which would allow trees to remove carbon from the atmosphere. Likewise, **afforestation**, the process of planting trees where no forest has previously existed, can increase carbon capture. And if we stopped filling in or draining wetlands and instead allowed new wetlands to forms, wetlands could fulfil their role as carbon sinks. These are long-term processes, but some governments and non-governmental organizations have instigated such activities.

Arguably, the government that can claim to plant the most trees in the world is China. In 1981, China passed a law mandating that one tree should be planted every year by every student over the age of 11. They continue to fund and implement forest improvement programmes but, as is the case with many countries, they also get outside financial help for such projects, notably from the World Bank. Just as with any project, if it is to be successful, an afforestation programme needs follow-up and maintenance. The programmes that work the best include provision for irrigation and for educating the local population about the need to preserve forests.

Nature of Science

As with many scientific endeavours, there is active debate when it comes to the best way to achieve carbon sequestration. Some experts advise countries to plant non-native species of trees that will grow faster than local trees and therefore capture more carbon more quickly. Others say that rewilding is the best approach, bringing back the native species that once flourished in an area. For wetland restoration or the introduction of carbon sinks such as peat bogs, once stagnant water has been introduced or reintroduced, plant species tend to arrive by natural means and start carbon sequestration rapidly, notably in tropical zones, for example the Eastern Congolian swamp forests, which contain peatlands in the Congo River basin that cover an area of 145,000 square kilometres and contain an estimated 30 billion tonnes of carbon.

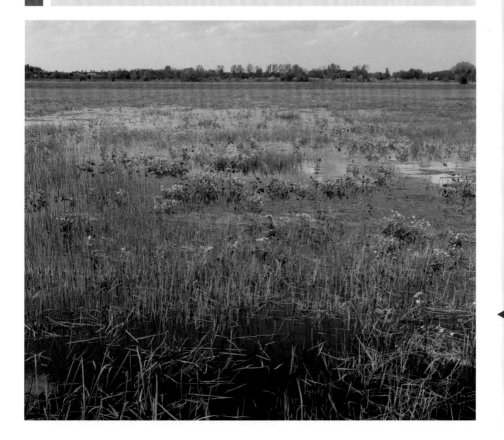

Wetlands can be used as carbon sinks. Instead of draining it to use the land for agriculture or urban development, this wetland in Poland has been set aside as a natural reserve where human activity is minimized.

The US Fish and Wildlife Service reported in 1990 that over 40 million hectares of wetlands had been destroyed since 1770, which was half of all the wetlands in the country. There are now protection policies in place to make sure that any wetlands that are destroyed are replaced. Restoring or introducing wetlands involves excavation (digging out) of the land to provide low-lying zones where rainwater can accumulate. Sometimes the types of plants that are best adapted to wet soil need to be deliberately introduced, such as marsh grasses or, in the case of peat bogs, *Sphagnum* mosses. Active management might also need to include the removal of invasive species of plants, but sometimes just leaving natural colonization and succession to do their work is enough once the water starts to accumulate.

Although many human activities tend to damage ecosystems and exacerbate climate change, efforts have been made to restore land to its natural state, improve biodiversity and bring the ecosystem back into balance. One of the best ways of doing this is to minimize the impact of human activities in an area. Nature reserves are places set aside by governments or private landowners to let nature do what it does best: optimize biodiversity and reach a sustainable equilibrium. By reducing access to motorized vehicles and prohibiting land use, natural restoration can take place, and sometimes it occurs over a surprisingly short period of time.

Gorse (*Ulex europaeus*) ▶

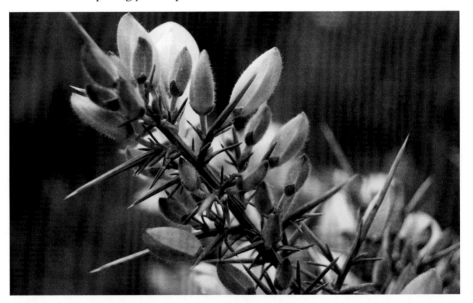

Gorse (*Ulex europaeus*) is a small thorny plant that produces yellow flowers. It was introduced to New Zealand by Europeans during British colonization of the islands. The plant quickly covers any open land such as abandoned farmland or pastures. Although it is considered an invasive species in New Zealand today, it has one advantage that favours the succession process: as it grows, its dense vegetation thins out, leaving spaces between branches for young saplings of native plants and trees to take hold and mature. The saplings benefit from the nurse canopies maintained by the gorse as it holds the soil in place, blocks the wind and shades the ground to reduce evaporation of water. As they grow in this nursery environment, native plants can eventually replace the gorse by growing above it, shading it out and taking its place.

This phenomenon has been observed over a very short period of time in the Hinewai Reserve in New Zealand. The reserve was started in 1987 and currently hosts many native plants that have not grown in the area for many decades. Although trees covered the region before humans lived there, farming in recent centuries has greatly reduced the original vegetation. Since the establishment of the reserve a few decades ago, a rapid regeneration of native plants is taking place. Today, if you visit the reserve or look at its website, you can see that native *kānuka* trees and podocarp trees such as the *tōtara* are repopulating the area.

What processes determine the distribution of organisms on Earth?

Guiding Question revisited

What are the drivers of climate change?

In this chapter you have learned that:

- human activity is the main driver of climate change
- activities such as burning fossil fuels or deforestation add carbon dioxide to the atmosphere
- methane is also a greenhouse gas and is added by farming methods such as raising cattle and rice farming
- there are positive feedback loops in global warming
- melting snow and ice decreases albedo so that the water underneath warms up more than usual, in a positive feedback loop
- in addition, warming oceans causes an accelerated release of carbon dioxide
- warmer temperatures mean that permafrost and peat also release carbon dioxide
- some forests may reach a tipping point, and rather than being carbon sinks they may become a carbon source.

Guiding Question revisited

What are the impacts of climate change on ecosystems?

In this chapter you have learned that:

- changes in ocean currents and atmospheric temperatures can lead to extreme weather events, drought and forest fires
- when the climate changes there are impacts on habitats
- warmer polar habits means that there is less ice to form breeding grounds for polar species
- changes to ocean currents can alter the timing of nutrient upwelling, which disrupts marine food chains
- an increase in carbon dioxide in oceans causes acidification and warming, which leads to bleaching of coral reefs and subsequent collapse of ecosystems
- species may change where they live, for example, montane birds may move further upslope.

Exercises

Q1. Which gases are contributing to the anthropogenic runaway greenhouse effect?

    I   Carbon dioxide.

    II   Ozone.

    III Methane.

  A I and II only   **B** I and III only

  C II and III only   **D** I, II and III

Q2. The El Niño Southern Oscillation modifies which of the following?

    I.   Nutrient upwelling.

    II.   Precipitation near the coasts.

    III. Primary production in marine ecosystems.

  A I and II only.   **B** I and III only.

  C II and III only.   **D** I, II and III.

Q3. Which Arctic animals are seeing their habitat modified by melting sea ice?

    I.   Polar bears.

    II.   Walruses.

    III. Emperor penguins.

  A I and II only.   **B** I and III only.

  C II and III only.   **D** I, II and III.

Q4. Explain the positive feedback loop related to climate change and methane production in regions with permafrost.

Q5. Distinguish between how a garden greenhouse works and how the greenhouse effect on Earth works.

Q6. A scuba diver returns to her favourite coral reef only to find it empty of life and all the corals turned white. She asks you if you know what this phenomenon is: what do you tell her?

D4 Practice questions

1. Which statements are characteristics of alleles?

    I.   Alleles differ significantly in number of base pairs.

    II.   Alleles are specific forms of a gene.

    III. New alleles are formed by mutation.

  A I and II only.   **B** I and III only.

  C II and III only.   **D** I, II and III.

(Total 1 mark)

2. What is a direct consequence of the overproduction of offspring?

  A Individuals become more adapted to the environment.

  B They will be subject to intraspecific competition.

  C They will diverge to produce different species.

  D They will suffer mutations.

(Total 1 mark)

3. What would restrict evolution by natural selection if a species only reproduced by cloning?

  A Too few offspring would be produced.

  B Mutations could not occur.

  C The offspring would show a lack of variation.

  D The offspring would be the same sex as the parent.

(Total 1 mark)

4. Some lice live in human hair and feed on blood. Shampoos that kill lice have been available for many years but some lice are now resistant to those shampoos. Two possible hypotheses are:

Hypothesis A	Hypothesis B
Resistant strains of lice were present in the population. Non-resistant lice died with increased use of anti-lice shampoo and resistant lice survived to reproduce.	Exposure to anti-lice shampoo caused mutations for resistance to the shampoo and this resistance is passed on to offspring.

Discuss which hypothesis is a better explanation of the theory of evolution by natural selection.

(Total 3 marks)

5. Explain how evolution by natural selection depends on mutations.

(Total 4 marks)

6. Extensive areas of the rainforest in Cambodia are being cleared for large-scale rubber plantations. Distinguish between the sustainability of natural ecosystems such as rainforests and the sustainability of areas used for agriculture.

(Total 3 marks)

7. Global warming has changed both the thickness and surface area of sea ice of the Arctic Ocean as well as the Southern Ocean that surrounds Antarctica. Sea ice is highly sensitive to changes in temperature.

Scientists have calculated a long-term mean for the surface area of sea ice in the Arctic and in the Southern Ocean around Antarctica. This mean value is used as a reference to examine changes in ice extent. The graph shows the variations from this mean (zero line) over a period of time.

(a) State the trend in the surface area of sea ice in the Southern Ocean around Antarctica. (1)

(b) Distinguish between changes in the surface area of sea ice in the Arctic and Antarctica. (2)

(c) Discuss the data as evidence of global warming. (3)

Adélie penguins (*Pygoscelis adeliae*) are only found in Antarctica and need sea ice for feeding and nesting. Biologists are able to deduce how these penguins have responded to changes in their environment for the last 35,000 years, as the Antarctic conditions have preserved their bones and their nests. The image on the next page is a map of Antarctica and the surrounding Southern Ocean. It shows the trends in the length of the sea ice season (days of the year when sea ice is changing) and the sites of nine Adélie penguin colonies.

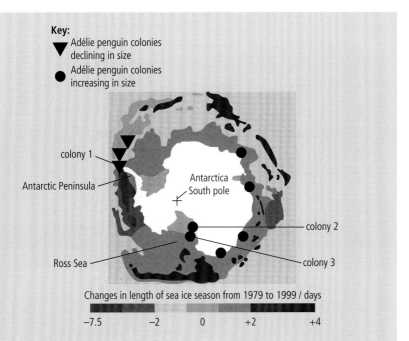

Key:
▼ Adélie penguin colonies declining in size
● Adélie penguin colonies increasing in size

colony 1

Antarctic Peninsula

Antarctica
South pole

colony 2

Ross Sea

colony 3

Changes in length of sea ice season from 1979 to 1999 / days

−7.5   −2   0   +2   +4

(d) Describe the trends in the length of the sea ice season around the Antarctic Peninsula and in the Ross Sea. (2)

The graphs shows the changes in penguin population in three of the colonies shown on the map.

(e) Analyse the trends in colony size of the Adélie penguins in relation to the changes in the sea ice. (3)

(Total 11 marks)

8. Several greenhouse gases occur in the atmosphere. Carbon dioxide (CO_2) is one of them but so are methane (CH_4) and oxides of nitrogen (NO_x). Why are oxides of nitrogen classed as greenhouse gases?

    A   They trap some of the long-wave radiation emitted by Earth's surface.

    B   They prevent short-wave radiation from reaching Earth's surface.

    C   They dissolve in rainwater to produce acid rain.

    D   They are only produced by human activity, whereas CO_2 and CH_4 are also produced naturally.

    (Total 1 mark)

9. The oceans absorb much of the carbon dioxide in the atmosphere. The combustion of fossil fuels has increased carbon dioxide ocean concentrations. What adverse effect does this have on marine life?

    A   Heterotrophs consume more phytoplankton.

    B   Phytoplankton have increased rates of photosynthesis.

    C   Corals deposit less calcium carbonate to form skeletons.

    D   Increased pH reduces enzyme activity in marine organisms.

    (Total 1 mark)

10. Describe the relationship between the rise in the concentration of atmospheric carbon dioxide and the enhanced greenhouse effect.

    (Total 5 marks)

Theory of Knowledge in biology

Theory of knowledge (TOK) is a way of thinking about our thinking. We ask ourselves questions about knowledge, about how we acquire, refute or confirm what we know. This chapter will guide you through how to think about TOK from a biologist's perspective in particular, and from the natural sciences' perspective in general. We can ask ourselves questions including "What makes biology different from the other natural sciences such as chemistry or physics in the way we acquire knowledge?" and "In what ways are the natural sciences different from other areas of knowledge: the human sciences, history, mathematics or the arts?" TOK is meant to challenge your brain and push you outside your comfort zone. Enjoy the ride.

An astronomer, a physicist and a mathematician are on a train going to a conference in Edinburgh, Scotland. Out of the window, they see a solitary black sheep.

- Astronomer: That's interesting, sheep in Scotland are black.
- Physicist: It would be more prudent to say that *some* of the sheep in Scotland are black.
- Mathematician: To be more precise, we can say that in Scotland there exists at least one field in which there is at least one sheep, which is black on at least one side.

What does this story reveal about scientific observations, hypotheses and conclusions? What does it reveal about the nature of each of the disciplines represented?

Is biology less exact than physics or mathematics? If there had been a biologist on board the train, what would they have said about the sheep?

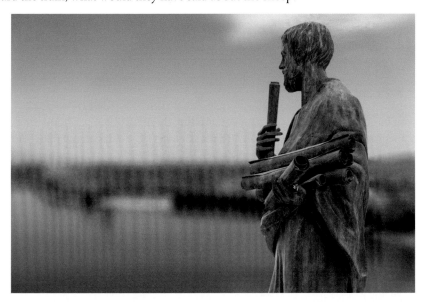

Theory of knowledge is different from your other classes because, instead of asking about facts and skills in a particular subject, it is a critical thinking toolbox for asking how we know what we know.

A paradigm is a way of seeing the world, and a paradigm shift is when new information transforms that view.

An antique microscope ▼

What is this chapter all about?

This chapter contains ideas, quotes, anecdotes, case studies and many unanswered questions. With TOK, it is important to develop your own ideas and arguments. Start with a **knowledge claim** such as "all sheep in Scotland are black" or a **knowledge question** such as "What role does imagination play in the pursuit of knowledge in the natural sciences?"

The study of biology is filled with objects that should help you see connections to TOK. For your TOK exhibition, you are asked to pick three such objects. Many examples are included in this chapter, and the microscope is one of them. The microscope represents a **paradigm shift** in biological thinking because it opened up a whole new category of life: organisms that occupy the microscopic world. A paradigm is a way of thinking about something, and a paradigm shift is when something new completely changes the old way of conceptualizing things. The microscope allowed biologists to see for the first time that there is a parallel universe of life living right under our noses.

Knowledge claim: a statement that someone declares as being true from their perspective. Upon verification, it may be confirmed, but a knowledge claim can also be refuted.

Knowledge question: a question that asks how we know something, how we acquire, confirm or refute knowledge. Knowledge questions are open-ended (they do not have a single correct answer), contain TOK concepts and vocabulary and, ideally, can be applied to a wide variety of situations in the real world.

Debates

Consider the following two knowledge claims about the nature of all human beings on Earth.

> A: We are all the same.

> B: We are all different.

Use your biological knowledge to support or refute these two claims. Choose one, and try to imagine someone saying to you, "That's not true! How can you say that?" How would you respond to that person?

Now try these two statements.

> X: Biology is a collection of facts about nature.

> Y: Biology is a system of exploring the natural world.

Use your critical thinking. Critical thinking is characterized by reflective inquiry, analysis and judgement. Ask yourself: "Should I believe this?" "Am I on the right track?" "How reliable is this information?" In short, you are deciding whether or not you should accept something as valid. If you think it is an easy, quick decision, then you are not treating the question in the way that you should.

Critical thinking

Is the statement valid? What is its source? Is the person who said it reliable? Do they have a **bias** that I should know about? Bias refers to a type of prejudice whereby a person gives an unfair preference to one perspective over another, rather than giving a balanced argument.

Coming back to the pairs of statements above, what would lead someone to believe one or the other? For each pair, could it be possible that both statements are valid? Or are they necessarily mutually exclusive? What about the following two statements?

- There is only one scientific method that is universal throughout the world: only by following the same method can scientists reach the same results and conclusions.
- Different scientists and different cultures in different regions of the world use different versions of the scientific method to obtain valid results and conclusions.

Is there just one scientific method? Do all scientists always follow the same steps?

TOK knowledge framework

One of the most important skills students are asked to develop in the IB programme is **analysis**. Analysis is the process of breaking apart something complex in order to see how it works. In biology, "lysis" refers to splitting, as in *cytolysis* or *hydrolysis*, words that are based on the ancient Greek word for "loosening" or "releasing". When you analyse something, hopefully you can pull out or release some meaning.

When you want to "analyse" something, where should you start? How should you break a concept up into its constituent parts? There is no single answer, but the TOK framework is a useful tool for analysing knowledge questions. When analysing a particular scenario, you are not expected to address each of the four points below exhaustively, but this framework can be a good place to start.

This framework is a useful tool for analysing knowledge questions.

In the TOK knowledge framework:

Perspectives refer to someone's point of view. It might be the perspective of an individual or it might be that of a group, such as molecular biologists or neo-Lamarckists.

Ethics refers to questions of responsibility. In TOK we ask how we know if it is morally right or wrong to genetically modify crops or to experiment on animals.

Scope concerns what questions the natural sciences can and cannot answer. The scientific method is only good at answering scientific questions, for example.

Methods and tools gets us to think about how scientists produce, acquire or refute knowledge. What is the scientific method and how is it different from the approach that historians or mathematicians use in their area of expertise?

Try the framework out on the sheep example at the beginning of the chapter. There may be some aspects of the framework that apply nicely, and others that do not fit well. Throughout this chapter there are a number of case studies; as practice, it is worth analysing them based on the knowledge framework. This is good practice for your essay writing because you are encouraged to incorporate ideas from the framework in your TOK essay.

What is knowledge?

Consider the knowledge questions below.

- What counts as knowledge in biology?
- How does biological knowledge grow?
- What are the limits of knowledge in biology?
- Who owns biological knowledge?
- What is the value of knowledge in biology?
- What are the implications of having or not having biological knowledge?
- Is there one way that is best for acquiring knowledge in biology?
- Where is biological knowledge? Is it a "thing" that resides somewhere: is it in books, in your head, in a computer database?

Look at the following images. Use the list of questions on the previous page and your critical thinking to evaluate whether some or all of these are valid as scientific knowledge. For example, can mythology count as scientific knowledge? To help guide your discussion, consider the following key concepts in TOK:

Evidence   Certainty   Truth   Interpretation   Power   Justification
Explanation   Objectivity   Perspective   Culture   Values   Responsibility

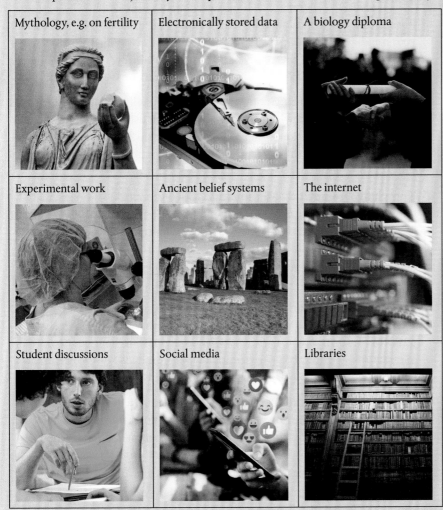

Mythology, e.g. on fertility	Electronically stored data	A biology diploma
Experimental work	Ancient belief systems	The internet
Student discussions	Social media	Libraries

How do we know?

An example of scientific knowledge in biology is: "The organelle in a plant cell that is responsible for photosynthesis is the chloroplast". How could you verify this? How can you be sure that there is not another part of the cell that performs photosynthesis? Is it a falsifiable idea? Such questions are **second-order questions**. They are not about the thing we want to know, they are about how we know it. In your IB Diploma subject exams and internal assessments, you are asked lots of first-order questions, but in TOK you need to focus on second-order questions.

Critical thinking does not mean you criticize everything and refuse to ever accept anything as valid. It means you are aware of questions of validity. You are not being negative, you are just being inquisitive and prudent.

Case study 1: Is there life on Mars?

This is not a simple question. Despite several visits by space probes, it has been extremely difficult to find conclusive evidence on Mars that can lead scientists to declare that there is life on its surface. And yet the search continues. One piece of compelling evidence that there was once life on Mars comes from a meteorite found in Antarctica that The National Aeronautics and Space Administration (NASA) claims came from Mars and contains fossils of bacteria.

If you apply your critical thinking, some knowledge questions should pop into your mind. How do we know that this chunk of rock is really from Mars? How does NASA know that the "fossils" are from bacteria? How certain are we that they were not formed from non-living chemical reactions?

Knowledge framework: Scope The question about life on other planets falls within the realm of knowledge that can be acquired using the natural sciences. Decades of data collection from probes sent to the surface did not reveal the presence of life, but is this an indication that there is no life on Mars, or does it show the limitations of our technology?

Knowledge framework: Methods and tools Although no humans have gone to Mars, researchers collect photos and soil samples, as well as measure the temperature and composition of the atmosphere, using various technologies on robotic probes. NASA analyses the data and proposes hypotheses. Subsequent missions are sent to collect more samples and test those hypotheses.

Knowledge framework: Ethics How do we decide that it is acceptable to spend millions of dollars on space missions when there are still so many problems to solve here on Earth? At what point do we accept that the amount of pollution here on Earth, in space and on Mars produced by the rockets and probes is worth the knowledge we gain from such missions?

Knowledge framework: Perspectives Such debates will have different points of view depending on which stakeholder you ask. NASA is likely to view the missions differently compared to environmentalists or groups fighting for social justice. Biologists will argue that it would be difficult to imagine any greater discovery in human existence than finding life on another planet.

From this specific example of life on Mars, two more general questions arise. Is it possible to really "know" the truth? Is information absolute or relative?

Are you an empiricist or a rationalist?

Empiricism = the belief that our senses allow us to acquire knowledge
Rationalism = the belief that reason allows us to acquire knowledge

Case study 2: Babies born on a full moon

Knowledge framework: Perspectives

Ask an experienced midwife "Are more babies born on a night when there is a full moon?" and the chances are she will say yes. You would have no reason to challenge her: she is the expert. She is an eyewitness to this phenomenon.

But knowledge questions arise. Is this verified knowledge or is it opinion? How does she know? Is it just a feeling, an intuition, a belief? Or is this knowledge claim based on carefully analysed statistics comparing birth numbers with a lunar calendar?

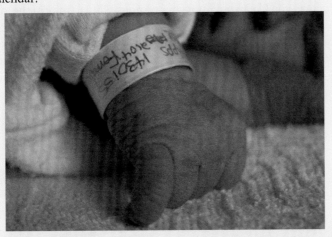

As it turns out, statistics do not support this knowledge claim. The evidence from maternity ward numbers does not show a correlation between births and the full moon. So what is going on? Is the midwife lying? In fact she is probably the victim of something we all are susceptible to: **confirmation bias**. Confirmation bias happens when we only remember the times when something confirmed our beliefs, and ignore the times when something refuted them. In the case of the midwife, on a busy night she might look out the window, see a full moon, and cry out to her colleagues, "See? I was right! More babies on nights when there is a full moon". Two weeks later, on another busy night, she looks out the window and what does she see? No moon at all. It is unlikely that she will now tell her colleagues, "Sorry, I was wrong: it's a busy night and yet there is no full moon". It is more likely that she will forget this negative result and only remember the positive result, thereby showing a bias for confirmation.

Should we tell her she is wrong? It could be argued that she is not hurting anyone and that it is lots of fun to have these sayings in our culture. Having shared beliefs unites people and strengthens a sense of community and belonging. Is it better to be right or to belong?

Confirmation bias happens when we only remember the times when something confirmed our beliefs and ignore the times when something refuted them.

TOK journal

Do you keep a TOK journal? You should. Write down any time one of your teachers mentions TOK, or any time you see a situation that could lend itself well to a knowledge question. For example, on a field trip you might be asked to determine how polluted a body of water is. You do some chemical tests and you catch some larvae that you identify as being sensitive to pollution. But then, as a good biologist and good TOK student, you start asking yourself how certain you can be that your measurements will allow you to conclude that the body of water is "clean" or "safe" (**Scope**, **Methods and tools**). What if you found that the water was contaminated by a local industry that makes products you like to buy? Is it your responsibility to inform the local authorities (**Ethics**)? What if you get different results and draw different conclusions compared to your classmates, or if you think that it does not matter what the data shows, it is more important that we keep industry productive in the region for financial success and access to local products (**Perspectives**)? How do you decide whose results are more valid?

Biology fieldwork should get you thinking about knowledge questions.

Catching a cold

Despite the biological evidence that colds are caused by viral infections, many people believe that you can catch a cold from being exposed to low temperatures or changes in humidity.

Who is right? Where does the truth lie? For something to be considered "true", does it have to be formally proven using a scientific method? Is a profound conviction that something is true good enough to make it valid? If one person believes that something is true, does that make it true or does there have to be a certain number of believers before the idea can be considered true? What if my grandmother says it and she learned it from her grandmother? Does that make it a valid claim?

Theories versus laws

A **law** is a generalization used to describe a phenomenon. The laws of thermodynamics describe the flow of energy in a food chain, for example. We make measurements and observations and realize that energy cannot be created or destroyed, only transformed from one type (e.g. light energy from sunlight) to another (e.g. chemical energy in sugar) and another (e.g. heat energy released during cellular respiration reactions). This is the concept of conservation of energy. It is a law of nature. But there is nothing in the law that attempts to explain the phenomenon. Laws only describe. They are not falsifiable because they make no contestable claims.

A **theory** is an explanation of a phenomenon. It can start out as a hypothesis, but to qualify as a theory it must be thoroughly tested and have ample evidence to support it, with little or no evidence to refute it. A theory is testable and can be refuted if enough evidence builds up against it, or can be replaced if a new theory does a better job of explaining the phenomenon. Robust theories are those that have withstood many attempts to falsify them. Darwin's theory of natural selection to explain evolution, for example, is one of the most tested and robust theories ever published in biology. For over a century and a half, it has withstood countless attempts to refute it. If someone finds one exception that goes against a theory but there is still overwhelming evidence for it, we do not abandon the theory.

The word "theory" can be problematic because different groups of people use it in different ways. Sometimes when someone says, "My theory is that this is caused by …", they are really referring to a hypothesis. They are suggesting a possible cause for a phenomenon. When it is used this way, it is sometimes qualified as "just a theory" meaning that it is, in fact, just a hypothesis. How can the language of science contribute to or detract from clarity when communicating knowledge?

A law describes a phenomenon and makes no attempt to explain it. Laws are not contestable because they make no claims. A scientific theory is a proposed explanation. A theory is contestable, and by challenging and testing it, the theory can be confirmed or refuted. A theory can be modified over time, or can be replaced with a better theory as new evidence appears.

Phrenology

The **pseudo-science** of **phrenology** claimed that the shape of a person's skull, and the bumps and indentations on it, determined a person's intelligence, personality and talents. A pseudo-science is a discipline that is presented as if it is a science but does not follow the rules of scientific inquiry and scrutiny. For example, in science, if a large body of evidence refutes an idea, the idea should be abandoned. Phrenology persisted for decades despite mounting evidence that it was not confirmed by observations and experiments. More controversially, it was used by some to justify the superiority or inferiority of "races" of humans.

How do you think it was demonstrated that the "laws" of phrenology were not, after all, scientifically valid? What needed to be done to disprove them? How can the social context affect what is studied and how? If scientists today wanted to use genome analysis to compare "races" the way phrenology did, how would the scientific community or the rest of society react? The word "race" has been used freely to refer to different groups of humans for hundreds of years. Linnaeus used the term in the 1700s. Although we still use the concept in words such as "racism" or "racist", scientists very rarely if ever use the term "race" when referring to subgroups of humans. This is an example of a **historical development**. It also shows how language changes over time as ideas about what the words represent change.

The term historical development in TOK refers to a change in methods and tools in an area of knowledge. For example, the use of the microscope to discover bacteria was an astounding historical development. More recently, genome sequencing has helped us see evolutionary relationships with much more detail. Do not confuse a historical development with a historical event, such as the signing of a treaty or the first time humans walked on the Moon.

▲
The study of phrenology has been discredited.

Tongue map

Tongue map

- ● **bitter**
- ● **sour**
- ● **sweet**
- ● **salty**

The idea of a tongue map, with zones where certain flavours are sensed, has been discredited.

As students and teachers, what do we claim to know about biology? Are we justified in making such claims? How?

What experiences have you had that give you insight concerning these issues? Consider the following example.

For generations of students, the idea of a "tongue map" (i.e. certain zones of the tongue relate to certain tastes) was propagated by biology textbooks, and taste-test investigations were suggested as laboratory work in schools. It has since been shown that all parts of the tongue can taste sweet, sour, bitter and salty.

There must be no barriers for freedom of inquiry. There is no place for dogma in science. The scientist is free, and must be free to ask any question, to doubt any assertion, to seek for any evidence, to correct any errors.

Robert Oppenheimer
1949

Art and imagination

Is there a place for creativity and emotional expression in science? Are there any parallels between biology and the visual arts? Could it be argued that just as an artist sees the natural world in their own way, so a scientist sees nature in their own way? Or, on the contrary, are science and the visual arts diametrically opposed ways of interpreting nature? Was someone like Ernst Haeckel, who drew visually striking drawings of what he observed, just as much a visual artist as a biologist?

Plate 32. *Rotatoria*, from Haeckel's *Kunstformen der Natur* (*Artforms of Nature*), published in 1904, showing members of the genus Rotatoria as observed under a light microscope.

Case study 3: Spirit/soul

Knowledge framework: Scope

In 1907, Dr Duncan MacDougall conducted experiments to determine whether people lost mass after death. His results seemed to suggest that they did, and led him to the conclusion that the human soul weighs 21 g. As his experiments (some of which did not give conclusive results) were carried out with scales of questionable accuracy, and he had only six subjects, his conclusions are widely criticized and are not taken seriously by the scientific community today.

Will questions about souls always remain beyond the capabilities of science to investigate or verify? Why has no one repeated this experiment in over a century? What do you think the reaction of the religious community would be if scientists repeated MacDougall's experiment?

Decisions, decisions ...

Should experiments be performed to answer fundamental questions, or should they only be done if they have a useful application in our everyday lives?

Who should decide which research pursuits are of the most value? Who should decide on how funding is distributed, or the prioritizing of the use of laboratory space and resources? Universities? Governments? Committees of scientists? Should taxpayers be allowed to vote on which research projects receive public funding?

Should research about a tropical disease such as malaria be paid for by tax money from non-tropical countries?

Is there an end?

Is scientific knowledge progressive? Has it always grown? Imagine a graph with scientific knowledge on the y-axis and time on the x-axis. How would you draw the graph? Would it be a curve or would it be linear? Is it always increasing? What units would you use? Could the graph ever go down? In other words, could scientific knowledge ever be lost (maybe because of the outbreak of war, a laboratory burning down, a brilliant scientist dying)?

Could there ever be an end to science? If there was an end, what would be the consequences?

> *Science knows no country, because knowledge belongs to humanity, and is the torch which illuminates the world.*
>
> Louis Pasteur

Doctor, which drug treatment is best for me?

How do doctors know which medication is best for their patients? One way is for them to keep up with the latest breakthroughs and developments published in scientific and medical journals. Doctors put their faith in these prestigious peer-reviewed journals and, because they do not have the time or the budget to do all the clinical trials themselves, they trust that the researchers doing the work are following sound practice. One problem is that a large percentage of these studies are being funded by the companies that make the drugs and, according to epidemiologist Dr Ben Goldacre's 2012 book *Bad Pharma*, it is common practice in the pharmaceutical industry to use a wide variety of tricks and manipulations to make a new drug look good in clinical trials. One trick follows this type of pattern: a company sets up a 2-year trial to test a new drug and then, after only 6 months, it decides to stop the trial and publish the data because the numbers show that its drug is performing well. This is advantageous to the company because it saves money (trials are very costly), and it reduces the chances that participants develop side effects or show negative results. Doctors reading about the clinical trials will never be informed, however, that the trials were stopped early. Another trick is to not report in the published study any participants who dropped out of the trial because they felt ill from side effects. By only mentioning the people who stayed in the study, they can report that, at the end of the trial, none of the participants complained of any major side effects. In short, Goldacre claims that the studies being published are not showing all the data and that, in order for doctors to decide whether a drug is safe to prescribe, they need to see both the positive and the negative results. What knowledge questions does Goldacre's book raise about the highly competitive pharmaceutical industry (that earns hundreds of billions of dollars annually)? Do the practices he denounces sound like the kinds of things your biology teacher encourages you to do in your laboratory investigations? If you were a scientist working at one of these companies, and you decided to complain and point out that some of the trials seemed unfair, what do you think your boss's reaction would be? If a company decided to publish the positive as well as the negative results of its drug trials, what do you think might happen to the sales of its drugs? Lastly, as doctors find out more about the practices followed, what will happen to their faith in the data presented by the medical journals? What kind of critical thinking or TOK questions should doctors apply when they pick up a medical journal and read about the latest breakthroughs in drug research?

The placebo effect

One of the ways that scientists test a drug is to compare it with a **placebo**. A placebo contains no active ingredients: it is often just a sugar pill. To find out if a new drug is effective, one group of volunteers in a study is given the drug and another group is given the placebo. Neither group knows whether it is taking an active pill or a placebo. Surprisingly, even in the group taking the placebo, there are patients who report that they feel better. This is called the **placebo effect**.

A placebo is a helpful tool in understanding the efficacy of drug treatments.

▼

According to Goldacre, researchers studying the placebo effect have observed that the following things have a positive influence on how effective the patients thought the pill was:

- the doctor was wearing a white laboratory coat
- there were diplomas on the wall in the doctor's office
- the doctor sat down and listened attentively to the patient.

The medical community is essentially unanimous on the validity and power of the placebo effect, and yet the mechanism of how it works is poorly understood. For example, astounding as it may seem, placebos seem to have an effect even when people are told that they are receiving a placebo. Some participants still feel better, even when they are aware that they have not been given any active drugs.

Inhabitants of industrialized countries often discredit herbal medicine and healers in indigenous peoples. And yet, those same critics may very well accept the effectiveness of the placebo effect: an effect that appears to be produced essentially by ritual (laboratory coat, diplomas, attentive listening). What knowledge questions are raised by this puzzling effect? What does it say about the limits of modern medicine?

Models

The double-helix shape for DNA and the fluid mosaic membrane are examples of models that were created in order to explain observed phenomena. Are such models just inventions of our imagination? If so, how is it that they can be used to make predictions or explain natural phenomena?

> *All models are wrong, but some are useful.*
>
> George E. P. Box
> (innovator in statistical analysis)

Who's right?

Among all the points of view that are available to you in the classroom, at home, in the media, on websites, how do you know which to trust?

Religion in an age of science

In what ways could someone's cultural or religious background influence their acceptance of certain scientific theories?

There was a time when scientists hesitated to publish their works out of fear of the church's reaction. Have the tables turned? Are there religious writers who fear scientific criticism if they publish their ideas?

If a student refuses to answer questions about evolution by natural selection on an IB exam because of their religious beliefs, should they get any marks?

In 1663, the Roman Inquisition condemned Galileo for defending the idea that the planet Earth goes around the Sun, and he remained imprisoned for nearly a decade before he died. In 2010, the Catholic Church formally apologized for Galileo's condemnation.

Ockham's razor

Simply put, the **principle of Ockham's razor** states that, all other things being equal, the simplest explanation should be preferred. This is reflected in the idea of **parsimony**: seeking out the least convoluted solution. Scientists take this principle very seriously and yet some aspects of science seem to be extremely complex. Is there a conflict here?

Limits of perception

Can we, here on Earth, possibly know of worlds beyond our own? Can we know what the distant past was like, or what the distant future will hold? Or are we like a frog at the bottom of a well trying to understand what the ocean might be like?

> *You cannot speak of the ocean to a well frog …*
>
> Chuang Tzu Taoist text
> (written more than 2,000 years ago)

The eye is not a camera

A fun classroom activity is to have someone unknown to the students barge in during a lesson, say something, take something off the teacher's desk, and leave, after which the teacher asks each student to write down a description of the person, and what they said and did. The students' observations are often remarkable in their diversity, and the activity demonstrates how human perception is notoriously bad at picking up crucial details, and notoriously good at filling in missing information. "The eye is not a camera" is a good example of a knowledge claim that can be explored and discussed after such an observation activity.

Do we see with our eyes or with our brain?

Although we associate eyes with vision, truly seeing something means interpreting the signals that arrive at the photoreceptors in the eye. This interpretation is done in the brain, so in fact, do we see with our eyes or our brain?

Another knowledge claim on this theme also relates to eyewitnesses: "all memories are reconstructed memories". Have you ever had a story in your family that was told time and time again for years, and then one day you found out that the event in question never actually happened? And yet, you could swear that you remember the event clearly. Or have you ever watched a video recording of something you experienced and thought to yourself, "That's funny, I don't remember it being like that: my memory of that event is very different". Such examples put into question the validity of eyewitness testimonies in a court of law. Given that we know that our memories and observations can trick us, should we trust an eyewitness's account as irrefutable evidence during a trial? Are such testimonies reliable enough to put defendants in jail or to sentence them to death? This theme is explored well in Sydney Lumet's 1957 film *Twelve Angry Men*.

We were wrong, here's the real story ...

In palaeontology, it seems that every time a new hominid fossil is dug up, we must redraw the human family tree. If you search the internet for human phylogeny, you will probably find that few sources agree with each other. Likewise, every few years nutrition experts change their minds about dietary advice.

Does this frequent revision give credibility to science, or does this make science less credible?

Perspective 1: it is important for scientists to be able to modify ideas as new evidence is revealed. This is how science grows and progresses and, without such a system, we would be intellectually stuck.

Perspective 2: why can so-called experts not make up their minds? One year they say one thing, and then a year or two later they say "Oh, we were wrong, here's the real story".

Archaeopteryx

Here is an object that should help you see connections between TOK and the real world: *Archaeopteryx*, one of the most famous fossils in the world. It has some features of a dinosaur, such as reptile teeth and a bony tail, but it also has some bone structures similar to a bird and the most bird-like feature of all: feathers. It did not take long for observers to jump to the conclusion that *Archaeopteryx* is the "missing link" between dinosaurs and birds. Can we be so sure that this fossil is the transition between the two? Are physical features enough to base such a decision on? What kind of evidence would give more credibility to this claim?

Many palaeontologists shy away from phrases like "missing link": what features of this use of language make the term unscientific? One way to approach this question is to examine the **assumptions** and **implications** of the phrase "missing link". It assumes species like dinosaurs and birds have enough similar features to say the latter descended directly from the former. In other words, you could link the various transformations together over millions of years into an unbroken chain. This implies that if we find an organism that is half-and-half, it must be one of the links in the chain. Can you see the limitations of jumping to conclusions like this? One perspective to

Assumption: an idea that is likely to be true and presumed to be true despite the fact that we are not certain that it is true. Assumptions are based on previous observations and experience and should be reasonable. Assumptions are often tacit, meaning that they are not stated explicitly. You have to read between the lines to find them.

Implication: a logical consequence. If the assumption turns out to be true, then this logical consequence will be true, too.

consider is given by palaeontologists who say, "What if *Archaeopteryx* is on a branch of the tree of life that led nowhere and all its descendants are now extinct?" It is as plausible as the hypothesis that it is the "first bird".

▲
Archaeopteryx, a fossil with features of both dinosaurs and birds. Can we call it a missing link in evolution? It depends on your perspective.

Here is an example.

- Observation: I see only white swans.
- Assumption: All swans are white.
- Implication: The next swan I observe will be white and I will never observe a swan of a different colour.

Assumptions are often a necessary starting point in an argument but we need to be careful to recognize that we are only presuming that they are true, and we should not be surprised when they sometimes turn out to be false. You are encouraged to consider assumptions and implications in TOK.

Nature of science?

For centuries, it was firmly believed that rats, maggots and mould sprang from rotting meat and vegetable matter. This was called spontaneous generation. It took tireless experiments by Louis Pasteur and others to refute this idea and demonstrate that the rats, maggots and mould came from the surrounding environment.

The end of spontaneous generation

The idea of spontaneous generation has been shelved as unscientific. It has no value as biological knowledge, but it does have historical value and it helps to illustrate how science works.

This is a good example of an original hypothesis that was disproved and falsified by experimentation. It can be argued that, in order for something to be considered valid as scientific knowledge, it has to be verifiable. If experiments show that the results do not support the hypothesis, or even refute it, the idea is falsified. This assumes that the experiment is repeatable. Other scientists should be able to do the same experiment and get similar results.

Case study 4: Science and government

Knowledge framework: Perspectives

Trofim Denisovich Lysenko was a Soviet biologist who opposed the ideas of Mendel and Morgan concerning genetics. Instead, he promoted the Lamarckian idea that acquired characteristics could be passed on from one generation to the next. Under Stalin, he was promoted to a high-ranking post in agronomy and given his own scientific journal to publish his ideas. The agricultural techniques he developed were used to feed the Soviet population and the Red Army. Once Stalin and Khrushchev were no longer in power, however, his methods were widely criticized and his theories attacked for lack of scientific validity. An inquiry revealed that, in order to retain his powerful position and promote his ideas, he had intimidated and removed scientists who questioned his theories. He was finally fired from his post at the Institute of Genetics in 1965 and his reputation was crushed. What does this story reveal about the influence of politics on scientific theories? In what ways does it reveal scientific bias? How do we know that Lysenko's critics were not simply trying to push their own opposing political agenda?

Unprovable assumptions?

Does biology make any assumptions that are impossible to prove? Consider this knowledge claim: all events in nature are caused by physical phenomena.

In other words, every natural event can be explained by the interactions between atoms and molecules. Is such a statement provable? If we find enough examples of instances where this is true, can we proceed by **induction** that it is true for all phenomena? Induction, or inductive reasoning, is when we look at many examples of a phenomenon and try to come up with a general pattern. This seems reasonable, and yet the philosopher David Hume criticized induction, saying that there is no logical reason to assume that it is the case. Consider Karl Popper's quote about swans, which illustrates clearly the problem of induction. Interestingly, because the claim can be tested and refuted, it makes it a scientific claim.

> *No matter how many instances of white swans we may have observed, this does not justify the conclusion that all swans are white.*
>
> Karl Popper
> 1992

For a long time, Europeans thought there were only white swans. Black swans do exist. They are native to Western Australia.

▼

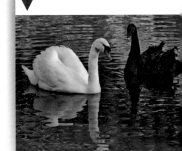

579

Scientific science

To what extent is there an overlap between biology and the social sciences? Are the latter "less scientific"? Consider psychology, sociology, anthropology and economics.

Science vocabulary

Does scientific language and vocabulary have a primarily descriptive or interpretive function? Consider the following expressions.

- Natural selection
- Concentration gradient
- Artificial intelligence

Wikis

Online wikis are filled with user-generated content on a wide range of subjects, including scientific ones. Wikis have been created for scientists to upload their latest laboratory findings. In what ways is this useful to scientists wanting to publish their results? In what ways is this useful to the general public? In what ways does this go against the very nature of peer-reviewed scientific publications, which is the norm today for sharing experimental results? For example, are such wikis just as valid as traditional scientific journals? How does the scientific community look upon these? Speaking of which, is there any such thing as a "scientific community"? Who belongs to it and how do we decide? When a specialist says something controversial and the media says that "the scientific community" does not agree with them, does that imply that they are not part of the scientific community? What about a wiki or a scientific journal for failed experiments? By seeing a list of published failures, would that not save researchers time by not repeating the same mistakes? Or, could it be that, if another scientist reads what one team thought was a failure but sees it for what it really is, a breakthrough, would that not help science advance?

Prediction is very difficult, especially about the future.

Niels Bohr
(a Danish physicist who helped us
understand how atoms work) 1970

580

Case study 5: Herbal medicine

Knowledge framework: Methods and tools

If a traditional treatment using a medicinal plant was tested in the laboratory and found to be effective, would that make the knowledge of the plant's properties more true? What if a single study disproved the plant's effectiveness as a medical treatment? Would that erase thousands of years of traditional knowledge? Turmeric, for example, has been used in herbal medicine for thousands of years to relieve digestive problems and inflammation. It is also widely used in savoury culinary dishes. Multiple scientific studies have been carried out to test its medicinal properties but none has conclusively seen any medical benefits. Have some types of knowledge been devalued by the perceived primacy of modern science? Conversely, today, some conventional Western medicine practitioners prescribe treatments that are categorized as complementary and alternative medicine, such as herbal remedies. What evidence would doctors need to be convinced that a herbal remedy is effective?

Seeing is believing: but what if you cannot see?

There is a story about a small group of blind men who encounter a tame work elephant, a creature none of them has ever had contact with before.

- One blind man touches the elephant's side and says "It's like a wall".
- Another grabs the end of its tail and says "It's covered in long hairs".
- Another feels a leg and says "Elephants are round and vertical like a pillar".
- A fourth holds his ear and says "It's like a sail".
- A fifth holds the animal's trunk and exclaims "Elephants are like snakes".

None of the men is wrong but no one is completely correct. This story illustrates how easy it is to jump to conclusions before having all the evidence. In science, is it possible to have all the evidence of any particular phenomenon?

Case study 6: Science deniers

Knowledge framework: Perspectives

In an editorial to the *Wall Street Journal* in March 2021, astrophysicist Neil deGrasse Tyson wrote about how surprised he was at the number of people who were sceptical of science when it came to dealing with the COVID-19 coronavirus pandemic: "If the enterprise of science were some new fangled, untested way of knowing, one might empathize with these sentiments. But the people who battle against science are the same ones who, for instance, wield and embrace their pocket-sized smartphones, which merge state-of-the-art engineering, mathematics, information technology and space physics". He noted that, for certain aspects of their lives, people trusted science, but for others they used non-scientific ways of thinking, even when considering science-related topics such as the immune system. He mentioned the influences of religion, culture and politics, and postulated that scientists were being out-competed by more savvy communicators from other domains. Is it the responsibility of knowers to share their knowledge with the public in a clear and convincing way? Should scientists be held responsible for not being better communicators?

Case study 7: CRISPR Cas9 technology, somatic cells versus germ-line cells

Knowledge framework: Ethics

In general, researchers working on gene-editing techniques in humans focus on somatic cells. They want to modify some tissue or an organ in a single individual. Modifying germ-line cells is much more controversial. Few researchers think it is a good idea to pursue knowledge about modifying the human species by intervening at the germ-cell level. The fear is that this would lead to "designer babies" with certain "desirable" traits, such as being taller, having a certain eye colour or being more intelligent. This raises many controversial issues and may lead to a return of eugenics, the practice of trying to improve the genetic makeup of humans. We genetically edit plant genomes to give them the characteristics we want, so why not do the same with our babies? How is such a distinction made: what causes a group of scientists to agree that this is where they will draw the line and say "no"?

Can you think of other situations in which a researcher might say, "Sorry, I refuse to do that on ethical grounds. Even though it is an interesting question to pursue, it goes beyond what I am willing to do."? How do scientists choose to draw such an ethical line between what knowledge is acceptable to pursue and what is not?

Perception

Which red circle is bigger? Judge using your eye first and then use a ruler to check your answer. What does this say about our perceptions and reality?

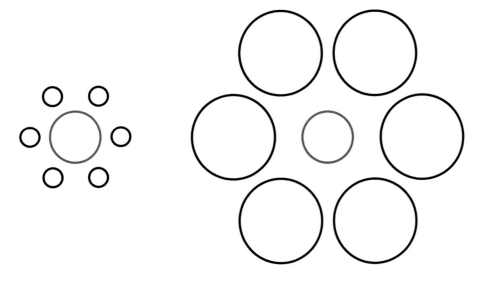

Science may be described as the art of systematic oversimplification.

Karl Popper
1992

What qualifies as an experiment?

Biology is an experimental science, but what constitutes an "experiment"? Do you have to have a hypothesis, controlled variables, a laboratory? What if you just have people filling out questionnaires? Is that an experiment? What about digging up fossils?

Theory versus myth

In what ways are theories and myths similar and different? Consider the similarities and differences when comparing and contrasting the two.

Is it based on well-substantiated facts? Is it passed on from generation to generation? Can it be modified over time? Can it be used to predict future events? Has it been tested repeatedly? Is it widely accepted as being true? Is it considered to be a supposition? Is it considered by many to be false?

Irrationally held truths may be more harmful than reasoned errors.

Thomas Henry Huxley

Biology and values

Do the ends justify the means? Consider the following domains of research in biology. What are the ethical issues associated with each?

- Gene therapy
- Vaccine tests
- Experimentation on human volunteers, notably prisoners
- Research involving human embryos

Nothing in this world is to be feared ... only understood.

Marie Curie

Marie Curie was the first woman to be awarded a Nobel prize, and the first person to be awarded two.

Science and technology

Is scientific knowledge valued more for its own sake or for the technology that it makes possible?

Reading your mind

With modern technology tracking everything we do on our computers and smartphones, it can be argued that the kind of privacy our grandparents had no longer exists. Can we at least say that our private and personal thoughts are still safe within our minds and cannot be tracked and monitored?

Functional magnetic resonance imaging technology (fMRI) allows researchers to see which parts of the brain are active when a person is thinking a specific thing or performing a specific task. This has led to the possibility of identifying thoughts or, as some call it, "mind reading". For example, researchers have shown a series of images to participants and recorded the patterns that show up on the fMRI scanner for each image. Later, they pick an image at random and show it to the participant while they are still in the scanner. A computer can match the current brain scan pattern with one of the patterns observed before and determine which image the person's brain is perceiving. Experts claim that they can use this technology to see whether someone is lying or to see whether someone recognizes a crime scene that they claim they have never been to. Marketing agencies are interested in seeing how the brain reacts to different advertising campaigns.

▲
Some companies claim they can use fMRI technology to identify peoples' thoughts or intentions.

Some major knowledge issues and knowledge questions arise from this. How can we know if such claims are valid? How do we test them and decide if the scanner and computer are accurate? How can we decide if the evidence collected in this way is credible enough to be used legally in court as evidence? Could complex thought patterns be identified, such as musical creativity or cruel intentions? Who should decide whether such experimentation and exploration into our private thoughts should be pursued or banned? Would you want a scan done of your thoughts?

Inaccessible worlds

Some scientific fields of exploration have only been possible since suitable technology has been invented, for example genetic engineering has only existed since the technological developments of the 1970s and 1980s. Could there be problems with knowledge that are unknown now because the technology needed to reveal them does not yet exist? Remember that, despite the fact that bacteria are all around us, we were not able to see them until the microscope was invented in the 1600s. Perhaps there are other phenomena that we simply cannot observe because no one has invented an apparatus to detect them yet.

Is there any science that can be pursued without the use of technology?

> *The most important discoveries will provide answers to questions that we do not yet know how to ask …*
>
> John Bahcall
> (commenting on the Hubble
> Space Telescope's capabilities)

> *My business is to teach my aspirations to conform themselves to fact, not to try and make facts harmonize with my aspirations.*
>
> Thomas Henry Huxley

Internal assessment

The scientific investigation

This chapter is not meant to replace the section in the official IB subject guide for biology about the internal assessment (IA) component of the course, the **scientific investigation**. Please be sure to read that first and consult it regularly, so that you keep in line with the criteria and the requirements. This chapter aims to help you understand the criteria and give you some guidance about how to fulfil them.

The IA component is worth 20% of your final grade and it is the only one in biology that assesses all four of the assessment objectives (AO):

AO1	Demonstrate knowledge
AO2	Understand and apply knowledge
AO3	Analyse, evaluate, and synthesize
AO4	Demonstrate the application of skills necessary to carry out insightful and ethical investigations

In other words, this is your opportunity to shine. Be sure to apply what you have been learning about the topic you choose to investigate and the laboratory skills and analytical techniques you have been acquiring during the course. And always keep in mind safety, ethics, environmental issues and academic integrity. These are all **transferable skills**: skills you can use later in a variety of contexts in your future studies and/or career.

Ways of obtaining your data

There are multiple ways of gathering the data you will need.

1. Laboratory work: carrying out an experiment you have designed. This allows you to manipulate a variable in controlled conditions.
2. Fieldwork: collecting data in streams, meadows woodlands or an urban park, for example. This allows you to select a variable such as sunlight or temperature to see if it affects the organisms you are studying.
3. Group data: analysing and modelling data collected by a group, either in the laboratory or during a field trip, and recorded in a shared spreadsheet. This allows you to choose one of the variables and see how it affects another in the spreadsheet. Each student needs to present different sets of data extracted from the spreadsheet.
4. Database: extracting and analysing data from a database such as a medical or environmental data set. This allows you to use data you would not otherwise be able to obtain on your own, such as numbers of patients with a particular tropical disease or statistics on deforestation in a distant country.
5. Simulation: using a computer model that allows you to change the parameters to see what happens. This could allow you to do things you would not be able to do in the laboratory or in the field, such as performing a genetics experiment on fruit flies or introducing new predators to an ecosystem.

The process from start to finish

Below is an overview of the steps to consider in the IA process in order to complete your scientific investigation.

1. Advance preparation: acquiring skills, learning how to design an investigation, learning the use of various laboratory techniques, databases, simulations or fieldwork techniques, practice data processing and writing a conclusion and evaluation.
2. Launch of the scientific investigation: criteria understood, calendar outlined.
3. Explore and brainstorm: initial research and narrowing down a topic followed by the formulation of a research question (RQ) and research about the methodology.
4. Plan and test: designing and trying out the method.
5. Data collection: this can be in the form of laboratory work, fieldwork, work with a database or with a computer simulation.
6. Data processing and analysis: calculations, graphs, statistical tests.
7. Writing, part 1: draft writing, including guidance and feedback from your teacher.
8. Writing, part 2: perfecting the draft of the report to produce the final submission, maximum word count 3,000 words. (See the official IB Biology guide for what is included in the word count.)

The criteria your work will be assessed on are listed here and total 24 marks (each is equally weighted).

- Research design: maximum 6 marks
- Data analysis: maximum 6 marks
- Conclusion: maximum 6 marks
- Evaluation: maximum 6 marks

Some parts of the work, notably the design of the method and the data collection (steps 4 and 5 listed in the overview above), can be done in small groups but most of the work is individual, notably the initial research, developing an RQ, the data processing, and all the writing involved in producing the report. Collaborative work can be done with up to three students as long as each has a different RQ and uses different data. Two examples of how collaborative work might be done are shown below.

Collaborative work: example 1

Two students want to use human volunteers for their investigations and decide it would be a good idea to work as a team. The first student has chosen a pupil reflex investigation and the second is looking at pulse rates in response to different types of exercise. They design their respective investigations collaboratively, each giving the other advice and suggestions, including how they are going to recruit their volunteers. They carry out the investigations collaboratively, asking each volunteer to take part in the first investigation and then the second. The students help each other with the setup and the recording of data. Each student uses the results collected to write their own report individually.

Collaborative work: example 2

Three students use nets to collect invertebrates from a stream during a class field trip. They assist each other with using the nets, identifying the invertebrates using dichotomous keys, and measuring the dissolved oxygen and flow rate in lotic (faster flowing) and lentic (slower moving) zones of the stream. Other measurements, such as water temperature, pH and light levels, are recorded using probes. They share all their raw data and qualitative observations but each student has a different RQ. For example, one might look at the relationship between dissolved oxygen levels and the presence of three species of invertebrates, another might look at the relationship between flow rate and the presence of all the species observed, and another could see how the invertebrates' anatomy is adapted to water speed. Each student has a separate RQ, each presents different raw data tables using the data they collected collaboratively, and each writes up a separate report individually.

In a similar fashion to the second example, large data sets collected by an entire class during fieldwork or laboratory work can be used as a database but the resulting investigations should be considered a database investigation. Each student presents only the data they have used. See the official IB Biology guide for more information on collaborative work.

Getting started

Once your teacher has launched the IA process, you need to come up with some possible ideas. Your early ideas may be rejected by your teacher because of safety or environmental reasons, or maybe just because the logistics will not work in the time given.

Some students start with a big topic, such as digestion, and narrow it down, for example to one enzyme being affected by one factor, after which they need to decide on what method to use. The danger with that approach is that not all themes or big topics will lead to an investigation that can be carried out. A student might be fascinated by parasites but the study of that topic in the laboratory raises serious logistical and ethical questions. It could be a database analysis instead of laboratory work, but it is not always easy to find a freely accessible database on a specialized topic.

Other students start with a particular method that they enjoyed, such as chromatography, and then work from there. If you are not sure where to start, one way is to make a list of the investigations you have already done and think about the techniques you are familiar with. Applying a familiar technique to a new situation is a great way to get started.

Example 1: applying a familiar technique to an unfamiliar situation

Lysa enjoyed an experiment using a delivery tube to measure the volume of gas produced when hydrogen peroxide was added to liver samples. She decided to apply the same measuring technique to yeast cells when feeding them different types of sugar.

Example 2: combining two familiar techniques

Matteo had already used the eyepiece graticule on a microscope to measure cell sizes in a previous laboratory, and in another microscope session while looking at living pondweed cells he observed cytoplasmic streaming of the chloroplasts. He thought, why not put the two together and measure the speed of cytoplasmic streaming? All he had to do then was come up with an independent variable. A bit more research on what factors activate or hinder cytoplasmic streaming, and he was on his way.

Example 3: database

Elif preferred research over laboratory work. She used the UniProt protein database and some online genomic databases, such as The National Center for Biotechnology Information (NCBI), to test different hypotheses about common ancestry in primates by comparing mutations in humans and non-human primates.

Example 4: trying to invent a technique unfamiliar to the student (unwise)

Fred wanted to measure stress levels in students who were given difficult mathematics problems to solve. His teacher asked, "Have we ever measured stress in the laboratory? Do we have a stress detector that you can point at people to measure their stress levels?" There is no such thing, and stress is subjective, so it was going to be very challenging to find a reliable measurement technique. Fred's teacher asked him to go back and look at things they had measured in previous experiments and come up with a more realistic idea, so he started to look into factors that influence heart rate.

Some questions to ask yourself when brainstorming topics and ideas for methods can include the following.

- Is the main focus on biology? Does it connect with a topic in the syllabus? *It is okay for the investigation to have some overlap with another subject, such as chemistry, psychology or geography, but the main focus should be on biology.* Where in the investigation is the living organism or the organic molecule(s) made by living organisms? *If there is not a connection to an organism, it is not biology.*

- Is it safe, ethic and environmentally friendly? Does it break any rules from the IB, my school or laws in my country?

- Does it involve a technique I am already familiar with? *If so, it will probably go more smoothly than if you try to invent a new technique no one has ever tried.*

- Can I actually measure (or count) the dependent variable? Do I have access to the necessary equipment to obtain number values that I will need later for data processing?

- Is it doable in the time frame on my school's calendar?

- Is it a relatively simple, straightforward idea that has lots of little details I can change or control? *Those work the best.*

- If I need human volunteers to participate, will the investigation follow the IB and my school guidelines for using human subjects?

- If I am using animals such as snails or mealworms, am I following the guidelines for the ethical use of animals in the laboratory?

- If it is fieldwork, are the logistics going to work? For example, can I carry all the equipment I need to the site?
- Will I be able to collect enough data and do a sufficient number of trials to get reliable data?
- If I am getting my numbers from a database or computer simulation instead of producing the data myself, is the source I am using going to give me usable data?
- Is the topic of interest to me and is it worthy of investigation?

Once you have narrowed down a topic and have some ideas of what you might measure and how, you will need to continue your background research. Reread the section in your textbook and in your notes that deal with your chosen topic. This will help you select key biological vocabulary to search for, such as "osmolarity", "membrane integrity", "patellar reflex" or "phototropism". With guidance from your teacher and your school or local librarians, try to find a variety of resources.

A general online search using a search engine is often not as productive as a search within a curated collection of resources that a librarian can direct you to. Ask your teacher if the science department has any books on your topic, or if there are resources at school or online that are specifically for high school-level biology. Use the research papers found in the list of sources at the bottom of Wikipedia pages to help guide you too. Many are flagged as being open access to the public without cost. If some are locked behind a paid subscription, ask your school librarian if the school has a subscription. Keep a list of all the books, articles and websites you access, including the date. That way, not only can you find them later, you can include the ones you use in your list of cited sources at the end of your report.

We will now examine each of the four criteria one at a time. Questions will help guide you.

It is mandatory to submit your IA work to complete your course. Failing to produce and upload a laboratory report for biology will result in no grade being awarded for the subject, even if you get fantastic results on all your exams. A missing grade could jeopardize your diploma, so be sure to give this work all the time and effort its needs.

Criterion 1: Research design

Writing an RQ

- If you are really stuck and do not know where to start, one formula for an RQ is to ask "What is the influence of X on Y?", where X and Y are factors or variables that can be measured, controlled, modified or counted.
- Be as precise as possible, even if it means that the RQ is quite long.
- The RQ should contain precise and focused words to describe both your independent and dependent variables (see below). For example, "patellar reflex times in adolescents from 16 to 18 years of age" is more focused than "knee reflex".
- If you are using any living organisms, or products from living organisms, such as seeds from a certain plant, give the most precise name you can and give the scientific name as well (e.g. *Pisum sativum* for garden peas).
- Even if your RQ is the title of your investigation, be sure to restate it clearly early on in your report. It is helpful if you state, "The research question is …".

Types of variables to consider

- The **dependent variable** is what you will be measuring as the results of your investigation. It is what changes in the experiment because of the manipulations of the experimenter. One way to think of the dependent variable is as "nature's answer": it is how the natural world's laws respond to your RQ.

- The **independent variable** is what is changed on purpose by the investigator to see effects it will produce. It is what you are testing to find out what happens. It should be the only thing that is different from one part of the experiment to another. For example, in an experiment testing the effect of different amounts of fertilizer on the growth of bean plants, a range of five different concentrations of fertilizer would represent the independent variable. Everything else must be the same: the type of plant, type of soil, age of plants, light conditions, etc. The one thing that you can vary on purpose is the concentration of fertilizer.

- The **controlled variables** are the things that are kept the same in all parts of the experiment in order to be sure that the experiment is fair. Controlled variables ensure that the independent variable really is solely responsible for any changes recorded. There is no need to make an exhaustive list, just be sure to identify the controlled variables that would most dramatically affect the results in an undesirable way.

- Remember: do not confuse **controlled variables** with **the control**. The control of an experiment is a variant of the experiment that is set up in order to have something to compare the other results with.

Writing about the context surrounding your RQ

- In addition to describing your RQ, have you given your reader some background information to put it into context? What are the main biological ideas surrounding your topic?

- What inspired you to pursue this investigation? Why is it of interest to you?

- What properties or laws of nature will you be investigating? How does it connect to what you have been studying?

- Have you explained why you chose what you did for your independent variable and your dependent variable?

- Why have you chosen this method for answering the RQ compared to other possible methods?

- For database or computer simulation investigations, how did you select your sources? What decisions did you make to select the data or the conditions for the simulation?

Writing a step-by-step method

When writing your method, take inspiration from methods your teacher has already introduced you to, or think about the recipe you use to prepare your favourite dessert. You should be as precise and concise as possible. Here are some things to consider.

- Could your method be read by someone else and be fully understood by that person?

- Have you clearly described how the independent variable is integrated into the steps? How about the dependent variable?

- For the controlled variables, have you explained how they will be controlled? If it is impossible to control one or more of them, have you described a method for monitoring them?

- In selecting the materials, apparatus and glassware, have you chosen the equipment with a degree of precision that is appropriate for your investigation?

- For glassware such as beakers and flasks, be sure to indicate the volume in millilitres (ml). If you just ask for test tubes, the standard size will be given but be aware that there are some with smaller or wider diameters.

- If the glassware is going to be heated, think of what you might need when moving it once it is hot, such as wooden pinchers or metal tongs.

- If the experiment involves cutting something, do not forget to ask for a knife (or scalpel if necessary).

- For chemical solutions, you must be precise about the concentration (in % or in moles per litre) as well as the volume (in ml) that you will need.

- Think about the materials used to transport things: the manipulation of liquids will probably require the use of pipettes or syringes, the manipulation of powdered chemicals will require a spatula, and, if you need to weigh the powder, how will you put it on the balance? Did you remember to ask for a balance?

- If you ask for any electronic probes (for temperature, light, humidity, etc.), be sure to ask for an interface for connecting them to your computer, or a data-logging device.

- Thermometers come in multiple forms, including glass, electronic and temperature probes. Be sure to state clearly what kind you need.

- If an experiment needs to be saved overnight from one lesson to the next, did you ask for a tray or a box to keep the samples in? Is it labelled?

- Will the methodology you are planning result in sufficient numerical data so that techniques of analysis such as standard deviation can be used? Clearly state how many different experimental variations (scope, range) you will have and state how many trials are needed for sufficient data to be collected.

- Have you explained how you have modified a standard method and made it your own design? For details such as the concentrations you chose for the independent variable, did you explain why you chose those particular increments? If you got your inspiration from outside sources, have you cited them?

- Have you mentioned safety, environmental and ethical concerns? Not all investigations will require comments about all three, so adapt your report accordingly.

- Is your step-by-step method concise? In other words, check that you have not repeated yourself and that you have given all necessary information, but not more.

Level descriptor for top marks

Here is what is expected if you want to get full marks for Criterion 1: Research design.

- *The research question is described within a specific and appropriate context.*

- *Methodological considerations associated with collecting relevant and sufficient data to answer the research question are explained.*

- *The description of the methodology for collecting or selecting data allows for the investigation to be reproduced.*

Criterion 2: Data analysis

This criterion asks to what extent your report provides evidence that you have presented, processed, analysed and interpreted the data in such a way that a conclusion can be reached that is in line with the proposed RQ.

- Have you selected and recorded raw data, including the uncertainties and appropriate qualitative data? Is everything you have chosen to put in the results directly relevant to the RQ?

- Have you selected an appropriate method for analysing the data? The Skills for biology chapter explores mathematical tools for analysing and graphing data, as well as statistical tests you can use.

Setting up effective raw data tables

- Give the table a number and a title (e.g. Table 2: Pea seed characteristics).

- Set up the rows and columns in an orderly way to facilitate interpretation, for example values that have been measured using the same tool, such as a thermometer, should be aligned in the same column.

- In the headings of each column, put three things: the name of what was measured, the appropriate units, and the degree of precision.

- Put only numbers in each box (cell) of the table, no units, and be sure to have only one value in each box of the table. Do not include symbols such as ± or ≈ in the cells with the raw data. An exception is the negative sign: that can be used.

- The number of decimal places after the decimal point should be in accordance with the degree of precision, for example if a thermometer is precise to ±0.5°C, then all the numbers in the column should end in .0 or .5 and not have any more or any fewer decimal places after the decimal point (even for 0.0).

- How reliable are the accuracy and precision of your results? (See Figure 1) Are there other uncertainties in the measurements that you can point out?

- Align the decimals, even when there are negative signs in front of some of the numbers.

- Is there a clearer or more concise way of showing the results?

- Have you followed all the conventions for presenting graphs and tables?

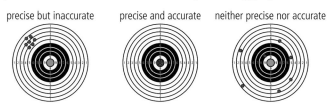

precise but inaccurate   precise and accurate   neither precise nor accurate

Figure 1 Accuracy is how close an investigator gets to measuring the accepted value that is reliable and verifiable. Precision is how close the data points are to each other. Precise measuring instruments do not give different values each time the same thing is measured. It is possible to measure something very precisely (getting the same results each time) but be very inaccurate, as seen in the first illustration of the target. This might happen if a balance was not set to zero after placing a recipient on it, giving the mass of the substance being weighed but also including in that value the mass of the container, thus falsifying the measurement.

Types of data to consider: quantitative and qualitative, raw and processed

Figure 2 shows qualitative data in purple. Such data cannot be expressed in numbers. Raw quantitative data are shown in green, and processed quantitative data are shown in orange.

Figure 2 Your results should show all three types of data: qualitative, quantitative and processed data.

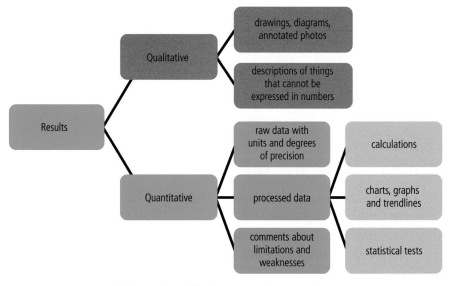

- Have you successfully analysed the data in a relevant and appropriate way?
- Is the analysis accompanied by consideration of the uncertainties?
- Do you have enough data to carry out sufficient data processing? Data processing ideally includes mathematical and statistical work as well as graphical work to support a valid conclusion fully. (Note: only graphing raw data is not considered to be data processing.)
- Do you need to use a null hypothesis test to demonstrate whether or not the differences you see in the data are statistically significant?
- For any processing you have done, have you made it clear what steps were taken?
- For any statistical tests, have you justified why this test was chosen over others?

Level descriptor for top marks

Here is what is expected if you want to get full marks for Criterion 2: Data analysis.

- *The communication of the recording and processing of the data is both clear and precise.*
- *The recording and processing of data shows evidence of an appropriate consideration of uncertainties.*
- *The processing of data relevant to addressing the research question is carried out appropriately and accurately.*

Criterion 3: Conclusion

This criterion measures how well you have answered the RQ. You will need to show that the conclusion is entirely in line with your RQ and has been described fully. In addition, the conclusion must be justified by the data you have collected, as well as justified through relevant comparison with the accepted scientific context.

Students are encouraged to do research to find similar investigations and see how their results and conclusions compare with those of other scientists who have published their findings. Such comparisons do not necessarily have to be quantitative but full citations of any sources used are required. If a comparison with a similar investigation is not possible, there could at least be a comparison with the current scientific understanding of the theories or laws governing the phenomenon being investigated. Always cite your sources.

Some things to consider when writing a conclusion

- Have you interpreted the analysis to form a conclusion?
- Is your conclusion relevant to the purpose of the investigation?
- Have you compared your conclusion to accepted scientific theory and given references?
- Have you explained how the data that was collected has answered the RQ stated earlier in the report?
- Have you explained how the results either confirmed the hypothesis or refuted the hypothesis? Use the expressions "confirmed by the data" or "refuted by the data" rather than "right" or "wrong". The latter two terms should be reserved for ethical arguments in science.
- Have you described any unexpected results: were there any outliers in the data, or any surprises?
- Have you commented on how much the measurement uncertainties may have influenced the results? In other words, how confident are you that the measurements gave results you can consider to be reliable?
- Have you explained what can be learned from the results? This is where you can usually connect the theory from class with your laboratory work. When possible, compare your first-hand data with literature values (secondary sources).

Level descriptor for top marks

Here is what is expected if you want to get full marks for Criterion 3: Conclusion.

- *A conclusion is justified that is relevant to the research question and fully consistent with the analysis presented.*
- *A conclusion is justified through relevant comparison to the accepted scientific context.*

Criterion 4: Evaluation

This criterion requires you to be reflective about the method of your investigation and to suggest possible improvements. Here are some things to ask yourself in your evaluation.

- Have you mentioned some of the strengths? Have you looked back at the list of controlled variables to get inspiration: were they, in fact, controlled?
- Do you feel the approach that you chose was an effective one to answer the RQ? Or should you have chosen a different method?

- Have you discussed the limitations and/or likely sources of error in the method?
- Have you discussed the reliability of the data? What weaknesses and limitations did you see in the data collection?
- Have you evaluated the level of impact of the sources of error? Are they minor, major or moderate?
- Have you explained how the source of error might generate unexpectedly high or low results?
- Have you commented on the range or spread of data? Do the data points follow a clear trend, for example, or are they very irregular and difficult to interpret?
- Have you assessed your sample size? It is big enough to have confidence in the conclusion?
- Have you commented on any assumptions that are being made? For example, if you used seeds for a germination experiment and all the seeds were from the same source, is it safe to assume they had similar genetics or ages or were kept in the same conditions? If those assumptions turn out to be incorrect, they could have an unexpected effect on the results.
- Have you suggested relevant and feasible modifications to the method? Have you explained how these would help improve the method? These must be things you could realistically do in a high school laboratory, so do not suggest that multi-million dollar laboratory equipment be used next time.
- Have you demonstrated that you understand the implications of the conclusion? For example, just because you saw a correlation in your results between the independent and dependent variables, does that mean that one causes the other to change?
- Have you suggested relevant and feasible extensions to the investigation? During investigations, often new questions arise. An extension is an idea for a future investigation to answer such new questions.

Level descriptor for top marks

Here is what is expected if you want to get full marks for Criterion 4: Evaluation.

- *The report explains the relative impact of specific methodological weaknesses or limitations.*
- *Realistic improvements to the investigation, that are relevant to the identified weaknesses or limitations, are explained.*

General questions to ask about your report overall

Have you communicated the focus, process and outcomes of your investigation clearly?

You need to write in such a way that you can communicate effectively to your reader.

- Have you written your report on the investigation in a concise, clear and logical format?
- Can your written explanation of data analysis be easily followed?

- Are your graphs, tables and images unambiguous? Photos should be annotated, have a legend describing what should be observed, and the source should be cited. If it is your own photo, put "author's photo" or "investigator's photo" as the citation.
- Is subject-specific notation used throughout?
- Have you used subject-specific terminology throughout?

Formatting checklist before submitting the final version

A downloadable version of this checklist is available on this page of the eBook.

- ☐ Check your document format: The IB only accepts Word, RTF or PDF. PDF is preferable as there is a lower risk of losing formatting.
- ☐ Check the overall word count is under 3,000 words. Make sure to include the word count in the report. The following are not included in the word count: charts and diagrams, data tables, equations, formulae and calculations, citations/references, bibliography, headers.
- ☐ Do you have a title? There is no need for a separate title page. You may use your RQ as the title, but this is not always practical as some are rather long.
- ☐ Include your IB candidate code (e.g. xyz123). Other than this code, the document should be anonymous. None of the following should be on any pages of the document: your name, your school's name, your teacher's name or anything else that could take away the anonymity of your work.
- ☐ Group work: if you collaborated with other students to collect data, their IB candidate codes should be listed too.
- ☐ Font: use a standard font and font size.
- ☐ Make sure the pages are numbered.
- ☐ Tables, graphs, photos and illustrations should all have titles or legends stating what they are, and any sources should be appropriately cited.
- ☐ Citations: cite all referenced works. Check with your school to see if there is a preferred citation format. Otherwise, you are allowed to use any format you like as long as it is a recognized standard that is used in the academic world, and as long as you stick to the same format throughout the document.
- ☐ Footnotes: using footnotes to explain ideas or define terms is seen as an attempt to get around the word limit. Put the ideas directly in your main text instead.
- ☐ Make sure you have a References or Works cited section: this is mandatory, but only works that are actually cited within the report can be in the list. Even if you got your inspiration from other works and think they are interesting, they do not belong in this list. You could add a separate section called Further reading or Additional sources, but only if you have space.

The process is long and challenging but, if you choose a topic you are really interested in, it will also be extremely rewarding. The ability to collect information, connect it to a larger context, analyse it, draw a conclusion and evaluate the process, are skills you will be able to apply for many years to come, whether at home, school or work.

Think about your IA work in the IB like an exam that you get to take home, work on for as long as you like, submit once to the teacher for feedback, and then get back to correct it before the final submission. IA work should take priority over other homework assignments because it counts directly towards your diploma. The points you earn will already be in your pocket as you walk into the final exams.

Skills in the study of biology

This chapter presents three sets of **tools** (experimental techniques, technology and mathematics) followed by three **inquiry skills** (exploring and designing, collecting and processing data, concluding and evaluating). You will need these to complete the biology course successfully but also to navigate today's world, which, more than ever, is filled with data and ideas that are presented as scientific but sometimes are not.

The approaches to learning (ATLs), shown in the grey boxes, should be integrated into the skills needed in biology, shown in the coloured boxes.

Skills in the study of biology

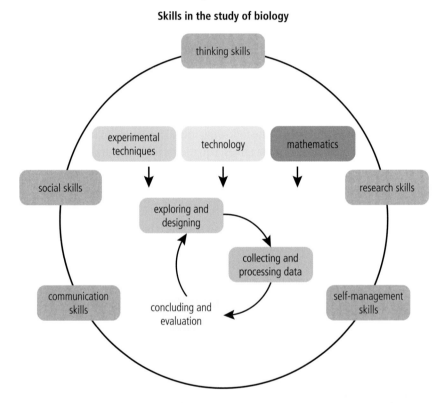

Tool 1: Experimental techniques

Addressing safety of self, others and the environment

When performing experiments in the laboratory, top priority should be given to safety. You should be mindful of preventing accidents such as burns, cuts, splashes and poisoning. Whenever possible, materials should be chosen that will not harm yourself, others or the environment. Students should never grow microbes without knowing what strains are present and must ensure that the colonies could not pose any health threats. Chemicals should be chosen in accordance with local safety and environmental rules. For example, if you need acidic buffers for an investigation, consider how much is necessary so that there is minimal waste, and think about how the chemicals will be disposed of at the end of the experiment. If you are using aquatic plants, choose a species that is non-invasive and consider what will happen to the plant at the end of the investigation. Evaluate whether using potentially toxic or dangerous materials is truly necessary. Be sure to consider ethical issues as well, such as the treatment of any animals you might use, respecting the privacy of human volunteers, and the impact of collecting samples from the field.

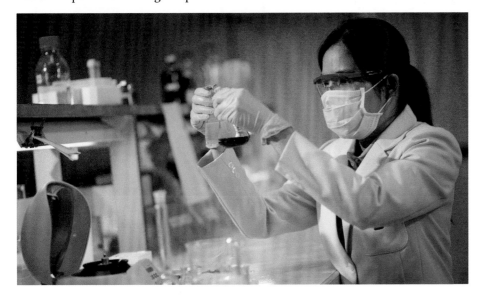

The first three rules of laboratory work: (1) safety, (2) safety and (3) safety.

Measuring variables

You should be able to measure basic variables such as mass, volume, time, temperature and length. All measurements are approximations, so you should also write down the degree of precision of your measurements. For example, when determining the mass of an object using a balance, look at the specifications of the balance, which will be in accompanying documentation, on a label on the balance itself or on the manufacturer's website. It might say "d = ±0.01 g" for example, which means it is precise to one one-hundredth of a gram. This means you can include up to 2 decimal places after the decimal point for each measurement.

When measuring volume, use the lowest part of a meniscus on a graduated cylinder and check whether the cylinder has a degree of precision printed on it. If not, you can estimate the degree of precision by looking at the smallest measurement interval and dividing that by 2. So if the graduated cylinder has graduations 2 ml apart, the degree of precision can be declared as ±1 ml, which means you cannot include anything after the decimal point when expressing your measurements in ml. However, a graduated cylinder with lines 1 ml apart can have a degree of precision that is ±0.5 ml, which would mean that all your measurements should end in either .0 or .5. If you need a more precise volume measurement, a syringe, burette or volumetric pipette would be better. When measuring time, ask yourself if you should take into account the reaction time to push the stopwatch button on and off. For temperature, glass thermometers are less precise but more practical to use than data logging temperature probes. Always read a glass thermometer straight on, never at an angle. For measuring length, ask yourself how precise the ruler is and whether you need to consider the fact that you are actually making two approximations, one at one end of the object being measured and another at the other end. If the ruler is precise to ±1 mm, maybe your measurement is only precise to ±2 mm. For high precision measurements of length, consider using callipers.

Be as precise as you can. Read the bottom of the meniscus by looking at the graduated cylinder straight on. In this example, you should get a measurement of 7.0 ml.

One of the basic foundations of science is the skill of observation. Be sure to write down what you observe. Table 1 indicates which parts of the course will help you practise the types of observation skills you need.

Table 1 Skills covered by the syllabus

Skill	Example	Relevant section of the syllabus
Counts	Counting plants found in a quadrat, or bubbles per minute to determine the rate of photosynthesis.	C4.1.3 and C1.3.7
Drawing annotated diagrams from observation	Drawing and annotation of cell organelles, distribution of tissues in stems and roots.	A2.2.11, B3.2.9 and B3.2.10
Making appropriate qualitative observations	Observations of tropic responses in seedlings: although measuring the angle is feasible, sometimes only qualitative descriptions are possible.	C3.1.17
Classifying	Working with dichotomous keys.	A3.1.2

Applying techniques

Table 2 on the following page lists some techniques that are useful in biology and gives a possible application of each. You might consider using one or more of these in your internal assessment work. The last column indicates which part of the syllabus covers the application of a technique: look at the relevant chapters for further details.

Technique	Description and purpose	Relevant section of the syllabus
Paper or thin layer chromatography	Using a solvent to separate mixtures into their constituent parts according to their polarity. Useful for separating pigments in leaves.	C1.3.4
Colorimetry or spectrophotometry	Measuring how much light of specific wavelengths can pass through a solution. Useful for measuring how much pigment leaks out of plant cells, such as from red cabbage, if the membrane is weakened by ethanol.	B2.1
Serial dilutions	Adding a small volume of a solution to larger volumes of water in calculated increments in order to reduce the concentration. Example: 10% > 1% > 0.1%. Useful for counting free floating cells under a microscope that would be too numerous to count directly from a sample.	A2.2.2
Physical and digital molecular modelling	Building models of sugars or amino acids using plastic, paper or metal, or with computer software. Useful for visualizing molecules that are otherwise invisible.	B1.2.1
Using a light microscope and eyepiece graticule	Viewing a slide under a microscope or using an eyepiece graticule as a ruler to measure the size of objects observed at a microscopic scale. Useful for both qualitative observations of cells and tissue and for comparing sizes.	A2.2.2
Preparation of temporary mounts	Preparing a microscope slide by sampling tissue, placing it on a slide and staining it. Useful for observing cells that are still alive in order to observe movement such as cytoplasmic streaming.	A2.2.2
Identifying and classifying organisms	Dichotomous keys can be used to identify organisms. Useful for determining the family, genus or species of an unknown organism while doing fieldwork.	A3.1.2
Using random and systematic sampling techniques	During fieldwork, tools such as quadrats and transects can be used for taking random or systematic samples of an area. Useful for determining what plant species are present and where, as well as testing a hypothesis about the association of two species.	C4.1.3 and B4.1.4
Karyotyping and karyograms	Reading karyograms can help determine the karyotype of a person. Useful in detecting the sex of a future baby or the presence of any genetic anomalies such as an extra or missing chromosomes.	A3.1.7

◀ **Table 2** Techniques covered by the syllabus

Tool 2: Technology

Applying technology to collect data

Data logging using probes and sensors

Studying a forest, stream, grassland or marine environment poses some unique challenges. Abiotic factors, such as temperature, air humidity and light, can vary considerably, and could have an influence on what is being studied. Because they cannot be controlled, abiotic factors should be monitored and data should be collected.

A student using a hand-held data-logging device to measure the pH and temperature of a sample of water in a river.

Temperature probes connected to **data-logging devices** can automatically record temperatures at particular intervals. Such devices can be equipped with probes and sensors for:

- temperature
- light intensity
- relative humidity
- flow rate (to see how fast water is flowing)
- dissolved oxygen.

Data loggers can also include a global positioning system (GPS) to record the exact location of each measurement.

There are various modes that can be used to collect data, including real-time or intervals, for example taking a measurement every 5 m or recording measurements over a 10-minute interval. Many of these devices allow you to graph and analyse the data directly on the screen. This is especially useful when doing fieldwork without a readily available computer.

Databases

Sometimes it is impossible to collect your own data, or you may want to compare what you have found with other sources. This is where **databases** can be helpful. Examples include using data on wolves and elk in Yellowstone National Park (see Section C4.1.16 for predator-prey relationships) and analysis of data related to polio vaccination campaigns (see Section C3.2.17 about vaccines). When using a database, it is useful to consider the following.

- Is it possible to extract the data you need from this database? If the data is sortable and searchable, it will make your work a lot easier.
- The reliability of the source: is it an authoritative academic or scientific source?
- How up to date it is: is the information outdated and can you find more recent data elsewhere?
- Is it possible to compare more than one database to see if they agree or disagree?

If you decide to use a database for your internal assessment or for an extended essay in biology, be sure to use some of the ideas above to justify why you chose a particular database.

Models and simulations

Some experiments are too dangerous, too time-consuming or too costly to carry out in the laboratory; **computer simulations** of those experiments can be performed safely and in a time-saving fashion. Because variables can be manipulated within them, simulations are often used to predict an outcome or to find out what the optimum parameters are for a system. Experiments based on mating fruit flies, for example, would take many weeks and do not comply with the IB rules about the treatment of animals in experiments. However, computer simulations allow us to collect data and perform experiments virtually (see Figure 1). If you choose a simulation for your internal assessment or for an extended essay in biology, make sure it is possible to modify variables to get different outcomes. The simulation should also be accessible to your teacher and IB examiners for verification.

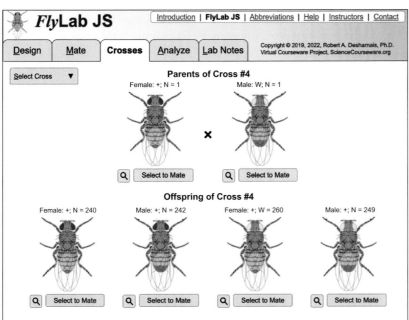

Figure 1 A simulation for mating fruit flies to see what genetic combinations are possible.

Applying technology to processed data

Spreadsheets

Make sure you know how to do the following with a **spreadsheet** program such as Excel, Numbers or LibreOffice Calc.

- Understand the system of identifying cells as A1, B2, C3, etc.
- Format: changing the format of the cell to match the type of data, such as number, date, percentage, text, time, scientific notation, etc.
- Format: changing the number of decimal places to correspond to the desired degree of precision.
- Insert: using maths operations by inserting an equals sign "=" followed by a formula, using "A1 + A2" to add, or "B3/B2" to divide, or "(A1 + A2 + A3)*B1" to combine more than one operation in the same formula.
- Insert: inserting predefined formulas such as sum, average, maximum or minimum, standard deviation, chi-squared, etc. As an example for Excel: typing "=max(A1:A100)" in cell A101 finds the maximum value between A1 and A100. Replacing the term "max" with "min" in the formula finds the minimum value.
- Converting a relative reference into an absolute reference by adding $, for example B2 does not behave the same way as $B2 or B$2 or B2 when the formula it is in is copied and pasted to another place on the sheet.

The worked example in the eBook shows how to use a spreadsheet program to calculate mean, mode and median from a data set.

Graphing

Make sure you know how to use a spreadsheet program to select a type of graph that will lead to useful analysis and insert a trend line and error bars. The worked example in the eBook will walk you through the steps needed and show you how to add labels so your graph is clear.

Just because you spend time making a graph look great does not mean it is worth including in a laboratory report. Sometimes it will inspire you to look for other patterns. In the example given in the eBook, further processing could be done by plotting soil temperature against light levels to see whether they are correlated.

Image analysis

Programs such as ImageJ, available through the National Institutes of Health, or LoggerPro from Vernier Software, have a function that allows you to measure things on a computer image or video. For ImageJ, once you have taken a photo and saved it to your computer, you can open it in ImageJ for analysis. Figure 2 shows an image of an iris and pupil taken during an investigation to measure pupil reflex after exposure to light. Tracing points around the iris allows the software to generate a measurement of area in arbitrary units. It is possible to convert this to centimetres if a scale bar is placed in the image, such as on the subject's face below the eye (e.g. a sticky note with a 1 cm scale bar drawn on it) and used to calibrate the tool.

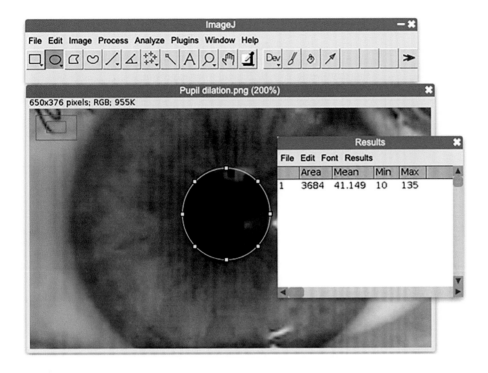

Figure 2 Image analysis software can help you generate quantitative data from images.

Tool 3: Mathematics

Applying general mathematics

You do not need to be a maths genius to be good at biology, but you do need to master several basic tools and techniques to understand other scientists' data and make sense of your own. Table 3 provides some examples of the kinds of mathematical tools used in the syllabus.

Table 3 Mathematical tools covered by the syllabus

Tool	Example	Relevant part of syllabus
Percentages	Punnett grids use percentages to express the chances that a genetic cross will produce certain traits in the offspring.	D3.2.2
Ratios	The surface area-to-volume ratio of an organism or cell decreases with increasing size, so bigger organisms will need to adapt their gas exchange system accordingly.	B2.3.6
Frequencies	How often something is found is its frequency and can be expressed as a percentage or decimal. Useful for comparing the number of times a mutation happens in different parts of a DNA sequence.	D1.3.5
Densities	Divide the number of things you are observing by the surface area to get the density. Useful under the microscope when determining stomatal density on a leaf.	B3.1.10
Proportions	Calculating how much of the whole a subcategory occupies. When studying an arctic ecosystem, the zones covered in landfast ice or sea ice can be expressed as a proportion.	D4.3.4

Tool	Example	Relevant part of syllabus
Scientific notation	Instead of writing out 3,500,000,000, for example, it is shorter to write 3.5×10^9. Useful when expressing tonnes of carbon in the carbon cycle.	C4.2.15
Approximation and estimation	It is not always possible to know an exact number, so we try to give a number that is as close as we can get. Useful in giving dates of the first living cells and estimations of population sizes.	C4.1.2
Calculate scales of magnification	The image size divided by the true size of an object in a microscope is its magnification. Useful in using scale bars to calculate magnification.	A2.2.2
Rates of change	By dividing quantities such as distance, mass or volume by time, it is possible to get the rate. Useful for transpiration rates in plants or enzyme reaction rates.	B3.1.9 and C1.1.8
Direct and inverse proportionality	If two variables are proportional to each other, they show a correlation. This can be positive (as X goes up, Y goes up) or negative (as X goes up, Y goes down). Useful for seeing connections in coronary heart disease statistics or anthropogenic causes of climate change.	B3.2.6 and D4.3.1
Percentage change or difference	Subtract the "before" value, V_1, from the "after" value, V_2, and divide by the "before" value, V_1, then multiply by 100 to get the percentage change. Useful for calculating the deforestation of the Amazon rainforest.	D4.2.3
Continuous and discrete variables	Continuous means that values do exist between data points (even if we did not measure them) whereas discrete means the data comes in separate categories and there is nothing in between. Height in humans is continuous whereas the ABO blood type has only four categories, A, B, AB and O, so it is discontinuous or discrete.	D3.2.14
The Lincoln index	Calculating the number of captured, marked and recaptured organisms. Useful in estimating the population size of a non-sessile (freely moving) organism.	C4.1.4

Measures of central tendency

You will be required to calculate measures of **central tendency** (mean, median and mode). The worksheet in the eBook explains how these measures are calculated.

Measures of dispersion

You will also need to be able to apply measures of dispersion (range, standard deviation, standard error and interquartile range). The worksheet in the eBook explains how these measures are applied.

Simpson reciprocal index

The **Simpson reciprocal index** can be used to estimate the biodiversity of a habitat. It takes into account the number of species found in an area and the number of individuals in each species. Table 4 on the following page shows the numbers of benthic invertebrates identified in the Triouzoune river.

Benthic invertebrates	Total
Plecoptera	2
Ephemeroptera	41
Trichoptera with cases	9
Trichoptera without cases	8
Megaloptera	1
Crustacea	8
Gastropoda	12
Diptera	7
Hirudinea	21
Oligochaeta	4
Total species found	**10**
Total number of organisms	**123**

◀ **Table 4** Data set 1: the numbers of benthic invertebrates identified in the Triouzoune river, France, June 2015

The formula for calculating the Simpson reciprocal index is:

$$D = \frac{N(N-1)}{\sum n(n-1)}$$

where n is the number of organisms that belong to one species and N is the total number of organisms. Applying this formula to data set 1 gives a value for D of 6.17. The higher this number, the more biodiversity is present. Lower numbers would suggest an unhealthy ecosystem.

Comparing the means and spread of data between two or more samples

Remember that in statistics we make inferences about a whole population based on just a sample of the population. Table 5 shows the results from an experiment growing bean plants in the sunlight and shade, and the data can be used to show how standard deviation is useful for comparing the means and the spread of data between two samples.

◀ **Table 5** Data set 2: the results of a bean plant experiment

Height of 10 bean plants grown in sunlight, in centimetres ±1 cm	Height of 10 bean plants grown in shade, in centimetres ±1 cm
125	131
121	60
154	160
99	212
124	117
143	65
157	155
129	160
140	145
118	95
Mean 131 cm	Mean 130 cm

Bean plants being grown for an experiment

Variance is a measure of variability. It gives a value to how much the numbers in a set of data vary from the mean.

When something is considered to be statistically significant, it means that there is a strong probability that it is not caused by chance alone. When something could be caused by chance, we say in statistics that the difference observed between two results is not statistically significant.

The standard deviation of the bean plants grown in sunlight is 17.68 cm, while the standard deviation of the bean plants grown in shade is 47.02 cm. Looking at the means alone, 131 cm and 130 cm, it appears that there is little difference between the two sets of bean plants. However, the high standard deviation of the bean plants grown in the shade indicates a very wide spread of data around the mean.

Significant difference between two data sets using a *t*-test

In order to determine whether the difference between two sets of data is a **statistically significant** difference, *t*-tests are commonly used. The **Student's *t*-test** compares two sets of data, for example the heights of bean plants grown in sunlight and the heights of bean plants grown in shade. Look at the top of Table 6, and you can see values for the **probability** (*p*) that chance alone could make a difference. If $p = 0.50$, it means the difference could be the result of chance alone 50% of the time.

Statistical significance refers to how probable it is that a relationship is caused by pure chance. If a relationship is statistically significant, it means that it is very unlikely that the relationship is caused by chance. We can also use this idea to see whether the differences between two populations are random or not.

If you reach $p = 0.05$, the probability that the difference is caused by chance alone is only 5%. This means that there is a 95% likelihood that the difference has been caused by something other than chance. A 95% probability is statistically significant in statistics. Statisticians are rarely completely certain about their findings, but they like to be at least 95% certain before drawing conclusions.

The formula used to compare two populations that are assumed to have equal variance is:

$$t = \frac{\overline{X}_1 - \overline{X}_2}{\sqrt{\left(\dfrac{(N_1 - 1)s_1^2 + (N_2 - 1)s_2^2}{N_1 + N_2 - 2}\right)\left(\dfrac{1}{N_1} + \dfrac{1}{N_2}\right)}}$$

$\overline{X}_1$ = the mean of population 1
$\overline{X}_2$ = the mean of population 2
N = sample size of the population
s = standard deviation

Note that you will not be asked this formula in exams: your graphing calculator can do it automatically.

If you put in the values from data set 2, you should get $t = 0.06$. You can use a table of critical *t*-values (Table 6) to find out what this number means. To do this, look at the left-hand column headed "Degrees of freedom", then look across to the given *t*-values. For a two-sample *t*-test like the one we are doing, the **degrees of freedom (d.f.)** are the sum of the sample sizes of the two groups minus two: $10 + 10 - 2 = 18$.

If d.f. = 18, we need to look at the row in the table of *t*-values that corresponds to 18. We see that our calculated value of *t* (0.06) is less than 0.69 in Table 6, indicating that the probability that the differences between the two populations of plants are due to chance alone is greater than 50%. In other words, we can safely declare that there is no statistically significant difference in the data collected from the bean plants in the sunlight and those from the shade. The differences are most likely due to chance. In order to be able to declare that our two populations show a level of 95% significance in their differences, we would need a *t*-value of 2.10 or more (look up d.f. = 18 and $p = 0.05$ (5%) in Table 6). Interpretations of such data processing can be a crucial addition to an effective conclusion in a laboratory report because just looking at numbers or graphs cannot tell us if the differences we see are statistically significant or not.

Degrees of freedom	Probability (*p*) that chance alone could produce the difference					
	0.50 (50%)	0.20 (20%)	0.10 (10%)	0.05 (5%)	0.01 (1%)	0.001 (0.1%)
1	1.00	3.08	6.31	12.71	63.66	636.62
2	0.82	1.89	2.92	4.30	9.93	31.60
3	0.77	1.64	2.35	3.18	5.84	12.92
4	0.74	1.53	2.13	2.78	4.60	8.61
5	0.73	1.48	2.02	2.57	4.03	6.87
6	0.72	1.44	1.94	2.45	3.71	5.96
7	0.71	1.42	1.90	2.37	3.50	5.41
8	0.71	1.40	1.86	2.31	3.37	5.04
9	0.70	1.38	1.83	2.26	3.25	4.78
10	0.70	1.37	1.81	2.23	3.17	4.59
11	0.70	1.36	1.80	2.20	3.11	4.44
12	0.70	1.36	1.78	2.18	3.06	4.32
13	0.69	1.35	1.77	2.16	3.01	4.22
14	0.69	1.35	1.76	2.15	2.98	4.14
15	0.69	1.34	1.75	2.13	2.95	4.07
16	0.69	1.34	1.75	2.12	2.92	4.02
17	0.69	1.33	1.74	2.11	2.90	3.97
18	0.69	1.33	1.73	2.10	2.88	3.92
19	0.69	1.33	1.73	2.09	2.86	3.88
20	0.69	1.33	1.73	2.09	2.85	3.85
21	0.69	1.32	1.72	2.08	2.83	3.82
22	0.69	1.32	1.72	2.07	2.82	3.79
24	0.69	1.32	1.71	2.06	2.80	3.75
26	0.68	1.32	1.71	2.06	2.78	3.71
28	0.68	1.31	1.70	2.05	2.76	3.67
30	0.68	1.31	1.70	2.04	2.75	3.65
35	0.68	1.31	1.69	2.03	2.72	3.59
40	0.68	1.30	1.68	2.02	2.70	3.55
45	0.68	1.30	1.68	2.01	2.70	3.52
50	0.68	1.30	1.68	2.01	2.68	3.50
60	0.68	1.30	1.67	2.00	2.66	3.46
70	0.68	1.29	1.67	1.99	2.65	3.44
80	0.68	1.29	1.66	1.99	2.64	3.42
90	0.68	1.29	1.66	1.99	2.63	3.40
100	0.68	1.29	1.66	1.99	2.63	3.39

You will find a second worked example in your eBook.

Using units, symbols and numerical values

You should be familiar with the units shown in Table 7; some are SI (Système International) base units, commonly referred to as the metric system, while others are not. See Table 7 for examples.

Table 7 Basic units ▶

Measurement	SI base units	Other examples
Length	metre (m)	
Time	second (s)	days, months, years
Amount of substance	mole (mole)	
Temperature	kelvin (K)	°C
Mass	kilogram (kg)	tonne
Luminous intensity	candela (cd)	
Illuminance	–	lux
Volume	–	litre (L)

Units can have prefixes added to them to make bigger or smaller units, such as millilitres, gigatonnes, nanoseconds and micrometres. Table 8 summarizes some of the prefixes used.

Table 8 Common prefixes for units ▶

Prefix	Base 10	Decimal	Name
giga	10^9	1,000,000,000	billion
mega	10^6	1,000,000	million
kilo	10^3	1,000	thousand
centi	10^2	100	hundred
milli	10^{-3}	0.001	thousandth
micro	10^{-6}	0.000 001	millionth
nano	10^{-9}	0.000 000 001	billionth

All measurements are approximations. Be sure to express the degree of precision by choosing how many decimal places are used. For example, if the balance you are using is precise to ±0.01 g, all your recorded measurements should have two decimal places after the decimal point. Sometimes temperature probes will give results such as "20.83510226°C". This does not mean that we should keep all those decimal places after the decimal point. Find out what the true precision is from the documentation provided by the manufacturer of the device.

Processing uncertainties

Why are decimal places important? If you want to show that there is a temperature increase during an experiment and the readings you get on your electronic temperature probe are 20.8351022°C for the before value and 20.8416491°C for the after value, it might look like the temperature has increased. But because the difference is in the second decimal place and is beyond the degree of precision of the temperature probe (±0.1°C) both values are, in fact, 20.8°C. So no increase can be declared.

This is why it is crucial to declare what the degree of precision is for your measurements. This can be included in the headings of data tables, as shown in Table 5 for data set 2, measured with a degree of precision of ±1 cm. It is recommended that in your reports you comment on the limitations of the measurement uncertainties and what impact they can have on the results of your investigations.

Error bars

Error bars are a graphical representation of the variability of data. Error bars can be based on several different values, such as the range of data, the standard deviation or the standard error. Notice the error bars representing standard deviation on the bar chart in Figure 3 and the graph in Figure 4.

The value of the standard deviation above and below the mean is shown extending above and below the top of each bar of the chart. As each bar represents the mean of the data for a particular tree species, the standard deviation for each type of tree can be different, but the value extending above and below a particular bar will be the same for normally distributed results. The same is true for the line graph in Figure 4. As each point on the graph represents the mean data for each day, the bars extending above and below the data point are the standard deviations above and below the mean.

Figure 3 Rate of tree growth on an oak–hickory dune in 2004–05. Values are represented as the mean ±1 SD of 25 trees per species.

Figure 4 Mean population density ±1 SD of two species of *Paramecium* grown in solution.

Regression models and coefficient of correlation

When scientists measure something, often they are looking to see whether they can demonstrate that the phenomenon is following a law of nature. Sometimes laws of nature follow patterns that can be expressed in mathematical equations. For example, when measuring the light that a leaf might use for photosynthesis, a scientist knows that the intensity of the light varies according to the distance to the light source. In Figure 5, the graph on the left illustrates the "pure" mathematical law for light intensity and distance from the light source. On the right is the same graph superimposed with measurements taken in a laboratory. Because of any number of things, including limitations in the equipment and human error, the laboratory measurements do not fit the mathematical model perfectly.

Figure 5 A model of what the data should show (on the left) and the actual collected data (dots on the right), of which only one point is actually where it was expected to be.

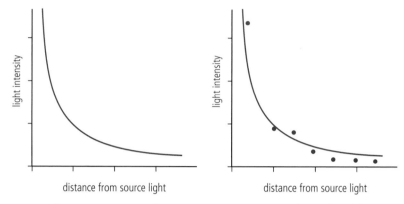

Now imagine the opposite. You take some measurements and wonder if there is some kind of mathematical equation that could act as a model of your data. You make a scatter plot and then see if there is a trend line that fits the data reasonably well. You might start with a straight line because that is the simplest relationship between two variables. This is called a simple **linear regression model**. But if that does not fit the data well, you could try other regression models that are not straight lines. Fortunately, statistical functions in a calculator, or a spreadsheet program on a computer, can do this for you in an automated fashion.

Figure 6 Three examples of data that have been modelled with a linear regression. The R^2-value is calculated to see how closely the linear regression model matches the data.

How can we know if the trend line's regression model is the best one for the data we have collected? The **squared correlation coefficient**, R^2, also called the **coefficient of determination**, is used to see how well a regression model matches the data collected. A value of $R^2 = 0$ means the regression model does not fit the data at all, whereas a value of $R^2 = 1$ means a perfect fit. Note that R^2 cannot be a negative number. Figure 6 shows some examples of R^2 values calculated using Microsoft Excel for three data sets and their trend lines.

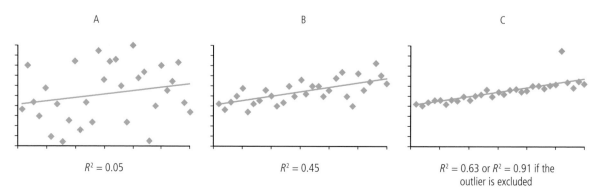

Notice what happens to the R^2-value as the variability of the data points is reduced from A to B to C. This reveals that the regression model shown by the trend line matches the observed data better and better. Graph A's regression line suggests that there is very little evidence of an agreement between the regression model and the data, whereas B and C show a stronger fit. Notice what happens in graph C: there is clearly an outlier at the top right. Fortunately, the investigator identified it as being a result of an error during the experiment. It can safely be ignored, and therefore the value of 0.91 can be used for analysis purposes. You are encouraged to used trend lines and R^2-values in your data processing, in order to better analyse the data you collect.

In addition to simply seeing whether the data points follow a predictable pattern, a regression model can be used to predict values that were not measured. Knowing the equation of the line or the curve allows us to add hypothetical values and get a prediction from the model. For example, changes in the human population in the coming decades can be predicted based on a regression model of current trends in the population. When using a regression model for prediction purposes, the R^2-value can help give a sense of how reliable the prediction will be. For example, predicting an outcome using graph A above would be extremely unreliable. However, using C's regression model would be more likely to give reliable results.

Correlation does not mean causation

Observing that wilting occurs in our house plants every time the soil is dry is a simple **correlation**, but a controlled experiment measuring wilting at different frequencies of watering provides us with evidence that the lack of water is the **cause** of the wilting. Experiments provide a test that shows cause. Observations without an experiment can only show a correlation. Also, in order for there to be evidence of causality, there must be a mechanism to explain how one phenomenon can cause the other. Knowing the properties of osmosis and turgidity in plant cells would explain the causality associated with the correlation, thus giving it greater scientific plausibility.

When using a mathematical correlation test, the value of the **correlation coefficient**, r, is a measure of the degree of linear relationship or **linear dependence** between two variables. This can also be called the Pearson correlation coefficient. The value of r can vary from +1 (completely positive correlation) to 0 (no correlation) to −1 (completely negative correlation). For example, we can measure the size of breeding cormorant birds to see whether there is a correlation between the size of males and females that breed together (Table 9).

Pair number	Size of female cormorants/cm	Size of male cormorants/cm
1	43.4	41.9
2	47.0	44.2
3	50.0	43.9
4	41 .1	42.7
5	54.1	49.5
6	49.8	46.5
$r = 0.88$		

- Trend lines are useful for seeing whether there is an overall pattern or tendency in the data points.
- The R^2-value, the coefficient of determination, is useful for seeing if the trend line matches the data points closely or not. It indicates how good the model is. The closer it is to 1, the better the model. Values close to 1 reveal that there is a strong correlation between the x- and y-values.
- If the regression model fits the data well, it can be used to predict values that were not measured.

A cormorant (*Phalacrocorax carbo*)

Table 9 Cormorant size data

Correlation does not necessarily mean causation. Just because two things show a relationship and have a strong *r*-value, does not mean one causes the other. To jump from one to the other requires a scientific explanation outlining the mechanism that causes the relationship.

The *r*-value of 0.88 shows a positive correlation between the sizes of the two sexes: large females mate with large males. However, correlation is not causation. To find the cause of this observed correlation requires experimental evidence. There may be a high correlation, but only carefully designed experiments can separate causation from correlation. Causality requires that the mechanism of exactly how X causes Y needs to be demonstrated. For example, the mathematics here does not explain whether it is the males choosing the females or the females choosing the males. Correlation says nothing about the direction of the influence.

Expected versus observed values: first application of the chi-squared test for goodness of fit

The **chi-squared (χ^2) test for goodness of fit** calculates how close our observed results are to the expected values. Chi is the Greek letter χ and is pronounced like the word "sky" without the s at the beginning.

In the activity accessed from this page of your eBook, we will use the χ^2 test to compare our observed results with what we can theoretically calculate the results should be (the "expected" results). To use this statistical test it is important to note down carefully all the observed results (O) and the expected results (E). In the case of genetics exercises, for example, the expected results would be the proportions of phenotypes as determined by a Punnett grid, such as 25%/50%/25% or 25%/75%, although it is important to use the actual numbers of offspring rather than percentages or ratios. Setting up a table to keep track of the numbers is helpful.

For more about using the chi-squared test for association between two species, see Section C4.1.15.

Graphing

Scientists use graphs extensively because they are useful tools for presenting data and identifying relationships that might otherwise remain hidden. Graphs are instrumental in analysing data, and if you know how to make accurate and appropriate graphs, your conclusion and evaluation will be greatly enhanced.

The most common forms of graphs you are expected to be able to use are:

- bar charts
- histograms
- line graphs
- scatter plots.

Occasionally, you may also need to use pie charts or box-and-whisker plots. Not all graphs are plotted using data. Sometimes you might be asked to sketch a graph rather than plot one. This means that you do not need to have numbers on your axes, only labels. Examples include sketching graphs of enzyme activation energy or the effect of temperature on photosynthesis. Sketches are assessed on the shape of the graph and the labels.

In your eBook, you will find descriptions of the graphs listed above.

Data analysis exercises

Both for your internal assessment work and in data-based sections of exams, you will be required to interpret sets of data presented either as tables or as graphs. Being able to extract scientific information from data is a key skill in biology.

The first thing to look for on a table or a graph is a title. When titles are not available, often the text before or after the tables and graphs will reveal some key information about what they are showing. The next clues to look for in order to interpret the data correctly are labels and units in the headings of tables, or labels and units on the axes of graphs. In both cases, the labels are often the dependent and independent variables of the investigation that generated the data. Knowing these will help you reach conclusions about the investigation. The units might be familiar to you, such as grams, millilitres or °C, but sometimes they are units you have never heard of. In such cases, do not panic, just be sure to include those unfamiliar units in your answers and in your analysis. The same goes for arbitrary units, which are sometimes used to avoid using confusing units.

Next, look at the scales on the axes of graphs. Do they show regular intervals (10, 20, 30, 40) or is there an atypical scale, such as a logarithmic scale (1, 10, 100, 1,000)? If two graphs are being compared, do they use the same scales and the same maximum and minimum values? If not, be careful how you compare the two because they may look the same but in fact be very different.

In the eBook, you will find a worked example of the analysis of a graph.

Throughout this book, there are examples of past paper questions, and some of them have graphs or tables of numbers that need to be interpreted and analysed. Be sure to practise analysing them because that is what you will be asked to do in exams.

◀ **Table 10** Applying skills

Skill	Example	Relevant part of syllabus
Interpolate values from a graph	To estimate or predict values between measured data points. Useful in estimating an enzyme rate at pH 5 even though you only have data points on your graph for pH 2, 4, 6, 8 and 10.	C1.1.9
Extrapolate values from a graph	To estimate or predict values beyond measured data points. Useful in predicting future atmospheric CO_2 concentrations based on current trends in the Keeling Curve.	C4.2.20
Quantifying energy flow	Represent energy flow in the form of food chains, food webs and pyramids of energy. Useful for showing the dependencies of species on others for energy.	C4.2.11
Pedigree charts	Construct a diagram to show which parents produced which offspring over many generations. Useful in tracing how genetic traits get passed down and the likelihood of a trait being passed on to the next generation.	D3.2.13

Inquiry 1: Exploring and designing

Here we will briefly introduce the three inquiry skills you need in biology; they are covered in more depth in the Internal Assessment chapter.

Exploring

It is hoped that when you are learning about topics in biology, you will have a desire to know more about certain concepts that spark your curiosity. You are encouraged to demonstrate independent thinking in your exploration as well as a good amount of initiative and insight. Insight is when you have an "A-ha!" moment and you see something in a new light. If you do things right, you should see a new connection or something will gain new meaning.

When researching a topic for investigation, a considerable amount of background research is necessary. It is unlikely that you will find what you are looking for just using a search engine. Talk to your teacher, and ask your school or local librarian if there are resources available on your topic.

Once you have done sufficient background research, you are ready to formulate a research question. This might take the form of "What is the effect of X on Y?" where X and Y are measurable quantities, at least one of which is connected in some way to a living organism. A hypothesis should be formulated based on your research, past experiences and perhaps some initial trials. A hypothesis must be stated as a claim rather than as a question. It is one possible answer to your research question.

Designing

Once you have a research question, you can design your method. You are encouraged to show some creativity in the design, implementation or presentation of the investigation. You are discouraged from trying to invent a brand new investigation using techniques no one has ever tried before. It is best to apply techniques you have used in the past but in a different way. There is more than one way of obtaining data for an investigation in biology, for example:

- a hands-on experiment in the laboratory
- fieldwork
- a database
- a computer simulation
- surveys.

You will need to identify and justify your variables.

- The independent variable is the thing you change on purpose in order to see what will happen.
- The dependent variable is the results of the investigation: what is measured or counted at the end.
- The controlled variables are a list of conditions that should stay the same in order to make the experiment a fair test, but also to show that it is only the independent variable that is causing the changes in the results and nothing else.

Have you consulted your school librarian or local librarian about what resources are available on your topic?

To justify the variables, say why it is a good idea to have chosen them. You will need to decide how much of a range you want to test with your independent variable. For example, if you are investigating the effects of different coloured light on photosynthesis, how many different colours are you going to test? What range would provide a graph that could answer the research question? For the dependent variable, you will need to decide how many trials are required in order to get a satisfactory result. With only two trials for each colour, for example, it would be difficult to know if one of them has been affected by an unexpected source of error. On the other hand, you are limited in time so you need to find a balance between the size of the range and the number of trials. When you choose a variable, explain why it was chosen. Say why it is important to keep controlled variables the same and explain what would happen if you did not control them.

Once you have made these decisions, you are ready to write your method. This should be written step-by-step with enough clarity and detail that any other student could pick it up, follow the instructions and get similar results as you.

For more information on how to design a laboratory investigation, see the chapter on internal assessment.

Controlling variables

In the laboratory and during fieldwork, some probes and sensors require calibration. Think of calibration in the following way: the difference between a glass of water and a graduated cylinder is that one has markings on the side that were set by comparing volumes to a known volume. Without the calibrated markings, you can only say if one volume is relatively bigger than another but not by how much.

If you have declared that temperature is one of your controlled variables, you are going to have to try your best to keep the temperature constant. In ideal conditions, all the environmental conditions for your investigation should be maintained at a steady state.

Sometimes an investigation requires measuring a sample of a population because measuring the whole population would be unrealistic, for example counting the number of dandelions in a field that is 20,000 m². We usually take **random samples** or **systematic samples**. If you do that, you should justify why one is a better choice than the other and this will depend on the nature of your research question. If, during your fieldwork, in a field that is 20,000 m² you counted dandelions in 200 sampled metres, is that enough? It represents only 1 in 100 square metres of the field's surface area, or 1%. But if that is all you can do in the time given, that will have to suffice. One temptation when performing a random sample is to favour zones that are easy to access, such as those on flatter or drier ground and with fewer obstacles. Unfortunately, this does not make the sampling truly random. If you do run into situations that could qualify as **sampling errors**, be sure to mention them in your report.

If you are testing the effects or influence of something, it is a good idea to set up a **control**. This is a version of your experiment without the characteristic you are using as an independent variable. For example, when testing the effect of different concentrations of salt solutions on seed germination, the version of the experiment that has no salt added at all is the control. Some investigations, notably in fieldwork, do not lend themselves to having a control but if it is possible, it is often useful.

For more information on how to keep controlled variables constant, see the chapter on internal assessment.

▲
A method for observing the rate of photosynthesis by capturing oxygen gas bubbles produced by an aquatic plant. When designing an investigation, careful attention must be paid to every detail and your choices should be explained.

Inquiry 2: Collecting and processing data

Collecting data

Quantitative results are those that can be expressed with numbers (e.g. temperature, number of plants, volume of solution) whereas **qualitative observations** are those that cannot.

The best laboratory investigations and fieldwork will have both qualitative and quantitative observations. Things like colour, texture and smell can be described with words. An annotated drawing of cells observed under a microscope, or a sketch of a zone studied along a river bank, are also qualitative observations.

When collecting data, it is important to be careful about being precise and getting sufficient results in order to be able to process the data. If you notice that there are some difficulties during the process and something is not going as planned, be sure to note it down so you can talk about it in your report. The data points collected at that stage will be less reliable but perhaps still exploitable. If you can modify the method to avoid repeating the problem, you should. This is why it is best to trial an experiment before undertaking data collection to answer the research question.

Investigations can be carried out in the field as long as the particular aspects of safety, ethics and environmental issues associated with working in nature are addressed.

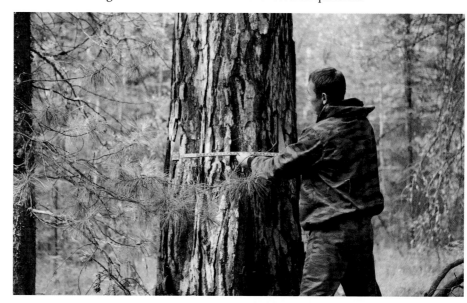

Processing data

There are several types of **data processing**, such as the mathematical operations, graphing or statistical analyses explained in the tools section earlier in this chapter. Be sure to pay close attention to how precise you can be and how confident you are of the reliability of the data points you use.

Here are some possible questions to ask yourself to make sure that your data processing and comments about your data are complete and helpful in reaching a conclusion.

- Is my data categorical, meaning it is in categories? (type A blood/type O …)
- Is my data normally distributed? (Meaning if graphed, it follows a Gaussian curve in the shape of a bell. See the central tendency worksheet linked to from page 895 of the eBook.)
- How representative is my sample of the population?

- Is my data ordinal: does the order matter (e.g. a questionnaire asking people to answer "always", "sometimes", "rarely", "never")?

- Is my data continuous or discrete?

- Do I have a way of showing the range of my data?

- Is the data well-suited for calculating the standard deviation?

- Do I have a way of determining outliers in the data? (Be sure to state how you determined outliers and what you decided to do with them.)

- If I see a correlation, am I careful about explaining whether there is a causation? (If so, explain how.)

- In my report, have I made comments about how confident I am of my results?

- Have I considered any confounding variables? (These are things that were not measured but which, in the end, had an unexpected influence on the results.)

- Should I do any hypothesis testing using my null hypothesis (H_0)?

- Should I calculate the regression on a regression line, R^2?

Interpreting results

Think about what comments you could make about your tables of quantitative results or your qualitative results. Say if you see any patterns that emerge or connections whereby qualitative and quantitative results agree or disagree with each other. Look at your graphs and comment on any trends they show. If one of your variables is supposed to show a relationship with another, comment on whether or not this is the case. If there are some data points that are very unreliable because something went wrong, what did you decide to do with them? How confident are you concerning your results? Comment on the accuracy, precision, reliability and validity of your results. For example, did the data points gathered from multiple trials of the same part of the experiment give similar results? Is the trend or relationship you are seeing supposed to be there?

Inquiry 3: Concluding and evaluating

Concluding

Now you are ready for the conclusion. The conclusion is an opportunity to give the answer to the research question. It says what can be learned from the results. You should say whether your hypothesis was supported or refuted by the results you obtained. You should include an answer to your research question and you should support your conclusion with examples from your results. How do you know if the conclusion you have reached is a valid one? You need to compare your results to the scientific context of your topic. For example, you could compare your results to what others have found previously in a similar investigation. It does not have to be identical to yours. You can do some research to find out if the outcome of your investigation can be explained by the current scientific understanding of your topic. In other words, say if your results fit with theories scientists have published. Cite your sources.

Evaluating

To evaluate something means to determine something's value or worth. An evaluation is an opportunity to comment on what might not have gone as you expected and suggest what to do differently next time. If your hypothesis was confirmed by the data, assess how fully it was confirmed or not. For example, were you way off or were you close to what you expected?

All investigations have their limitations. Obviously, if we had more time to collect results we could get more data, but that is not necessarily possible. Were there things on your list of controlled variables that turned out to be impossible to control? Did they have an influence on the reliability of the data? Were there any assumptions that need to be addressed? For example, if you assumed that an experiment left overnight was not subjected to any temperature changes in the laboratory, how sure can you be of that assumption? Be sure to consider the impact of uncertainties on the conclusions. Think about how confident you are of the precision of your measurements and decide if the precision is enough to give you confidence in your conclusion. If a result shows a difference that is ten times larger than the degree of precision, you can be more confident that the results are not simply due to the lack of precision of the measuring device than if the difference was smaller.

Comment on any sources of error that gave unreliable results. Some might be **random errors** and some might be **systematic errors**. Brainstorm all the possible sources of error you can think of and then put them in order based on which ones would have the biggest impact. Then focus on the ones that are specific to your investigation rather than ones that could be true for any investigation. For example, "maybe the steps of the method were not followed correctly" is too generic.

For each source of error, describe what the source is and then comment on whether it would tend to give unreliably high results or unexpectedly low ones. In a good evaluation, sources of error should be assessed for how severely they affect the data. Is the source of error minor, moderate or major? How would you fix the problem? Suggested improvements to the method or the types of equipment used should be realistic in terms of time available, resources you can access and safety. Some suggestions might solve one problem but create another one, such as being more time consuming.

For more specific details on what to put in a laboratory report, see the chapter on internal assessment.

Random errors are ones that happen by chance. For example, a temperature probe might be accurate to ±0.1°C and fluctuate between 21.7°C and 21.6°C but it is not because the temperature is changing, it is just the device. Systematic error would be if the temperature probe was defective or improperly calibrated and gave results that were always 2°C too low no matter what.

Advice on the Extended Essay

Introduction

One of the requirements of the IB Diploma is to write an Extended Essay. An Extended Essay is an in-depth study of a limited topic within a particular subject area. It provides the opportunity to carry out independent research within a subject of your choice. Biology is a subject often selected by students for their Extended Essay. It is a popular subject because many of the topics studied in IB biology stimulate further research ideas. Laboratory work carried out in the IB biology course also provides a basis for student ideas involving possible research questions. A good research question is essential for Extended Essay success.

General guidelines for all Extended Essays

The following guidelines apply to Extended Essays in all subjects. You should:

- not exceed the upper word limit of 4,000 words, not including the bibliography, contents page, appendices, or any labelling or captioning of graphs, diagrams, illustrations or tables
- spend around 40 hours working on your essay
- be assigned a teacher with Extended Essay training who will act as your supervisor and provide general guidance for the project
- pay attention to your school's set deadlines (your essay will be assessed externally using published criteria but your school will set its own deadlines)
- ensure your essay represents a unique approach to addressing a specific research question
- work independently
- select a topic and research question that is of interest to you, and be certain to show that interest in your writing
- complete the Reflections on Planning and Progress Form (RPPF) at the end of the process, which focuses on your final insights and methods evaluation.

General guidelines for biology Extended Essays

The following guidelines apply specifically to biology Extended Essays.

- A biology Extended Essay should involve biological theory and concepts. It will be assessed on the nature of biology.
- Most successful biology Extended Essays involve some sort of independent, hands-on experimental work along with literature-based research.

- A detailed procedure representing the exact steps carried out in any experimental work must be given. This detailed procedure is known as the **protocol**.

- Biology Extended Essays may involve data collected through experimentation, survey, microscopic observations, fieldwork or some other appropriate biological approach. It is essential that a proper analysis of this primary data is presented.

- Students taking an experimental approach to the biology Extended Essay must also consult and reference secondary data resources.

- Some biology Extended Essays do well when they are mostly literature based. These literature or secondary data-based Extended Essays in biology should include a unique analysis of raw data generated by reputable protocols and procedures.

- A successful Extended Essay will clearly illustrate access to sufficient data or information to effectively research the topic selected.

- The main body of the biology Extended Essay should centre on an argument or evaluation based on the primary and/or secondary data presented.

- Any research involving organisms must be ethical. Any animal research must follow the IB guidelines concerning the use of animals in IB World Schools.

- IB states clearly that biology topics dealing with symptoms and treatment of particular human diseases are very rarely successful Extended Essays.

- Topics dealing with ethical issues, different general approaches to medical treatments, and surveys involving attitudes or opinions concerning science research, are rarely successful biology Extended Essays.

- Extended Eessays based on experimental or practical work at a laboratory outside school must have a cover letter submitted with the essay detailing your role in the protocol design. This letter must also specifically describe any guidance received. This is especially important when the research is done at a university or research institution. For safety and/or academic honesty reasons, some schools do not allow students to work outside the school, so check with your teacher.

Suggested steps towards a successful essay

The following steps are a suggested way in which to approach an Extended Essay in biology.

I. Initial research and planning

(a) Decide on your subject of interest.

(b) Think of potential research questions.

(c) Make sure your research question has a "biological" focus. It must directly relate to an organism in some specific way.

(d) Meet with your supervisor to discuss your proposed topics and research questions. One of the most important functions of the supervisor is to help develop a proper research question for the Extended Essay.

II. **Continued research involving your chosen research question**

   (a) Research should involve a survey of the topic literature, keeping a detailed account of the sources from which ideas and/or data are used.

   (b) Plan your procedure for any experimental work.

   (c) Discuss your proposed research and procedure with your supervisor.

   (d) It may be necessary to refine your topic and research question as more information is gathered. Proper focus of the research question is essential.

III. **Experimental work and data collection**

   (a) Make certain your experimental procedure is safe and ethical in the opinion of your supervisor before beginning the procedure.

   (b) Arrange for all necessary equipment, chemicals and specific needs before beginning the experimental work. This may involve sources outside your school. Be certain all sources of materials outside your school are acceptable to your supervisor.

   (c) It is extremely important to consider the independent variables, the dependent variables, and the controlled variables in your procedure or procedures.

   (d) An essential part of all experimental work is to ensure an adequate sample size. It is important to discuss sample size with your supervisor.

   (e) Control groups and experimental groups must be carefully considered.

   (f) A plan should be in place for recording the raw data before the procedure begins. Qualitative data and quantitative data should both be considered in the data collection and recording stage.

   (g) Processing and presentation of data is an essential part of the experimental work. Careful consideration should be given to tables, graphs and statistical tests so that data will allow meaningful and proper conclusions to be reached.

   (h) If taking a non-experimental approach to the Extended Essay, it is essential there is sufficient secondary data to research the topic effectively. It is also important with this approach to use the secondary data and manipulate or analyse it in an original way.

   (i) Your supervisor may give general suggestions throughout this experimental work.

IV. **Writing the essay**

   (a) Your essay should have a structure that allows for an acceptable and appropriate presentation. An acceptable Extended Essay organization is as follows.

   - Title page
   - Table of contents
   - Introduction with research question stated early and clearly
   - Hypothesis and explanation of hypothesis
   - Background information
   - Presentation of variables
   - Materials used

- Protocol of experimental procedures
- Data collection and presentation
- Data analysis
- Evaluation
- Conclusion
- Bibliography
- Appendix (This is optional, and may include details of protocols, raw data, or any calculations using the raw data. It is important to note that the essay should be sufficient without the presence of an appendix.)

(b) A first draft should be submitted to your supervisor so that general directions may be given for writing the final draft.

(c) The first draft should be checked against the IB marking criteria by you and your supervisor.

(d) The bibliography style should be one used at your school. There is not a specific form of bibliography to use. It is important that some reference in the essay is made to each bibliography source provided. Information about any online sources used must be appropriate and complete.

V. Final draft

(a) Make changes generally suggested in the first draft by your supervisor.

(b) Proofreading is essential.

(c) Double check your final essay against the "presentation" criterion in the Extended Essay marking criteria.

(d) Arrange a meeting with your supervisor to submit your final essay. Your supervisor should go over the final essay with you, making certain the major sections have been included.

The Extended Essay criteria and advice for each criterion

Criterion	Advice
Criterion A: Focus and method – 6 marks possible	• Explain the topic in a clear and focused manner. • The research question must be effectively stated early in the paper. • The research question should lend itself to discussion and even debate within the Extended Essay. • A sound research question can only be presented when appropriate sources are utilized and properly cited. • The research question should be utilized to formulate a hypothesis, or hypotheses, that can be tested. • Methods used in the Extended Essay should be obviously well planned and allow gathering of data that is relevant to the research question. • Methods used should involve controls and allow adequate data collection.

	• It is essential the research question and the method are biological in nature.
	• If an investigation is conducted in an external laboratory, you must clearly demonstrate your understanding of the methods and materials utilized. In this situation, you should also clearly present *your role* in choosing and applying any methods and materials utilized.
Criterion B: Knowledge and understanding – 6 marks possible	• The essay must demonstrate a thorough understanding of the topic. • Your essay should flow in a logical way towards the development of a proper conclusion concerning the research question. • Clearly demonstrate in the Extended Essay that you understand all aspects of the essay. • Your analyses should represent an obvious understanding of the topic and research question. • Sources utilized in the Extended Essay must be appropriate and contribute to analyses and the conclusion of the essay. • It is essential the terminology used in the essay is accurate, focused and relevant. • Any technical terms used in the essay should be explained and used appropriately within the text. • Symbols, equations, significant digits and SI units should be utilized throughout the essay.
Criterion C: Critical thinking – 12 marks possible	• Be certain to present a convincing argument in your Extended Essay. • All arguments or data presented must relate logically to the research question. • It is wise to present alternative views as you develop your research question. • Carefully analyse all of the sources used in your essay. • Evaluate all aspects of the argument/experiment for appropriateness. • All data must be analysed. This analysis may involve mathematical transformations, statistical analysis, tables and/or graphs. • Tables and graphs should be presented logically and appropriately. Each table and graph used in the essay must relate to the research question and to the conclusion. • You must comment on the quality and quantity of the secondary sources and data used. • Limitations, validity and reliability of data should be critically commented on within the text of the essay.
Criterion D: Presentation – 4 marks possible	• Proper structure and layout are essential for high marks with this criterion. • The structure and layout of the essay should add to the presentation and development of the argument. • Scientific and annotated representations of equipment and experimental setups should be present and clear. • Procedural steps should be appropriately summarized. • Irrelevant details should be minimized. • Clarity of diagrams, graphs and tables is essential. They should add to the effective communication essential in the development of the analyses towards a logical conclusion.

	• Raw and processed data tables must be clearly displayed in the most appropriate form.
	• Any mathematical processes used in analysing raw data should be illustrated/explained clearly.
	• The essay must not exceed 4,000 words. It is essential to include the following in your Extended Essay: title page, table of contents, page numbers, appropriate illustrations, proper citations and bibliography, and appropriate appendices, if needed.
	• Graphs, figures, calculations, diagrams, formulas and equations are not included in the word count.
	• Examiners will not assess any material presented after the 4,000 word upper limit.
Criterion E: Engagement – 6 marks possible	• The examiner will utilize your RPPF after the assessment of the essay to determine the marks achieved for this criterion.
	• Your reflections should explain how and why the topic was determined.
	• Your reflections should clearly demonstrate how methods and approaches were determined in developing the Extended Essay.
	• Reflections should demonstrate knowledge gained by performing this activity. It is essential that you explain suggested changes in further work on the topic chosen.

Final advice for your Extended Essay

1. Be cautious concerning plagiarism. Presenting someone else's ideas or work as your own is plagiarism. Be certain to give proper credit to all ideas or work that has been used in any way in your Extended Essay.
2. Your title page should include:
   - the title of the essay
   - the research question
   - the subject for which the essay is registered
   - the word count, within the 4,000 limit.
3. Neither your name nor your school's name should appear on the title page or any page headers.
4. The introduction must present a strong reasoning for pursuing a conclusion to the presented research question, which should be stated clearly and early. Why the research question is significant for your Extended Essay should also be stated.
5. Any experimental procedures must be clearly and appropriately presented in a way they can be easily replicated.
6. Ethical and safety factors must be thoroughly addressed. It must be obvious that ethical and safety factors have been seriously considered in pursuing the research question.
7. Pages should be clearly numbered. The title page is not numbered. Sections of the essay should be clearly and appropriately labelled.
8. Footnoting and the bibliography must be proper and consistent. Sources in the bibliography not specifically used in the paper should be minimal.

9. All visual presentations must be clear, labelled appropriately, and must pertain to the research question and conclusion.

10. The conclusion must be clearly related to the research question. Limitations to the conclusion should be discussed. It is suggested that you present a brief plan for possible further development of a conclusion to your research question.

11. Any appendices used must be appropriate. Large tables of raw data collected are best included in an appendix. However, a representative sample of collected raw data should be included in the core of the essay in a data table.

12. Be certain the essay is based on suitable biological topics. Stay away from psychology- or medicine-focused topics.

13. Remember that a strong evaluation includes a comparison of your results with that of the literature. This evaluation may be addressed at several points in the essay besides the conclusion. An evaluation only presented in the conclusion is often rather shallow and ineffective.

14. Be certain your Extended Essay does not duplicate any other work being submitted for the Diploma programme. The Extended Essay should not take the form of an internal assessment. The Extended Essay assesses the ability to analyse and evaluate scientific arguments.

Viva voce and the RPPF

The completion of your Extended Essay is signified by the viva voce (concluding interview). This is a 10–15 minute interview with your supervisor. It provides an opportunity to reflect on successes and what has been learned.

After completing the viva voce, you must complete your RPPF. This is your final reflections on the topic, methods, analyses and conclusion of the Extended Essay.

Enjoy your research.

abdominal muscles 175, 176
abiotic factors 200–3, 212, 513–14
absorption spectra 261–3
acetylcholine 281, 294
acetylcholinesterase 281
acinar cells 156
acquired characteristics 516–17
actin filaments 42
action potentials 275–6, 277–9, 292
action spectra 261–3
activation energy 239–41
active transport 143, 146, 276
adaptation
    herbivores and plants 222–3
    plants for photosynthesis 227–8
    predators and prey 223–6
adaptation to environment 197–211
    abiotic factors 200–3
    ecosystems and communities 205–7
    hot desert adaptations 207–8
    limiting factors on populations
        201–3
    marine ecosystems 199–200, 203–4
    plants 198–200, 206, 207, 209
    salty environments 198–200
    and survival 515–16
    tropical forest adaptations 209–10
adaptive radiation 95, 206
adenine 17, 19, 22–3
adenosine diphosphate see ADP
adenosine triphosphate see ATP
adhesion 7–8
adipose tissue 127, 503
ADP (adenosine diphosphate) 237,
    248–9
adrenal glands 289, 298
adrenaline 289, 298, 502
aerobic respiration 214, 251–3, 381
afforestation 555
agriculture 530–1
AIDS 319–20, 412
air 10
albedo 542–3
alcohol 21
alcoholic fermentation 251
algae 215, 532
alimentary canal 303–4
allele frequency 519
alleles 474–5
    co-dominant 477
    difference from genes 475
    dominant and recessive 472, 476–7,
        478–9, 481

multiple 480–2
production 419
Punnett grids 472–4
allelopathy 356–7
α (alpha) cells 498
alveoli 174–5
Amazon rainforest 209, 522–3, 524–5
amino acids 14, 21, 366
    condensation reactions 116
    essential 133
    genetic code 23, 24–5, 401, 402–3
    hydrolysis and condensation 115–17
    peptide bonds 132–3
    sequences 71–2, 82
    structure 131–2
    translation process 399–400
    variety of possible peptide chains
        133–4
amphipathic molecules 128, 140
amylopectin 120, 121
amylose 120, 121
anabolic reactions 236, 364
anaerobic respiration 214, 250–1,
    252–3
analogous structures 86, 87
animal breeding 83
animal cells 40, 47, 49–50, 149
    cytokinesis 426–7
    water movement 444–5
animal tissue banks 106
anthers 463
anthropogenic species extinction 97–9
antibiotics 37, 320–2
    allelopathy 356–7
    antibiotic resistance 321–2
    penicillin 356–7
antibodies 313–14
anticodons 400
antidiuretic hormone (ADH) 299
antigens 123–4, 311, 313, 314–17
aorta 300–1
aposematism 226
aquaporins 145, 443
aquatic habitats 10–11
aqueous solutions 4, 8–10
archaea 36, 218, 375, 544
Archaeopteryx 578–9
arteries 183, 184–6
    carotid arteries 301
    narrowing of coronary arteries
        187–8
arterioles 183
artifacts 34

artificial insemination 105–6
artificial pollination 471–2
artificial selection 83
asexual reproduction 452–3
asteroids 372
atmosphere 257, 376–7, 378–81,
    538–40
    carbon cycle 374–5
    future carbon dioxide levels and
        photosynthesis 267–8
ATP (adenosine triphosphate) 21–2
    active transport 146
    ATP cycle 248–9
    and cell respiration 249–53
    chromosome movement 429
    electrons for producing 365, 366
    function 247–8
    in neurons 276, 277
    production 44, 50
    structure 237, 247–8
autonomic nervous system (ANS)
    185, 289–91, 301, 497
autosomal recessive diseases 478–9
autotrophs 216, 257, 364–5, 381
axons 274–5
    depolarization 276–7
    motor neurons 293–4
    myelinated/non-myelinated 277–8,
        295
    at resting potential 276–7

B-lymphocytes 313, 316–17
bacteria 36, 357, 512
    antibiotic resistance 321–2
    binary fission 39
    cell structure 37–8, 39
    chromosomes 39
    Gram-positive/Gram-negative 38
    at hydrothermal vents 365, 366
    photosynthetic 215
    root nodules and nitrogen fixation
        346
    saprotrophs (decomposers) 217
    Winogradsky column 525–6
baroreceptors 300–1
basal bodies 47
behaviour 504, 517–18
behavioural adaptations 224–5, 226
benthic zone 541
β (beta) cells 498
binary fission 39, 426
binding sites 313–14
binomial nomenclature 59–60

biodiversity 93–109, 523, 607–8
  change over time 95–6
  conservation 104–7
  crisis 100–4
  ecosystem loss 99–100
  and human activities 97–100, 102–7
  loss of keystone species 527–8
  surveys 100–2
  types of 93–5
bioinformatics 71
biological species concept 61–2
biomagnification 532–4
biomass 265, 372–4
biomass (organic waste) 377
biomes 197, 204–10, 373, 545
biotic factors 213
birds 501, 514, 516, 547
  aquatic birds 11
  cryptic coloration 517–18
  and dinosaurs 578–9
  range migration 552
bivalent 434, 437
blood
  aqueous solution 9
  capillaries and chemical exchange
    183–4
  circulatory system 183, 184–8
  clotting 310–11, 485–6
  lymphocytes 307, 313
  monitored by chemoreceptors 301,
    302
  oxygen transport 290
  pH 301, 302
  plasma proteins 310
  platelets 310, 311
  red blood cells 51, 163, 164, 444–5
  regulation of blood glucose
    498–500
  transfusions 315
  transport of hormones 289
  transport of waste products 290
  white blood cells 163, 164, 307, 310
  see also haemoglobin
blood pressure 185, 301
blood types 123–4, 315, 481–2
blood vessels 183–8, 300–1
BOD (biochemical oxygen demand)
  532
body processes 289–90
body systems
  coordination 287–8
  feedback control 300–2, 305
  hierarchy of subsystems 288–9

integration 287–9, 304–5
  negative feedback control 287
  see also endocrine system; nervous
    system
bone marrow 161–2
bonobos 89–90
box-and-whisker plots 491–3
brain 273–4, 290–1, 293–4
  coordinating movement 296–7
  and endocrine system 299–300
  see also hypothalamus
brainstem 291
breathing 175–7
brome grass 350
bronchioles 175
buoyancy 10, 11

cacti 207–8
calcification 553
calibration 32–3
camouflage 226
cancer
  cells 160, 162
  and mutations 397, 414–15, 417, 418
capillaries/capillary beds 183–4
capillary action 7–8, 189
capsule 37
capture–mark–release–recapture
  method 336–7
carbohydrates 111, 114
  chains in cell membranes 147–8
  conjugated 123–4
  disaccharides 111, 116
  energy storage 111, 119–21
  monosaccharides 111, 118–19
  polysaccharides 111, 119–22
carbon compounds 112–15
  chemical energy 361
  ring structure 113
  synthesis by autotrophs 364–5
carbon covalent bonds 113
carbon cycle 374–80
carbon dioxide
  atmospheric concentration 378–80,
    540
  carbon cycle 374–5
  dependence on atmospheric oxygen
    and carbon dioxide 380–1
  diffusion 141–2
  FACE experiments 267–8
  future levels in atmosphere 267–8
  increased production with exercise
    302

photosynthesis 257, 263, 264, 266,
    267–8
  release by decomposition of organic
    matter 541, 542–3
  release during combustion 376–8,
    544–5, 546
  role in climate change 538–40
carbon fixation 364
carbon footprint 531
carbon sequestration 554–7
carbon sinks/sources 376, 539, 542,
  543, 546
carbonic acid 301, 302, 553
carboxyl group 116, 124–5, 132
carnivores 220–1
carotenoids 258
carrier (genetics) 475, 486
carrier proteins 145
carrying capacity 337–8, 513
catabolic reactions 236
causation 350, 539, 614–15
cell-cell recognition 123
cell division 425–6, 439
  binary fission 39, 426
  chromosome movement 429–30
  cytokinesis 426–7
  cytoplasm division 427
  DNA condensation 429
  DNA replication 428
  nuclear division 428
  reduction division 433–5
  stem cells 161
  see also meiosis; mitosis
cell fractionation 153–4
cell plate 426–7
cell reproduction 159–60
cell research 152–4
cell respiration 178, 247, 255
  aerobic 251–3
  anaerobic 250–1, 252–3
  and ATP 249–53
  blood glucose levels 498–9
  electron transport chain 252
  energy release 367
  glycolysis 250–1, 252
  heat loss to environment 371
  Krebs cycle 251–2
  link reaction 251
  rate of 253–5
cell signalling 160
cell specialization 159–67
  and cell size 163–7
  differentiation process 159–60, 166

muscle fibres adaptations 165
stem cells 159, 160–3
surface area-to-volume ratios 166–7
cell structure 29, 35–6, 53–5
animal cells 40, 47, 49–50
atypical cell structure 50–2
comparisons 48, 49–50
drawing 52–3
eukaryotic cells 40–7, 49–52
organelles 40, 41–7
plant cells 41, 46, 47, 49–50
prokaryotic cells 36–9
cell theory 29
cell walls 37–8, 121, 444–6
cells
biochemistry 9
diploid cells 64
haploid cells 64
and microscopes 30–5
origin and evolution 4
size and specialization 163–7
specialized functions 50–2
surface area-to-volume ratio 166–7
cellulose 121–2
central nervous system (CNS) 273–4,
292, 293, 303
centrioles 47
centromere 65, 66, 67, 428
centrosomes 47, 429
cerebellum 291, 296
cerebrum 291, 293–4
channel proteins 145
chemical adaptations 223, 225
chemoautotrophs 218, 257, 364–6
chemokines 412
chemoreceptors 300–1, 302
chemosynthesis 218
chi-squared tests 351–3, 615
chimpanzees 66–7, 68, 89–90, 220,
221
chlorophyll 215, 258–60
chloroplasts 46, 258
cholera 308
cholesterol 128–9, 149
chromatids 428, 429, 431, 437
chromatin 45, 429
chromatography 153, 258–60
chromosome movement 429–30
chromosome numbers 64, 433–4,
435–6
chromosomes 38, 45
bacterial 39
before cell division 428, 429–30

centromere 65, 66, 67
daughter chromosomes 39, 426
extra 435–6
homologous 434, 437–8, 453–4
human chromosome 2 evolution
66–8
karyograms 65–6
meiosis 428, 433–8
mitosis 428, 431–2
random orientation 438
recombinants 437
sex chromosomes 484–5, 486
shape 66–7
cilia 47, 310
circadian rhythm 297–8
circulatory system 183, 184–8
cisternae 43
classification systems 58, 96
binomial nomenclature 59–60
biological species concept 61–2
morphological classification 59
cleavage furrow 426–7, 432
climate change 538–58, 560–3
anthropogenic causes 538–40
carbon dioxide and methane
538–40, 541, 542–4, 546
ecosystem collapse 553–4
global warming 542–5, 547, 550
mitigation by carbon sequestration
554–7
ocean current change 548–50
polar habitat change 547
positive feedback loops 542–5, 546
range shifts 550–2
scepticism and critics 541
tipping points 546
tolerance levels 523
climatic conditions 207
clones 317
codons 400, 402–3
cohesion 6–7
cohesion-tension theory 189
collagen 134, 185
collision theory 242, 244
combustion 376–8
common ancestors 63, 66–7, 84–5, 96
communities 197, 205–7, 343
companion cells 51
compartmentalization 38, 40, 45, 152
advantages 155–7, 158
and functions of organelles 155
competition 513
deterring 356–7

endemic and invasive species 348,
349–50
interspecific 229, 345, 348–50
intrasexual 518
intraspecific 344, 515–16
competitive exclusion 229
complementary base pairs 19, 22–3,
388–90, 396–7
computer simulations 604
condensation reactions 17, 18,
115–16, 122, 124–5
confirmation bias 570
Congo River 89–90
conjugated organic molecules 123–4
conservation of biodiversity 104–7
consumers 216, 361–3, 366–8
continuous variation 58, 489–90
contractile vacuole 445
cooperation 344
coral bleaching 347, 553–4
coral reefs 203, 347, 355, 553–4
coronary arteries 187–8
coronary heart disease 188–9
coronavirus see COVID-19
corpus luteum 459
correlation 350, 539–40, 614–15
correlation coefficients 189, 279,
613–15
cortex 191–2
cotyledons 467
courtship behaviour 517–18
covalent bonding 5, 112–13
non-polar 124
nucleotides 15–17
COVID-19 323, 583
pandemic data analysis 326–7
vaccines and immunity 324–5
cows 222
CpG sites 417
cranial nerves 291
crenation 444–5
CRISPR-Cas9 gene editing 583
cristae 44
crops 83–4
cross-pollination 463, 464–5, 474
crossing over 434, 437–8
cryogenic electron microscopy 34
cryosphere 543
cryptic coloration 517–19
cuticle 10
cystinuria 145
cytokines 316
cytokinesis 426–7

cytology 29
cytoplasm 35, 36, 38, 42
  compartmentalization 156–7
  division 427
  translation 156
  water-soluble substances in 9
cytosine 17, 19, 22–3, 417
cytoskeleton 42
cytosol 36, 42

Darwin, Charles 25, 81, 84
  theory of evolution 80, 509–10, 516
data 605–16, 621
  analysis 594–5, 616
  collection 587, 603–4, 619
  comparing datasets 608–10
  evaluating 325–7, 621
  graphing 605, 612, 613, 615
  image analysis 605–6
  public availability 540
  qualitative/quantitative 33
data logging devices 603
databases 604
daughter cells 426–7, 435
DDT (dichloro-diphenyl-
  trichloroethane) 533–4
de-extinction 88
decomposers 217, 364, 370, 381, 541
decomposition of organic matter 532,
  542–5
deductive reasoning 30, 488
deforestation 99–100, 103, 524–5, 546
dehydration 446
denaturation 238, 243
dendrites 274–5
deoxyribonucleic acid see DNA
deoxyribose 17, 19, 21, 388
dependent variables 267
deserts 205, 207–8
detritivores 364
diabetes 499–500
diaphragm 176
dicotyledonous root tissues 192
dicotyledonous stem tissues 191–2
diet 133, 478
  and blood glucose levels 498, 499
  and dentition 218–21
  of great apes 220, 221
diffusion 141–2, 144, 145, 146–7
  neurotransmitters across synaptic
  cleft 280–1
digestion 115, 117–18, 419
dinosaurs 578–9

dipeptides 132
diploid cells 64, 433, 470
diploid number 64
disaccharides 111, 116
discontinuous variations 58, 490
disease 307–29
  adaptive immune response 312,
  314–18
  antibiotics 320–2
  antibodies 313–14, 316
  antigens 313, 314–17
  bacterial infections 320–2
  defence against infection 309–10
  diabetes 499–500
  genetic disorders 404, 409–11, 418,
  478–80, 485–7
  herd immunity 325, 326–7
  HIV and AIDS 319–20, 412–13
  incidence in populations 307
  infection control by phagocytes 312
  innate immune response 311, 312
  opportunistic infections 319
  pathogens 307–9
  primary immune response 318, 324
  public health and infection control
  308
  secondary immune response 318
  transmission 319
  vaccines and immunization 323,
  324, 325
  zoonotic diseases 322–3
  see also cancer; viruses
diversity of organisms 57–8
DNA 14–15, 25–7
  amplifying and separating 390–3
  in a cell 35–6
  chloroplasts 46
  comparison with RNA 20–1
  complementary base pairs 19, 22–3,
  388–90
  eukaryotic cells 44, 45
  and ionizing radiation 414–15
  mitochondrial 44, 71–2
  nitrogenous bases 17–18, 19
  nucleosomes 45
  nucleotides 15–18
  plasmids 39
  satellite DNA 67
  storage of genetic information 23–4
  structure 19–20, 388
  sugar-phosphate bonding 16–17
  telomeric DNA 67
  triplet codes 23, 24

DNA condensation 429
DNA polymerase 390
DNA profiling/fingerprinting 154,
  392–3
DNA proofreading 415
DNA replication 23, 321, 387–94,
  422–3
  before cell division 428, 437
  complementary base pairing 389–90
  errors 404, 415
  polymerase chain reaction (PCR)
  391
  role of helicase and DNA
  polymerase 390
  semi-conservative 388–90
DNA sequences 24–5, 74
  base pairs 69
  comparing genomes 71–2
  endangered species 107
  evidence for evolution 81–2
  non-coding sequences 416
  personalized medicine 75
DNA templates 395, 397
DNA testing 420
drought 544, 546

Earth 3–4, 359
ecosystem stability 522–37
  biomagnification 532–4
  eutrophication 532
  human impacts 522
  investigating with mesocosms
  525–6
  keystone species 527–8, 535–6
  plastic pollution 534–5
  rewilding 105, 535–6
  sustainability of natural ecosystems
  522–5
  sustainable agriculture 530–1
  sustainable resource harvesting
  528–30
  tipping points 524–5
ecosystems
  abiotic factors 200–1, 212
  biomass 372–4
  biomes and communities 205–7,
  373
  biotic factors 213
  carbon sinks/sources 376
  conservation 105
  ecosystem collapse 527–8, 553–4
  ecosystem diversity 94
  ecosystem loss 99–100

energy loss 370–2
energy pyramids 368–9
energy sources 360–1, 364, 371–2
energy transfers 361, 362, 363, 371–2
  food chains and webs 362–3, 367–70, 372–3
  marine ecosystems 198–200, 203–4
  nutrient cycles 364, 381–2
  open systems 359–60
  recovery after natural disaster 339–41
  recycling of chemical elements 381–2
  regeneration and restoration 105, 554–7
  species biodiversity 94
  trophic levels 367–70, 372–3
  see also ecosystem stability
ectothermic animals 500
EDGE of Existence programme 107
egg cells 163–4, 433
eggs 454, 455, 460–1, 462, 516
ejaculation 455, 460
El Niño 548–50
elastin 185
electrical signals 273, 283, 288, 291
  see also nervous system; neural signalling
electromagnetic spectrum 260
electron microscopes (EMs) 30, 32, 34, 52–3
electron transport chain 252
electrons 5
elephants 63
embryos 462, 467
emergent properties 289
endangered species 97, 105–8
endergonic reactions 240, 249
endocrine system 287, 289–90, 299–300, 497–8
endocytosis 43, 141, 156–7, 311, 312
endometrium 457–8, 460
endoplasmic reticulum (ER) 42–3
endosperm 463
endothermic reactions 240
endotherms 127, 173, 501
energy 237, 256–7, 382, 523
  flow through food chains 362–3, 367–70
  heat energy 371
  loss between trophic levels 370
  loss from cell respiration 371

pyramids 368–9
  release by cell respiration 367
  from the Sun 360–1, 364, 371–2
energy sources 360–1, 364, 365–7, 376–8
energy storage compounds 111, 119–21, 126, 127
enteric nervous system (ENS) 303–4
enzyme-catalysed reactions 239–45
enzymes 9, 156, 235–6, 246
  activation 239–41
  active sites 238, 239, 241, 242
  as catalysts 235, 239–40
  cell respiration 250
  denaturation 238, 243
  hydrolysing 43, 117–18
  induced-fit model 239
  in lysosomes 43
  mechanism of enzyme action 242
  naming 118
  smooth ER 43
  in synaptic clefts 281
epidemiological studies 436
epidermis 191–2, 309
epinephrine (adrenaline) 289, 298, 502
epiphytes 227–8
errors
  DNA replication errors 404, 415
  error bars 612
  measurement precision 600, 611–12
  random/systematic 621
  sampling errors 335, 618
erythrocytes see red blood cells
ethics 255, 569, 583, 585–6
  animal breeding 83
  DNA testing 420
  stem cell research 163
eukaryotes 70–4
eukaryotic cells 36, 40–7, 49–50
  atypical for specialized functions 50–2
  comparison with prokaryotic cells 48
eutrophication 532
evolution 24–5, 79–81, 90–2
  convergent 86–7, 206
  divergent 87
  evidence for 81–6
  homologous and analogous structures 84–6, 87
  Lamarckism 80, 516–17
  modern synthesis theory 80

see also natural selection; speciation
exergonic reactions 240, 249
exothermic animals 173
exothermic reactions 240
extended essay 622–8
extinction 88, 95, 96, 97–9, 513
extracellular fluids (ECF) 447
extreme weather events 544

F1 generation 472–3
F2 generation 473
facilitated diffusion 145, 146–7, 498
facultative anaerobes 214
Fallopian tubes 461
fatty acids 124–5, 126
feedback control 287, 296, 300–4, 305
  see also negative feedback; positive feedback
fenestrated capillaries 184
fermentation 251
fertilization 454, 455
  flowering plants 462, 463, 465, 474
  humans 455, 460–1
  self-fertilization 462, 474
  in vitro fertilization (IVF) 462
fertilizers 531
fibrous proteins 310
fish 63, 214, 518–20
fishing 528–9
flagella 38, 47
flamingos 516
flowering plants 349–50, 462–7, 471–4
fluid mosaic model 140, 148–9
fluorescent dyes/stains 34, 154
follicle cells 458–9
follicle-stimulating hormone (FSH) 458
food chains and webs 362–3, 367–70, 372–3, 548–9
forest 207
forest biome 207, 209
forest fires 376–7, 525, 544–5, 546
forests 99–100, 524–5, 529–30, 545–6, 554–5
fossil fuels 378
fossils 68, 95, 219, 578–9
foxes 207–8, 228–9
frameshifts 412
freeze fracture 34
frogs 209–10, 212, 213, 225, 226
frugivores 220
fruits 465–6

functional groups 113–14
fungi 214
   cell structure 49–51
   mutualism with plants 346–7
   saprotrophs (decomposers) 217, 364

Galapagos Islands 87–8, 348
gametes 64, 159, 433, 454–5, 470
   nuclear division 428
   Punnett grids 472–4
gas exchange 171–82
   comparison of plant and mammal
      182
   concentration gradients 172–3
   gills 172, 173
   leaves 178–9
   lungs 174–7
   between organism and environment
      171–2, 178
   surfaces 172–5
   transpiration 179–80
gel electrophoresis 154, 391–2, 393
gene pools 94–5, 480–1
genes 14, 21, 69–70, 159–60
   difference from alleles 475
   expression 397
   inheritance 485
genetic code 23–5, 159, 401–3
genetic crosses 471–4
genetic diversity 70–2, 94–5, 428,
    437–8, 523
genetic engineering 76, 401
genetic information 23–4
genetic traits 406–7
genetic uniqueness 24–5
genetic variation 419, 511–12
   and meiosis 428, 437–8, 453–4,
     511, 512
   plants 464–5
genomes 68–9
   diversity of eukaryote genomes
     70–2
   Human Genome Project 69–70
   sizes 70–1, 73–4
   whole genome sequencing 74–6
genomics 71
genotypes 475
   multiple alleles 480, 481
   pedigree charts 487, 488
   Punnett grids 472–4
genus 60
geographical isolation 87–8, 89–90
germ cells 417–18, 583

germ line 418
germplasm 106
gills 172, 173
glandular cells 299
global warming 542–5, 547, 550
globular proteins 134, 238
glucagon 498
glucose 119–21
   alpha and beta 121–2
   blood glucose levels 498–500
   cell respiration 250–1
   for generating body heat 503
   hydrolysis and condensation 116,
     117, 122
glycerol 124–5, 140
glycocalyx 148
glycogen 120–1, 498–9
glycolipids 147–8
glycolysis 250–1, 252
glycoproteins 123–4, 147–8, 149, 314
goats 348
Golgi apparatus 43–4
gonadotropin-releasing hormone
    (GnRH) 458, 459
gorillas 220, 221
gorse 556
Graafian follicles 458–9, 462
Gram staining 38
grana 46
grassland 207
Great Barrier Reef 94
greenhouse effect 538–9, 540
greenhouse gases (GHGs) 538–9,
    543–4
grey matter 292
guanine 17, 19, 22–3, 417
guard cells 178

habitats 197–8, 337–8, 500
haemodialysis 147
haemoglobin 51, 82, 134
   sickle cell disease 404, 409–10
haemophilia 485–6
halophytes 447
haploid cells 64, 433, 434, 470
haploid number 64, 470
hares 353–4
heart 187–8, 300–1
helicase 390
herbaceous plants 228
herbal medicine 576, 582
herbivores 220–1, 222–3, 345
herd immunity 325, 326–7

heritable characteristics 80, 83–4,
    516–17
hermaphroditic plants 462
heterotrophs 216, 257, 366–7, 381
heterozygotes 472
heterozygous genotype 472, 475
hexose monosaccharides 119
histones 45, 134, 429
HIV 319–20, 412–13
holozoic nutrition 216
homeostasis 496, 504–5
   body temperature control 127,
    500–4
   negative feedback control 287, 299,
    497
   role of hormones 299, 498–9
Hominidae 219–21
homologous chromosomes 434,
    437–8, 453–4
homologous structures 84–5, 86, 87
homozygous genotype 472, 475
hormones 288, 289–90, 297–300
   epinephrine (adrenaline) 298
   for IVF treatment 462
   melatonin 297–8
   and menstrual cycle 457–60
   negative feedback 299, 459
   regulating blood glucose levels
    498–500
   steroids 9, 128–9
   for thermoregulation 503
hot desert biome 205, 207–8
human activities
   and biodiversity 97–100, 102–7
   and climate change 538–40
   and ecosystem loss 99–100
   and rate of extinctions 95, 97–9
   sustainable 528–31
Human Genome Project 69–70, 73
humans
   chromosomes 65–8, 428, 433,
    435–6, 438
   genome 68–70, 73, 74, 75–6
   population growth 102–4
   stem cell niches 161–2
   teeth and diet 220, 221
   thermoregulation 501–4
   see also human activities
humus 364
hunting 225
Huntington's disease 411, 487
hydrogen 113
hydrogen bonds 5, 441

adhesion 7–8
cohesion 6–7
DNA structure 19
proteins 134
role in transcription 396
hydrolysis 43, 115–16, 117–18, 248–9
hydrophilic molecules 8–9
hydrophobic molecules 8, 9–10
hydrostatic pressure 446
hydrothermal vents 365, 366
hydroxyl group 21, 117, 122, 124–5, 132
hyperglycaemia 499–500
hypertonic solutions 144, 442, 443
hyphae 50–1, 346–7
hypocotyl 467
hypothalamus 299–300
    menstrual cycle regulation 458, 459
    osmoregulation 299
    thermoregulation 501–2
hypotheses 572, 580, 617
hypotonic solutions 144, 442, 443

ice cores 539
ice melt 542, 547
iguanas 87–8
immune system 307, 311–18
immunofluorescence 34
independent variables 267
inductive reasoning 30, 488, 580
infrared radiation 538–9
inheritance 470–95, 506–7
    blood types 481–2
    codominance 477, 483
    combinations of alleles 474–5
    continuous variation 489–90
    dominant and recessive alleles 476–7
    genetic crosses in flowering plants 471–4
    genetic disorders 478–80, 485–7
    genotypes and phenotypes 475–8
    haploid gametes and diploid zygotes 470
    incomplete dominance 483
    Mendel's pea plant experiments 471–4
    polygenic 489
    Punnett grids 472–4
    recessive genetic conditions 478–80
    sex chromosomes 484–5, 486
    sex-linked traits 485–6
    single-nucleotide polymorphism (SNP) 480–1

insects 222, 463–4
inspiration (breathing in) 176
insulin 134, 401, 498–9
    and diabetes 499–500
    resistance 500
integral proteins 142–3, 149
intercostal muscles 175, 176
internal assessment (IA) 587–98
    assessment criteria 591–7
    collaborative work 588–9
    conclusion and evaluation 596–7
    data analysis 594–5
    data collection 587
    research question 591–3
    topic choice 589–91
interneurons 294, 295
interphase 437
interspecific relationships 229, 344–54
intracellular fluids (ICF) 447
intraspecific relationships 344
intravenous (IV) fluids 447
invasive species 104, 348, 349–50, 556
involuntary control 303
IPBES reports 101
isolation 87–90
isotonic solutions 144, 442–3, 447
IUCN Red List 97, 101–2, 107
IVF (in-vitro fertilization) 163, 462

Japanese encephalitis 323

kangaroos 80
karyograms 65–6
karyotypes 65–7
Keeling Curve 378–80
keratin 134
keystone species 527–8, 535–6
kidney cells 145
kidney dialysis 147
kinetochores 431
Krebs cycle 251–2

lactic acid fermentation 251
lactose 117, 419
ladder diagram 19
Lake Victoria, East Africa 63
Lamarckism 80, 516–17
leaching 531
leaves 7, 10, 178–81
legacy carbon 546
leucocytes see white blood cells
lianas 227

lichens 336, 345
life 3–4
    functions of 29, 48–9
    tree of life 96
light 227–8, 260–1
light microscopes 30–1, 32–3, 34–5
lignin 190
limiting factors 201–3, 265–7, 338
Lincoln index 336
linear regression model 613
link reaction 251
Linnaeus, Carolus 59
lipase 134
lipids 36, 111, 112, 114
    cholesterol 128–9, 149
    conjugated 123, 124
    energy storage 111, 126, 127
    phospholipids 124–5, 127–8, 140
    solubility 124, 127
    triglycerides 111, 124–7
    see also phospholipid bilayers
liver cells 498
Lucy (hominid fossil) 68
lungs 52
    alveoli 174–5
    bronchioles 175
    concentration gradients 173
    gas exchange 174–5
    ventilation 175–6, 301–2
    volume 177
luteinizing hormone (LH) 458
lymphatic system 313–14
lymphocytes 307, 313–14
lynx 353–4
lysosomes 43, 156–7

macromolecules 115–17
malaria 410
mammals
    body temperature control 501–4
    see also specific types
mangroves 199–200
marine ecosystems 198–200, 203–4, 548–50
    intertidal area 527
    population controls 355–6
marsupials 86–7
mass extinction events 97, 513
measurements 4, 245, 600–1
mechanoreceptors 293
medical treatments 447, 575–6
    herbal 576, 582
    organ transplants 148, 314–15, 447

stem cell research 162–3
using human genome 75–6
vaccines 323, 324, 325
see also antibiotics
medulla 291
meiosis 426, 428, 433, 453–4, 462
and genetic variation 428, 437–8,
453–4, 511, 512
non-disjunction 435–6
phases 434–5
melatonin secretion 297–8
melting points 126
membrane potential 276
membrane transport 139, 143–6
active 143, 146
carrier and channel proteins 145
diffusion 141–2, 144, 145, 146–7
facilitated diffusion 145, 146–7
osmosis 144–5
passive 143–5, 146
selectivity in membrane
permeability 146–7
membranes 139, 150–1
diffusion across 141–2
fluid mosaic model 140, 148–9
fluidity 149
glycoproteins and glycolipids 147–8
integral and peripheral proteins
142–3
mitochondria 44
partially permeable 144
permeability 146–7
phospholipid bilayer 128, 139–40
proteins 36, 123, 145
Singer-Nicolson model 140
structure 139–41, 147–50
memory cells 316, 318
Mendel's pea plant experiments 471–4
menstrual cycle 457–60
menstruation 458, 459
mercury 533
meristematic tissue 161, 467
mesocosms 525–6
messenger RNA (mRNA) 21, 156,
395
codons 402–3
translation process 398–400, 403
metabolic pathways 250–1, 252
metabolism 115, 117, 236, 246
metaphase plate 431, 434
methane 375, 538–9, 544
methanogens 544
micrometers 32

microplastics 534–5
microscopes 29, 30–5, 42, 565
microscopy techniques 32–3, 34
microtubules 47, 429–30, 431
migration 341, 500
"missing link" 578–9
mitochondria 44, 251–2
mitochondrial DNA 44, 71–2
mitosis 426, 428, 462
for embryo development 461
phases 430–2
mitotic spindle 431, 432
mixotrophic nutrition 216–17
moas 98
models 342, 576, 604
molecular motions 241
molecules 3–13, 111–30
amphipathic 128
bonding 112–14
carbon compounds 112–15
categories 112, 114
condensation and hydrolysis 115–18
conjugated 123–4
energy storage 111, 119–21, 126,
127
macromolecules 115–17
monomers 111, 115–17
polar 119
structural 121–2
see also carbohydrates; lipids;
nucleic acids; proteins
monoculture 100, 103
monohybrid cross 472
monosaccharides 111, 116, 117,
118–19
monounsaturated fatty acids 126
morphogens 160
morphological classification 59
mosses 556
mother cells *see* parent cells
motor cortex 296
motor end plates 294
motor neurons 164–5, 274, 293–4,
295–6
motor proteins 429–30, 431
Mount Saint Helens, USA 339–40,
341
movement 295–7
mucous membranes 309–10
multicellular organisms 49
multipotent stem cells 162
muscle cells 498
muscle fibres 165

muscles 21
cell structure and function 51–2
coordinating movement 296
and motor neurons 293–4
muscular activity and body
temperature 501
neuromuscular junctions 281, 294
pain reflex arcs 295–6
mutagens 413–15
mutation hotspots 417
mutations 397, 404, 406–21
DNA replication errors 404, 415
exposure to ionizing radiation 413,
414–15
genetic variation 419–20, 511
hereditary/non-hereditary 417–18
insertions and deletions 406,
411–13
location 416–17
mutagens 413–15
point mutation 404, 406–10
randomness 416–17
substitutions 406–10
types of 406
mutualism 345–7
mycorrhiza 346
myelin sheath 275, 277–8, 295

natural resources 528–30
natural selection 25, 509–21, 559
abiotic factors 513–14
adaptation and survival 515–16
competition 513, 515–16
and genetic mutation 417, 419
intraspecific competition 515–16
mechanism for evolution 80, 81,
509–10
modelling 518–20
overproduction of offspring 513
selection pressures 513–14, 517–20
sexual selection 517–18
sources of variation 511–12
struggle for survival 513
nature reserves 105, 536, 556–7
negative feedback 287, 299, 497
control of blood glucose 498–9
menstrual cycle 459
populations 338–9
thermoregulation 500–3
nerve cells 52, 164–5
nerve fibres 277, 294–5
nerve impulses 274–8, 279
nerves 274, 294–5

nervous system 287, 289–90, 304
  autonomic nervous system (ANS) 185, 289–91, 301, 497
  central nervous system (CNS) 273–4, 292, 293, 303
  control of peristalsis 303–4
  conveying information 293–4
  coordinating movement 296–7
  enteric nervous system (ENS) 303–4
  feedback control 300–2
  motor neurons 293–6
  pain reflex arcs 295–6
  sensory neurons 293, 294, 295–6
  unconscious processes 292
neural cell body 274–5
neural signalling 273, 283–5
  action potentials and nerve impulse 275–6
  neurotransmitter release 280–1
  postsynaptic potential 281
  role of neurons 273–5
  speed of nerve impulse 277–9
  synapses 280–1
neuromuscular junctions 281, 294
neurons 146, 164–5
  depolarization 276–7
  interneurons 294, 295
  motor neurons 164–5, 274, 293–4, 295–6
  myelinated/unmyelinated 277–8, 295
  pain reflex arcs 295–6
  postsynaptic 280–1
  presynaptic 280–1
  resting potential 276–7
  role in signalling 273–4
  sensory neurons 274, 293, 294
  structure 274–5
neurotransmitters 274, 280–1, 304
niches
  ecological 212, 228–30
  fundamental and realized 228–9
  stem cell niches 161–2
nitrogen 113, 381
nitrogen fixation 346
nitrogenous bases 14, 17–18, 19
  combinations and genetic information 23–4
  complementary base pairs 19–20
nocturnal animals 297
nodes of Ranvier 277, 295
non-disjunction 435–6

non-polar covalent bonds 5
non-polar molecules 9–10
nuclear division 428
nuclear envelope 45
nuclear power 415
nucleic acids 21–2, 112, 114, 324
  see also ATP; DNA; RNA
nucleoid region 39
nucleoli 46
nucleosomes 45
nucleotides 15–16, 21, 35–6
  combinations 23
  condensation reactions 116
  DNA replication 389–90, 391
  drawing 17–18
  structure of ATP 247–8
  structure of DNA and RNA 18–20, 388
  sugar-phosphate bonding 16–17
nucleus 45–6, 155–6
nutrient cycles 364, 381–2, 523
nutrient leaching 531
nutrient upwelling 548–50
nutrition 214–18
  see also diet

obligate aerobes 214
obligate anaerobes 214
observation 577–8, 601
ocean acidification 553–4
ocean currents 548–50
Ockham's razor 577
oestradiol 128–9, 458, 460
omnivores 220–1
oocytes 458
oogenesis 427
orangutans 220, 221
orchids 346–7
organelles 40, 41–7, 50, 152, 155–8
  functions 155, 156–7
  isolating for research 152–4
organic compounds see carbon compounds
organisms 36, 40
  adaptation see adaptation; adaptation to environment
  binomial nomenclature 59–60
  development 159–60
  genetic information see genomes
  producers/consumers 216, 361–3, 364, 366–8
  types of 49
  variation between 57–8, 68–9

see also classification systems
organs 49, 288
  integration 289–90
  transplants 148, 314–15, 447
  see also specific organs
osmoreceptors 299
osmosis 144–5, 443–4
ovarian cycle 460
ovaries 456, 458
ovulation 457, 458–9, 460
ovules 462, 463, 465
oxidation reactions 247, 365–6, 367
oxygen
  aerobic respiration 214
  covalent bonds 113
  dependence on atmospheric oxygen and carbon dioxide 380–1
  diffusion 141–2
  photosynthesis by-product 257, 263, 264

P generation 472–3
pain reflex arcs 295–6
palisade mesophyll 178
palm oil 100
pancreas 498, 499, 500
pandemics 325, 326–7
paradigm shift 510, 565
parasites 345
parent cells 426–7, 437
pathogens 157, 324, 345
pea plants 471–2
peat 377, 542
pedigree charts 486–8
penguins 514, 547
penicillin 356–7
pentadactyl limbs 84–5
pentose monosaccharides 118
pentose sugars 15–17, 18, 21, 116, 118
peptide bonds 116, 132–3
peptidoglycan 37
percentage change 325–6
percentage difference 325–6
peripheral nerves 274
peripheral proteins 142–3, 149
peristalsis 303–4
permafrost 543–4
perspiration 502
pH
  blood pH 301, 302
  effect on enzyme-catalysed reactions 243–4, 245

effect on protein structure 134
phagocytes 307, 311, 312, 314
phagocytosis 43, 156–7
pharmaceuticals testing 575–6
phenotypes 69, 476–8, 481
  continuous variation 489–90
  intermediate and dual phenotypes 483
  sex-linked traits 486
phenotypic plasticity 477–8
phenylketonuria (PKU) 478–9
pheromones 223
phloem 51, 178, 183, 191–2
phosphate groups 116, 124–5, 127–8
  ATP and ADP structure 237, 247–8, 249
  DNA and RNA structure 15–17, 388
phospholipid bilayers 139–40
  as a barrier 141
  fluid mosaic model 148–9
  formation 127–8, 141
phospholipids 124–5, 127–8, 140
phosphorus 113, 531
photoautotrophs 257, 364, 365
photosynthesis 46, 119–20, 215–16, 256–68
  absorption and action spectra 261–3
  equation 257
  limiting factors 265–7
  pigments 258–60, 261, 263
  rate of 178, 263–7
  reverse of aerobic respiration 381
phrenology 572
phylogeny and phylogenetic trees 74, 82, 86, 88
physical adaptations 224, 226
phytoplankton 541
phytotoxins 223
pili 38
pineal gland 297
pituitary gland 299–300
placebo effect 575–6
plant cells 41, 46, 47, 49–50
  cytokinesis 426–7
  water movement 144–5, 446–7
plants
  adaptation for protection 222–3
  adaptation to environment 198–200, 206, 207, 209
  allelopathy 356
  biodiversity conservation 106
  carnivorous 206

flowering plants 462–7, 471–4
  function and parts of a flower 464
  genetic crosses 471–4
  genetic variation 464–5
  mutualism with fungi 346–7
  poisons 223
  selective breeding 83–4
  sexual reproduction 462–7
  water transport 7, 189–92
  see also trees
plasma 9, 134
plasma membrane 35, 36, 37–8
  bilayer 36, 128
  binary fission 39
  egg and sperm 461
plasmids 39
plasmolysis 446–7
plastic pollution 534–5
platelets 310, 311
pluripotent stem cells 162, 163
point mutation 404, 406–10
polar habitat change 547
polar molecules 5
polarity of bonds 5, 119
pollen grains 462–3, 465
pollen tube 463
pollination 462–5, 471–2, 474
pollution 103, 531, 532–5
polymerase chain reaction (PCR) 391
polymers 15–17, 116–18, 119–22
polypeptides 131, 132–3
  examples of 134
  hydrolysis and condensation 116, 117
  synthesis 398–400, 403
  variety 133–4
polysaccharides 116, 117, 119–22
polyunsaturated fatty acids 126
population growth curves 339–42
populations 62–3, 333–4, 357–8
  abiotic limiting factors 201–2
  carrying capacity 337–8, 513
  chi-squared test of association 351–3
  control of 353, 354–6
  density-dependent control factors 338–9, 353–4, 514
  density-independent factors 513–14
  deterring competition 356–7
  endemic and invasive species 348
  estimating size 334–7
  genetic diversity 94–5
  growth 102–4, 338–42

heritable characteristics 80
  incidence of disease 307
  interspecific competition 348–50
  interspecific relationships 344–54
  intraspecific cooperation/ competition 344
  isolated 87–90
  negative feedback 338–9
  positive feedback 338, 339
  predator-prey relationships 353–4
  sources of genetic variation 511–12
positive feedback 541
  climate change 542–5, 546
  menstrual cycle 458
  populations 338, 339
postsynaptic potential 281
predator-prey relationships 345, 353–4, 518–19
predators 223–6, 536
prey 223–6, 353–4, 518–19
primary consumers 367–8
primary production 373, 374, 546, 549
primates 66–7, 68, 89–90, 218–21
primroses 349–50
producers 216, 361–3, 364–5, 367
progesterone 459, 460
prokaryotes 214
prokaryotic cells 36–9, 48
propagules 200
protein synthesis 395–405
  mutations 404
  producing polypeptide chain 403
  transcription 395–7
  translation 398–400, 403
proteins 9, 112, 114, 131–5
  carrier proteins 145
  in cell membranes 142–3, 147–9
  channel proteins 145
  conjugated 123–4
  cytoskeleton 42
  denaturation 134
  effect of pH and temperature 134
  hydrogen bonds 134
  hydrolysis and condensation 115–17
  peptide bonds 132–3
  pump proteins 146
  shape 134
  structure 134, 404
  synthesis 39, 43, 132–4, 156
  types of 143
  variety of possible peptide chains 133–4
proxies 539

puberty 451, 457
pulse rate 186
pump proteins 146
Punnett grids 472–4
pyruvate 250, 251

rabies 322–3
radiation 414–15
radicle 467
rainforest 99–100, 209, 522–3, 524–5
random orientation 438
random sampling 334–6
range migration 550–2
receptors 290–1, 299
　baroreceptors 300–1
　chemoreceptors 300–1, 302
　osmoreceptors 299
　pain receptors 295–6
　sensory receptors 292–3
recombinants 437
red blood cells (erythrocytes) 51, 163,
　　164, 444–5
　see also haemoglobin
reductionist approach 152
reflex arc 292
reforestation 554–5
reproduction 451–69
　asexual reproduction 452–3
　differences between sexual and
　　asexual 452–3
　eukaryotic cells 46
　flowering plant sexual reproduction
　　462–7
　gametes 454–5
　genetic variation in populations 512
　hormonal regulation of menstrual
　　cycle 457–60
　hormones 462
　male and female gametes 455
　meiosis and genetic variation 453–4
　reproductive systems 456–7
　see also fertilization
reproductive isolation 88
reproductive potential 516
respiration 141–2
respiratory centres 301–2
respirometers 253–4
rewilding 105, 535–6
Rhizobium bacteria 346
ribonucleic acid see RNA
ribose 17, 21, 118, 119
ribosomal RNA (rRNA) 21, 39,
　　398–400

ribosomes 21, 39, 42, 43, 44
RNA
　comparison with DNA 20–1
　complementary base pairing 400
　microscopy techniques 34
　nitrogenous bases 17–18
　nucleotides 15–18, 82
　ribosomal RNA (rRNA) 21, 39,
　　398–400
　structure 18
　sugar-phosphate bonding 16–17, 18
　synthesis 395–7
　vaccines 324
　see also messenger RNA (mRNA);
　　transfer RNA (tRNA)
RNA polymerase 395–6
roots 192, 467
　buttress roots 209
　mangroves 199, 200
　root nodules and Rhizobium bacteria
　　346
ruminants 222, 223

S phase 428
sampling methods 334–7
sand dunes 198–9
saprotrophic nutrition 217
saprotrophs 217, 364
satellite DNA 416
saturated fatty acids 126
scanning electron microscope (SEM)
　　34
Schwann cells 277, 295
sea bass 528–9
sea oat (grass species) 198–9
seals 11–12, 98, 127
secondary compounds 223
secondary consumers 367–8
secondary metabolites 356–7
secondary production 374
seeds 350
　dispersal 465–6
　germination 466–7
　inhibited germination 356
　seed banks 106
selective breeding 83–4
self-incompatibility mechanism 465
self-pollination 462, 474
self-renewal 161
sensory gating 292
sensory neural pathways 292–3
sensory neurons 274, 293, 294, 295–6
sexual selection 517–18, 519

shivering 503
sickle cell disease 404, 409–10
sieve tube elements 51
Simpson reciprocal index 607–8
Singer-Nicolson cell membrane model
　　140
single-celled organisms 24–5
single nucleotide polymorphism (SNP)
　　69, 409, 480–1
sinoatrial (SA) node 300
skin 309–10, 501–3
sleep patterns 297–8
snails 282, 477–8
snakes 223
snow 545–6
SNP see single nucleotide
　　polymorphism
sodium chloride 441
sodium-potassium pump 146, 276–7
soil 8, 364, 542–4
soil erosion 530–1
solar radiation 542–3
solvation 441
solvents 4, 8–10, 440–1
somatic cells 417–18, 583
somatic tissue 106
Southeast Asia 99–100
spatial habitat 212–13
speciation 62–3
　adaptive radiation 95
　and biodiversity 95–6
　divergence of isolated populations
　　87–90
　geographical isolation 89–90
　reproductive isolation 88
species 57, 76–7
　biological species concept 61
　chromosome numbers 64
　classification and naming 58–62
　distinguishing from populations
　　62–3
　endangered 97, 105–8
　endemic 348, 349–50
　evolutionary relationships 74–6
　extinction 88, 95, 96, 97–9, 513
　invasive 104, 348, 349–50, 556
　morphological classification 59
　variation between 70–2, 73–4
　variation within 57–8, 70–2
species diversity 94
specific heat capacity 10, 11
sperm 454, 455, 460–1
sperm cells 52, 163–4, 433

pinal cord 273, 274, 292, 295
spinal nerves 291, 292, 294–5
spirometer 177
spongy mesophyll 178
spreadsheets 605
squirrels 229
standard deviation 609, 612
taphylococcal infections 322
starch 116, 119–20, 121
statistical significance 609
stem (plant) 191–2
stem cells 159, 160–3
steroid hormones 9, 128–9
stigma 463
stomata 7, 178–9, 180–1
striated muscle fibres 165
stroma 46
substrate concentration 244
sugar-phosphate bonding 16–17, 18
sulfur compounds 526
supercoiled DNA 429
superovulation 462
surface area-to-volume ratio 166–7,
    171–2
surface tension 6
surfactant 175
survival 515–16
sustainability 522–5, 528–31
Svedberg units 43
symbiosis 346
synapses 280–1
synaptic cleft 280, 281
synaptic terminal buttons 274–5,
    280–1
systematic sampling 335
systems theory 359–60

T-lymphocytes 313, 316, 319
t-test 609–10
taiga 205, 207, 545–6
Taq polymerase 391
target tissue 289, 299
Tasmanian tiger 86–7
taste 480–1
taxonomy 58, 60
teeth 218–21
telomeres 67
temperature
    effect on enzyme-catalysed reactions
        243, 245
    effect on photosynthesis rate 266
    effect on protein structure 134
    global temperatures 540

internal body temperature 496,
    500–4
template strand 395
tension 7
terrestrial biomes 204–7
tertiary consumers 367–8
testosterone 128–9
theory of knowledge (TOK) 564–86
    assumptions and implications
        578–9, 580
    case studies 569–70, 574, 580, 582,
        583
    confirmation bias 570
    creativity 573
    critical thinking 565–6, 569
    debates 565–6
    discredited ideas 572–3
    empiricism and rationalism 569
    ethics and values 569, 583, 585–6
    historical development 572
    indigenous knowledge 576, 582
    journal 571
    knowledge framework 566–7
    knowledge questions/claims 565–71
    language and communication 581,
        583
    methods and tools 569, 574, 575–6,
        579–81, 584
    perception and observation 577–8,
        582, 584
    perspectives 569, 570, 576–9, 580,
        581, 583
    religion and science 574, 576
    scope of biology 569, 573–4
    technology 585–6
    theories vs laws 571–2
    theories vs myth 584
therapeutic cloning 162
thermal conductivity 10, 11
thermocycler 391
thermoreceptors 501
thermoregulation 500–4
thoracic cavity 175–6
thylakoids 46
thymine 17, 19, 22–3
tipping points 524–5, 546
tissues 288
    adipose tissue 127, 503
    animal tissue banks 106
    dicotyledonous leaf 179
    dicotyledonous root 192
    dicotyledonous stem 191–2
    for gas exchange 172

gonadal 129
plant tissue 179, 191–2
triploid tissue 463
tolerance 213
tools and inquiry skills
    concluding and evaluating 620–1
    controlling variables 618
    data collection 603–4, 619, 624
    data processing 605–16, 619–20
    errors and uncertainties 611–12,
        618, 621
    experimental techniques 599–602
    graphing 605, 612, 613, 615
    interpreting results 620
    mathematics 606–16
    measuring variables 600–1
    precision and accuracy 600, 611–12
    technology 603–6
    topics and investigation design
        617–18
tortoises 348
totipotent stem cells 162
transcription 129, 156, 395–7
transducin 406–7
transduction 292–3
transects 202
transfer RNA (tRNA) 21, 398–400,
    403
translation 156, 398–400, 403
transmission electron microscope
    (TEM) 34
transpiration 7, 179–80, 189
transplants 148, 314–15, 447
transport 183–93
    blood and circulatory system 183–8
    lymphatic system 313–14
    water in plants by transpiration 7,
        189–92
    see also membrane transport
tree of life 96
trees 227, 550–1
    black cherry trees 529–30
    dipterocarp trees 99–100
    kapok tree 209
    mangrove trees 199–200
    see also forests
trinucleotide repeat expansion 411
triplet codes 23, 24, 400–1
triploid tissue 463
trophic cascade 527–8
trophic levels 367–70, 372–3
tropical forest biome 209–10
tuberculosis 323

tundra 205, 207
turgor pressure 446–7
twins 24, 438

uncertainties 611–12
unconscious processes 292
unicellular organisms 48–9
    see also prokaryotic cells
unipotent stem cells 162
units 611
uracil 17, 20
uterine cycle 460

vaccines 323, 324, 325
vacuoles 47, 156–7
vascular bundle 191–2
veins 183, 184–5, 186–7
ventilation rate 301–2
vesicles 43–4
    cell plate 426
    cortical granules 461
    at synapses 280, 281
viruses 15, 314, 571
    capsid 314
    HIV and AIDS 319–20, 412–13

treating 320, 321
vaccines 324
zoonotic diseases 322–3
    see also COVID-19
viscosity 10, 11

Wallace, Alfred Russel 25, 80, 510
walruses 547
water 3–13
    adhesion with other substances 7–8
    capillary action 7–8
    cohesion 6–7
    eutrophication of body of water 532
    hydrophilic/hydrophobic substances
        8–10
    as medium for life 3–4
    molecules 5–8, 116, 440–1
    physical properties 10–11
    as a solvent 4, 8–10, 440–1
    surface tension 6
water movement 440–9
    effect of cell walls 444–6
    imbibition by seeds 467
    medical applications of isotonic
        solutions 447

osmosis 144–5, 443–4
    in plants 7, 9, 189–92
    and solute concentration 442, 447
    water potential 440
waxy cuticle 178
wetlands 555–6
white blood cells (leucocytes) 163,
    164, 307, 310
white matter 292
wings 86
Winogradsky column 525–6
wolves 225, 527–8
woolly mammoth 63

X chromosome 484–6
xylem 7, 9, 178, 183, 190–2

Y chromosome 484–5, 486
yeast 214, 427

zoonotic diseases 322–3
zoos 105–6
zooxanthellae 347
zygotes 64, 159–60, 164, 470